Planetary Science: The Geological Perspective

Planetary Science: The Geological Perspective

Editor: Max Hall

MURPHY & MOORE
www.murphy-moorepublishing.com

Murphy & Moore Publishing,
1 Rockefeller Plaza,
New York City, NY 10020, USA

Visit us on the World Wide Web at:
www.murphy-moorepublishing.com

ISBN: 978-1-63987-430-9 (Hardback)

Cataloging-in-Publication Data

Planetary science : the geological perspective / edited by Max Hall.
p. cm.
Includes bibliographical references and index.
ISBN 978-1-63987-430-9
1. Planetary science. 2. Solar system. I. Hall, Max.
QB601 .P53 2022
523.4--dc23

Table of Contents

Preface

This book was inspired by the evolution of our times; to answer the curiosity of inquisitive minds. Many developments have occurred across the globe in the recent past which has transformed the progress in the field.

The scientific study of planets including Earth, moons, and planetary systems as well as the processes of their formation is known as planetary science. It studies objects, aiming to determine their composition, dynamics, formation, interrelations, and history, ranging in size from micrometeoroids to gas giants. Planetary science is a sub-field of astronomy and earth science. It encompasses many disciplines such as cosmochemistry, atmospheric science, oceanography, planetary geology, hydrology, glaciology and exoplanetology. Planetary geology deals with the planetary bodies near the Earth such as the Moon, and the two neighbouring planets, Venus and Mars. This book unravels the recent studies in the field of planetary science. While understanding the long-term perspectives of the topics, it makes an effort in highlighting their impact as a modern tool for the growth of the discipline. This book is a vital tool for all researching and studying this field.

This book was developed from a mere concept to drafts to chapters and finally compiled together as a complete text to benefit the readers across all nations. To ensure the quality of the content we instilled two significant steps in our procedure. The first was to appoint an editorial team that would verify the data and statistics provided in the book and also select the most appropriate and valuable contributions from the plentiful contributions we received from authors worldwide. The next step was to appoint an expert of the topic as the Editor-in-Chief, who would head the project and finally make the necessary amendments and modifications to make the text reader-friendly. I was then commissioned to examine all the material to present the topics in the most comprehensible and productive format.

I would like to take this opportunity to thank all the contributing authors who were supportive enough to contribute their time and knowledge to this project. I also wish to convey my regards to my family who have been extremely supportive during the entire project.

Editor

1

Towards scaling laws for subduction initiation on terrestrial planets: constraints from two-dimensional steady-state convection simulations

Teresa Wong* and Viatcheslav S Solomatov

Abstract

The strongly temperature-dependent viscosity of rocks leads to the formation of nearly rigid lithospheric plates. Previous studies showed that a very low yield stress might be necessary to weaken and mobilize the plates, for example, due to water. However, the magnitude of the yield stress remains poorly understood. While the convective stresses below the lithosphere are relatively small, sublithospheric convection can induce large stresses in the lithosphere indirectly, through thermal thinning of the lithosphere. The magnitude of the thermal thinning, the stresses associated with it, and the critical yield stress to initiate subduction depend on several factors including the viscosity law, the Rayleigh number, and the aspect ratio of the convective cells. We conduct a systematic numerical analysis of lithospheric stresses and other convective parameters for single steady-state convection cells. Such cells can be considered as part of a multi-cell, time-dependent convective system. This allows us a better control of convective solutions and a relatively simple scaling analysis. We find that subduction initiation depends much stronger on the aspect ratio than in previous studies and speculate that plate tectonics initiation may not necessarily require significant weakening and can, at least in principle, start if a sufficiently long cell develops during planetary evolution.

Keywords: Convection; Stress; Lithosphere; Subduction; Plate tectonics

Background

Plate tectonics is central to many aspects of the geology and evolution of terrestrial planets. While Earth is the only planet where plate tectonics is observed, its driving mechanism and timing of initiation are still poorly understood. Subduction is thought to be the fundamental process for plate tectonics initiation because the slab pull of subducting slab contributes most to the forces that drive plate movements. On the Earth, initiation of subduction is greatly facilitated by tectonic forces associated with plate motions *already occurring elsewhere* (Mueller and Phillips 1991; Hall et al. 2003). Various models for subduction initiation has been proposed (e.g., McKenzie 1977; Turcotte 1977; Ogawa 1990; Mueller and Phillips 1991; Kemp and Stevenson 1996; Toth and Gurnis 1998; Stern 2004; Solomatov 2004a; Ueda et al. 2008; Nikolaeva et al. 2010), many of which involve existing plate boundaries or weak zones. Incipient subduction zones are often found near transform faults or fracture zones because of their physical weakness (e.g., Mueller and Phillips 1991; Gurnis et al. 2004).

On one-plate planets such as Venus and Mars, the absence of plate tectonics is likely to be due to the difficulty of subduction initiation in the absence of forces due to plate motions. In other words, the problem of plate tectonics initiation can be viewed as the problem of the very first occurrence of subduction. Due to the high sensitivity of viscosity to temperature, the lithosphere acts as the cold rigid thermal boundary layer that has a very high strength. On these planets, mantle convection is likely to be in the stagnant lid regime (e.g., Morris and Canright 1984; Fowler 1985; Solomatov 1995). One possible mechanism for the very first episode of subduction is due to the lithospheric stresses generated by mantle convection (Ogawa

*Correspondence: twong@levee.wustl.edu
Department of Earth and Planetary Sciences, Washington University, 1 Brookings Drive, 63130-4899 St. Louis, MO, USA

1990; Fowler and O'Brien 2003; Solomatov 2004a). The magnitude of these stresses is relatively small compared to the lithospheric strength suggested by laboratory and field observations (e.g., Kohlstedt et al. 1995; Gurnis et al. 2004), and thus, it is usually believed that to initiate subduction, some weakening mechanisms must be present in the lithosphere.

Much effort has been devoted to understand the weakening mechanisms of the lithosphere. Several studies showed that the frictional shear stress resisting subduction at transform faults and fracture zones have to be less than 10 MPa for subduction to occur (Toth and Gurnis 1998; Hall et al. 2003; Gurnis et al. 2004). Stress drop estimates from earthquakes also indicate that fault strength may be approximately 10 MPa (Kanamori 1994; Kanamori and Brodsky 2004). Models are able to describe global reduction in the lithospheric strength, as well as localized weak zones such that plate-like features can be generated from mantle convection in a self-consistent manner (e.g., Trompert and Hansen 1998; Moresi and Solomatov 1998; Tackley 2000a; Bercovici 2001; Branlund et al. 2001; Regenauer-Lieb et al. 2001; Regenauer-Lieb and Kohl 2003; Regenauer-Lieb et al. 2006; Korenaga 2007; Landuyt et al. 2008). Various approaches have been used to deal with the creation of weak zones (Bercovici 2001; Bercovici and Ricard 2005; Landuyt et al. 2008; Branlund et al. 2001; Regenauer-Lieb and Kohl 2003). The two-phase damage theory with a grain-size dependent rheology was developed to explain the formation of weak plate boundaries and track the evolution of deformation (e.g., Bercovici and Ricard 2005; Landuyt et al. 2008; Bercovici 2012). Some studies suggested that water might play an important role in the localization of deformation (Regenauer-Lieb et al. 2001; Regenauer-Lieb and Kohl 2003; Regenauer-Lieb et al. 2006). Water also weakens the lithosphere by lowering the activation energy (Regenauer-Lieb et al. 2001; Regenauer-Lieb and Yuen 2004; Regenauer-Lieb et al. 2006) and increasing the pore fluid pressure (Kohlstedt et al. 1995).

One approach to quantify the weakening of lithosphere is to set a yield value to the rheology of the lithosphere to simulate brittle behavior (Fowler 1993; Trompert and Hansen 1998; Moresi and Solomatov 1998; Richards et al. 2001; Tackley 2000a,b; Fowler and O'Brien 2003; Solomatov 2004a; Stein et al. 2004; O'Neill et al. 2007; Stein and Hansen 2008). The yield stress can be regarded as a simplification of mechanisms that describe the strength of the lithosphere. Convection with yield stress is usually categorized into three regimes: mobile lid regime, transitional regime with some episodic failure, and stagnant lid regime (Moresi and Solomatov 1998; Tackley 2000b; Stein et al. 2004). Stein and Hansen (2008) further subdivided the transitional regime into episodically mobile and stable plate mobilization regimes. To access the conditions of a planet to have plate tectonics, some researchers presented regime diagrams in terms of Rayleigh number, viscosity contrast, and yield stress (e.g., Stein et al. 2004; O'Neill and Lenardic 2007).

A number of studies attempted to derive scaling relations for convective stresses and yield stress to extrapolate to planetary conditions (e.g., Moresi and Solomatov 1998; Fowler and O'Brien 2003; Solomatov 2004a,b; O'Neill et al. 2007; Valencia and O'Connell 2009; Korenaga 2010b; Van Heck and Tackley 2011; Stamenkovic and Breuer 2014). Yet the accurate description of these convection-induced stresses inside the lithosphere and thus the yield stress is lacking.

This study seeks to understand the stress distribution of the steady-state convecting cell with respect to various convective parameters using the pseudoplastic rheology as a first step. The goal of this study is to find a scaling law for the lithospheric stress (hereafter referred as lid stress) and the critical yield stress, which is the highest yield value at which the stagnant lid could be mobilized. Note that an alternative and, perhaps, more intuitive approach would be to assume that the yield stress is known and to try to figure out under what dynamic conditions it can be reached. However, (a) the 'normal' yield stress is so high that it is nearly impossible to reach and (b) given the uncertainties in the weakening mechanisms and thus the actual magnitude of the yield stress, it should be treated as an unknown.

In this study, we first examine the stress structure in steady-state stagnant lid convection and explore scaling relationships between convective parameters especially in relation to aspect ratio to develop a scaling theory for lid stress and critical yield stress. We then compare the theoretical scaling laws with numerical results. In addition, we investigate the accuracy of the Frank-Kamenetskii approximation for modeling the initiation of plate tectonics.

Rheology

Viscous creep governs the flow in the mantle as it has high temperatures and low stresses. It can be described by a constitutive relation (Hirth and Kohlstedt 2003), which is an Arrhenius function of temperature T, activation energy E, pressure P, and activation volume V with power law dependences on stress τ (second invariant of stress tensor), grain size d, water fugacity f_{H_2O}, and an exponential function of melt fraction ϕ:

$$\eta = A\tau^{1-n}d^m f_{H_2O}^r \exp(-\alpha\phi)\exp\left(\frac{E+PV}{RT}\right), \qquad (1)$$

where A and α are constants, R is the gas constant, and m, n, and r are exponents for grain size, stress, and water fugacity, respectively. Depending on the temperature, grain size, stress, pressure, and composition, the dominating deformation mechanism in the mantle would be different (Karato and Wu 1993; Karato et al. 1995; Hirth and Kohlstedt 2003). In the lithosphere, the major factor controlling the viscosity is temperature. Thus, the viscosity function to investigate subduction initiation is often written as:

$$\eta = A' \exp\left(\frac{E}{RT}\right). \tag{2}$$

Frank-Kamenetskii approximation

Many numerical studies used use a relatively low viscosity contrast to observe plate behavior, which has limited applications to realistic planetary situations. Moresi and Solomatov (1998) investigated the convective regimes with viscosity contrast ranging from 3×10^4 to 3×10^7, and in Tackley (2000a), the viscosity contrast was limited to 10^4, whereas Richards et al. (2001) and Stein and Hansen (2008) used viscosity contrast on the order of 10^5. The viscosity contrast across the terrestrial lithosphere is many orders of magnitude higher.

The low-viscosity contrast is used because high-viscosity contrasts are difficult to treat in numerical calculations (Moresi and Solomatov 1995). Thus, the Arrhenius function is often approximated by the Frank-Kamenetskii function, which reduces the viscosity contrast by many orders of magnitude compared to Arrhenius viscosity function. This makes the problem of convection with strongly temperature-dependent viscosity more computationally tractable.

Frank-Kamenetskii approximation originated from the combustion theory. Frank-Kamenetskii pointed out that since the activation energy E is large, we can consider the rate of reaction only in a narrow range of temperature around the combustion temperature (Frank-Kamenetskii 1969). The equation for the rate of reaction is similar to the strain rate in the constitutive relations, which also has an Arrhenius form $\exp(-E/RT)$. Since convection mostly takes place in the interior of the cell where the temperature is close to the interior temperature T_i, we use the same approximation by expanding the exponent E/RT in the Arrhenius form so that the viscosity can be expressed as an exponential function of temperature only:

$$\eta = B \exp(-\gamma T), \tag{3}$$

where B and γ are constants. In the interior:

$$\eta_{i,\mathrm{Arr}} = \eta_{i,\exp}, \tag{4}$$

$$\left(\frac{d\eta_{i,\mathrm{Arr}}}{dT}\right)_{T=T_i} = \left(\frac{d\eta_{i,\exp}}{dT}\right)_{T=T_i}, \tag{5}$$

where η_{Arr} and $\eta_{\exp}$ are the interior viscosities of the Arrhenius function and that of Equation 3 (hereafter referred as exponential viscosity), respectively. Equations 4 and 5 give γ in terms of activation energy and interior temperature:

$$\gamma = \frac{E}{RT_i^2}. \tag{6}$$

This method of expanding the terms in the exponent preserves the interior viscosity and the change of viscosity with temperature close to T_i, where convection actively takes place. Some studies expanded the terms inside the exponents differently (e.g., King 2009). However, it is important to use Equations 4 and 5 to ensure the asymptotic accuracy of Frank-Kamenetskii approximation (Morris 1982; Morris and Canright 1984; Fowler 1985; Frank-Kamenetskii 1969).

Frank-Kamenetskii approximation was shown to be sufficiently accurate for the interior of the convective layer with large viscosity contrast (Solomatov and Moresi 1996; Ratcliff et al. 1997; Reese et al. 1999). Recent studies have examined convection with Arrhenius rheology and suggested slightly different scaling laws compared to convection with Frank-Kamenetskii viscosity (Korenaga 2009; Stein and Hansen 2013). Here, we assess the accuracy of the Frank-Kamenetskii approximation in predicting the values of critical yield stress.

Pseudoplastic rheology and plastic yielding

The brittle behavior of the lithosphere can be simplified with a viscoplastic rheology that causes yielding when the convective stresses exceed a plastic yield stress τ_y (Moresi and Solomatov 1998; Trompert and Hansen 1998; Tackley 2000b; Fowler and O'Brien 2003). The yield stress can be defined by Byerlee's law (Byerlee 1978):

$$\tau_y = \tau_0 + \mu \rho g z, \tag{7}$$

where τ_0 is the yield stress at the zero hydrostatic pressure, μ is the frictional coefficient, and $\rho g z$ is the hydrostatic pressure. Viscous deformation occurs according to Equation 3 when stresses are less than the yield stress. Above the yield stress, the deformation follows a plastic flow law defined by a non-linear effective viscosity:

$$\eta_{\mathrm{eff}} = \frac{\tau_y}{\dot{e}}, \tag{8}$$

where $\dot{e}$ is the second invariant of the strain rate tensor. The yield stress defines a change on deformation mechanism based on the second invariant of the deviatoric stress tensor, which corresponds to the Von Mises yield criterion. In this study, we consider two types of yield stress: a constant yield stress τ_y or a depth-dependent yield stress with a constant gradient τ_y'.

Methods

Formulation of the problem

Equations of thermal convection

The equations of thermal convection of an incompressible fluid in Boussinesq approximation and infinite Prandtl number are as follows:

$$\frac{\partial u_i}{\partial x_i} = 0, \quad (9)$$

$$\alpha\rho g_i T' - \frac{\partial p'}{\partial x_i} + \frac{\partial \tau_{ij}}{\partial x_j} = 0, \quad (10)$$

$$\frac{\partial T'}{\partial t} + u_i\frac{\partial T'}{\partial x_i} = \kappa\frac{\partial^2 T'}{\partial x_i^2}, \quad (11)$$

where ρ is density, p' and T' are pressure and temperature perturbations, g_i is the gravity vector, α is the thermal expansivity, $\kappa = \frac{k}{\rho c_p}$ is the thermal diffusivity, k is the thermal conductivity, and c_p is the isobaric specific heat. τ represents the elements of the stress tensor according to the following equation:

$$\tau_{ij} = \eta\left(\frac{\partial u_i}{\partial x_j} + \frac{\partial u_j}{\partial x_i}\right), \quad (12)$$

where i and j, are indices of the coordinate axes.

The boundary conditions are as follows. For a cell with only base heating, the top and bottom surfaces are isothermal. The temperature of the top surface as T_0 and that of the bottom surface as T_1. The temperature difference is $\Delta T = T_1 - T_0$. The vertical boundaries are thermally insulated. All surfaces are free-slip. The velocity normal to a cell boundary is zero.

Non-dimensionalization

The above equations are non-dimensionalized as follows:

$$\bar{x_i} = \frac{x_i}{d},\ \bar{u_i} = u_i\frac{d}{\kappa},\ \bar{\eta} = \frac{\eta}{\eta_1},\ \bar{\tau} = \tau\frac{d^2}{\kappa\eta_1},\ \bar{t} = t\frac{\kappa}{d^2},\ \bar{T} = \frac{T}{\Delta T}, \quad (13)$$

where d is the layer depth, t is time, η_1 is the reference viscosity (at the bottom of the convective layer), and ΔT is the temperature drop across the layer. The Rayleigh number can then be used to characterize the system:

$$Ra = \frac{\alpha\rho g\Delta T d^3}{\kappa\eta_1}. \quad (14)$$

The Arrhenius viscosity is non-dimensionalized as:

$$\bar{E} = \frac{E}{R\Delta T}, \quad (15)$$

$$\bar{\eta} = \eta_{r,\mathrm{Arr}}\exp\frac{\bar{E}}{\bar{T_0} + \bar{T'}}, \quad (16)$$

while the dimensionless exponential viscosity is as follows:

$$\bar{\eta} = \eta_{r,\exp}\exp(-\theta\bar{T'}), \quad (17)$$

where the pre-factors $\eta_{r,\mathrm{Arr}}$ and $\eta_{r,\exp}$ are chosen to ensure that the viscosity is equal to unity at $\bar{T'} = 1$ and the Frank-Kamenetskii parameter θ is the non-dimensionalized form of the constant γ (Equation 6):

$$\theta = \gamma\Delta T. \quad (18)$$

In this case, the viscosity contrast is characterized by only one parameter (θ):

$$\Delta\eta = e^{\theta}, \quad (19)$$

The non-dimensional yield stress is as follows:

$$\bar{\tau_y} = \bar{\tau_0} + \bar{\tau'}_y\bar{z}, \quad (20)$$

where

$$\bar{\tau_0} = \frac{d^2}{\kappa\eta_1}\tau_0 \quad (21)$$

is the non-dimensional yield stress at the surface and

$$\bar{\tau'}_y = \frac{\rho g d^3}{\kappa\eta_1}\mu \quad (22)$$

is the non-dimensional yield stress gradient. The dimensionless pseudoplastic viscosity is as follows:

$$\bar{\eta}_{\mathrm{eff}} = \frac{\bar{\tau_y}}{\bar{\dot{e}}},\ \bar{\dot{e}} = \frac{d^2}{\kappa}\dot{e}. \quad (23)$$

In the following discussion, all parameters are assumed to be non-dimensionalized and the bar sign will be dropped. The non-dimensional forms of Equations 9 to 11 are as follows:

$$\frac{\partial u_i}{\partial x_i} = 0, \quad (24)$$

$$RaT'\mathbf{e}_i - \frac{\partial p'}{\partial x_i} + \frac{\partial \tau_{ij}}{\partial x_j} = 0, \quad (25)$$

$$\frac{\partial T'}{\partial t} + u_i\frac{\partial T'}{\partial x_i} = \frac{\partial^2 T'}{\partial x_i^2}. \quad (26)$$

where $\mathbf{e}_i$ is a unit vector in the direction of gravity.

Matching Arrhenius viscosity and exponential viscosity in non-dimensional form

To compare the two viscosity laws, the Arrhenius viscosity and the exponential viscosity are matched according to Equations 4 and 5:

$$\eta_{\mathrm{r,Arr}}\exp\frac{E}{T_i + T_0} = \eta_{\mathrm{r,exp}}\exp(-\theta T_i), \quad (27)$$

$$\frac{E}{(T_i + T_0)^2}\eta_{\mathrm{r,Arr}}\exp\frac{E}{T_i + T_0} = \theta\eta_{\mathrm{r,exp}}\exp(-\theta T_i). \quad (28)$$

Equations 27 and 28 would yield:

$$\theta = \frac{E}{(T_i + T_0)^2}. \quad (29)$$

T_i differs from the bottom temperature T_1 by a rheological temperature difference ΔT_{rh}, which is on the order of θ^{-1}.

Equation 29 shows that there are various combinations of E and T_0 that would give the same θ, and they would result in different Arrhenius viscosity contrasts:

$$\Delta\eta_{\text{Arr}} = \exp\frac{ET_i}{T_0(T_i + T_0)}. \quad (30)$$

Thus the ratio of Arrhenius viscosity contrast to exponential viscosity contrast $\exp\theta$ is as follows:

$$\frac{\Delta\eta_{\text{Arr}}}{\exp(\theta)} = \exp\frac{ET_i}{T_0(T_i + T_0)^2}. \quad (31)$$

Results

Steady-state convection

We use the finite element code CITCOM (Moresi and Solomatov 1995) to simulate convection in a $64a \times 64$ box, where a is the aspect ratio. Several high viscosity cases were ran with $128a \times 128$ resolution for more accurate results. All cases were run until they reached a steady state. We consider the range of parameters in which convection is in the stagnant lid regime (Solomatov 1995).

The structure of a steady-state convection cell is shown in Figure 1. Due to the temperature-dependent viscosity, the top part of a convective cell forms a stagnant lid and convection only penetrates into the lid by length of δ_{rh} - the rheological layer thickness (e.g., Solomatov 1995). A cold rigid lid, which is often defined by an isotherm, is naturally developed in the top part of the cell sloping downward to the downwelling end. The stress field (Figure 1, right) shows a stress boundary layer near the surface. This is consistent with the analytical solutions of Fowler (1985).

To compare the stresses in exponential and Arrhenius viscosity, we choose a range of T_0 and calculate their corresponding E that gives the same θ according to Equation 29 with $T_i \approx 1$.

Aspect ratio

The horizontally averaged profiles (Figure 2) show that the lid thickness slightly depend on the aspect ratio of the convective cell but the bulk of the temperature and viscosity profile does not vary much with the aspect ratio. However, the stress profile in small aspect ratio cells ($a = 0.25$) is distinctly different from that in larger cells ($a = 0.5$ to 1).

This difference is more apparent in the 2-D plots (Figure 3). In wider cells, the layer with highest stress (red) is approximately symmetric along the half-width of the cell, increasing in depth towards both edges and greater towards the downwelling edge. Below the surface stress boundary layer, the stresses in the middle of the lid are highest as they are not affected by the free-slip boundary conditions at the vertical edges. Although surface stress boundary layer is obvious in horizontally averaged stress profiles, the 2-D stress fields reveal that the surface stress are not always greater than that at depth. Figure 3 shows that at mid-width, it is possible that the surface stresses are lower than the interior. The high-stress region (orange to lime) roughly correspond to the cold lid shown in the temperature distribution, both having slopes towards the downwelling edge. In the narrowest cell however, this high-stress slope deviates from the thermal lid slope. There is a high-stress 'core' within the lid where the magnitude of stresses is close to that of the surface stress boundary layer. This implies that steady-state convection in small aspect ratios may be in another

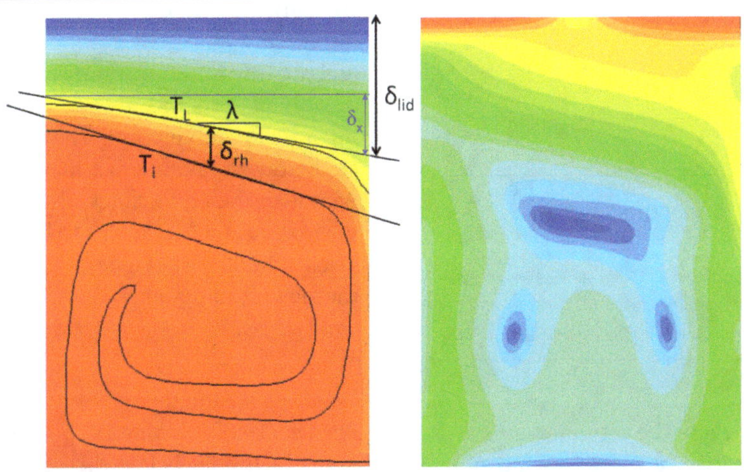

Figure 1 Temperature (left) and stress fields (right) of a steady-state convective cell. Color scale goes from high (red) to low (blue). The lid is defined by an isotherm T_L, and the interior temperature T_i is found by averaging the temperature of the convecting interior excluding boundary effects. For scaling purposes, the lid slope λ and rheological sublayer thickness δ_{rh} is taken at mid-width, whereas lid thickness δ_{lid} is extrapolated to the edge from the lid slope in the middle.

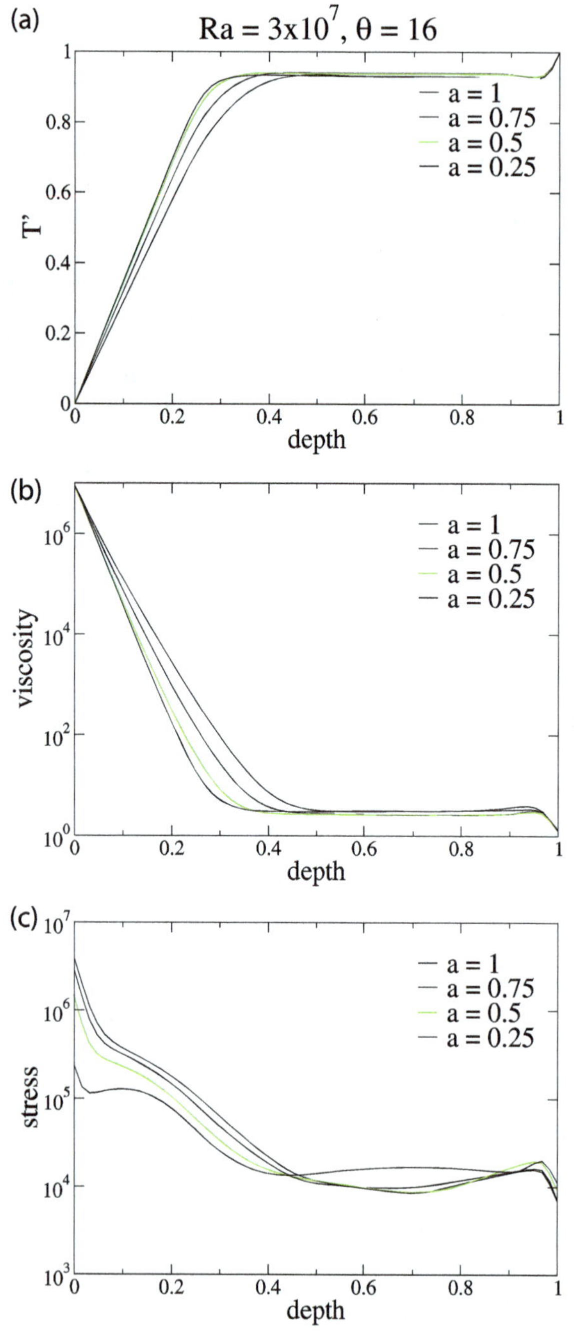

Figure 2 Comparison of **(a)** viscosity, **(b)** temperature, **(c)** stress profiles of exponential viscosities for $\theta = 16$, $Ra = 3 \times 10^7$, and varying a. The somewhat different stress profile of $a = 0.25$ suggests that the surface stress boundary layer is not the region with highest stress.

regime of plastic failure where the interior stresses reach the yield stress first, such that the plastic zone could initiate at depth while the surface may or may not be plastic, depending on the yield stress.

For cases with high-viscosity contrasts and larger aspect ratios, convection is localized to a small part of the cell, or there could be multiple convective cells (Figure 4). These cases might belong to the subcritical convective regime (Solomatov 2012), which has a different behavior from supercritical convection and therefore not in our scope of study. To stabilize one-cell flow, smaller aspect ratio cases are considered to investigate the dependence of critical yield stress on aspect ratio.

Rayleigh number

Increasing Ra reduces the lid thickness as well as the lid slope (Figure 5). The stresses are larger with higher Ra, as is expected with more vigorous convection.

Viscosity contrast

The effects of viscosity contrast on the interior profile are illustrated in the plots in Figure 6. The conductive lid becomes thicker and the interior temperature is closer with the bottom temperature with increasing θ. In the stagnant lid regime ($\theta > 10$), the stress boundary layer near the surface is more pronounced than in the transitional regime.

The interior viscosity, temperature, and stress for Arrhenius viscosity and exponential viscosity are close. We investigate a range of different Arrhenius viscosity values by varying T_0, noting that at $T_0 = 2.0$ is a rather high surface temperature. As T_0 decreases, the difference in viscosity contrasts calculated by the two viscosity laws becomes larger. However, the temperature and the stress profiles are similar, as shown in the horizontally averaged profiles in Figure 7. There is only a slight decrease in thickness of the stress boundary layer as the Arrhenius viscosity contrast increases. The 2-D stress distributions in Figures 8 and 9 reflect the small differences in the stress distribution as viscosity contrast increases, both in exponential viscosity and Arrhenius viscosity. This is contrary to the findings of Stein and Hansen (2013), which observed a distinctly thinner lid in Arrhenius viscosity and slightly different temperature and viscosity profiles. The small difference that we observe may affect the scaling laws for subduction. The vertical variation in stresses at the downwelling edge may be particularly important. Although the lid thickness is about the same, the contrast in stresses at $x = a$ seems to be greater at low T_0 or high-viscosity contrast (Figure 9). Since subduction occurs at the downwelling edge, this may influence the scaling of yield stress.

Lid stress scaling theory

Fowler (1985) obtained a polynomial expression for the stress in the lid below the surface stress boundary layer in large lid slope approximation, which allows a comparatively simple scaling relation for stress. In order to solve the equations of convection, Fowler took Ra and θ to be asymptotically large and assumed the magnitude of lid

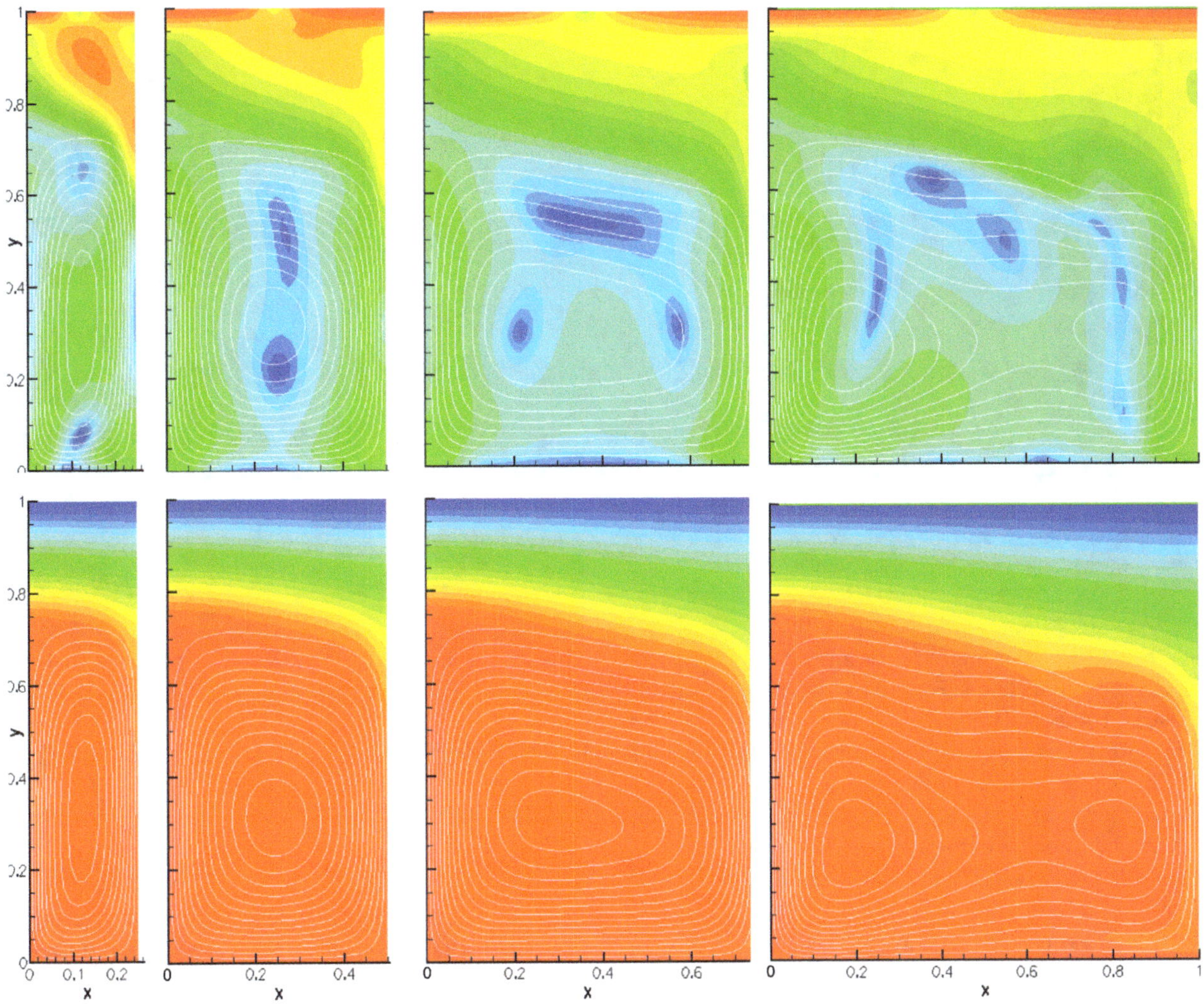

Figure 3 Stress (top row) and temperature fields (bottom row) of convecting cells with $Ra = 3 \times 10^7$, $\theta = 16$, and various a.

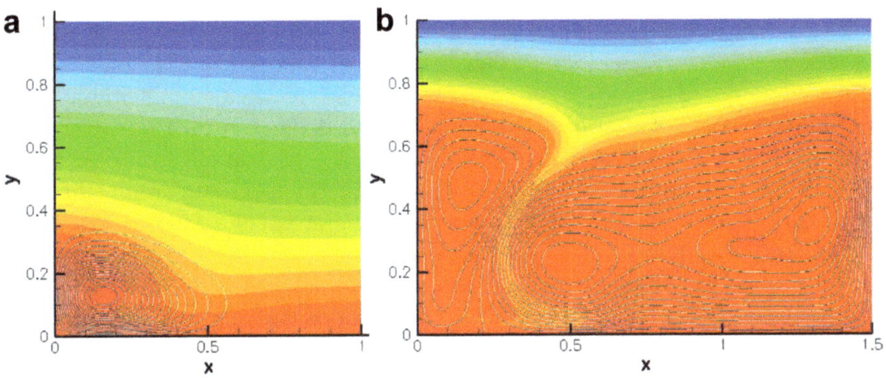

Figure 4 Localized convection and multiple subcells in steady-state convection. **(a)** $\theta = 19$, $T_0 = 0.6$, $a = 1$, $Ra = 1 \times 10^7$. **(b)** $\theta = 16$, $T_0 = 0.6$, $Ra = 3 \times 10^7$.

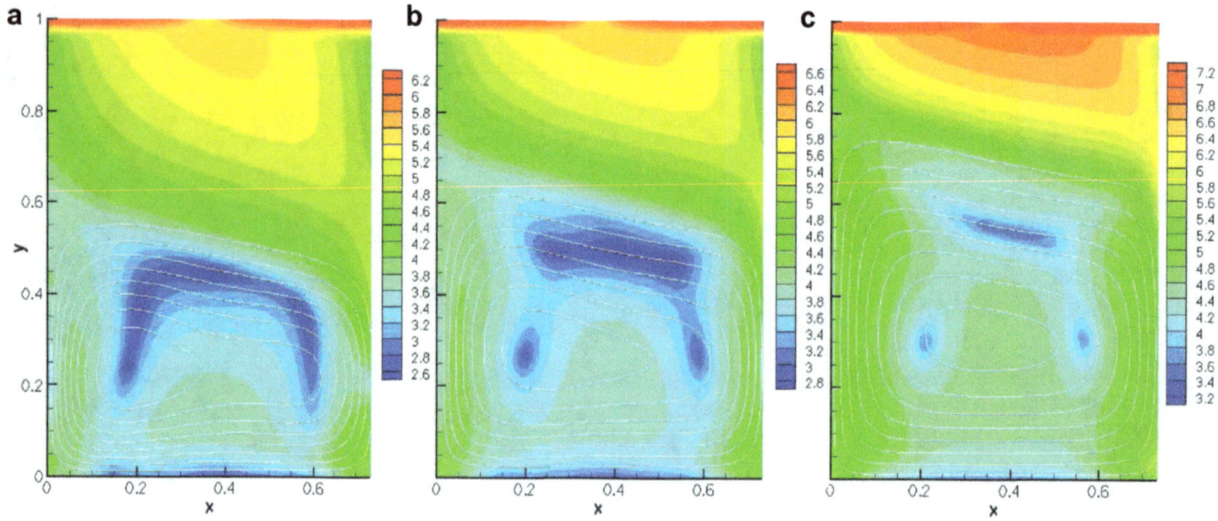

Figure 5 Stress structure of steady state cases with $\theta = 16, a = 0.75, T_0 = 0.8$. **(a)** $Ra = 10^7$. **(b)** $Ra = 3 \times 10^7$. **(c)** $Ra = 10^8$. The area of convecting interior becomes larger as Ra increases and the thickness of the stagnant lid decreases. Scale shows the values of $\log \tau$.

slope to be either on the order of lid thickness or rheological sublayer thickness. For the Earth and some smaller terrestrial planets as well as most numerical simulations, Ra may not be as high as the asymptotic theory require, and thus, some other theory may be needed for lid stress scaling. Moreover, as we find in this study, the lid slope does not follow either of these two end-member cases and thus needs to be scaled based on numerical simulations.

Fowler also found that the interior flow can be uncoupled from the rheological sublayer, which makes the problem setup akin to a viscous lid gravitationally sliding along a slope. We can therefore estimate the shear stress in the rigid lid τ_{lid} by considering the force balance on the lid (Figure 1):

$$\frac{d\tau}{dy} = \Delta\rho g_x. \tag{32}$$

For density changes due to temperature variations, the lid stress can be integrated from Equation 32:

$$\tau_{\text{lid}} = -\alpha\rho_0 \frac{dT}{dy} g \sin\lambda \frac{y^2}{2} + \tau_i, \tag{33}$$

where τ_i is the stress at the interior temperature T_i.

There are two assumptions that allow us to simplify Equation 33. The first is that τ_i is negligible since $\tau_i \ll \tau_{\text{lid}}$. The second is the small lid slope approximation. For small λ, $\sin(\lambda)$ is approximately equal to λ. The non-dimensional form of Equation 33 becomes:

$$\tau_{\text{lid}} = -\text{Ra}\frac{dT}{dy}\frac{y^2}{2}\lambda. \tag{34}$$

Thus, the lid stress is determined by Ra, dT/dy, λ, and y, which will be defined in the following discussion. This scaling is similar to that obtained from the analytical solutions of Fowler (1985).

Lid base temperature

The lid base is often defined by an isotherm:

$$T_L = T_i - C\theta^{-1} \tag{35}$$

where C is a constant. We determine the lid base from velocity profile (Solomatov and Moresi 2000) to find the constant C. We first find the greatest velocity gradient at a specific distance x from the upwelling edge of the cell. This velocity gradient is then extended to the depth at which velocity is zero, as shown in Figure 10. This depth defines the lid thickness. This process is repeated for all x (from 0 to a) to obtain the shape of the lid base across the convecting cell. This velocity gradient-defined lid base is then used to find the thermal lid base defined by Equation 35. The interior temperature T_i is found by averaging the temperature in the middle part of the interior to exclude the boundary effects. The constant C is determined by matching the lid thickness at mid-width ($x = 0.5a$) given by T_L and that defined by the velocity gradient (Figure 11). We choose the value at mid-width because in most cases, the two lid bases are closest around the middle of the cell for a large lateral extent. In some cases, especially for narrower cells, the temperature lid base and the velocity lid base may not match, so the mid-width serves as a reference point for consistency in defining the lid base.

Equation 35 allows us to determine the rheological temperature difference $\Delta T_{\text{rh}} = T_i - T_L = C\theta^{-1}$. It varies with Ra, θ, and aspect ratio, and the scaling exponents are summarized in Table 1.

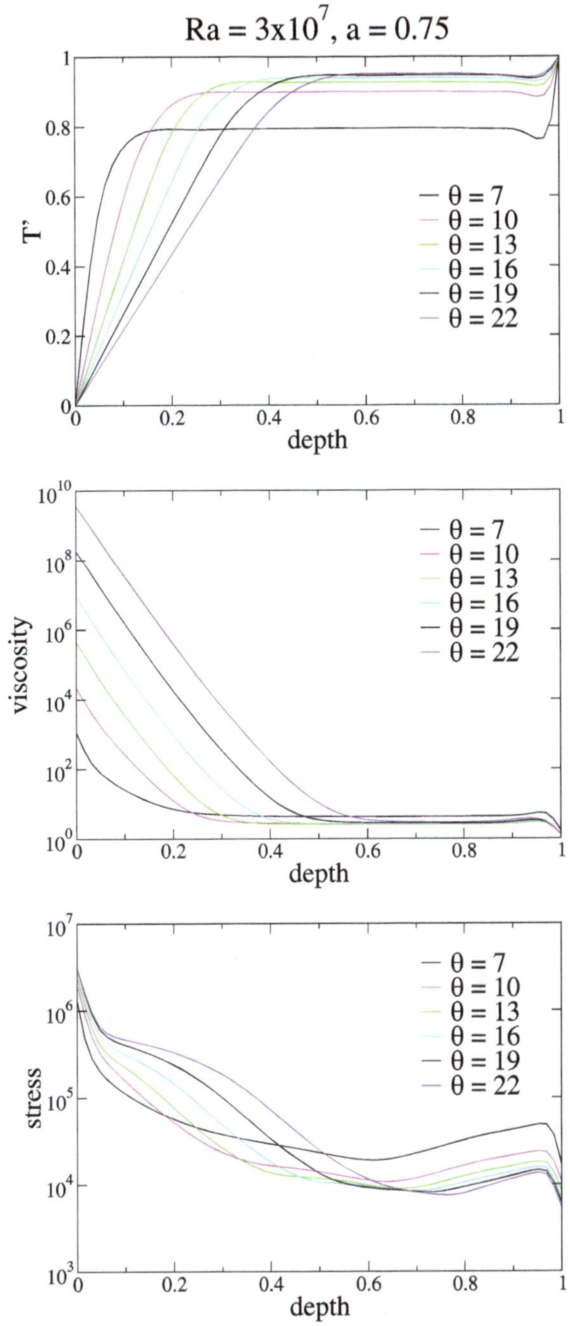

Figure 6 Temperature, viscosity, and stress profiles of $Ra = 3 \times 10^7$, $a = 0.75$, and varying Frank-Kamenetskii parameter θ.

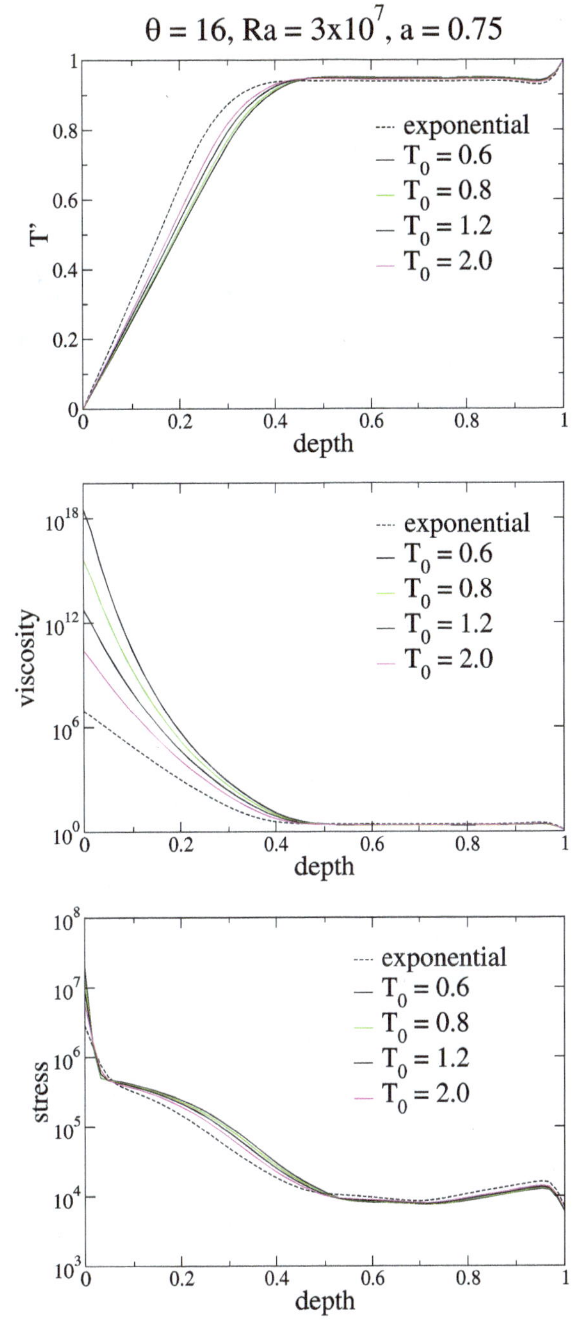

Figure 7 Comparison of temperature, viscosity, and stress profiles of Arrhenius and exponential viscosities for $\theta = 16$, $Ra = 3 \times 10^7$, $a = 0.75$, and various T_0.

Lid slope and lid thickness

Gravitational sliding requires a downward dipping slope λ as indicated in Equation 34. In larger aspect ratio cells, although the lid thickness varies horizontally, the lid slope is approximately constant in the middle portion of the cell. This is different for smaller cells, where the lid base could be some function of x instead of a straight line (Figure 12). For example, Fowler suggests that the lid base varies with $x^{0.4}$ (Fowler 1985). For consistency in our scaling analysis, the lid slope is taken to be the slope of the thermal lid base at mid-width.

To check whether the lid slope scales with lid thickness or rheological sublayer thickness (Fowler 1985), we look at the vertical drop δ_x (Figure 1) and define the lid slope as δ_x/a. If the lid slope scales with the lid thickness, then $\delta_x \sim \delta_{lid} \sim \mathrm{Nu}^{-1}$, and thus

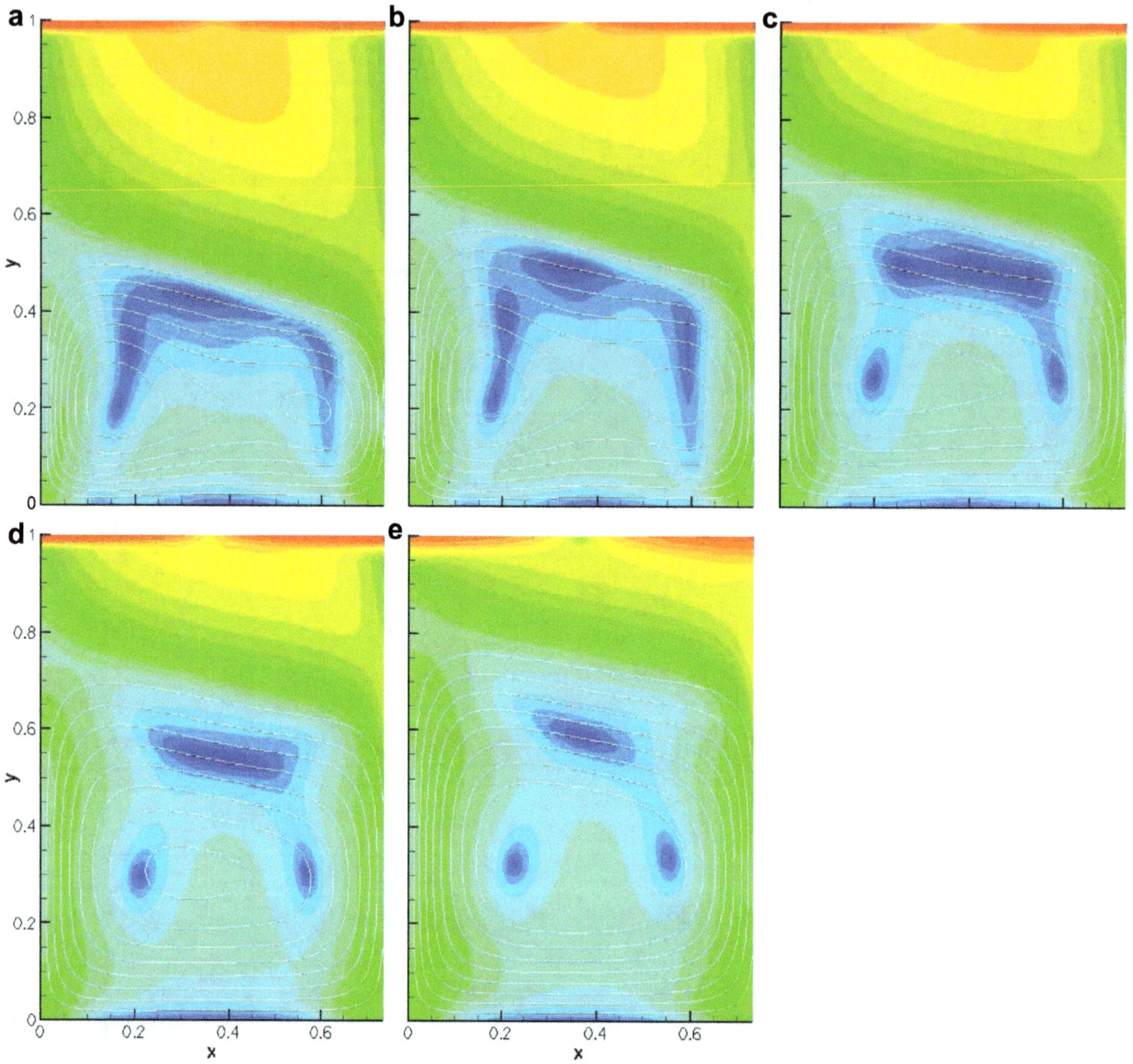

Figure 8 Stress structure of steady state cases with $Ra = 3 \times 10^7$, $T_0 = 0.8$, $a = 0.75$. **(a-d)** Arrhenius viscosity with **(a)** $\theta = 22$, **(b)** $\theta = 19$, **(c)** $\theta = 16$, **(d)** $\theta = 13$, and **(e)** exponential viscosity with $\theta = 13$. Color scale shows the values of $\log(\tau)$. White lines represent streamlines.

$$\delta_x/\delta_{lid} \sim Nu\delta_x \sim \text{constant.} \tag{36}$$

If the lid slope scales with δ_{rh}, then $\delta_x \sim \delta_{rh}$. As $\delta_{rh}/\delta_{lid} \sim \theta^{-1}$, thus $\delta_x/\delta_{lid} \sim \theta^{-1}$, or

$$\theta\delta_x/\delta_{lid} \sim Nu\delta_x\theta \sim \text{constant.} \tag{37}$$

We plot $Nu\delta_x$ and $Nu\delta_x\theta$ in Figure 13. Neither combination remains constant with θ, with $Nu\delta_x$ increases with θ and $Nu\delta_x\theta$ decreases with θ. This suggests that the lid slope is somewhere in between these two extreme cases. Therefore in deriving the scaling laws for stresses, we need to determine the dependence of lid slope on various convective parameters (Table 1).

Thermal gradient

The temperature is approximately a linear function of depth and the thermal gradient in the lid is about constant with depth. At mid-width ($x = 0.5a$), it is approximately equal to the Nusselt number, which is the non-dimensional horizontally averaged surface temperature gradient. Since we are looking at temperature changes from the interior to the bottom of the lid which includes the rheological sublayer, we also check the thermal gradient in the rheological sublayer $\Delta T_{rh}/\delta_{rh}$ to note any difference in the scaling relations. As before, we choose the values $\Delta T_{rh}/\delta_{rh}$ at the mid-width to exclude boundary effects for scaling purposes.

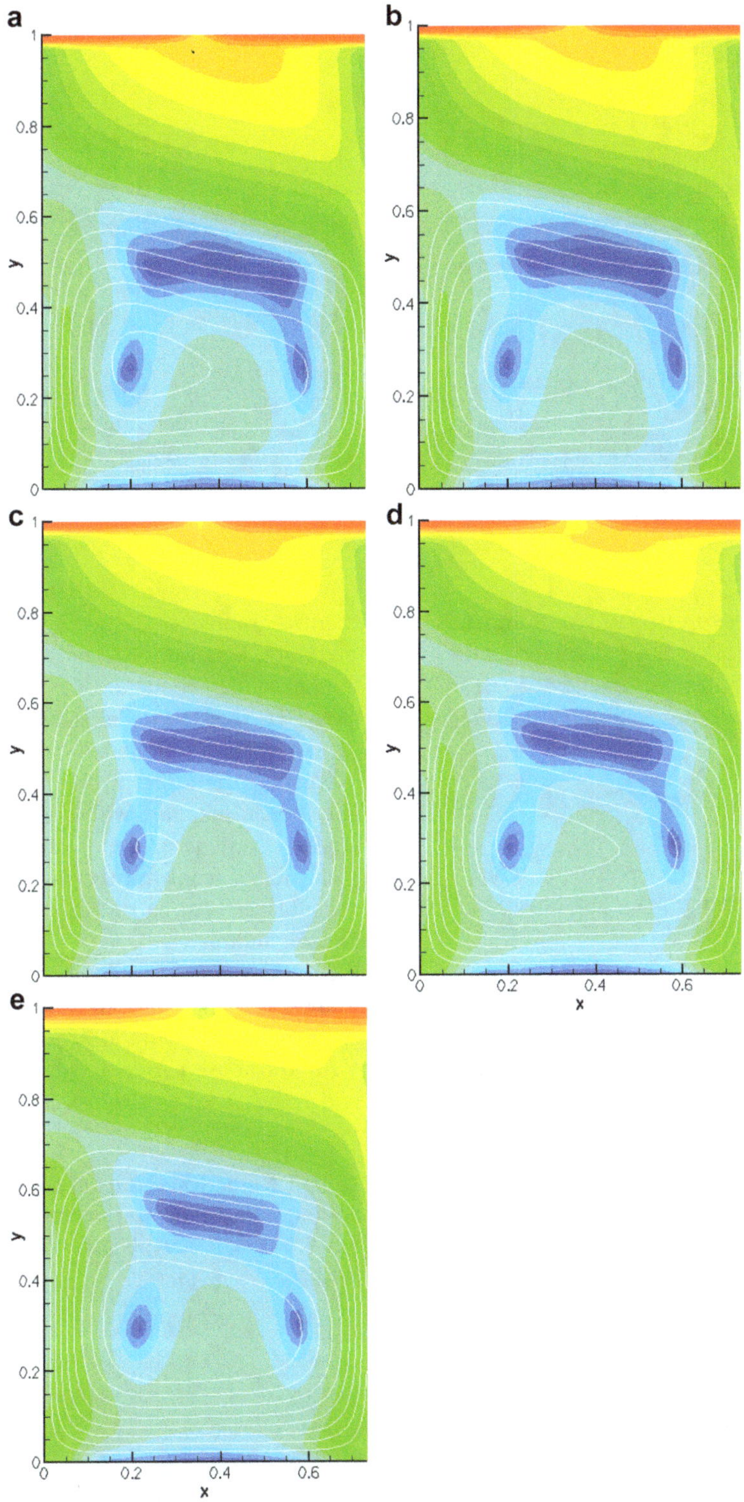

Figure 9 Stress structure of steady state cases with $Ra = 3 \times 10^7, \theta = 16, a = 0.75$. Cases **(a-d)** use Arrhenius viscosity with **(a)** $T_0 = 0.6$, **(b)** $T_0 = 0.8$, **(c)** $T_0 = 1.2$, and **(d)** $T_0 = 2.0$. Case **(e)** uses exponential viscosity. Color scale shows the values of $\log(\tau)$. White lines represent streamlines.

In previous theories, $\Delta T_{\rm rh}/\Delta T \sim \theta^{-1}$ and $\delta_{\rm rh}/\delta_{\rm lid} \sim \theta^{-1}$. The determination of C follows the description in the previous section on lid base temperature, and it is found to be dependent on aspect ratio and θ. Therefore, $\Delta T_{\rm rh}/\Delta T$ and $\delta_{\rm rh}/\delta_{\rm lid}$ will also have a dependence on a and θ, and their scaling relations are summarized in Table 1.

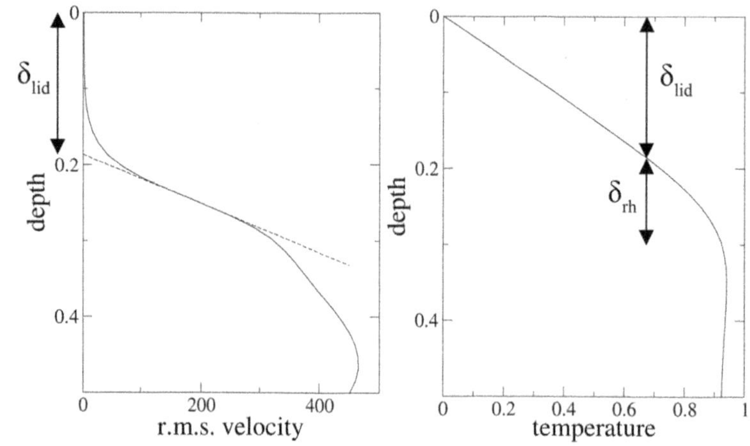

Figure 10 Velocity profile taken at mid-width of the cell ($x = 0.5a$). Dotted line is the linear extrapolation of the maximum velocity gradient, and the lid base is marked at the depth at which this line intersect with the vertical axis.

Stress scaling

Equation 34 can now be expressed in a non-dimensional form with the definitions of various parameters in the previous discussion. At the bottom of the lid at δ_{rh} from the interior, where the temperature difference is ΔT_{rh}, the stresses are as follows:

$$\tau_{\text{lid}} \sim \text{Ra}\frac{\Delta T_{\text{rh}}}{\delta_{\text{rh}}}\lambda y^2. \tag{38}$$

Further into the lid, the non-dimensional temperature gradient is the Nusselt number. Therefore Equation 34 can be alternatively scaled as:

$$\tau_{\text{lid}} \sim \text{RaNu}\lambda y^2. \tag{39}$$

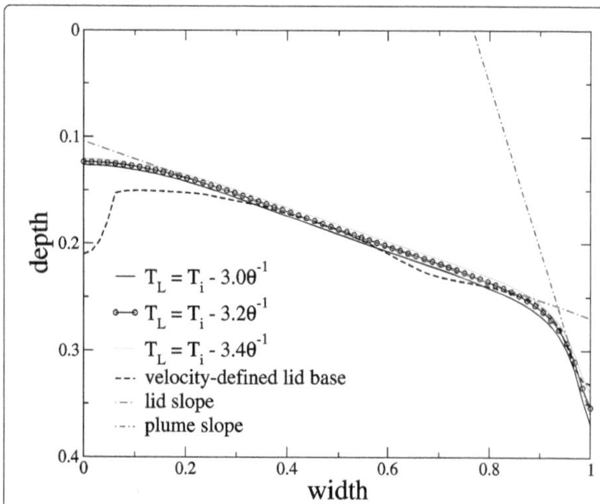

Figure 11 Top part of a convective cell showing various definitions of the lid base. The lid base defined by the velocity gradient is shown in dashed line, and it matches the temperature lid base defined at $T_L = T_i - 3.2\theta^{-1}$ at around the mid-point. The lid slope is estimated at the mid-point.

As shown in Figure 10, there is a slight difference in the thermal gradient in the lid and in the rheological sublayer; therefore, Equations 38 and 39 may result in slightly different scaling exponents. We check both scaling relations to see whether the stresses at the lid base and those in the lid can be scaled similarly.

We plot the stress profile according to Equations 38 and 39 and compare with that from numerical solutions (Figure 14). The prefactor of the stress as a function of y calculated from the thermal gradient in the rheological sublayer ($T_{\text{rh}}/\delta_{\text{rh}}$) is 5.9×10^6, whereas that from the Nusselt number is 9.6×10^6. This demonstrates that the theoretical stress profiles match fairly closely with the numerical one, and the best fit can be obtained with some small adjustments in the coefficient.

All the above parameters depend on Ra, aspect ratio a, and θ; thus, they can be expressed as $\text{Ra}^{\beta}a^{\zeta}\theta^{\alpha}$ where β, ζ,

Table 1 Numerical results of power law coefficients in scalings of different parameters with Ra, aspect ration *a*, and Frank-Kamenetskii parameter θ

Parameter	c in 10^c	Ra	a	θ
Nu	0.23 ± 0.07	0.24 ± 0.01	-0.17 ± 0.03	-1.17 ± 0.05
$\Delta T_{rh}/\Delta T$	0.25 ± 0.17	0.010 ± 0.02	0.39 ± 0.08	-0.84 ± 0.12
δ_0/d	-0.26 ± 0.10	-0.22 ± 0.01	0.27 ± 0.04	1.12 ± 0.06
δ_{lid}/d	-0.74 ± 0.06	-0.23 ± 0.01	0.12 ± 0.03	1.44 ± 0.04
$\delta_{\text{lid,max}}/d$	-0.62 ± 0.07	-0.19 ± 0.01	0.33 ± 0.03	1.26 ± 0.05
δ_{rh}/d	0.02 ± 0.33	-0.20 ± 0.04	0.60 ± 0.15	0.34 ± 0.21
Lid slope	-0.98 ± 0.13	-0.07 ± 0.02	0.14 ± 0.06	0.63 ± 0.09
τ_{lid}	0.36 ± 0.21	0.68 ± 0.03	0.92 ± 0.09	-0.15 ± 0.13
$\tau_{y,cr}$	0.40 ± 0.22	1.09 ± 0.03	1.67 ± 0.10	-1.49 ± 0.14
$\tau'_{y,cr}$	2.34 ± 0.32	1.34 ± 0.04	1.45 ± 0.14	-3.47 ± 0.21

$\Delta T_{rh}/\Delta T$ and τ_{lid} taken at mid-width.

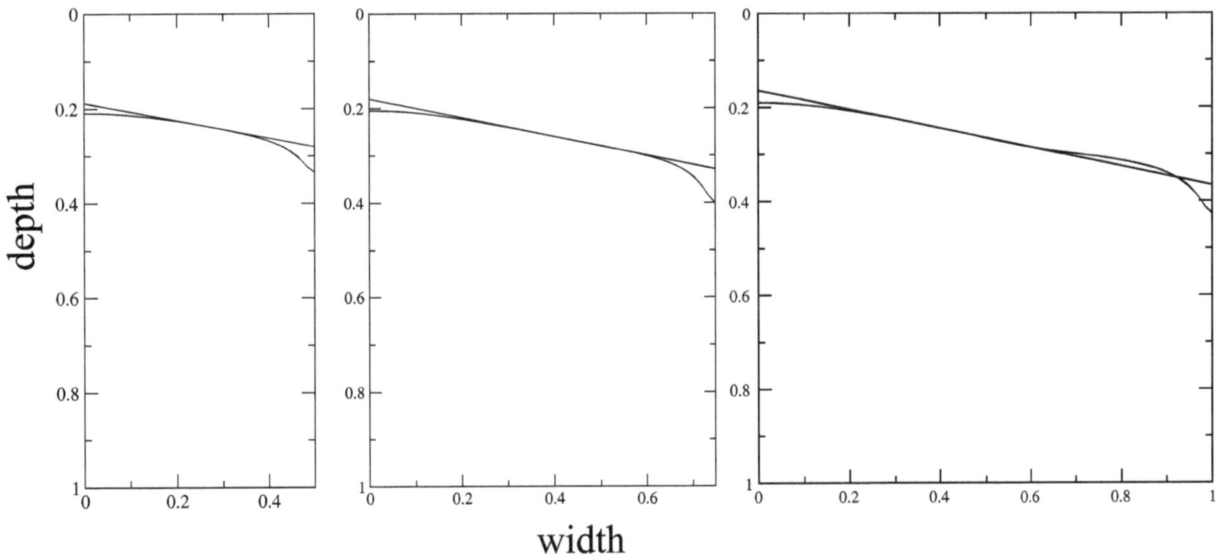

Figure 12 The slope of the lid base at various aspect ratios. The lid slope deviates from the linear approximation as the cell aspect ratio and viscosity contrast increase; therefore, a non-constant value of the lid slope may affect the scalings.

and α are scaling exponents. The results are summarized in Table 2.

Convection with yield stress

We use the steady-state solutions as the starting point before imposing a yield stress to simulate plastic yielding. For the yield stress gradient, a small cohesion term (surface yield stress) was introduced to stabilize the solution.

Regimes of convection with constant yield stress or constant yield stress gradient

When a yield stress is present, the regions with stresses that reach the yield value would have an approximately constant stress close to τ_y. These plastic zones develop first at the corners of the cell where the stresses are highest. As τ_y decreases, the plastic zones extend both in depth and horizontally, narrowing the width of the high-viscosity part of the lid (Figure 15). If yield stress is too high, the depth of the plastic zone is small or the plastic zone is entirely absent; thus, the stagnant lid does not fail. If the yield stress is sufficiently low, the plastic zone extends suffcieintly deep so that the stagnant lid is mobilized.

We examine the stress and viscosity profiles at various locations x to see how they change in the presence of a yield stress (Figures 16 and 17). At high yield stress or high yield stress gradient, the plastic zone only occurs at shallow depths and the bulk of the stress and viscosity profiles are unaltered from the stagnant lid state. The plastic zone extends deeper as the yield stress or yield stress gradient decreases. As the yield stress approaches the critical value, a small change in yield stress induces a change in plastic depth that is comparable to the change caused by an order of magnitude change in yield

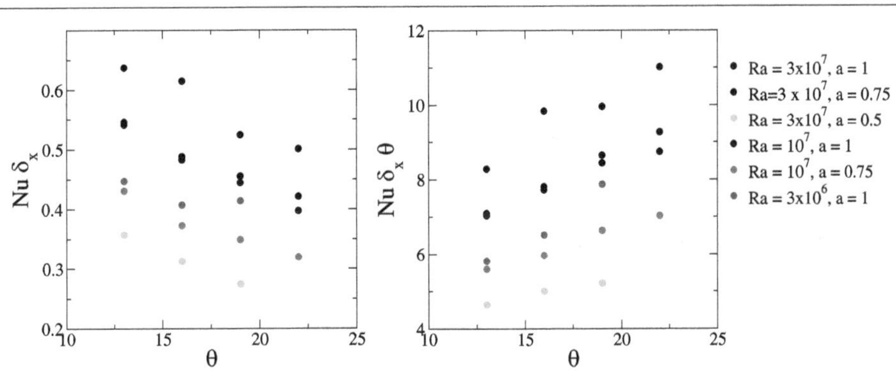

Figure 13 Plots of $Nu\delta_x$ (or δ_x/δ_{lid}) and $Nu\delta_x\theta$ (or $\theta\delta_x/\delta_{lid}$) as a function of θ.

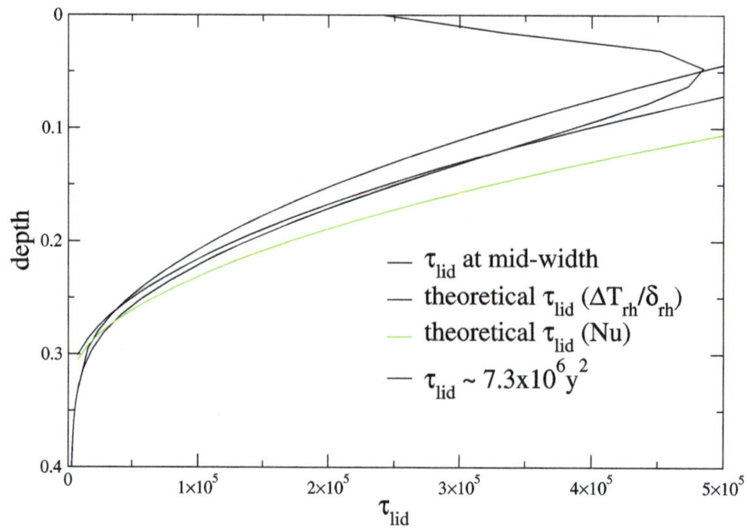

Figure 14 Comparison of stress profiles at mid-width obtained from numerical calculations and theory. Blue line represents the best fit to numerical solutions of stresses. The stress profile is taken at mid-width, where the stresses in the surface boundary layer is lower than the interior as shown in the 2-D stress fields.

stress when it is far from critical. Although subduction does not occur, this implies a change in convection regime from stagnant lid to some sort of transitional regime.

In this transitional regime, the yield stress slightly changes the interior dynamics as can be seen in various convective parameters of the interior region of the cell. The yield stress increases the lid slope whereas the lid thickness remains approximately the same. When the yield stress is slightly above the critical value, these changes caused by yield stress are negligible and convection remains in the stagnant lid regime. Therefore in deriving the critical yield stress scalings, we refer to the steady-state structure that has a yield stress just above the critical value, so that the scaling relations for various convective parameters (Table 1) from steady-state stagnant lid convection can still be used.

Time evolution of lid weakening and failure

When the lid fails, the surface velocity continuously increases and overturn occurs (Figure 18). Figure 19 shows the time sequence of stress structures before and during failure. When the surface velocity is still low compared to the bottom velocity (Figure 18, left), the variation in stress structure is not obvious. It is not until the velocity begins to increase drastically that the plastic yield zones from the two corners start to connect in the middle of the cell to form a plastic lid. The weak lid then becomes unstable and starts to subduct.

Table 2 Power law coefficients in scaling laws for stresses: numerical results vs theory

Parameter	Ra	a	θ	Method
τ_{lid}	0.68 ± 0.03	0.92 ± 0.09	-0.15 ± 0.13	Numerical
	0.73 ± 0.09	1.14 ± 0.29	0.146 ± 0.42	Theory (in terms of $\Delta T_{rh}/\delta_{rh}$)
	0.77 ± 0.12	1.18 ± 0.39	0.14 ± 0.56	Theory (in terms of Nu)
$\tau_{y,cr}$	1.09 ± 0.03	1.67 ± 0.10	-1.49 ± 0.14	Numerical
	0.94 ± 0.10	1.81 ± 0.32	-1.13 ± 0.47	Theory (in terms of $\Delta T_{rh}/\delta_{rh}$)
	0.97 ± 0.13	1.85 ± 0.42	-1.16 ± 0.61	Theory (in terms of Nu)
$\tau'_{y,cr}$	1.34 ± 0.04	1.45 ± 0.14	-3.47 ± 0.21	Numerical
	1.13 ± 0.11	1.48 ± 0.35	-2.39 ± 0.51	Theory (in terms of Nu)
	1.16 ± 0.14	1.52 ± 0.45	-2.38 ± 0.65	Theory (in terms of $\Delta T_{rh}/\delta_{rh}$)

$\Delta T_{rh}/\Delta T$ and τ_{lid} taken at mid-width.

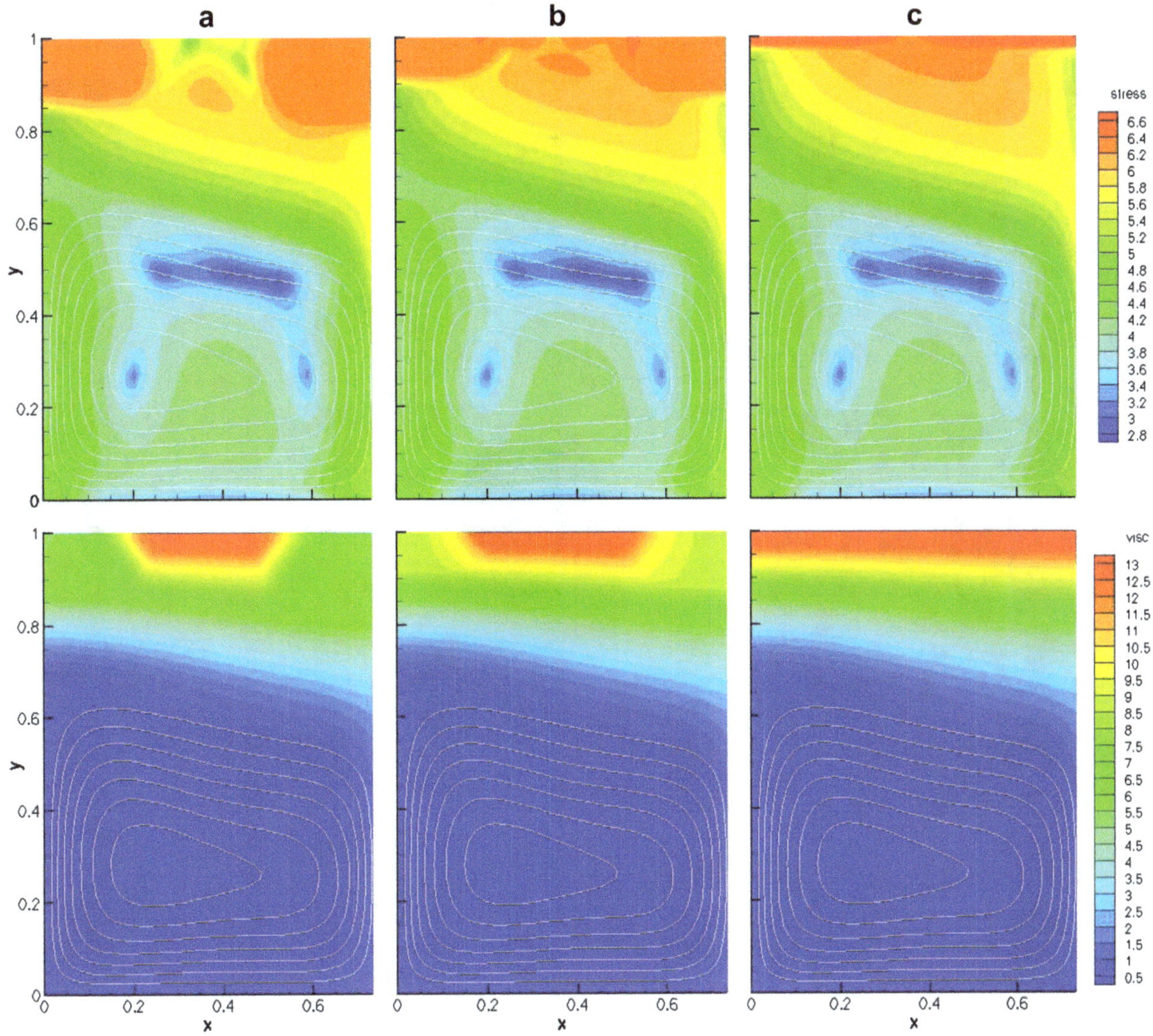

Figure 15 Stress (top) and viscosity (bottom) fields of case $Ra = 3 \times 10^7, \theta = 16, a = 0.75, T_0 = 0.8$. **(a)** $\tau_y = 6.7 \times 10^5$, **(b)** $\tau_y = 7 \times 10^5$, **(c)** $\tau_y = \infty$. Failure occurs at $\tau_y = 6.6 \times 10^5$, while the stagnant lid remains at higher τ_y. Color bars show $\log \tau$ values.

A possible contribution to the uncertainty in determining the critical yield stress is that at the vicinity of the critical value, the behavior may be difficult to interpret. In some cases that as the yield stress gets close to a critical value, the surface velocity increases slowly and it may take more than 10^5 timesteps to reach a point of overturn, whereas typically, it takes less than 10^4 timesteps to a drastic increase in surface velocity (Figure 18 right). This may be due to the behavior of dynamic system near a critical point.

Critical depth of plastic failure zone

For subduction to occur, the lithosphere has to be sufficiently weak so that it can be mobilized by stresses arisen from mantle convection. As seen in Figures 16 and 17, the lid remains stagnant if the plastic yield zone is small. Therefore, the question is how deep does this plastic zone have to penetrate for the lid to be mobilized? Previous theories proposed that the plastic zone has to penetrate through some critical depth $\delta_{\rm pl}$ defined by a critical temperature T_c:

$$\frac{\delta_{\rm pl}}{\delta_0} = \frac{T_c - T_0}{T_i - T_0}, \qquad (40)$$

where $\delta_0 = \delta_{\rm lid} + \delta_{\rm rh}$. The model of Fowler and O'Brien (2003) predicts that δ_{pl} is defined by the temperature that gives the interior viscosity. For Newtonian rheology, this means that the plastic zone has to extend through the base of the lid. Solomatov (1995,2004a) suggested that δ_{pl} only has to penetrate to the isotherm at which the viscosity

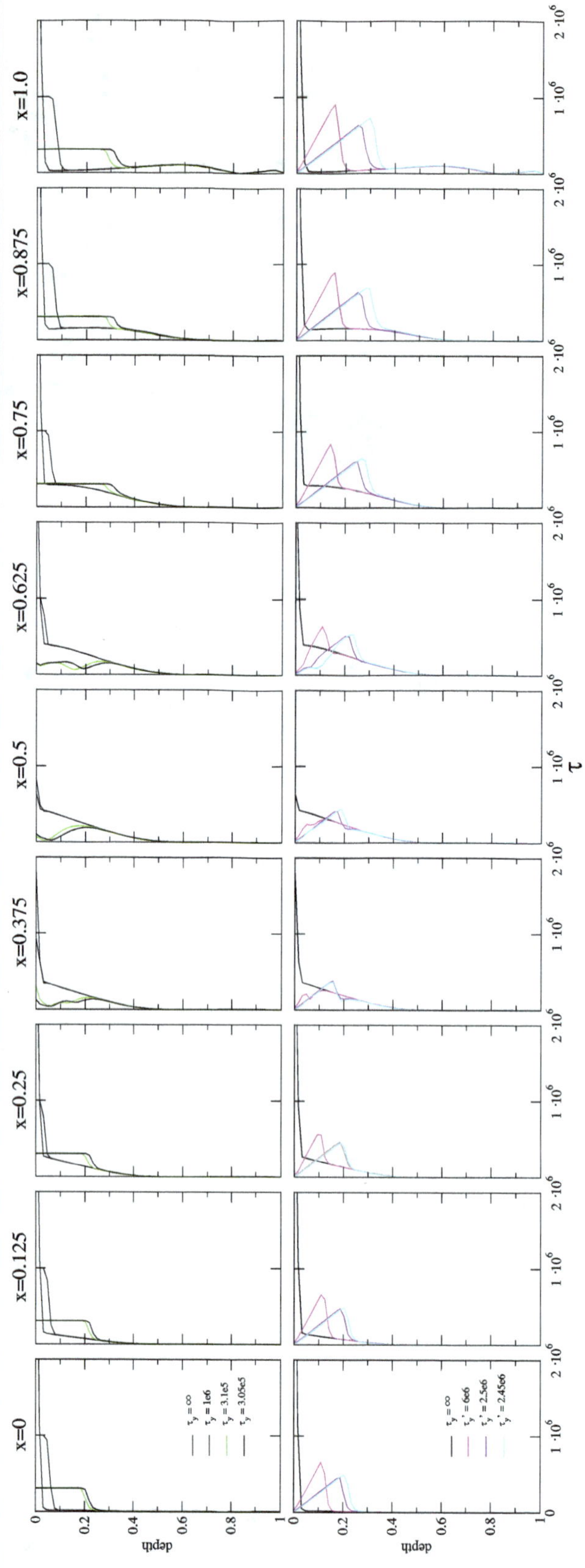

Figure 16 Stress profiles at different widths at various yield stress or yield stress gradient.

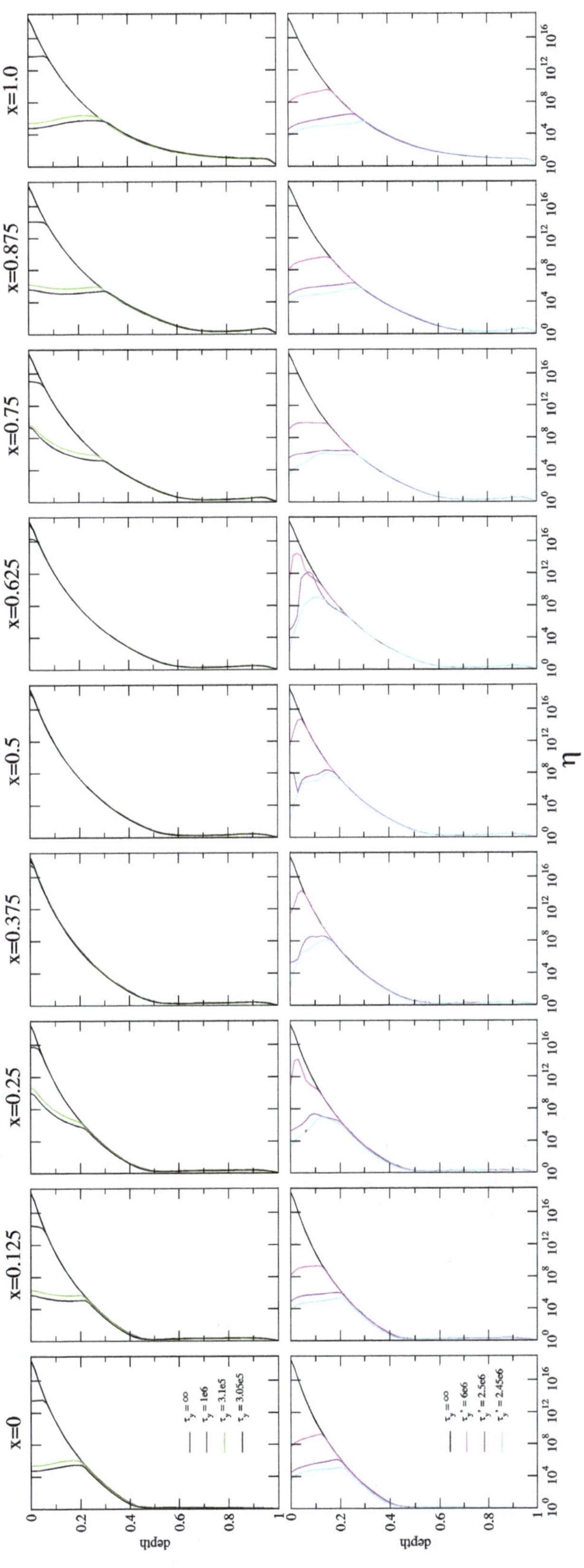

Figure 17 Viscosity profiles at different widths at various yield stress or yield stress gradient.

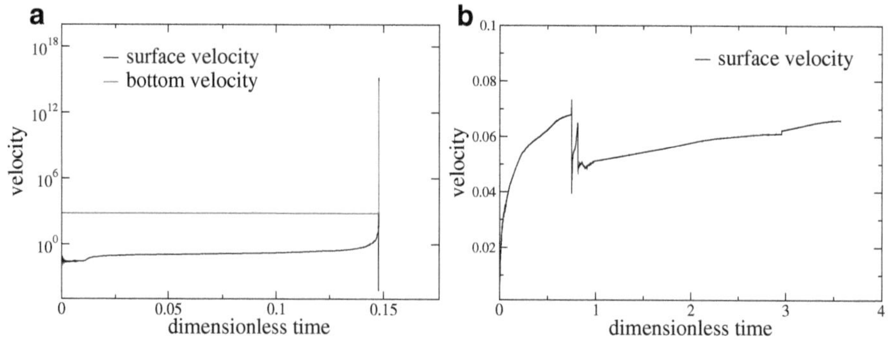

Figure 18 Surface velocities over time for $Ra = 3 \times 10^7, \theta = 16, a = 0.75$, **(a)** $T_0 = 0.8, \tau_y = 6.6 \times 10^5$, **(b)** $T_0 = 0.4, \tau_y = 6.1 \times 10^5$, bottom velocity ≈ 650. The surface velocity in (b) increases slowly. Although it has evolved for approximately 25 times as much as the period for (a) to fail, it is far from reaching the bottom velocity.

contrast with the interior viscosity is $e^{4(n+1)}$, where n is the stress exponent for non-Newtonian viscosity.

To examine these hypotheses, we look at the stress profile at the downwelling edge ($x = a$) to determine the depth of the plastic zone, as the stresses at this edge are the highest and this is where subduction starts (Figure 20). The depth of the plastic zone is defined by extent of the stress modified by the plastic flow law in Equation 7. The exact value of δ_{pl} and the stress at this depth are found by the intercept of the stress calculated from linear extrapolation from the top (where stress is determined by the yield stress) and exponential extrapolation from the creep regime just below the plastic depth. The values of δ_{pl} for constant yield stress cases and that for constant yield stress gradient cases are close. It is noted that the stress at depth δ_{pl} in constant yield stress gradient cases is about twice as much as that in constant yield stress cases. The force on the lid is $\delta_{pl}\tau_{y,cr}$ for constant yield stress and $0.5\delta_{pl}\tau'_{y,cr}$ for constant yield stress gradient. This implies that the force on the lid is about the same for both constant yield stress and yield stress gradient.

As we see from the effects of aspect ratio on steady-state stress distribution, the stresses in the interior become comparable or even exceed the surface stresses as the aspect ratio and Ra decrease and viscosity contrast increases. This means that the plastic zone may not propagate from the top but also develop at depths in the lid, and the plastic zone does not span the whole top part of the cell. (Figure 21). This may represent another regime of lid failure. These cases are thus excluded from our scaling analysis.

The depth of the plastic zone is determined from drop in stress by the yield stress. We note that the zone of reduced viscosity due to the yield stress may correspond to the zone of reduced stress as shown in Figure 20. However, this does not always hold, especially for cases with higher θ. Figure 22 shows that the transition of the plastic zone to creep flow may not correspond to the sharp change in the viscosity. This means that the viscosity at the plastic depth and the maximum viscosity are different, since the reduction in viscosity is not only determined by the yield stress but also by the strain rate.

To find out whether there is a critical viscosity contrast in lithospheric failure, we examine both the viscosity contrast at plastic depth $\Delta\eta_{\text{pl}}$ and the maximum viscosity contrast $\Delta\eta_{\text{max}}$. The maximum viscosity needs to be determined by extrapolation as the resolution near the point where the viscosity is maximum is not high enough to resolve sharp changes in stress and viscosity. The point of maximum viscosity is found by extrapolating the values from both above and below the maximum point (e.g., Figure 22 right). The viscosity is extrapolated linearly from the two points above the maximum point, and below the maximum point, the viscosity is calculated from temperature. The intersection of these two curves determines the maximum viscosity and its depth. We found that these two viscosity contrasts are mostly within the same order of magnitude. Figure 23 shows that the maximum viscosity seems to depend on the original non-yielding viscosity contrast $\Delta\eta$. For exponential viscosity and Arrhenius viscosity cases at low non-yielding viscosity contrast, the maximum viscosity contrast $\Delta\eta_{\text{max}}$ increases with non-yielding viscosity contrast. This may be because they are close to transitional regime. At higher $\Delta\eta$, the increase in $\Delta\eta_{\text{max}}$ seems to decline with increasing $\Delta\eta$, but the spread of data prevents us from concluding that $\Delta\eta_{\text{max}}$ converges towards higher viscosities. As with the maximum viscosity contrast, $\Delta\eta_{\text{pl}}$ also does not display linearity or convergence clearly with either the non-yielding viscosity contrast or θ (Figure 24).

Since the critical viscosity contrast is neither a constant or a function of θ, we look at the depth of the plastic zone to derive scaling relations for the yield stress. We investigate the plastic depth δ_{pl} as a fraction of the lid thickness. For scaling purposes, the lid thickness

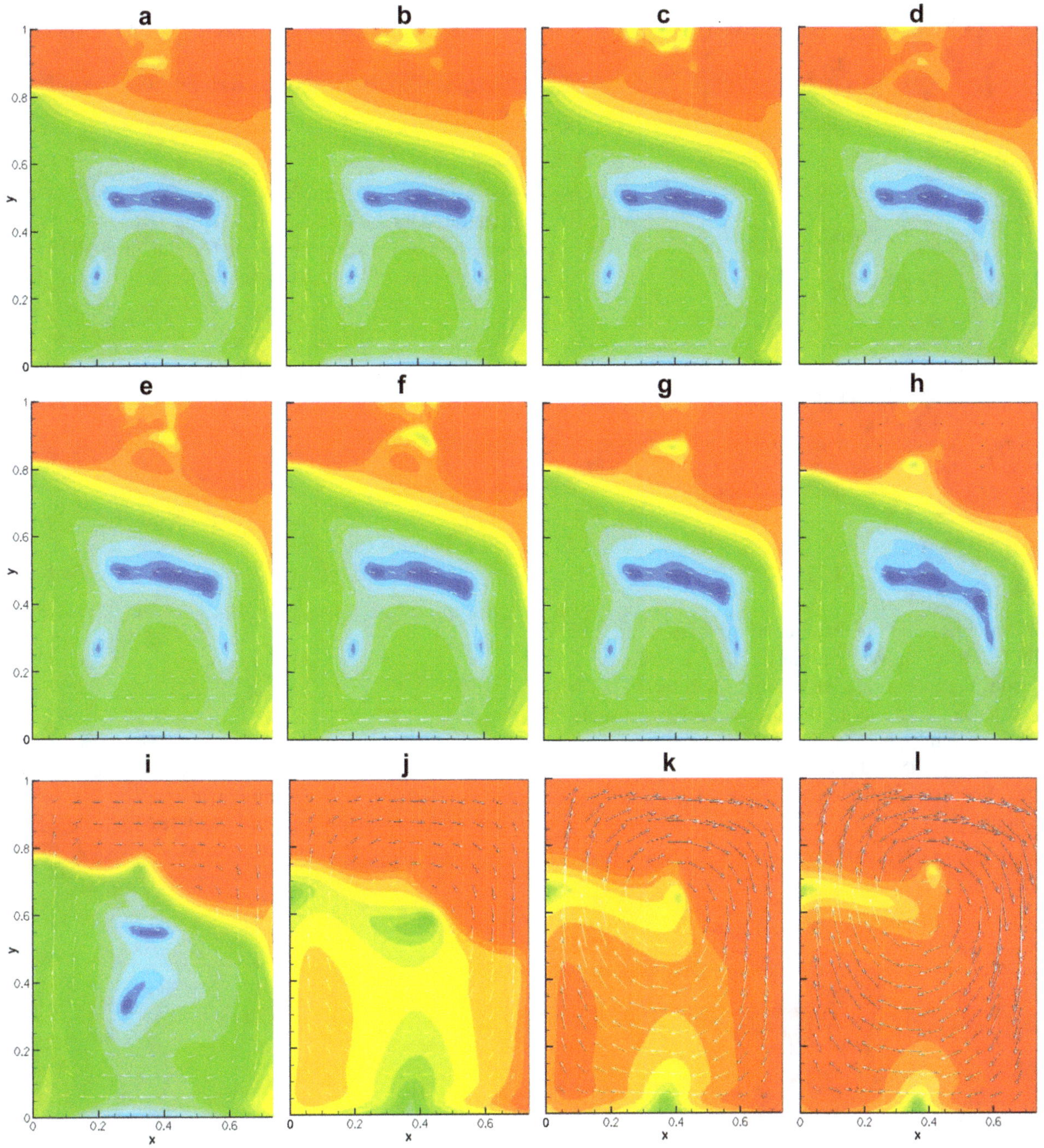

Figure 19 Snapshots of stress fields before and at the point of failure for case $Ra = 3 \times 10^7, \theta = 16, a = 0.75, T_0 = 0.8, \tau_y = 6.6 \times 10^5$. Time sequence goes from left to right and top to bottom. White arrows show velocities. **(a)** and **(b)** are close to beginning of simulation, **(c)** is at the mid-point between the start and failure, and **(d-i)** are right before overturning (depicted in **(j-l)**) occurs.

was previously defined at the middle of the convecting cell. Here, since δ_{pl} is defined at the downwelling edge, we have to determine a lid thickness at the edge $\delta_{lid,max}$. This is done by extrapolating the mid-width lid slope to the downwelling edge (Figure 10). As shown is Figure 24, the plastic depth is approximately 0.3 to 0.5 of the lid thickness. We take approximate values for our scaling relations rather than scaling these properties with convective parameters because the trends observed in Figure 24 maybe due to insufficient viscosity contrasts which place convection on the boundary of transitional regime, especially for $\theta = 13$ in which the viscosity is reduced to approximately 10^4 by the critical yield stress.

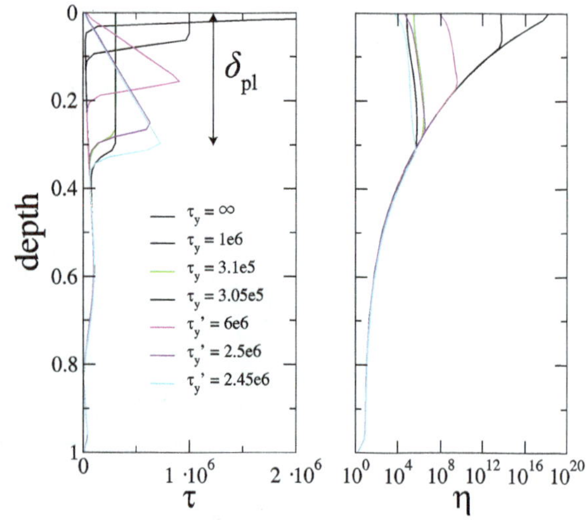

Figure 20 Depth of plastic zone determined by the yield stress (right) and the drop in viscosity due to the yield stress (right). Profiles taken at the downwelling edge ($x = a$) of the convecting cell.

To find the lid stress at $\delta_{\rm pl}$ using Equation 34, we also need to determine the distance of the base of the plastic zone from the convective interior $y_{\rm pl}$. The lid base is at $\delta_{\rm rh}$ from the interior, so we express $y_{\rm pl}$ in terms of $\delta_{\rm rh}$ to give a sense of distance in relation to rheological sublayer thickness. While there is a general trend of increasing of $y_{\rm pl}/\delta_{\rm rh}$ with θ, it is difficult to discern a correlation as $y_{\rm pl}/\delta_{\rm rh}$ fluctuates; thus, we take $y_{\rm pl} \approx 3\delta_{\rm rh}$ for our scaling relations.

The dependence of $\delta_{\rm pl}$, y_{pl}, and $\Delta\eta_{\rm pl}$ on Ra and aspect ratio are very weak and therefore assumed negligible.

Scaling for critical yield stress and critical yield stress gradient

Critical yield stress scaling theory

In deriving a theoretical scaling for the critical yield stress, we assume an approximate balance between the force generated by the shear stresses acting at the base of the lid and the normal stress acting on the side of the lid and we assume that the latter are largely dominated by the stresses in the plastic zone δ_y (Solomatov 2004a and Figure 25). Thus, we can express the yield stress as:

$$\tau_y \sim \tau_{\rm lid}\frac{a}{\delta_y} = {\rm Ra}\frac{dT}{dy}\lambda\frac{y_{\rm pl}^2}{2}\frac{a}{\delta_y}. \qquad (41)$$

In the case of a constant yield stress, gradient ($\tau_y = \tau_y' z$), τ_y' can be scaled as

$$\tau_y' \sim \tau_{\rm lid}\frac{a}{\delta_y^2} = {\rm Ra}\frac{dT}{dy}\lambda\frac{y_{\rm pl}^2}{2}\frac{a}{\delta_y^2}. \qquad (42)$$

In Equations 41 and 42, the yield stress is treated as the normal stress whereas the lithosphere stresses are shear stress. However, the magnitude of stresses is expressed in second invariant, and the yield stress in von Mises

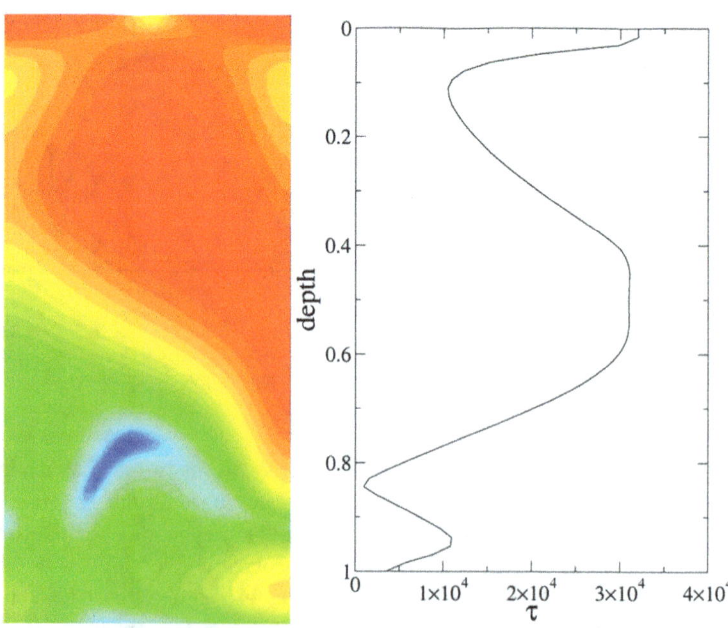

Figure 21 Stress field (left) and stress profile at the downwelling edge. $Ra = 3 \times 10^6$, $a = 0.5x5$, $\theta = 19$, $\tau_y = 3.2 \times 10^4$, exponential viscosity.

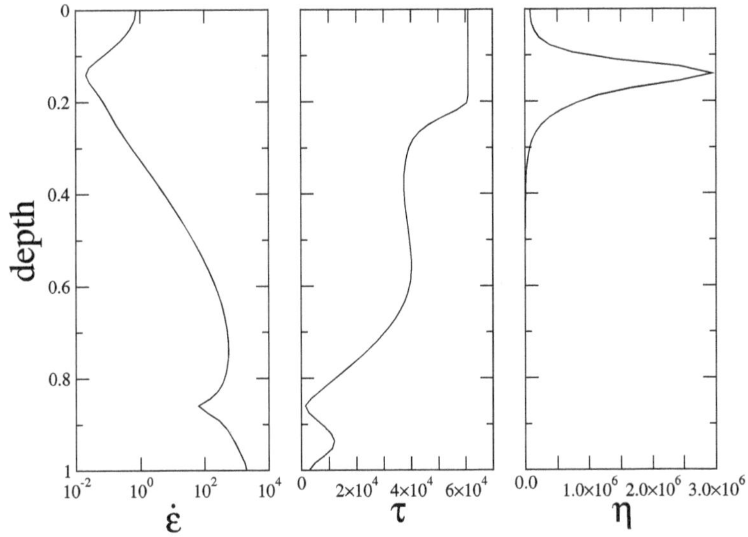

Figure 22 Strain rate, stress, and viscosity profile at the edge with $Ra = 3 \times 10^6$, $a = 0.75$, $\theta = 19$, exponential viscosity. The point of maximum viscosity may not correspond to the brittle-plastic transition.

criterion also put a limit the second stress invariant. To see which stress component contributes to the second invariant, we plot the normal stress and shear stress profiles in Figure 26. Inside the lid where plastic failure occurs, normal stress dominates, whereas shear stress exceeds normal stress below the plastic depth.

The critical yield stress is the stress at $\delta_y = \delta_{pl}$. The critical plastic depth δ_{pl} is taken to be 0.3 to $0.5\delta_{\rm lid}$, and y_{pl} is about 2 to $4\delta_{rh}$.

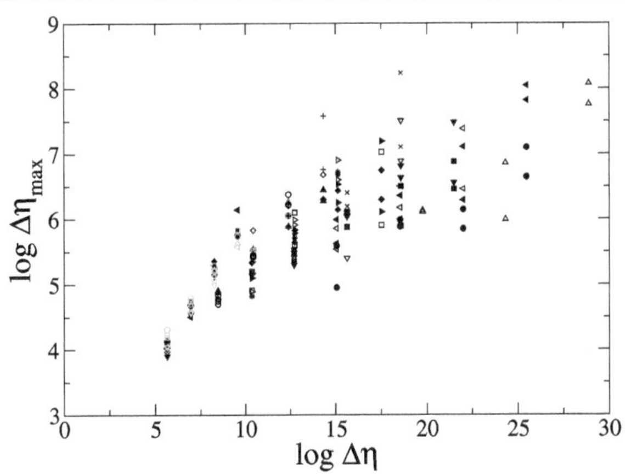

Figure 23 Maximum apparent viscosity in the convective cell at the point of lithospheric failure as a function of the viscosity contrast in the absence of yield stress. Black, exponential viscosity with constant $\tau_{y,cr}$; red, Arrhenius viscosity with constant $\tau_{y,cr}$; green, exponential viscosity with constant $\tau'_{y,cr}$; blue, Arrhenius viscosity with constant $\tau'_{y,cr}$.

$$\tau_{\rm lid} \sim C_1 {\rm Ra} \Delta T_{rh} \lambda \delta_{rh}, \tag{43}$$

$$\tau_{y,cr} \sim C_2 {\rm Ra} \Delta T_{rh} \lambda a \frac{\delta_{rh}}{\delta_{\rm lid}}, \tag{44}$$

$$\tau'_{y,cr} \sim C_3 {\rm Ra} \Delta T_{rh} \lambda a \frac{\delta_{rh}}{\delta_{\rm lid}^2}. \tag{45}$$

For $\tau_{\rm lid}$ expressed in terms of Nu (Equation 39), noting that ${\rm Nu} \sim dT/dy \sim \delta_{\rm lid}^{-1}$,

$$\tau_{\rm lid} \sim C_1 {\rm Ra Nu} \lambda \delta_{rh}^2, \tag{46}$$

$$\tau_{y,cr} \sim C_2 {\rm Ra Nu}^2 \lambda a \delta_{rh}^2, \tag{47}$$

$$\tau'_{y,cr} \sim C_3 {\rm Ra Nu}^3 \lambda a \delta_{rh}^2. \tag{48}$$

where C_1, C_2, C_3 ranges from 4 to 16, 8 to 53, and 16 to 178, respectively.

From the previous sections, since $\Delta T_{\rm rh}/\Delta T$, $\delta_{\rm rh}/d$, $\delta_{\rm lid}/d$, and λ are all scaled in terms of Ra, a, and θ with scaling exponents summarized in Table 1. $\tau_{\rm lid}$, $\tau_{y,cr}$, and $\tau'_{y,cr}$ can be scaled in terms of Ra, a, and θ. The results are listed in Table 2.

Numerical results for critical yield stress and yield stress gradient: Arrhenius vs. exponential viscosity

Figures 27 and 28 show that both $\tau_{y,cr}$ and $\tau'_{y,cr}$ decrease with increasing θ and total viscosity contrast and might converge to asymptotic values at high viscosity contrasts, although it is difficult to tell from our limited data. To estimate the accuracy of Frank-Kamenetskii approximation in the prediction of critical yield stress, we express the ratio of yield stress for Arrhenius viscosity to that for exponential viscosity $R_\tau = \tau_{y,cr,{\rm Arr}}/\tau_{y,cr,{\rm exp}}$ and similarly

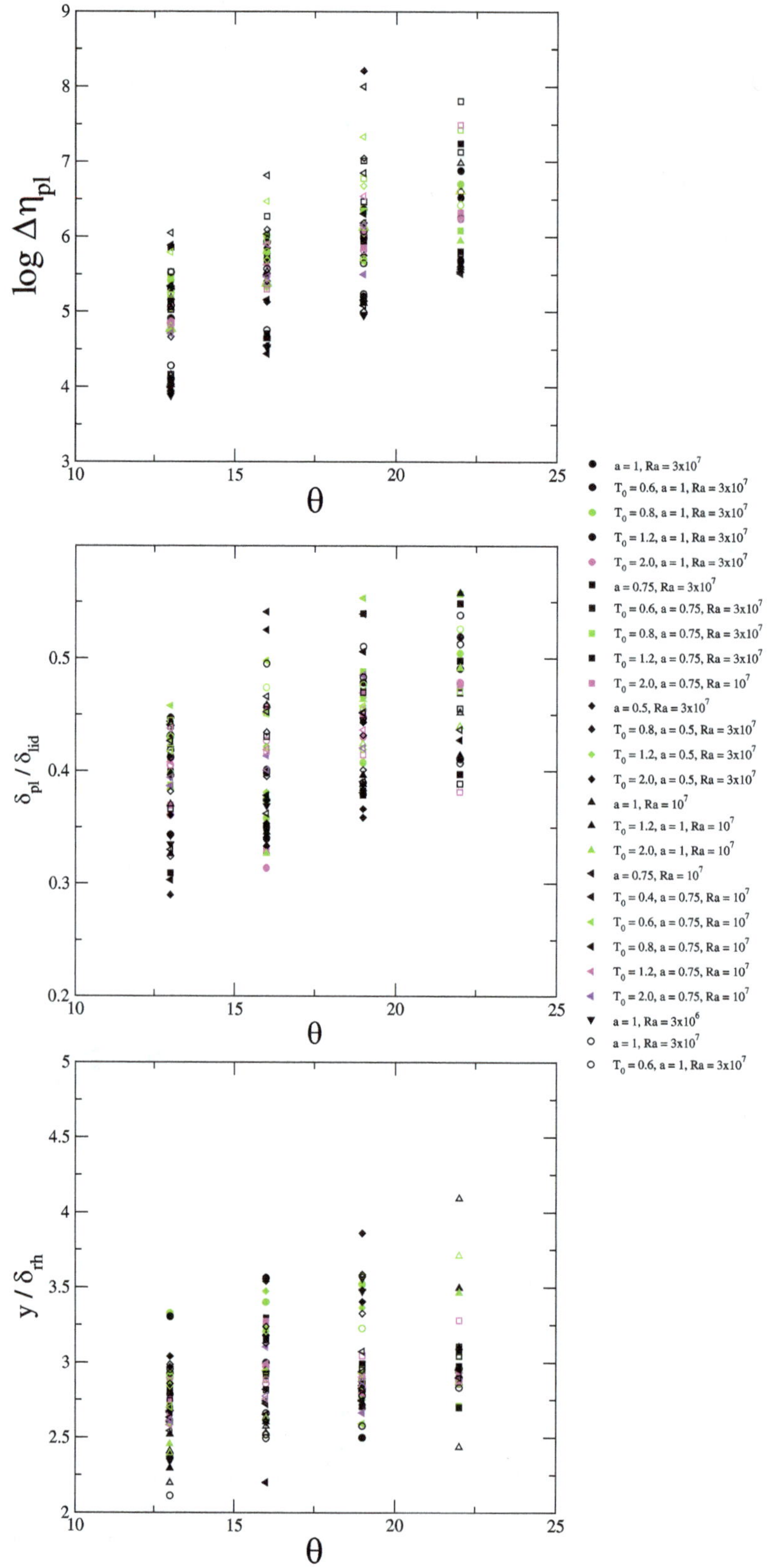

Figure 24 Viscosity contrast at plastic depth, fraction of lid that experience plastic failure, and the distance between the interior and plastic depth as a function of θ. Solid symbols represent constant $\tau_{y,cr}$ cases, while open symbols are constant $\tau'_{y,cr}$.

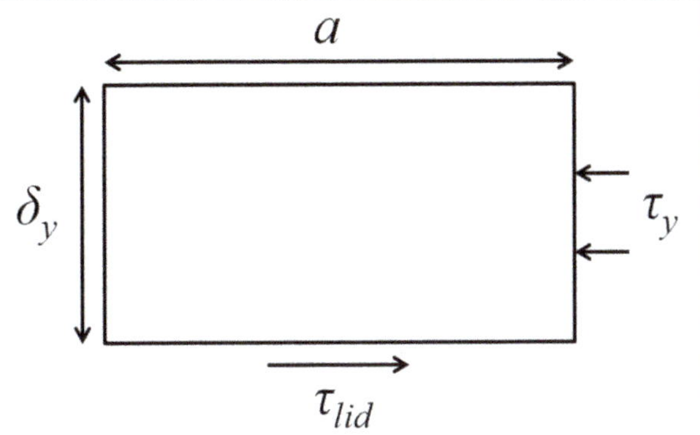

Figure 25 Schematic diagram of surface stresses on the plastic zone in the lid. The shear stress τ_{lid} acting on the base of the lid of horizontal length a is balanced by the normal stress τ_y acting on the side with depth δ_y, developed under free-slip boundary conditions.

for critical yield stress gradient with $R_{\tau'} = \tau'_{y,cr,\mathrm{Arr}}/\tau'_{y,cr,\exp}$ to look at the dependence of these ratios on θ (Figures 29 and 30). Both R_τ and $R_{\tau'}$ increase with θ and the values of yield stress for Arrhenius viscosity and that for Frank-Kamenetskii approximation get closer as the viscosity contrast increases, assuming that $R_\tau < 1$ and $R'_\tau < 1$ at all θ.

For the cases tested at resolution that is doubled, we find that the values for R_τ and $R_{\tau'}$ are within 5% difference. Therefore, $64 \times 64a$ resolution is sufficient for our single-cell steady-state convection analysis.

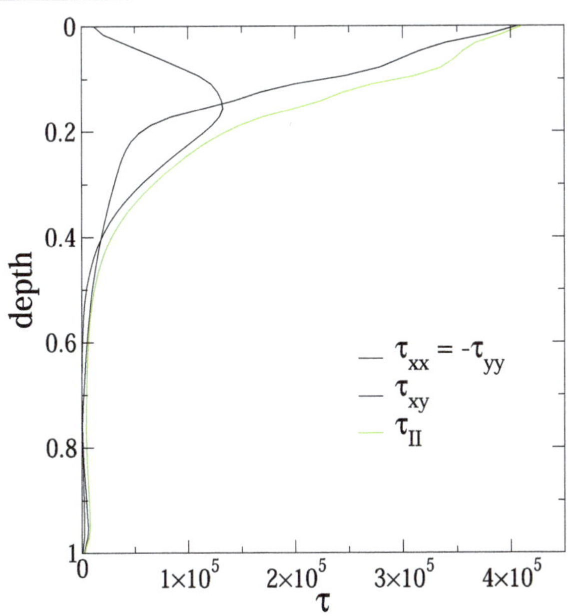

Figure 26 Horizontally averaged stress components and second invariant of stress as a function of depth.

To find the dependence of R_τ and $R_{\tau'}$ on the Arrhenius viscosity contrast $\Delta\eta_{\mathrm{Arr}}$, we consider the ratio of Arrhenius viscosity contrast to exponential viscosity contrast $(\Delta\eta_{\mathrm{Arr}}/\exp\theta)$. This ratio reflects the difference between Arrhenius viscosity contrast and exponential viscosity contrast: the larger the ratio, the greater the difference. Figure 31 shows that values of the two $\tau_{y,cr}$ approach each other as the Arrhenius viscosity gets closer to the exponential viscosity. It also suggests that R_τ and $R_{\tau'}$ are mainly determined by the Frank-Kamenetskii parameter θ but less sensitive to Ra and a. As the difference between the Arrhenius viscosity and exponential viscosity increases, both R_τ and $R_{\tau'}$ appear to approach some asymptotic value.

We also investigate the dependence of critical yield stress and yield stress gradient on aspect ratio (Figures 32 and 33). The positive power law coefficient (see Table 1) implies that with smaller aspect ratio, the critical yield value is lower and thus more difficult to reach the yielding criterion. Therefore, smaller cells are more stable. This may explain the phenomenon where overturning of the cold lid is observed once or a few times but then stabilized after reconfiguring into smaller cells.

Figures 34 and 35 and Table 1 show that $\tau_{y,cr}$ and $\tau'_{y,cr}$ are approximately proportional with Ra.

The plots in Figures 27, 28, 32, 33, 34, and 35 show that the scaling exponents could have a range of values, and the cases using Arrhenius viscosity may have a bit different values from those with exponential viscosities. Figure 36 shows that while the scaling exponent for Ra and a for exponential viscosity cases lie between the range obtained from Arrhenius viscosity cases, the exponent for θ varies with surface temperature T_0, Ra, and a. In general, the scaling exponent of θ increases with the viscosity contrast (i.e., lower T_0), a, and moderately with Ra. This may be related to the difference in the stress distribution in the lid in Arrhenius cases, especially towards the downwelling edge, as discussed previously. In the future, it will be worthwhile to look in more detail at the stress variations for the Arrhenius viscosity and to see if there are any relationships between R_τ, $R_{\tau'}$, and T_0 at even lower T_0 (i.e., at higher viscosity contrasts.)

Discussion and conclusions

Comparison with other studies of stress scaling laws

The studies dealing with convective stress scaling often aim to provide an expression for stress in terms of radius and mass of a planet to predict the likelihood of plate tectonics. They reached various conclusions (e.g., O'Neill et al. 2007; O'Neill and Lenardic 2007; Valencia and O'Connell 2007;2009; Korenaga 2010a; Van Heck and Tackley 2011; Stamenkovic and Breuer 2014). One of the main difficulties in deriving convincing scaling laws for

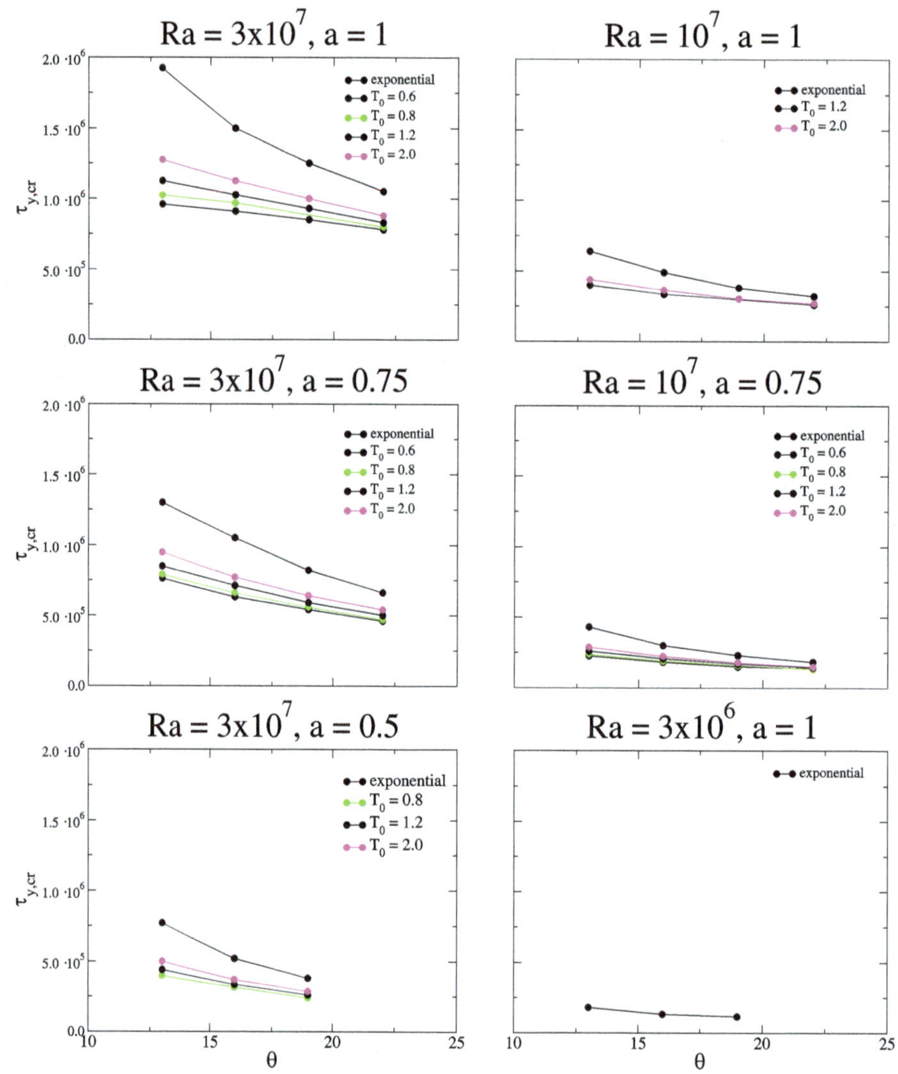

Figure 27 Critical yield stress $\tau_{y,cr}$ as a function of Frank-Kamenetskii parameter θ. The Arrhenius viscosity is calculated with the non-dimensional surface temperature T_0 and the activation energy E that gives the corresponding θ. Lower T_0 gives a higher viscosity contrast $\Delta\eta_{Arr}$.

plate tectonics initiation was a poor understanding of lid stresses and how they are related to lid failure. In the present study, we have addressed these issues using two-dimensional steady-state convective cell simulations. This is the simplest system to analyze, and yet even for this system, the derivation of scaling laws proved to be complicated and not well described by the existing asymptotic theories. Below, we discuss some differences between our study and previous studies and summarize our scaling laws in a dimensional form.

In some studies (e.g., Moresi and Solomatov 1998; Trompert and Hansen 1998; Tackley 2000b; Fowler and O'Brien 2003), the authors assumed that subduction occurs when the stresses in the convective interior exceed the yield stress. This means that subduction begins when not only the lid but also the interior of the convective cell fails. However, subduction initiation may not necessarily require the failure of interiors but instead may only require failure of just a small portion of the lid. The stresses in the lid are several orders of magnitude higher than the stresses in the interior, and also, they scale differently. Thus, the assumption regarding what part of the convective cell must fail in order for subduction to begin is critically important. In this study, we have investigated this assumption quantitatively, based on a detailed analysis of stresses and other parameters in the convective cell, and then formulated the critical conditions for subduction initiation.

In agreement with Fowler (1985), we have shown that the lid slope is a key factor in determining the stresses

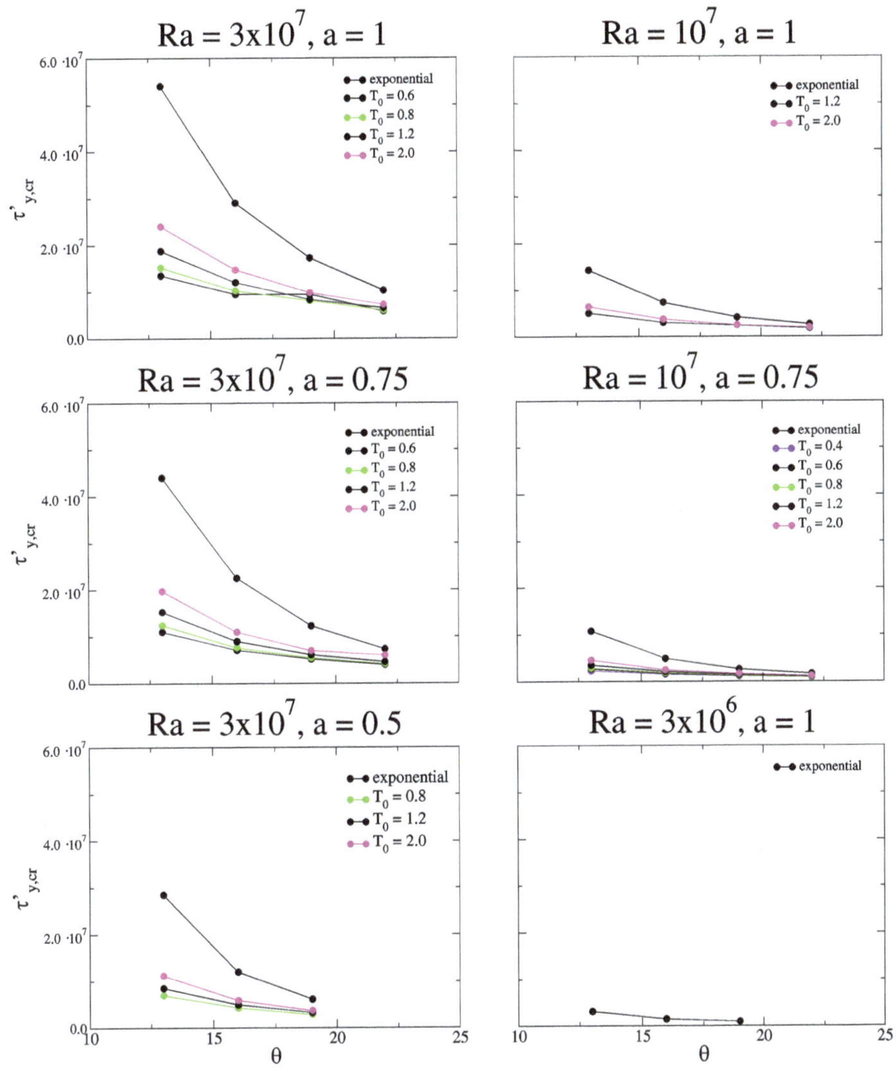

Figure 28 Critical yield stress gradient $\tau'_{y,cr}$ as a function of θ.

in the lid. However, our model has several important differences from Fowler (1985). The theoretical solution in Fowler (1985) is a similarity solution and does not take into the finite horizontal extent of the lid. Our model has vertical boundaries, and thus, the structure of the lid in our model is more complex. Also, the solution in Fowler (1985) is an asymptotic solution requiring very high values of parameters, such as Ra and θ, and a satisfaction of certain asymptotic conditions, which are not reached in our simulations and may not necessarily be reached on planets. Thus, our scaling laws are not asymptotic in this sense. Also, solutions in Fowler (1985) are obtained for two end-member cases, the large lid slope case, and the small lid slope case. We find that the lid slope behaves in a more complex way and is between these two end-member cases. We have determined a scaling law for the lid slope numerically and used it to derive the scaling law for the stresses in the lid.

Our analysis suggests that the stresses in the lid increase approximately as a square of the distance from the bottom of the lid (Equation 34 and Figure 14). This agrees with the asymptotic analysis of Fowler (1985) but is different from the stress distribution in Solomatov (2004a). In Solomatov (2004a), the stress distribution was more complex because the convective cell was heated from within rather than from the bottom and the internal heating affected the temperature-induced density distribution in the lid. At Rayleigh numbers higher than those reached in Solomatov (2004a), the lid is expected to become sufficiently thin so that the heat production inside the lid would be negligible compared to the heat flux at the base of the lid. Thus, we expect that for convection with internal heating, the

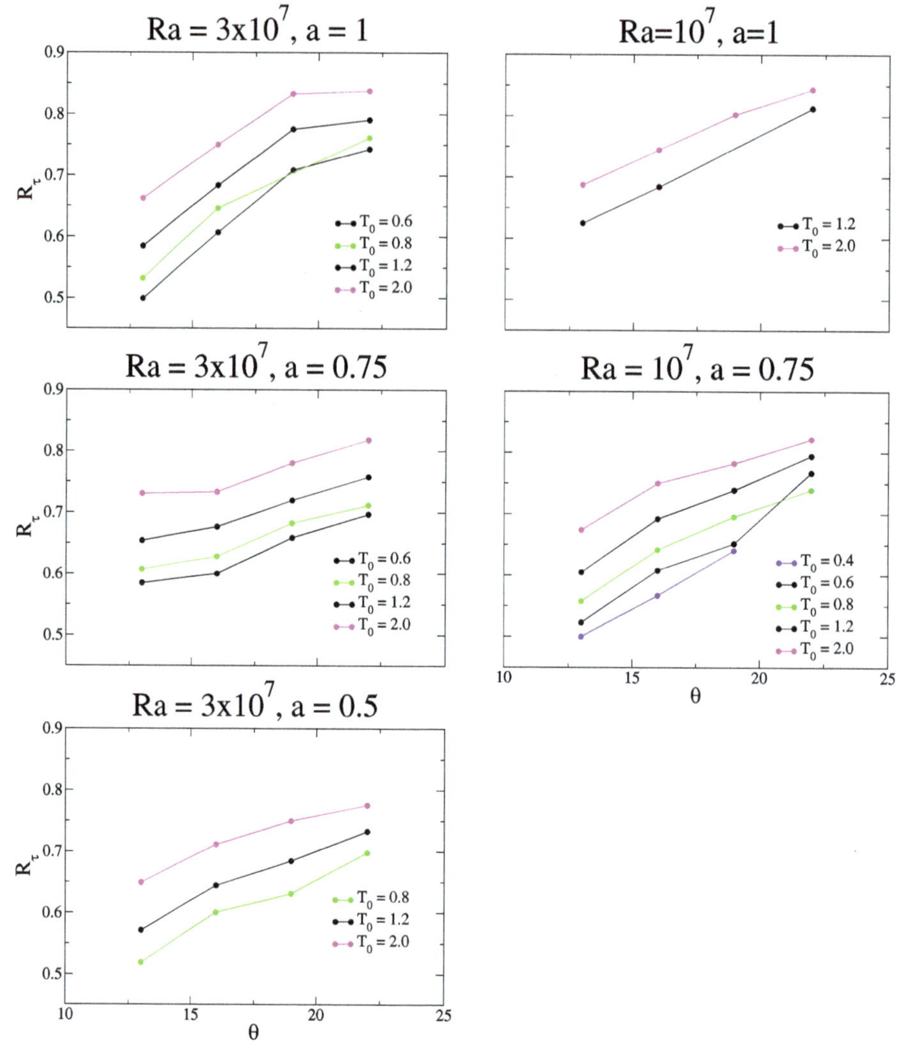

Figure 29 The ratio of Arrhenius yield stress to exponential yield stress R_τ as a function of Frank-Kamenetskii parameter θ.

stress distribution in the lid should approach the quadratic distribution that we observe for convection with bottom heating.

We find that subduction initiation requires that only a part of the lid undergoes plastic failure, roughly 0.3 to 0.5 of the total lid thickness. This generally agrees with the analysis in Solomatov (2004a) and confirms that the plastic failure does not have to extend all the way to the bottom of the lid as was assumed in Fowler and O'Brien (2003). However, unlike Solomatov (2004a), we determine the distance to the boundary of the plastic failure zone by measuring it from the base of the lid and scaling it in terms of the rheological boundary layer thickness. We find that such an approach is more appropriate because the mobility of the lid is largely controlled by the viscosity contrast between the zone of failure and the convective interior of the cell, which in turn is scaled with the rheological boundary layer thickness.

Estimates for the Earth

To compare our results with those obtained in Solomatov (2004a,b), we convert the critical yield stress $\tau_{y,cr}$ and critical yield stress gradient $\tau'_{y,cr}$ into their dimensional forms (Equations 20 and 22) and estimate the critical yield strength and the critical coefficient of friction μ for subduction initiation on the Earth.

The interior viscosity cannot be reliably estimated from the viscosity law alone and is usually determined from better constrained properties such as lithospheric thickness. Therefore, following Solomatov (2004a), we use the scaling law for δ_0 (Table 1) and present

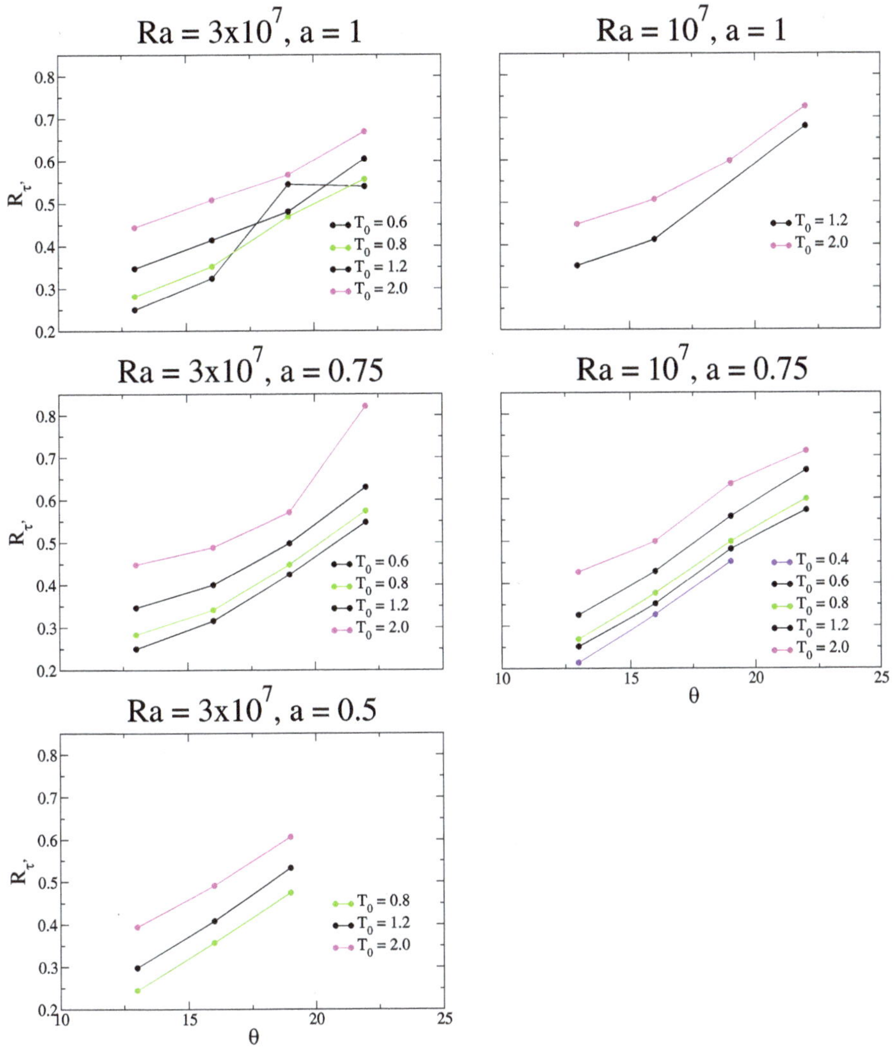

Figure 30 The ratio of Arrhenius yield stress gradient to exponential yield stress gradient $R_{\tau'}$ as a function of θ.

the results in terms of the thickness of the thermal boundary layer $\delta_0 \sim 100$ km instead of the mantle viscosity η_1.

The scaling law for the critical yield stress depends strongly on aspect ratio a. Previous studies have scaled the aspect ratio from half-space cooling of lithosphere (Korenaga 2010b; Stamenkovic and Breuer 2014) or estimated from numerical simulations (Solomatov 2004a,b), whereas it was assumed to be on the order of 1 in Valencia and O'Connell (2009). We use the horizontal width of the convective cells as $l_{\rm hor} = ad \sim$100 km as a very rough value to compare our estimates with those in Solomatov (2004a,b). This value was inferred from observational constraints on the present-day horizontal scale of sublithospheric convective structures (Solomatov 2004a).

Using the results in Table 1, we obtain that the dimensional critical yield stress for subduction initiation is as follows:

$$\tau_{y,cr} \sim 1.95\alpha\rho g \left(\frac{E}{RT_i^2}\right)^{-1.03} \Delta T^{-0.03}\delta_0^{-0.41} l_{\rm hor}^{1.78} d^{-0.37}. \quad (49)$$

For comparison, Solomatov (2004a,b) give:

$$\tau_{y,cr} \sim 13\alpha\rho g \left(\frac{E}{RT_i^2}\right)^{-2} \Delta T^{-1} l_{\rm hor}. \quad (50)$$

The critical coefficient of friction μ for subduction initiation is as follows:

$$\mu \sim 89\alpha \left(\frac{E}{RT_i^2}\right)^{-1.74} \Delta T^{-0.74}\delta_0^{-1.55} l_{\rm hor}^{1.87} d^{-0.32}, \quad (51)$$

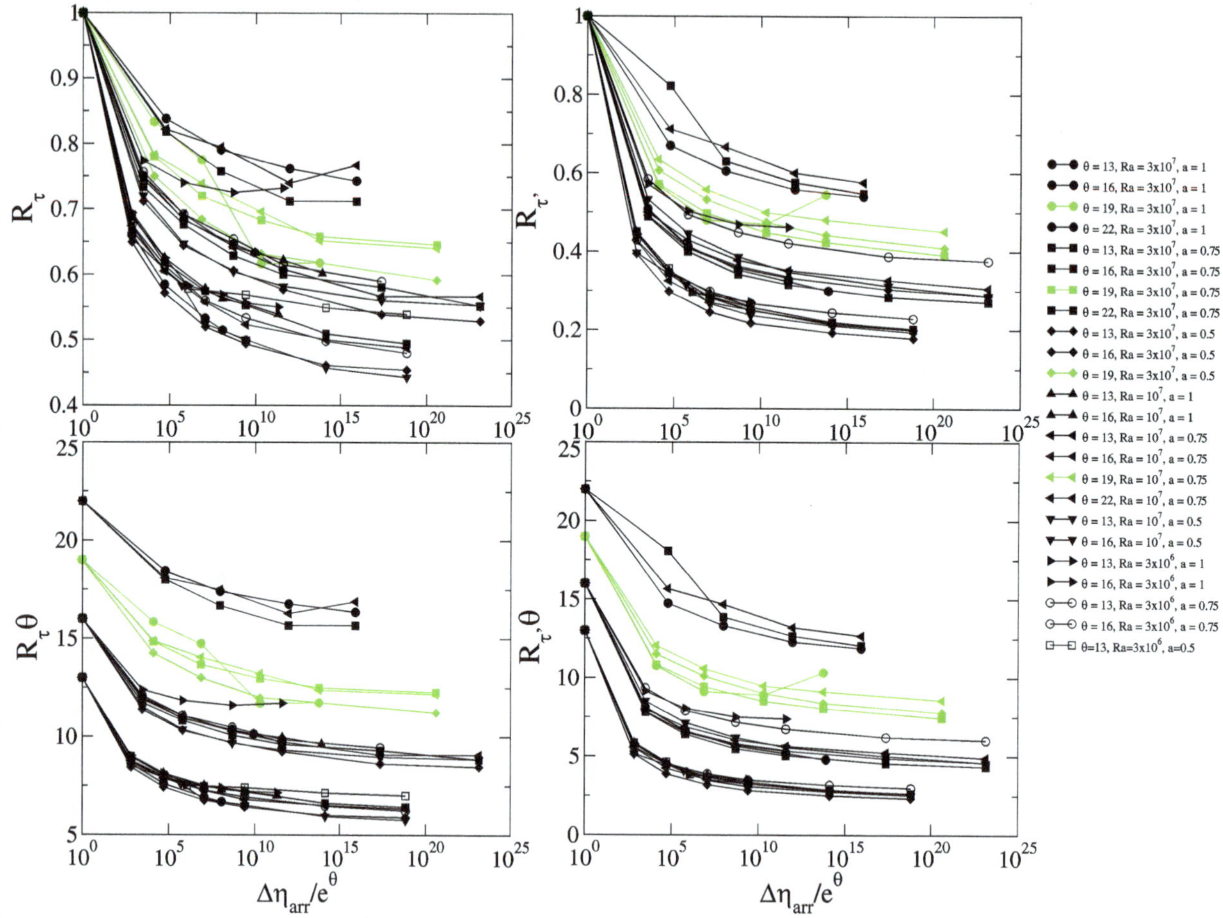

Figure 31 Top figures: ratio of yield stress for Arrhenius viscosity to that for exponential viscosity R_τ and $R_{\tau'}$ as a function of Arrhenius viscosity contrast normalized to exponential viscosity contrast $\Delta\eta_{Arr}/\exp(\theta)$. $\Delta\eta_{Arr}/\exp(\theta) \geq 1$, and it can go up to many orders of magnitude. Bottom figures: ratio of yield stress multiplied by θ. For cases with the same θ but various Ra and a, these ratios do not differ much, meaning that θ is the controlling factor for the difference in yield stress predicted by Arrhenius viscosity and exponential viscosity. Asymptotically towards high $\Delta\eta_{Arr}/\exp(\theta)$, R_τ and R'_τ are approximately proportional to θ.

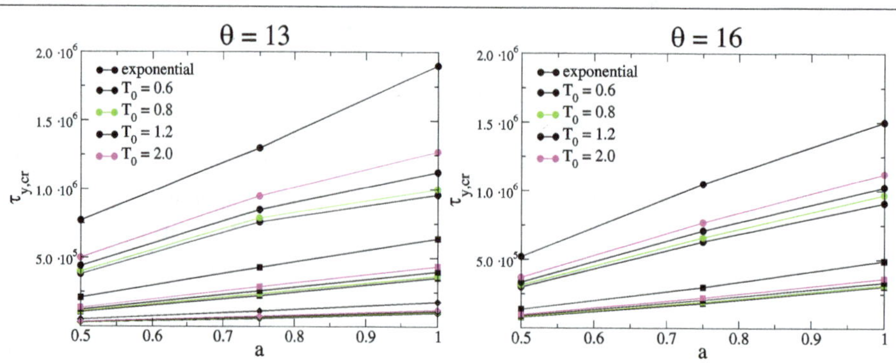

Figure 32 Critical yield stress $\tau_{y,cr}$ as a function of aspect ratio. Circles represent Ra = 3×10^7, squares Ra = 10^7, diamonds Ra = 3×10^6.

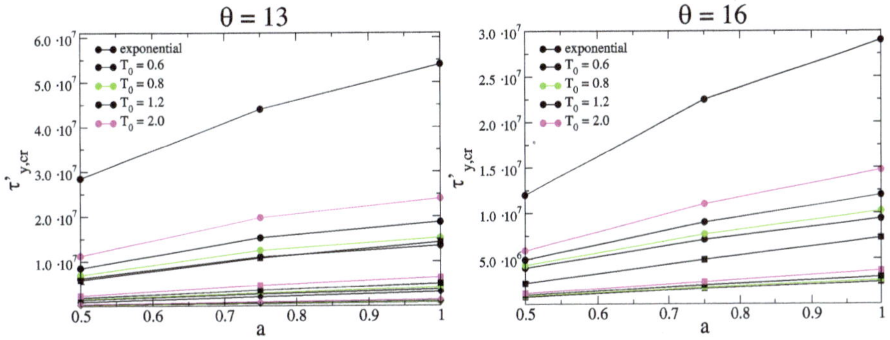

Figure 33 Critical yield stress gradient $\tau'_{y,cr}$ as a function of aspect ratio. Legends are the same as in Figure 32.

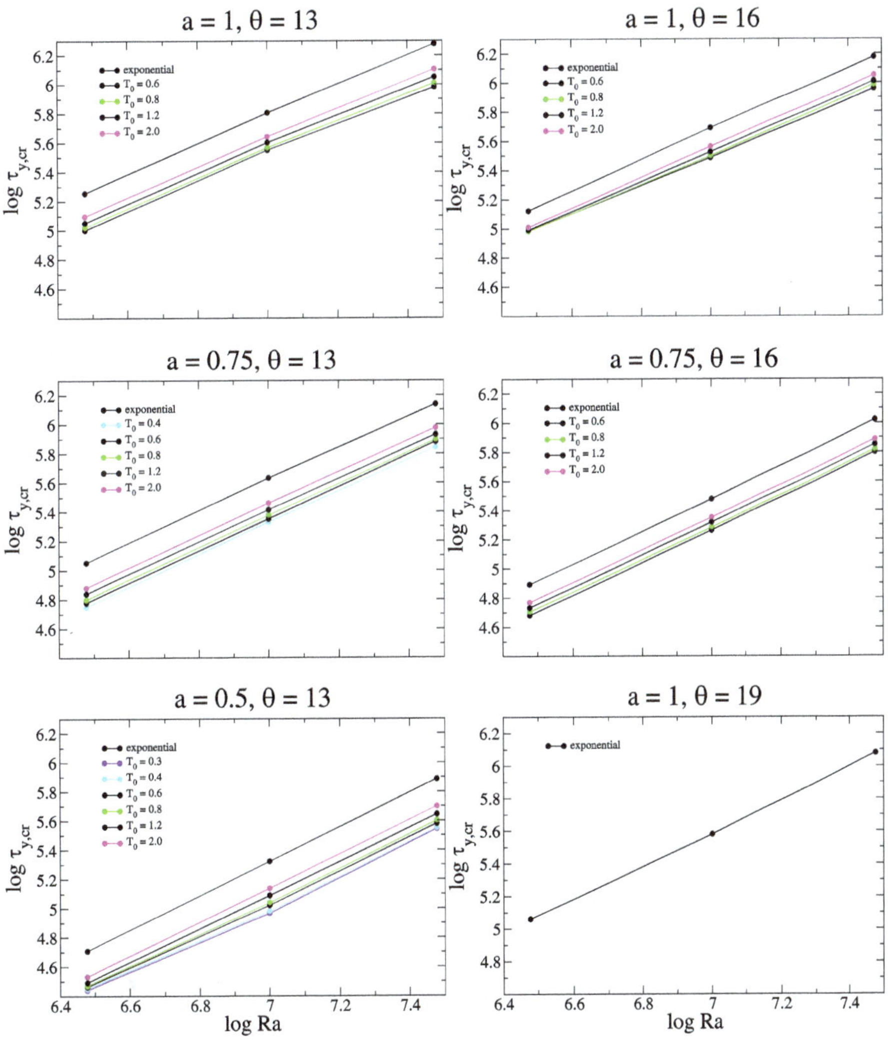

Figure 34 Critical yield stress $\tau_{y,cr}$ as a function of Rayleigh number.

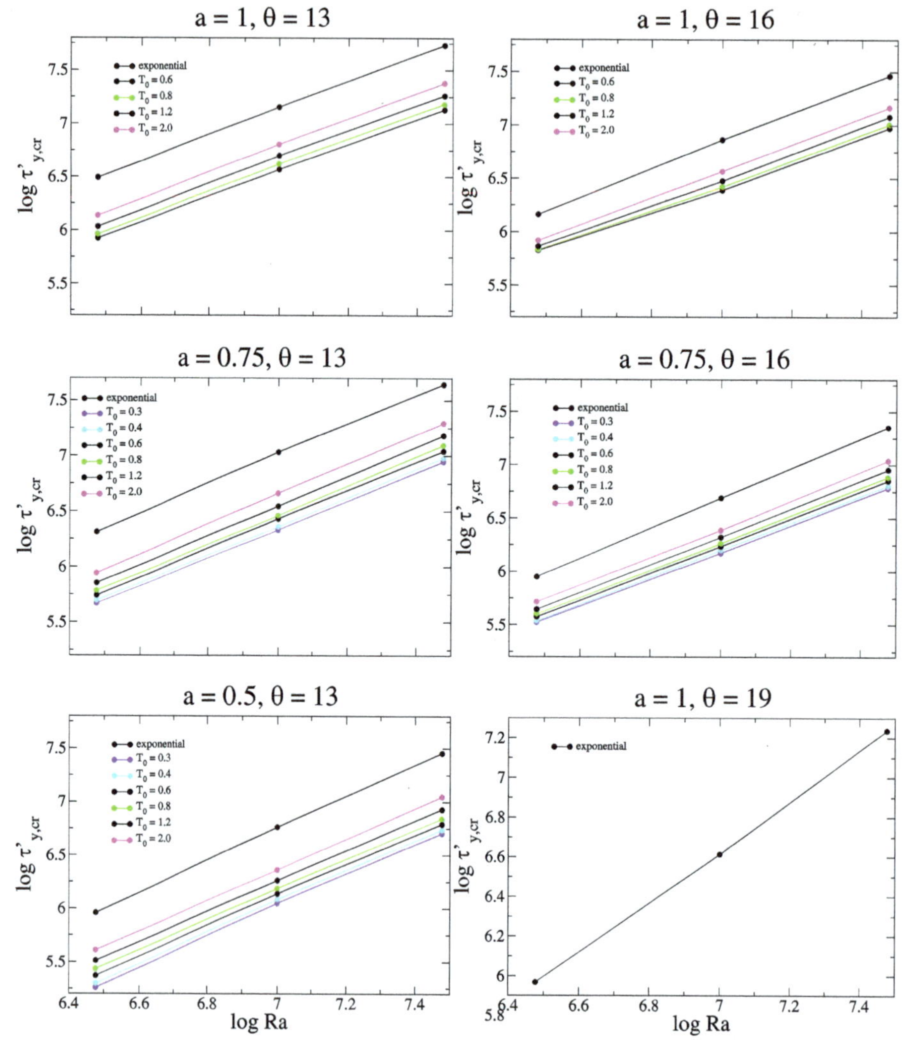

Figure 35 Critical yield stress gradient $\tau'_{y,cr}$ as a function of Rayleigh number.

and in Solomatov (2004a,b):

$$\mu \sim 50\alpha \left(\frac{E}{RT_i^2}\right)^{-2} \Delta T^{-1} \delta_0^{-1} l_{\mathrm{hor}}. \tag{52}$$

Using the typical values of various physical parameters (Table 3), we estimate that the yield strength for the Earth is 5 MPa which is of the same order of magnitude as 3 MPa obtained by Solomatov (2004a,b). To see how variations in various parameters may affect these estimates, it is useful to present the estimates in a different form. Our estimate (Equation 49) can be written as:

$$\tau_{y,cr} \sim 5 \left(\frac{100\ \mathrm{km}}{\delta_0}\right)^{0.41} \left(\frac{l_{\mathrm{hor}}}{100\ \mathrm{km}}\right)^{1.78} \left(\frac{500\ \mathrm{km}}{d}\right)^{0.37} \mathrm{MPa}, \tag{53}$$

and the estimate from Solomatov (2004a,b) (Equations 50) is as follows:

$$\tau_{y,cr} \sim 3 \left(\frac{l_{\mathrm{hor}}}{100\ \mathrm{km}}\right) \mathrm{MPa}, \tag{54}$$

from Equation 50.

Our estimates of the critical friction coefficient μ is 8×10^{-3}, which is a factor of 3 larger than 3×10^{-3} obtained in Solomatov (2004a,b). Our estimate can be written as

$$\mu \sim 0.008 \left(\frac{100\ \mathrm{km}}{\delta_0}\right)^{1.55} \left(\frac{l_{\mathrm{hor}}}{100\ \mathrm{km}}\right)^{1.87} \left(\frac{500\ \mathrm{km}}{d}\right)^{0.32}, \tag{55}$$

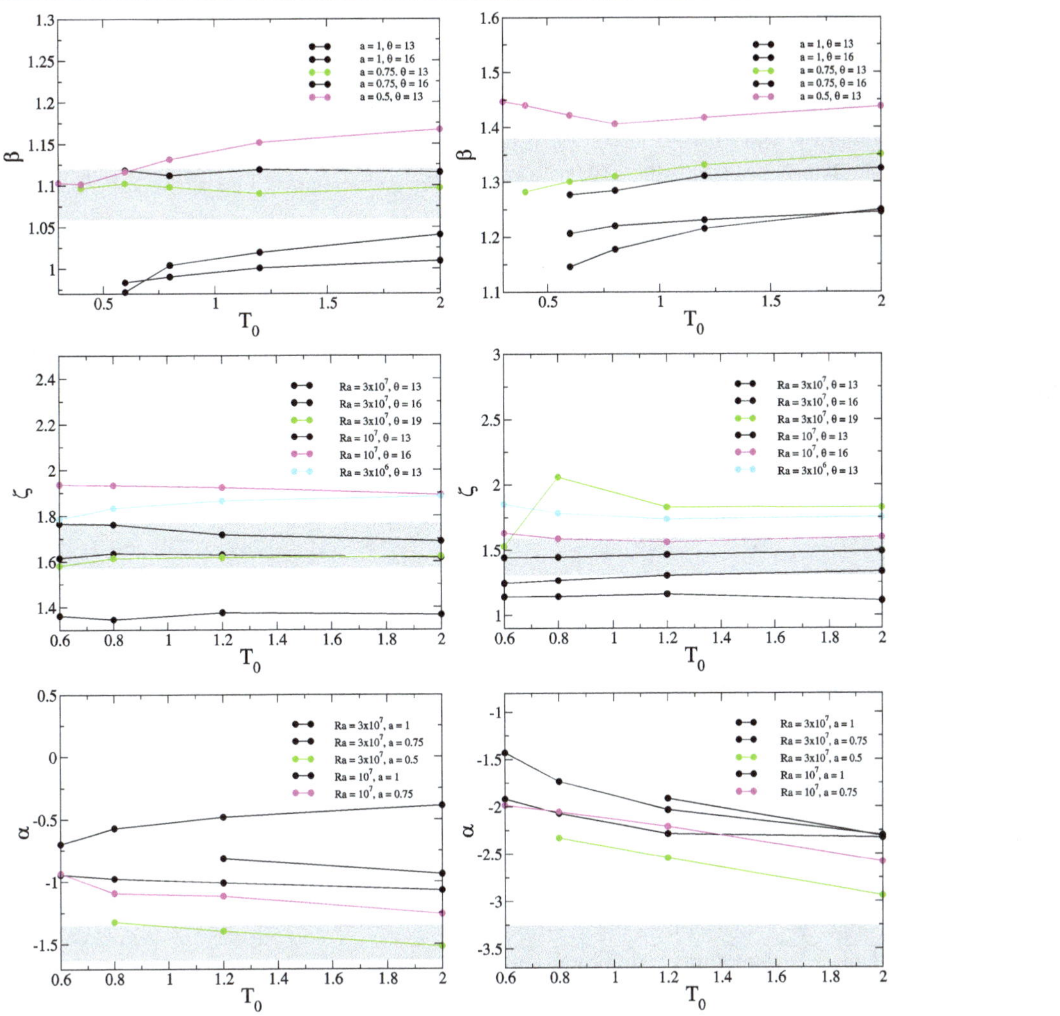

Figure 36 Scaling exponents in $Ra^{\beta}a^{\zeta}\theta^{\alpha}$ for τ_y (left three figures) and $\tau'_{y,cr}$ (right three figures) with varying T_0 in Arrhenius viscosity. The grey stripes represent the value obtained from exponential viscosity calculations, with the width determined by error bars.

Table 3 Parameters used to estimate $\tau_{y,cr}$ and $\tau'_{y,cr}$ for Earth as in Solomatov (2004a,b)

α	3×10^{-5}
κ	10^{-6} m^2 s^{-1}
δ_{lid}	100 km
k	3 W m^{-1} K^{-1}
E	430 kJ mol^{-}1
d	approximately 500 km
g	10 m s^{-2}
ρ	3,300 kg m^{-3}
T_0	300 K
T_i	1,700 K
l_{hor}	100 km

and the estimate from Solomatov (2004a,b) (Equation 50) is as follows:

$$\mu \sim 0.003\left(\frac{l_{\text{hor}}}{100\ \text{km}}\right)\left(\frac{100\ \text{km}}{\delta_0}\right). \tag{56}$$

If we take into account the fact that Frank-Kamenetskii approximation that we used to derive the scaling laws overestimate the critical yield stress and the critical friction coefficient (Figure 31), then both ours and the estimates in Solomatov (2004a,b) should be further reduced by a factor of 2 (Figure 31), depending on the values of the viscosity parameters and the Rayleigh number.

One major difference between our scaling laws and the scaling laws obtained in Solomatov (2004a) is a much stronger dependence of the critical yield stress and critical friction coefficient on the width of the convecting layer l_{hor} - they scale roughly as approximately l_{hor}^2 as opposed to the previous scaling approximately l_{hor}. This means that the critical values of the yield stress and friction coefficient would increase by 2 to 4 orders of magnitude if the width of the convective cells increased by 1 to 2 orders of magnitude (for example, in the past history of the Earth),and thus, at least in principle, could reach the experimentally observed values that are on the order of 1,000 MPa for $\tau_{y,cr}$ and μ ~0.6 to 0.85 (e.g., Byerlee 1978; Goetze and Evans 1979; Kohlstedt et al. 1995; Mei et al. 2010) and values constrained by loading models with *in situ* stress measurements of Hawaiian Islands (Zhong and Watts 2013) which are 0.25 to 0.7 for μ and 100 to 200 MPa for lithospheric stress. This implies that the chances of plate tectonics might be higher than we thought before. Time-dependent calculations and a more realistic formulation of the problem are required to better understand the implications of these results for plate tectonics initiation.

Uncertainties in stress scaling

The scaling laws derived here are applicable to Newtonian rheology; therefore, the activation energy for diffusion creep is used in our calculations. However, it should be noted that dislocation creep is probably the dominant mechanism in the lithosphere (Karato and Wu 1993). For the Earth, wet dislocation creep may be preferable (Solomatov and Moresi 2000) while for other terrestrial planets such as Venus might have dry lithosphere. To apply on a wider range of planets including icy bodies, scaling laws based on non-Newtonian rheology will be required.

In previous scaling theories, the lid slope is often considered to be small because the lid thickness is assumed to be relatively small. Even in the large lid slope end member case in Fowler's theory, δ_{lid} is assumed to be small relative to the thickness of the convecting layer.

However, our simulation indicates that the slope may be significant, so the derivations may need to be modified to take this into account.

Free-slip boundary conditions are often used in solving equations for thermal convection, but this restricts the vertical motion of the surface. Recent studies have used the free-surface boundary conditions, which is closer to natural surface condition as both normal and shear stress on the surface is reduced to zero (Zhong et al. 1996; Schmeling et al. 2008; Kaus et al. 2010; Crameri et al. 2012; Kramer et al. 2012). It maybe computationally expensive to implement this for the time being, but it could be worthwhile to explore its effect on scaling relations for stresses in the future.

Our numerical results show that $\tau_{y,cr}$ of Arrhenius viscosity approaches that of exponential viscosity as the Frank-Kamenetskii parameter θ increases. This enables us to use exponential viscosity law to extrapolate to high Arrhenius viscosity contrast conditions. Besides the Frank-Kamenestskii approximation, the viscosity contrasts can be reduced in other ways, one of which is to set a cutoff viscosity. The stress structure resulting from the cutoff viscosity will have to be examined. We can then compare the accuracy of these approximations and apply them to extrapolate the results to planetary parameters.

Our results generally support previous conclusions that in order for the convective regime on the terrestrial planets in the inner Solar System to change from stagnant lid convection to plate tectonics, the yield stress of the lithosphere should be much smaller (several MPa) than that predicted by laboratory experiments on rock deformation (hundreds of MPa as predicted by Byerlee's law). However, our results suggest a much stronger dependence of the critical yield stress on the horizontal width of the convective cells. This opens a possibility of subduction initiation even for the large, experimentally measured, lithospheric strength provided that a sufficiently long convective cell forms in a time-dependent mantle convection. In the future, it would be important to investigate the role of initiation conditions and statistical fluctuations of convective cells for the initiation of subduction in time-dependent convection.

Competing interests

The authors declare that they have no competing interests.

Authors' contributions

Both authors read and approved the final manuscript.

Authors' information

TW is a graduate student in Washington University in St. Louis and VS is her advisor.

Acknowledgements

The authors thank NSF for partial support and the two anonymous reviewers for their constructive comments to improve the manuscript.

References

Bercovici D, Ricard Y, Schubert G (2001) A two-phase model for compaction and damage 3. Applications to shear localization and plate boundary formation. J Geophys Res-Solid Earth 106(B5):8925–8939. doi:10.1029/2000JB90043

Bercovici D, Ricard Y (2005) Tectonic plate generation and two-phase damage: void growth versus grain size reduction. J Geophys Res-Solid Earth 110(B3):B03041

Bercovici, D, Ricard Y (2012) Mechanisms for the generation of plate tectonics by two-phase grain-damage and pinning. Phys Earth Planet Int 202–203:27–55. doi:10.1016/j.pepi.2012.05.003

Branlund J, Regenauer-Lieb K, Yuen D (2001) Weak zone formation for initiating subduction from thermo-mechanical feedback of low-temperature plasticity. Earth Planet Sci Lett 190(3–4):237–250

Byerlee J (1978) Friction of rocks. Pure Appl Geophys 116(4–5):615–626. doi:10.1007/BF0087652

Crameri F, Schmeling H, Golabek GJ, Duretz T, Orendt R, Buiter SJH, May DA, Kaus BJP, Gerya TV, Tackley PJ (2012) A comparison of numerical surface topography calculations in geodynamic modelling: an evaluation of the 'sticky air' method. Geophys J Int 189(1):38–54. doi:10.1111/j.1365-246X.2012.05388.x

Frank-Kamenetskii DA (1969) Diffusion and heat exchange in chemical kinetics. Plenum Press, New York. p 574

Fowler A (1985) Fast thermoviscous convection. Stud Appl Math 72(3):189–219

Fowler, A (1993) Boundary-layer theory and subduction. J Geophys Res-Solid Earth 98(B12):21997–22005

Fowler A, O'Brien S (2003) Lithospheric failure on venus. Proc Roy Soc Lond Math Phys Sci 459(2039):2663–2704

Goetze C, Evans B (1979) Stress and temperature in the bending lithosphere as constrained by experimental rock mechanics. Geophys J R Astr Soc 59(3):463–478

Gurnis M, Hall C, Lavier L (2004) Evolving force balance during incipient subduction. Geochem Geophys Geosystems (G^3) 5:07001. doi:10.1029/2003GC000681

Hall C, Gurnis M, Sdrolias M, Lavier L, Muller R (2003) Catastrophic initiation of subduction following forced convergence across fracture zones. Earth Planet Sci Lett 212(1–2):15–30

Hirth G, Kohlstedt D (2003) Rheology of the upper mantle and mantle wedge: a view from the experimentalists. Geophys Monogr Ser 138:83–105

Kanamori H (1994) Mechanics of earthquakes. Annu Rev Earth Planet Sci 22:207–237

Kanamori H, Brodsky E (2004) The physics of earthquakes. Rep Progr Phys 67:1429–1496

Karato S, Wu P (1993) Rheology of the upper mantle - a synthesis. Science 260(5109):771–778

Karato S, Zhang S, Wenk H (1995) Superplasticity in Earth's lower mantle - evidence from seismic anisotropy and rock physics. Science 270(5235):458–461

Kaus BJP, Mulhaus H, May DA (2010) Astabiliplumes algorithm for geodynamic numerical simulations with a free surface. Phys Earth Planet Int 181:12–20

Kemp DV, Stevenson DJ (1996) A tensile flexural model for the initiation of subduction. Geophys J Int 125:73–96

King S (2009) On topography and geoid from 2-D stagant lid convection calculations. Geochem Geophys Geosystems (G^3) 10(3):Q03002. doi:10.1029/2008GC002250

Kohlstedt, D, Evans B, Mackwell S (1995) Strength of the lithosphere - constraints imposed by laboratory experiments. J Geophys Res-Solid Earth 100(B9):17587–17602

Korenaga, J (2007) Thermal cracking and the deep hydration of oceanic lithosphere: a key to the generation of plate tectonics? J Geophys Res-Solid Earth 112(B5):05408. doi:10.1029/2006JB004502

Korenaga J (2009) Scaling of stagnant-lid convection with arrhenius rheology and the effects of mantle melting. Geophys J Int 179:54–170. doi:10.1111/j.1365-246X.2009.04272.x

Korenaga, J (2010a) On the likelihood of plate tectonics on super-earths: does size matter? ApJ 725(1):43–46. doi:10.1088/2041-8205/725/1/L43

Korenaga J (2010b) Scaling of plate tectonic convection with pseudoplastic rheology. J Geophys Res-Solid Earth 115:11405. doi:10.1029/2010JB007670

Kramer SC, Wilson CR, Davies DR (2012) An implicit free surface algorithm for geodynamic simulations. Phys Earth Planet Int 194–195:25–37

Landuyt W, Bercovici D, Ricard Y (2008) Plate generation and two-phase damage theory in a model of mantle convection. Geophys J Int 174(3):1065–1080. doi:10.1111/j.1365-246X.2008.03844.x

McKenzie DP (1977) The initiation of trenches: a finite amplitude instability. In: Talwani M, Pitman WCI (eds). Island Arcs, Deep Sea Trenches, and Back-Arc Basins, Maurice Ewing Series, Chap. 1. AGU, Washington, D. C. pp 57–61. doi:10.1029/ME001

Mei S, Suzuki AM, Kohlstedt DL, NA D, Durham WB (2010) Experimental constraints on the strength of the lithospheric mantle. J Geophys Res:B08204

Morris S (1982) The effects of a strongly temperature-dependent viscosity on slow flow past a hot sphere. J Fluid Mech 124:1–26

Morris S, Canright D (1984) A boundary-layer analysis of benard convection in a fluid of strongly temperature-dependent viscosity. Phys Earth Planet Int 36(SI):355–373

Moresi, L, Solomatov V (1995) Numerical investigation of 2D convection with extremely large viscosity variations. Phys Fluids 7(9):2154–2162

Moresi L, Solomatov V (1998) Mantle convection with a brittle lithosphere: thoughts on the global tectonic styles of the earth and venus. Geophys J Int 133(3):669–682

Mueller S, Phillips R (1991) On the initiation of subduction. J Geophys Res 96(B1):651–665

Nikolaeva K, Gerya T, Marques FO (2010) Subduction initiation at passive margins: numerical modeling. J Geophys Res 115:03406. doi:10.1029/2009JB006549

Ogawa M (1990) Perturbation analysis of convective instability of oceanic lithosphere and initiation of subduction zones. J Geophys Res-Solid Earth 95(B1):409–420

O'Neill C, Lenardic A (2007) Geological consequences of super-sized earths. Geophys Res Lett 34(19):19204

O'Neill C, Jellinek AM, Lenardic A (2007) Conditions for the onset of plate tectonics on terrestrial planets and moons. Earth Planet Sci Lett 261(1–2):20–32. doi:10.1016/j.epsl.2007.05.038

Ratcliff J, Tackley P, Schubert G, Zebib A (1997) Transitions in thermal convection with strongly variable viscosity. Phys Earth Planet Int 102(3–4):201–212

Reese C, Solomatov V, Moresi L (1999) Non-Newtonian stagnant lid convection and magmatic resurfacing on Venus. Icarus 139(1):67–80

Richards M, Yang W, Baumgardner J, Bunge H (2001) Role of a low-viscosity zone in stabilizing plate tectonics: implications for comparative terrestrial planetology. Geochem Geophys Geosystems (G^3) 2(2000GC00043):1040

Regenauer-Lieb K, Kohl T (2003) Water solubility and diffusivity in olivine: its role in planetary tectonics. Mineral Mag 67(4):697–715. doi:10.1180/0026461036740l2

Regenauer-Lieb K, Yuen D (2004) Positive feedback of interacting ductile faults from coupling of equation of state, rheology and thermal-mechanics. Phys Earth Planet Int 142(1–2):113–135. doi:10.1016/j.pepi.2004.01.00

Regenauer-Lieb K, Yuen D, Branlund J (2001) The initiation of subduction: criticality by addition of water?. Science 294(5542):578–580

Regenauer-Lieb K, Hobbs B, Yuen D, Ord A, Zhang Y, Mulhaus H, Morra G (2006) From point defects to plate tectonic faults. Philos Mag 86(21-22):3373–3392. doi:10.1080/14786430500375l5. Symposium on Instabilities Across the Scales, Cairns, Australia, Sep 14–17, 2004

Schmeling H, Babeyko AY, Enns A, Faccenna C, Funiciello F, Gerya T, Golabek GJ, Grigull S, Kaus BJP, Morra G, Schmalholz SM, van Hunen J (2008) A benchmark comparison of spontaneous subduction models–towards a free surface. Phys Earth Planet In 171:198–223. doi:10.1016/j.pepi.2008.06.028. Recent Advances in Computational Geodynamics: Theory, Numerics and Applications

Solomatov V (1995) Scaling of temperature- and stress-dependent viscosity convection. Phys Fluids 7(2):266–274

Solomatov, V (2004a) Initiation of subduction by small-scale convection. J Geophys Res-Solid Earth 109(B1):B01412. doi:10.1029/2003JB002628

Solomatov VS (2004b) Correction to "initiation of subduction by small-scale convection". J Geophys Res 109(B05408). doi:1029/2004JB003143

Solomatov, V S (2012) Localized subcritical convective cells in temperature-dependent viscosity fluids. Phys Earth Planet In 200–201:63–71. doi:10.1016/j.pepi.2012.04.005

Solomatov V, Moresi L (1996) Stagnant lid convection on venus. J Geophys Res-Planets 101(E2):4737–4753

Solomatov, V, Moresi L (2000) Scaling of time-dependent stagnant lid convection: application to small-scale convection on earth and other terrestrial planets. J Geophys Res-Solid Earth 105(B9):21795–21817

Stamenkovic V, Breuer D (2014) The tectonic mode of rocky planets: part 1 driving factors, models & parameters. Icarus In press. doi:10.1016/j.icarus.2014.01.042

Stein C, Hansen U (2008) Plate motions and the viscosity structure of the mantle - insights from numerical modelling. Earth Planet Sci Lett 272(1–2):29–40

Stein, C, Hansen U (2013) Arrhenius rheology versus Frank-Kamenetskii rheology - implications for mantle dynamics. G-cubed 14(8):2757–2770. doi:10.1002/ggge.20158

Stein C, Schmalzl J, Hansen U (2004) The effect of rheological parameters on plate behaviour in a self-consistent model of mantle convection. Phys Earth Planet Int 142(3–4):225–255

Stern R (2004) Subduction initiation: spontaneous and induced. Earth Planet Sci Lett 226(3–4):275–292

Tackley P (2000a) Self-consistent generation of tectonic plates in time-dependent, three-dimensional mantle convection simulations 2. Strain rate weakening and asthenosphere. G-cubed 1:2000–000043. doi:10.1029/2000GC0003

Tackley PJ (2000b) Self-consistent generation of tectonic plates in time-dependent, three-dimensional mantle convection simulations. G-cubed 1(8):1021. doi:10.1029/2000GC000036

Toth J, Gurnis M (1998) Dynamics of subduction initiation at preexisting fault zones. J Geophys Res-Solid Earth 103(B8):18053–18067

Trompert R, Hansen U (1998) Mantle convection simulations with rheologies that generate plate-like behaviour. Nature 395(6703):686–689

Turcotte DL (1977) Lithospheric instability. In: Talwani M, Pitman WCI (eds). Island Arcs, Deep Sea Trenches, and Back-Arc Basins, Maurice Ewing Series. AGU, Washington, D. C. pp 63–69. Chap. 1

Ueda K, Gerya T, Sobolev SV (2008) Subduction initiation by thermal-chemical plumes: numerical studies. Phys Earth Planet Int 171(1–4):296–312. doi:10.1016/j.pepi.2008.06.032

Valencia D, O'Connell RJ (2007) Inevitability of plate tectonics on super-earths. ApJ 670:45–48. doi:10.1086/524012

Valencia, D, O'Connell RJ (2009) Convection scaling and subduction on earth and super-earths. Earth Planet Sci Lett 286(3–4):492–502

van Heck HJ, Tackley P (2011) Plate tectonics on super-earths: equally or more likely than on earth. Earth Planet Sci Lett 310:252–261. doi:10.1016/j.epsl.2011.07.029

Zhong S, Watts AB (2013) Lithospheric deformation induced by loading of the hawaiian islands and its implications for mantle rheology. J Geophys Res 118:1–24. doi:10.1002/2013JB010408

Zhong S, Gurnis M, Moresi L (1996) Free-surface formulation of mantle convection - I. basic theory and application to plumes. Geophys J Int 127:708–718

2

Mars core structure–concise review and anticipated insights from InSight

George Helffrich

Abstract

This review summarizes the knowledge of Mars' interior structure, its inferred composition, and the anticipated seismological properties arising from its composition with particular focus on Mars' core. The emphasis on the core stems from the unusual morphology of the liquidus diagram of iron at moderate pressures when enriched in sulfur. From a fairly detailed liquidus diagram constructed from experimental studies, I identify a set of processes that could act within Mars' core: an iron "snow" from the core-mantle boundary's surface and a $Fe_{3-x}S_2$ "ground fog" forming at the base of the core. Depending on temperature and bulk sulfur composition, these could form an inner core or could stratify the outer core by enriching it in sulfur, or both. Core stratification could be one explanation for the extinction of Mars' magnetic field early in the planet's history, and I demonstrate the feasibility of this mechanism. The crystallization processes in the core could be observable in the seismic data that the future Mars geophysical mission, InSight, is planned to provide. The core size, the presence of an inner core, and the wavespeed profile of the outer core, whose radial derivative provides a proxy for changes in composition, are key observables to seek.

Keywords: Mars, InSight, Dynamo, Core, Fe-FeS phase relations, Stratification

Introduction

The Earth's density distribution is known principally through the eigenfrequencies of its whole-body oscillations excited by large earthquakes (Dziewonski and Anderson 1981; Masters and Gubbins 2003). The interior structure of Mars is not as well known as the Earth due to the lack of this type of seismic information. This was not for lack of trying, however. The Viking Lander instrument package included a seismometer, but it was not deployed on the ground. Its placement on the lander deck led to high noise levels due to wind-induced motion of the probe and thus low signal-to-noise levels for seismic signals (Lorenz and Nakamura 2013). Only one potential earthquake was recorded (by the Viking 2 lander; the Viking 1 instrument failed), limiting the seismicity to levels significantly lower than the Earth. The putative earthquake was shallow and located about 100 km from the lander (Anderson et al. 1977). The small event was not energetic enough to set the body of the planet vibrating; hence, no inferences on internal structure were possible.

Correspondence: george@elsi.jp
Earth-Life Science Institute, Tokyo Institute of Technology, 2-12-1-IE-1 Ookayama, Meguro-ku, Tokyo 152-8550, Japan

Consequently, physical parameters provided by astronomical and orbital measurements yield the only constraints on Mars' internal structure. According to Rivoldini et al. (2011), they are the bulk planet density 3934.97 ± 1.2 kg m^{-3}, the mean moment of inertia $I/\left(Mr^2\right)$ 0.3645 ± 0.0005, and the elastic tidal Love number k_2 0.159 ± 0.009. The k_2 value itself indicates that the core is at least partially liquid and in a range of 1520–1840 km (Yoder et al. 2003). Using these further properties to constrain internal structure requires a mineralogical and thermal model of the planet; hence, the estimates in Table 1 are model dependent.

Mars' core (radius 1600–1810 km according to Konopliv et al. (2006)) is about half the radius of Earth's (3480 km according to PREM (Dziewonski and Anderson 1981)), and the planetary radius (3389.5 km according to Seidelmann et al. (2002)) is also about half of Earth's (6371 km again according to PREM). Hence, the interior structure of Mars is roughly a half-sized scale model of the Earth (Fig. 1), and the core occupies an eighth of the planetary volume. Mars' core mass is about 0.24 of the planetary mass (Table 1), and Earth's core mass fraction is 0.32 (Stacey 1992). The pressure at the center of the

Table 1 Mars internal properties

Property	Model	Value	95% conf. range	
Core radius (km)	Median (cold)	1810	1701	1900
	Median (hot)	1784	1673	1879
	Average	1797		
Core mass fraction	Median (cold)	0.2404	0.2095	0.2683
	Median (hot)	0.2327	0.2171	0.2474
	Average	0.2366		
Core-mantle boundary P (GPa)	Median (cold)	19.0	17.8	20.4
	Median (hot)	19.4	18.1	20.0
	Average	19.2		

Source: Rivoldini et al. (2011)

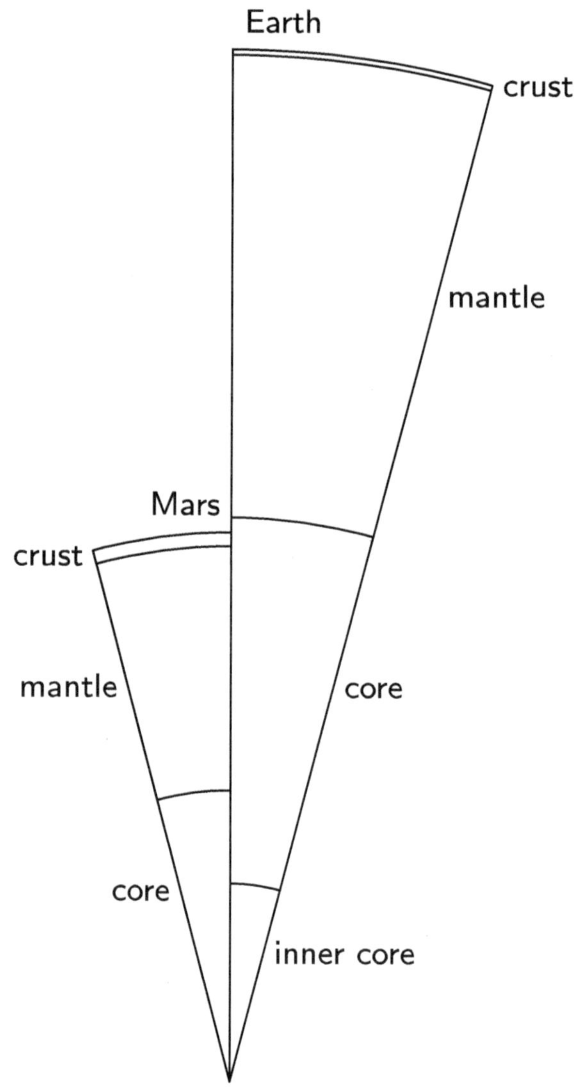

Fig. 1 Comparison of sizes and interior features of Earth and Mars. Mars is almost a half-size scale model of the Earth, except for its thicker crust (Rivoldini et al. 2011) and its probable absence of an inner core

planet 37–38 GPa (Rivoldini et al. 2011) is much lower than Earth's central pressure of 364 GPa (Dziewonski and Anderson 1981). The InSight mission, due for launch in May 2018 (originally March 2016) and to reach Mars in November 2018, is a lander designed as a geophysical observatory (Banerdt et al. 2013). Its instrument package includes a heat flow sensor, a seismometer, a magnetometer, wind and temperature sensors, and a radio transponder for lander range tracking that will supply geodetic data. Unlike the Viking missions, the seismometer is deployed separately on the surface next to the lander. Some quite basic seismic observations will significantly constrain compositional and structural models of Mars' interior. Here, I will review the key observations and what they will tell us about alternative interior structures. The review is concise, forfeiting a broad topical survey to focus instead on how the seismological information will elucidate core structure. The reviews of InSight mission seismology objectives found in Panning et al. (2015, 2017) cover aspects of marsquake location precision and silicate mantle structure.

Review

Composition

The composition of Earth's core, like Mars', is not directly observable. The core's seismic wavespeeds do provide significant constraints on composition due to the velocity-density systematics discovered by Birch (1952), who showed that there were systematic relations between the density and liquid seismic velocity depending on mean atomic weight. While the density-seismic velocity relation of the Earth's core, as revealed by its seismic wavespeed profile, was close to that experimentally determined for pure iron, it did not coincide. Thus, Birch estimated that there was ~10 wt% lighter material alloyed with the core's iron, an estimate not much improved after 60 years of further study. Denied this information about Mars' core, the inferences become more tenuous and model dependent, but comparison with Earth suggests the level of detail that will emerge once Mars is studied seismologically.

There are two main sources of compositional information: geochemical/cosmochemical and geophysical. The models underlying the use of the geophysical information provide a range of potential compositions for the core, as shown in Fig. 2. The models start with a bulk composition based on CI chondrites or the composition of SNC meteorites (believed to be Mars crustal fragments excavated by impacts on Mars), a mineralogical model for the composition as a function of pressure and temperature, and then attempt to match planetary mass, radius, and moment of inertia by varying a quasi-adiabatic temperature profile through the planet. Core metal is taken to be Fe-Ni alloy with added light elements S, C, Si, O, N, and H.

Rivoldini et al. (2011) models only considered sulfur as an alloying element. This is almost certainly wrong, but Rivoldini et al. (2011) argue for its likelihood, partly on the basis of the antipathy that S and Si have in iron (Kilburn and Wood 1997) and the fact that, at low pressure, O solubility in liquid metal is low (Rubie et al. 2004). (Systematic depletions of siderophile elements in the Earth's core, for example, suggest that it harbors significant amounts of P, Sb, Se, Te, Ag, and Ge, in addition to Ni and S and the possible, but thermodynamically unlikely, Si and O for these conditions (Rubie et al. 2004; Wood et al. 2006). These are in fact the bases for cosmochemical estimates of the Earth's core composition (Allègre et al. 1995; McDonough and Sun 1995)). Martian meteorites themselves suggest a core composition of ~11–15 wt% S (Dreibus and Wänke 1987; Lodders and Fegley 1997). Nevertheless, Rivoldini et al.'s (2011) modeling efforts revise this to 15–17 wt% S in the core. I consider the range to be, conservatively, 11–17 wt%. Earlier estimates for Mars' core from various sources yield a range of 14–17 wt% (and one as low as 10.5 wt%), yet all suggest much more than is believed to be in Earth's core: 1.7 wt% (Dreibus and Palme 1996).

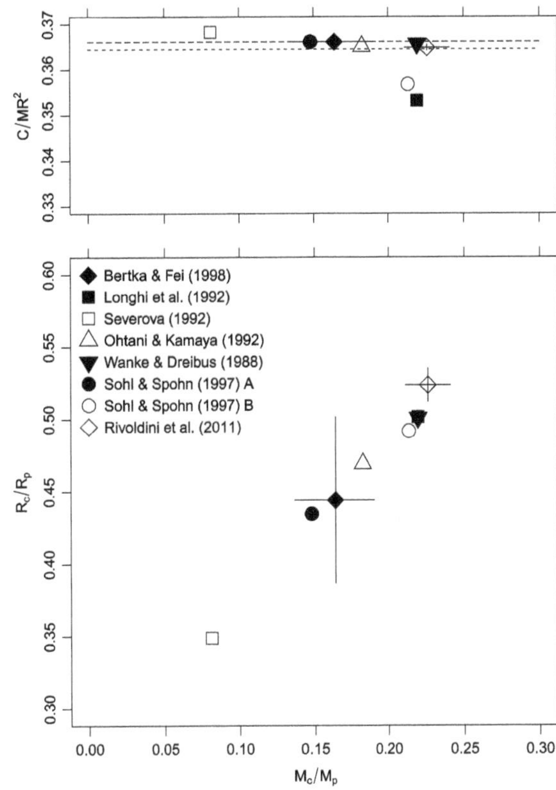

Fig. 2 Dependence of internal properties on compositional models for Mars. Top panel shows polar moment of inertia C/Mr^2 for each of seven models, characterized by their fraction of core mass to total planetary mass M_c/M_p. Bottom panel shows core radius as a fraction of planetary radius for each compositional model. Horizontal lines in upper panel are upper limit for C/Mr^2 (Esposito et al. 1992) and its recently determined value from the Mars Reconnaissance Orbiter mission (Konopliv et al. 2011)

Phase relations

Figure 3 shows the approximate melting relations in the Fe-FeS system derived from various sources (Fe melting curve (Nguyen and Holmes 2004), Fe-FeS melting relations (Fei et al. 1997; Fei et al. 2000; Morard et al. 2008; Buono and Walker 2015)). For the conditions in Mars' core, the first solids to crystallize are likely to be $Fe_{3-x}S_2$ if the composition is relatively rich in sulfur and Fe otherwise. In either case, Fe_3S will be the dominant crystalline phase in Mars' inner core after the core entirely solidifies and if it undergoes subsolidus equilibration. A naive reading of the lack of a present-day magnetic field (Acuña et al. 1999) is that an inner core is absent, incapable of releasing compositional buoyancy to drive a dynamo. A more complex view attributes the end of Mars' dynamo era to substantial stagnation of the core, leading to an annular dynamo (Takahashi and Matsushima 2006) becoming too thin to sustain activity (Olson and Christensen 2002; Christensen and Aubert 2006). The solids that do crystallize in the core lead to quite different outcomes compared to the Earth due to the difference in core composition. Figure 4 shows the conditions for core crystallization. If core CMB temperature T_{CMB} is less than 1300 K, the core will be completely solid. This is ruled out by value of the k_2 tidal Love number, placing a firm lower bound on T_{CMB}. If $T_{\mathrm{CMB}} > \sim 1550$ K, the core will be completely liquid. More interesting phenomena arise when $1300 \leq T_{\mathrm{CMB}} \leq 1500$ K. In all cases, Fe will crystallize at the top of the core, forming an iron "snow" (Stewart et al. 2007). Moreover, if $T_{\mathrm{CMB}} < 1450$ K, $Fe_{3-x}S_2$ will crystallize at the core's center and produce a phenomenon akin to ground fog on Earth. Unlike the water droplets making up fog, the solid is considerably more S-rich (~28 wt% S compared with the liquid's 14–16 wt% S) sinking and depleting the remaining liquid in S. If well-mixed, this will effectively foster further Fe crystallization because the remaining liquid will crystallize iron. If not well-mixed, the lowermost core will stratify due to Fe-rich liquid left there by $Fe_{3-x}S_2$ crystallization. The inverse will happen near the CMB: denser Fe will snow downwards in the core and leave the outermost liquid richer in S. If convection is not vigorous enough, the stratification could stabilize the core into compositionally distinct layers, in which convection occurs only in the steadily shrinking well-mixed central zone. This might quell dynamo activity and end Mars' magnetic era (Weiss et al. 2002; Williams and Nimmo 2004), like what may be occurring in Ganymede (Rückriemen et al. 2015).

The route to Mars' core stratification may be visualized with the help of Fig. 5. The figure shows the areas affected by the crystallization of solids as a function of temperature at the CMB. It is derived from Fig. 3 by fixing a CMB temperature and following an adiabat with $\gamma = 1.4$ (see Fig. 4) down into the core (upward in pressure) and plotting the S content at the liquidus for the different crystallizing phases (Fe and $Fe_{3-x}S_2$). For adiabatic temperature profiles initiated there at temperatures cooler than 1575 K, the top of the core will snow iron downwards (Stewart et al. 2007). At CMB temperatures less than ~1420 K, solid $Fe_{3-x}S_2$ will simultaneously crystallize at the bottom of the core. The combined effects will enrich the remaining core liquid in Fe, due to crystallization at the center, and to snow Fe downward from above due to the crystallization at the surface. If equilibrium conditions prevail, the snow would descend until it was dissolved back into the remaining liquid and the liquid would adopt an outwardly increasing S enrichment profile controlled by the Fe solubility along the adiabat. The isotherms of $Fe_{3-x}S_2$ crystallization show that the inner core size increases rapidly as the CMB temperature drops: for example, if the bulk S content is 15 wt%, the inner core will grow from 0 to 500 km in radius with a CMB temperature change of about 80 K. These phenomena will have a detectable effect on the seismological structure of the core, which is the next topic in this review.

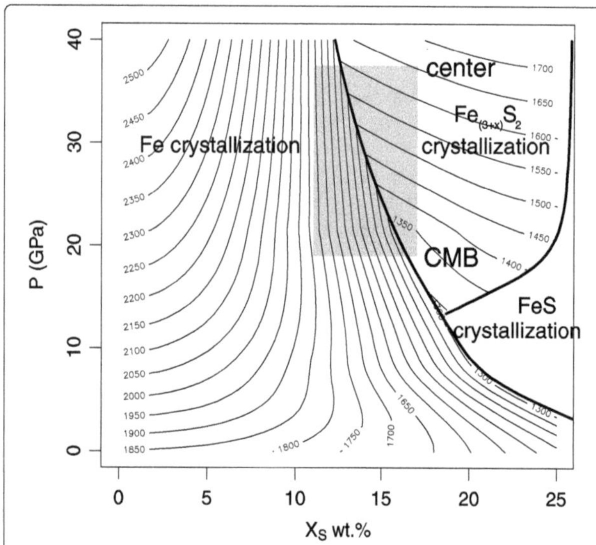

Fig. 3 Melting relations in the Fe-FeS system in the pressure and compositional range suitable for Mars' core. Isotherms show temperature (K) of appearance of first solid in the system depending on the composition and pressure. Thick lines delimit fields of first crystallizing solid: Fe, FeS, and $Fe_{3-x}S_2$ (Fe-FeS eutectic on left and incongruent melting of $Fe_{3-x}S_2$ on right). Gray box shows likely pressure and compositional range for core. Intersection of box with isotherms shows that for any crystallization of solids in the inner core, temperatures must be less than ~1650 K. First solids to crystallize will be $Fe_{3-x}S_2$ if in S-rich compositions and will be Fe if S-poor. When eutectic reached, crystallizing solids will vary depending on subsolidus phase diagram but will ultimately be Fe + Fe_3S or $Fe_{3-x}S_2$ + Fe_3S (Fei et al. 2000)

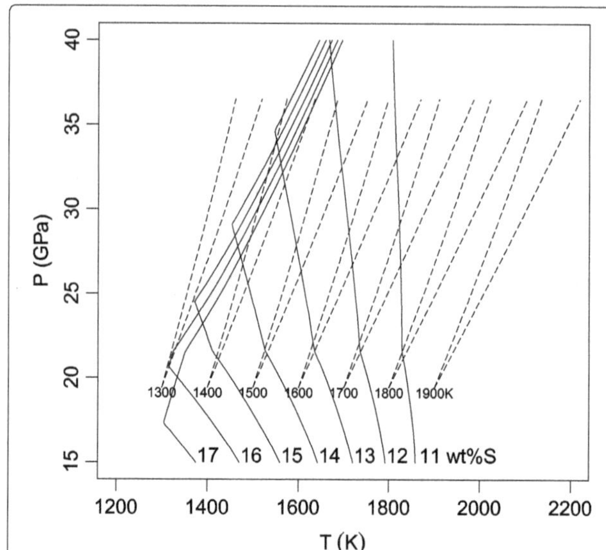

Fig. 4 Liquidus curves for a suite of fixed compositions in Mars' core (*solid lines*; these are iso-compositional sections through Fig. 3), superimposed with various adiabatic temperature profiles initiated at the core-mantle boundary (*dashed lines*; ranges shown for $1.2 \leq \gamma \leq 1.6$ for each labeled T_{CMB}) and extending through the core. Fe crystallizes where the liquidus curve slope is negative, whereas $Fe_{3-x}S_2$ crystallizes if the slope is positive (Fig. 3). For $T_{CMB} < 1300$ K, the core is solid, whereas if $T_{CMB} > 1450$ K it is liquid. At any $T_{CMB} < 1850$ K, the top of the core crystallizes an Fe 'snow'; if, moreover, $T_{CMB} < 1450$ K, the base crystallizes an $Fe_{3-x}S_2$ "ground fog." Adiabats calculated using $d\log(T)/dz = \gamma g(r)/V_P{}^2(r)$ with z the depth below the CMB using data from Rivoldini et al. (2011)

Seismology

For comparison, I focus on two seismological models of Mars' interior developed by Sohl and Spohn (1997) and by Rivoldini et al. (2011). The approaches they took to construct the models were similar, using meteorite-based compositional information to predict the mineral assemblages as a function of radius and hence the radial density profile, refining the internal temperature profiles until a match was achieved to the basic planetological data (radius, moment of inertia and k_2 Love number). As a side effect of these calculations, the models also provide seismic wavespeed profiles.

Sohl and Spohn (1997) essentially took a single compositional model due to Wänke and Dreibus (1988) and relied on either the chondritic Fe/Si ratio (1.71) to fix the bulk the mantle/core partitioning of iron (and thus the oxidation state of the mantle) or relied on matching the rotational moment of inertia, C, to its upper limit of $0.366 \times Mr^2$ as then known from available data. It was not possible to satisfy both constraints simultaneously, so they provided two solutions, A and B. I show here for comparison model A because its philosophy is most similar to those explored by the following alternative.

Rivoldini et al. (2011) used a Bayesean approach to solving the internal structure problem. They used updated astronomical and geodetic data for Mars and a wider suite of composition models advanced by meteoriticists as well as researchers who explored the phase relations in the silicates comprising Mars' mantle and the metal content of its core. Aggregates of solutions were developed that fit the observational data within each compositional model, and assuming temperature profiles through the planet that were "hot" and "cold." The statistics of the aggregate solutions provided confidence levels for each model, one of which is shown in Table 1.

I compare Sohl and Spohn (1997) model A with Rivoldini et al. (2011) "cold" profile in the Dreibus and Wänke (1985) compositional model because these represent similar geotherms and similar bulk compositions. Thus the essential differences are the advance in the state of physical knowledge and the methodology improvement (due to increased computing power) between the modeling efforts. The resulting wavespeed and density profiles are shown in Fig. 6. The main difference between them is their core size. With a chondritic Fe/Si ratio, their model B core (1667 km; not shown) is larger and similar to Rivoldini et al. (2011) core size, 1731 km. The two models' crust thicknesses are similar, ~100 km. The mantle seismic wavespeed profiles differ because Sohl and Spohn (1997) use a simplified peridotitic mantle model and first-order discontinuities at the transitions between the olivine polymorphs (olivine, wadsleyite, and ringwoodite), whereas Rivoldini et al. (2011) use a more complete mantle model and allow for two-phase regions through the polymorph transition intervals. There is no transition to bridgmanite in either model. Mantle densities are broadly similar. The smaller core of Sohl and Spohn (1997) model requires it to have a higher density to match the planetary mass. Hence, there is less light element in the core, and wavespeeds are correspondingly higher.

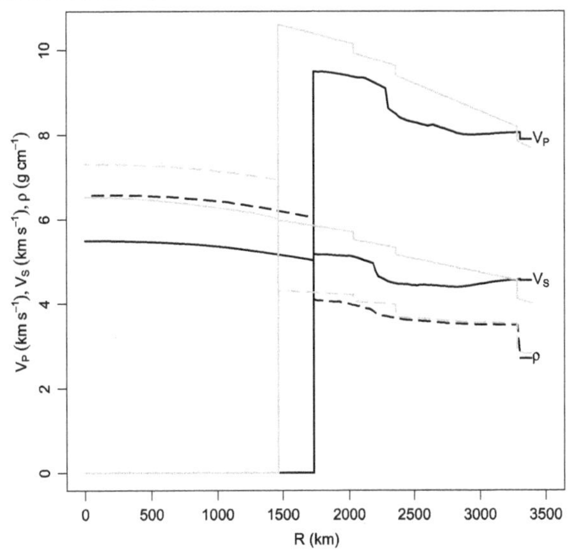

Fig. 6 Radial wavespeed and density profiles of two Mars internal structure models, Rivoldini et al. (2011) (R11DW black) and Sohl and Spohn (1997) (SS97a gray). The most pronounced difference between them is their core size. Other important seismological details are that R11DW's core wavespeed is lower than V_S at the base of the mantle, whereas SS97a's is slightly higher

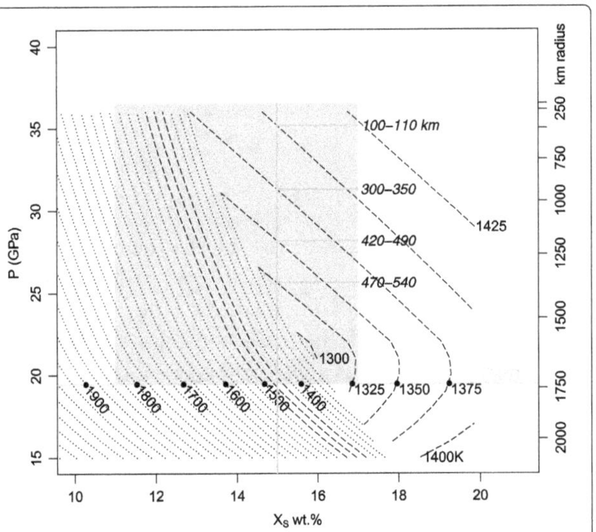

Fig. 5 Crystallization process of Mars' core. Contours show pressure (left scale) and radius (right scale) where crystallization takes place along an adiabat extending from the CMB (horizontal gray line, labeled with T_{CMB}). Dotted lines show Fe crystallization conditions, whereas dashed lines show $Fe_{3-x}S_2$ crystallization conditions. Gray box shows pressure/radius range of the core, and the anticipated compositional range of the core; range of radii of solid inner core crystallized from the liquid of 15 wt% S (vertical line) indicated by italic labels within the box. The dashed adiabatic temperature contours show that the core will grow rapidly with small changes in CMB temperature. The dotted Fe crystallization contours show that an iron 'snow' will form at the top of the core at temperatures below ~1575 K. The range of inner core radii arises from alternative compositions of the solid core: Fe_3S_2 or Fe_3S

To illustrate the first-order information that seismology will yield about Mars' core, I show the travel time curves for selected arrivals of interest to constrain internal structure. Figure 7 shows two diagrams for the Sohl and Spohn (1997) model, with and without an inner core. The model is meant only to be illustrative; the inner core is only a scaled version of Earth's inner core transplanted to Mars, with similarly scaled radius and V_P jump; a more realistic model will be examined below. The diagrams show a pronounced difference in travel times. The geometries of the individual curves are key structural information about the interior, as indicated in the figure caption.

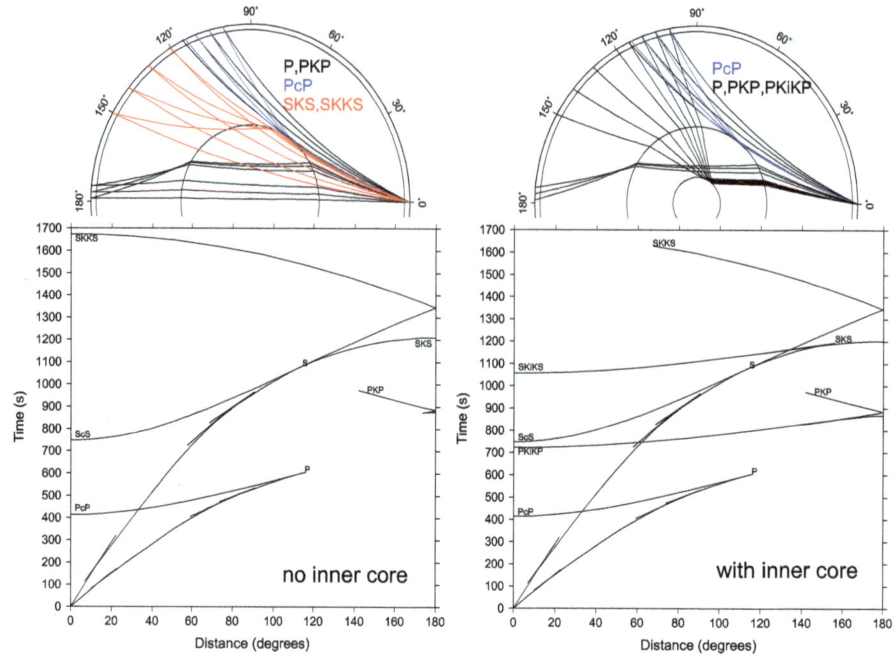

Fig. 7 Ray paths (top) and calculated travel times (bottom) in the Sohl and Spohn (1997) model A, with (right) and without (left) an inner core. The inner core's properties are only scaled version of the Earth's inner core and only meant to be indicative of the change in the travel time curves, and not otherwise physically justified. The horizontal axis is distance from an earthquake source at 0, measured in degrees of arc around the planet. The lag time until an arrival is seen is the time plotted on the vertical axis. Each type of seismic wave arrival traces a continuous path as a function of distance. Curves with positive slopes travel directly between the source and receiver (along the minor arc, to distances <180°). Curves with negative slopes travel the long way around the sphere (along the major arc to distances >180°). Each wave type is labeled with its name (see the "Appendix" section for nomenclature). The core's low wavespeeds with respect to the mantle cause a P-wave arrival shadow, the gap between P and PKP between ~120–~ 180° (top left). The main difference is that the inner core reflection PKiKP (and PKIKP) illuminates the shadow region when an inner core is present (top right). SKS and SKKS illuminate the core, but the ranges of the SKS (>160°) and SKKS (>300°) arrivals are truncated by the presence of an inner core

A comparison of the travel time curves and their associated ray paths provides further insights as to how different models lead to different travel time curves. Figure 8 compares Sohl and Spohn (1997) model A (SS97a) with Rivoldini et al. (2011) model based on the Dreibus and Wänke (1985) compositional model (R11DW). The first difference between the models is the near-constant V_P in the shallow mantle of R11DW. These lead to straight-line ray paths, and a slope for the P travel time curve that is less curved than the corresponding curve for SS97a. The second difference is that of core size. PcP (a core-reflected P-wave) arrives earlier when the core is larger. This constrains the core size, and hence the Fe/Si ratio of the planet. The final difference is the travel time curve for SKKS. Due to the lower core V_P compared to V_S at the base of the mantle, there is a shadow (gap) in the travel time curve for SKS and SKKS. The very small increase at the core-mantle boundary in the SS97a model eliminates the shadow (see Fig. 7) and leads to a "whispering gallery" effect at the top of the core (wave fronts may travel horizontally just under the boundary), illuminating its structure up the the outermost core surface. In contrast, with R11DW, the outermost outer core's structure will be less visible. Owing to SKS' high amplitude (Panning et al. 2017) and polarization in the vertical plane, it should be visible with the InSight instrumentation. This arrival's travel time curve will probably be the main way to determine core structure since the PKP amplitude is considerably lower.

Figure 9 shows the sensitivity limits of various core-related arrivals given the design criteria of the sensors on the InSight lander. One obvious difference is that of the two models, SS97a leads to more visible core arrivals. This is essentially due to the size of the velocity jump at the CMB: larger velocity jumps lead to stronger reflections (PcP, ScS) at short range, and more efficient propagation into the core (SKS, SKKS) at long range. It appears that short-range (5°–15°) ScS and ScP observations will give the best information about core size. However, longer-range ScP, ScS (20°–75°) and SKS or SKKS (>75°) may also provide valuable constraints on core size. Unfortunately, core arrivals are so weak that they require large earthquakes (in excess of 6.0 M_W) and low environmental noise for detection. Taylor et al. (2013) estimate from faults

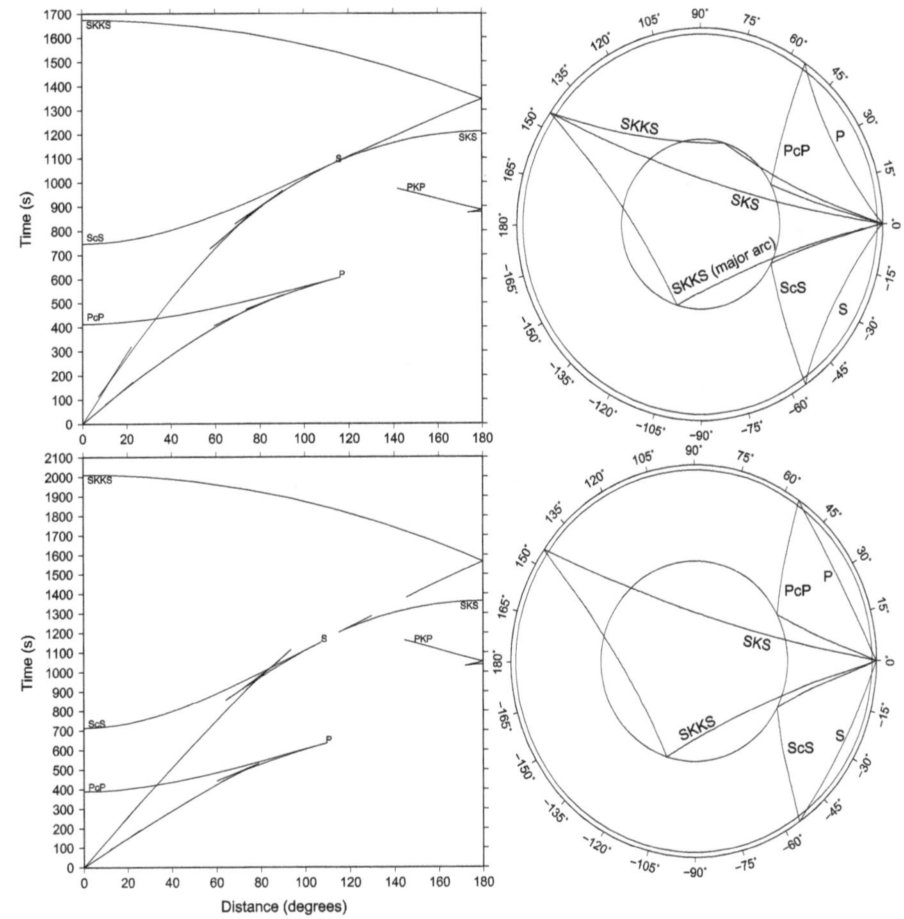

Fig. 8 Travel time curves (left) and ray paths (right) for two different Mars seismic models. (Top panels) Sohl and Spohn (1997) model A (SS97a); (bottom panels) Rivoldini et al. (2011) model based on the Dreibus and Wänke (1985) compositional model (R11DW). The diagram shows how the combination of core sizes and wavespeeds at the top of the outer core relative to V_S at the base of the mantle affect the gross form of the seismic wave arrivals. The right panels show rays emanating from a seismic source at 0° to stations at 55 and 145°. R11DW features a core shadow to SKS and SKKS due to its lower core wavespeeds, one effect of which is that minor arc SKKS disappears at 130°. The core size affects the travel time of the core reflected arrivals PcP and ScS

in Mars' Cerberus Fossae region that 0.2–500 detectable earthquakes of this size could occur per Earth year, suggesting that optimism is warranted for constraining the core's size. Cerberus Fossae is around 25° from the InSight landing site on Elysium Planita, putting the marsquake region at a range of relatively high sensitivity for ScS and ScP core reflections (Fig. 9).

Stratification of the core may result if a favorable combination of temperature and composition prevails (Fig. 5). The temperatures anticipated at the CMB for Rivoldini et al. (2011) models are between ~1700–2100 K, and for Sohl and Spohn (1997) models are ~1750–1950 K. These are too high for any inner core formation (Fig. 4). However, these models were constructed without including the mineralogy for $Fe_{3-x}S_2$ or Fe_3S forming the inner core, and the consequences of its crystallization for stratification. (Rivoldini et al. (2011) explored inner core sizes but assumed Fe as the only constituent solid.) Without exploring the resulting core structure's compatibility with the $I/(Mr^2)$ and k_2 constraints (but see the post-hoc justification later), I calculate the seismic signature of a stratified model with an inner core. Some simplifications are in order to avoid burdening the review with excessive details of thermophysical properties and equations of state; a brief summary appears in the "Appendix" section. I use the estimated thermophysical properties of Fe_3S_2 and Fe_3S (Saxena and Eriksson 2015) to calculate the density of the solid and compare it with liquid density estimates from previous core models (Rivoldini et al. 2011; Sohl and Spohn 1997). The densities at Mars' center are 7701 kg m^{-3} (Fe_3S_2) and and 8330 kg m^{-3} (Fe_3S) are greater than the liquid densities (6600–7300 kg m^{-3}), so the crystals would sink. The resulting inner core size depends on

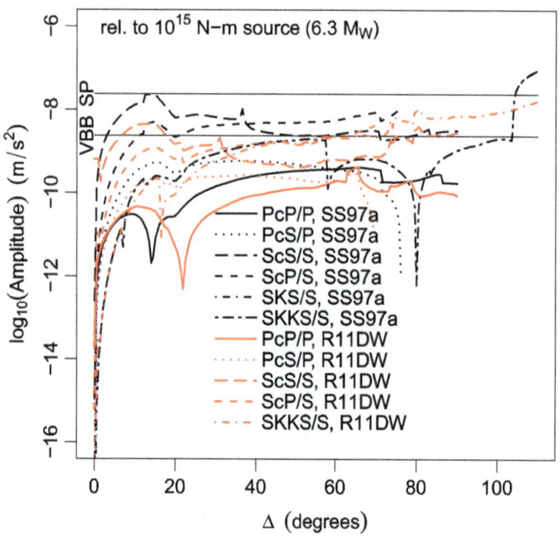

Fig. 9 Detection limits of selected core-related arrivals needed to determine core radius and outer core structure. Various curves show the expected acceleration levels expected for a 10^{15} N-m source recorded at various ranges (in degrees of great circle arc) from the seismometer. The sensitivity levels of the InSight SP and VBB instruments are shown as horizontal lines (Lognonné et al. 2014). For methods used, see the "Appendix" section

whether Fe_3S_2 or Fe_3S (or a mixture of both) is the stable high-pressure assemblage at Mars' center. The assemblage itself depends on both the precise phase relations (Fei et al. 2000; Saxena and Eriksson 2015) and the presence of other alloying elements such as Ni (Stewart et al. 2007), so a range of inner core sizes results from a single bulk core composition and CMB temperature (Fig. 5). Rivoldini et al. (2011) models found a range of core radii (of pure Fe) of 500 ± 500 and 600 ± 500 km depending on the CMB temperature, 1700 and 2100 K, respectively. For a core with 15 wt% S, the ICB radius may be as large as 500 km if Fe_3S_2, suggesting that a more sulfur-rich solid in the core is also capable of satisfying the $I/(Mr^2)$ and k_2 constraints.

To show the features resulting from a stratified core, I construct a core structural model that contains an inner core and a sulfur-rich layer at the top of the core from the liquidus relations shown in Fig. 5. The CMB temperature is 1350 K. At a bulk core composition of 15 wt% sulfur, the inner core size is 490 km. Because crystallization of Fe at this temperature deflects the core liquid's sulfur content to the adiabat, the topmost 200 km of the core will be stratified. Depletion of sulfur at the base through crystallization of Fe_3S_2 and at the top by Fe pumping down the adiabat reduces the bulk liquid S content slightly to 14.5 wt%. The liquid part of the core is 1111 km. Figure 10 shows the structure, travel time curves, and ray paths compared with R11DW.

The effect of stratification is the wavespeed and density reduction in the core in the 200 km nearest to the CMB. Density is affected most strongly: it is reduced by about 1% relative to the ambient core density. The effect on wavespeeds is much smaller, and serves to reduce them, marginally changing the ray geometry of SKS and SKKS at the CMB but not changing the form of the travel time curves. If the increased sulfur content raised wavespeeds above the shear wave speed at the base of the mantle, minor arc SKKS would re-emerge (see the SS97a model in Fig. 8), but this is not predicted by the core liquid model used (see the "Appendix" section). Panning et al.'s (2017) seismic wavespeed model brings mantle shear-wave speeds very close to core compressional wave speed, holding out the hope that the shallow core may be illuminated better by SKS and SKKS.

The Brunt-Väisälä frequency of the stratified layer is ~0.4 mHz (0.7 h period), suggesting strong stratification resistant to convection (Loper 2000). A simple calculation shows how difficult it would be to disrupt the stratification. The density change $\delta\rho/\rho$ due to a temperature change ΔT is $\alpha\Delta T$, whereas the density change due to a compositional gradient in the radial range Δr is $\rho^{-1}(d\rho/dr)\Delta r$. Equating these and taking the radial derivative, The temperature gradient required to eradicate the density gradient is therefore $dT/dr = (\alpha\rho)^{-1}d\rho/dr$. Using characteristic values $\alpha \approx 4.7 \times 10^{-5}$ K^{-1}, $\rho \approx 6400$ kg m^{-3} (Table 2), and $d\rho/dr \approx 0.875$ kg m^{-3} km^{-1} (Fig. 6) a temperature gradient of ~2.9 K km^{-1} is required. If the core's thermal conductivity is 30 W m^{-1} K^{-1}, this would entail a core heat flux of 87 mW m^{-2} or 0.88 TW, in the range of sub-lithospheric heat flux estimates of 15–40 mW m^{-2} (0.69–1.84 TW) (Ruiz et al. 2011). Hence, the core heat flux required to disrupt layering entails the unlikely requirement of zero radiogenic heat production in Mars' mantle. It is also a higher core heat flux than encountered in any of the post-plate tectonic Mars mantle evolution scenarios explored by Nimmo and Stevenson (2000) and Williams and Nimmo (2004).

The clearest evidence for stratification is the presence of an inner core. Figure 5 shows that crystallization at the base of Mars' core occurs only at low temperatures below ~1420 K. At these temperatures, Fe would already be snowing from the CMB region, forming the stratified layer. Hence, if an inner core exists, the core must be stratified (with the proviso that S is the major alloying element and present in the concentration range of 11–17 wt%). It is possible to stratify without an inner core if the CMB temperature is less than ~1850 K at the lowest expected sulfur content, 11 wt%, but this lies well below the limits found by geophysical modeling (Rivoldini et al. 2011). In the range 11–14 wt% sulfur, an effectively pure iron inner core could form below a substantially stratified liquid in

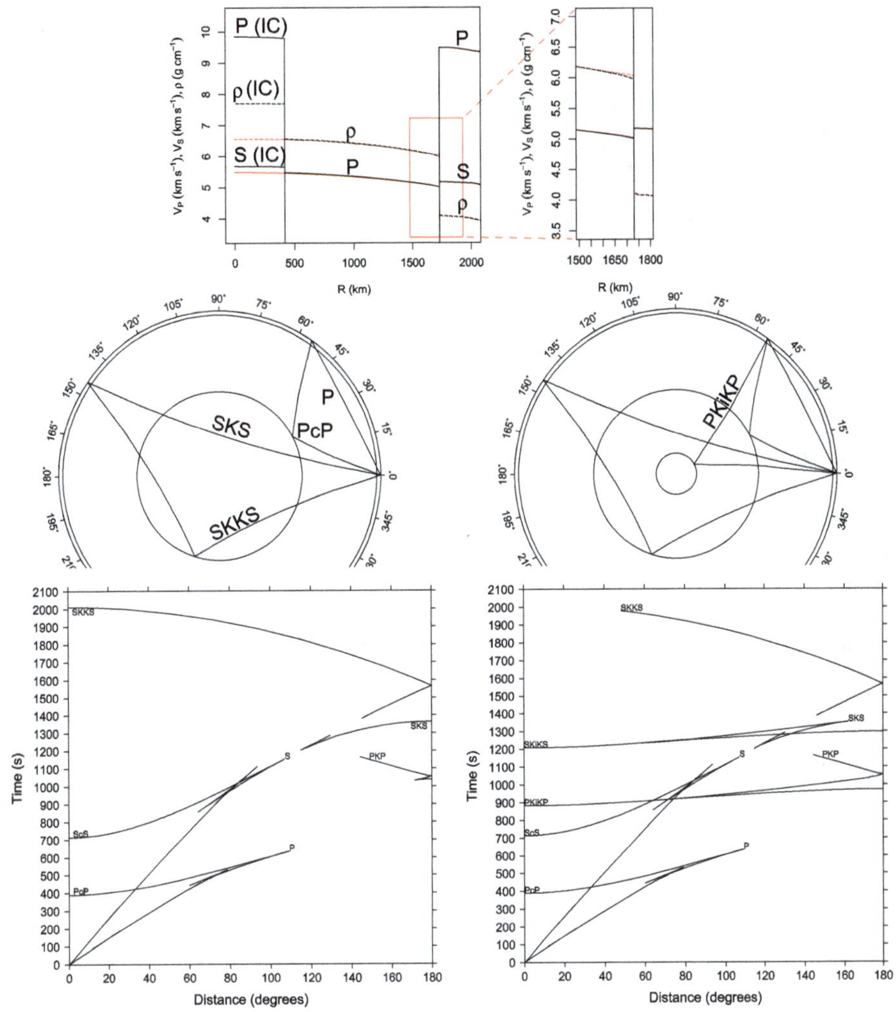

Fig. 10 Mars core structure when stratification and crystallization of an inner core occur simultaneously. Top panels show size of inner core (assumed Fe_3S_2) and slight perturbation to density and P wave speed at the top of the outer core when stratification enriches sulfur in the core, lowering both density ρ and P wave speed in the outer core relative to unstratified state (red lines). The profiles of P, S and ρ in the inner core are labeled with (IC) for clarity; the unstratified P and ρ profiles (red) continue smoothly to Mars' center. The inset (top right) of the boxed area near the CMB expands the density and P wavespeed profiles to emphasize their subtle structural changes. Bottom panels show ray paths and travel times for unstratified (left) and stratified (right) cases. The stratification itself is unobservable because P wavespeeds are already lower than S in the mantle, blocking minor-arc SKKS (Fig. 8, top). The only visible consequence of stratification is the presence of the inner core itself, indicated by the existence of PKiKP (and its extension to 180° as PKIKP)

the temperature range 1375–1575 K enriched in sulfur by up to 2.5 wt% over the bulk core content.

Between 1575–1850 K, there would be no inner core but a stratified outer core, whose stratification depth varies with temperature. Detecting the stratification could be difficult for the reason discussed above concerning the relative wavespeed of S at the base of the mantle to P in the liquid outer core.

Functioning dynamos in small planetary bodies appear to be a common outcome of the accretion process (Monteux et al. 2011). In Mars' case, dynamo activity ended early in the planet's history (Acuña et al. 1999). Perhaps this was mediated by impacts as, for example, Monteux et al. (2015) envisage, but internal processes might also curtail dynamo activity. I explore this possibility in the following paragraph.

Temperatures at the CMB are expected to be in the range of 1700–2100 K based on the internal planetary models cited earlier. In this range, core stratification is possible if the bulk sulfur content is less than 13 wt%. At the lowest temperature and lowest sulfur content (11 wt%), about half of the core's depth would be stratified (Fig. 5), and any dynamo would have to operate within a radius of ~1000 km. The heat flux required to operate a thermal dynamo would also have to be carried conductively through a thick stagnant layer. One can use dynamo scaling relations developed by Christensen and Aubert (2006) to decide whether dynamo extinction through a change in conductive heat transport is possible. The relations are strictly valid for annular dynamos, so I assume a small inner core of 100 km radius and a well-mixed liquid region extending to 1100 km radius. This roughly corresponds to a scenario where the T_{CMB} is 1700 K and the bulk sulfur content is 11 wt%; the top of the core will be enriched to 13 wt%. Using the thermophysical values given in Table 2, I calculate the amount of heat flux in excess of the adiabat to drive a dynamo by heat alone at the critical Rayleigh number, Ra^*_C. The details are relegated to the table for completeness; the result is simply that a heat flow excess Q_e of at least 1.8% over the adiabatic flow in the convective region is required. This heat flow must be carried through the stagnant layer conductively, along an unknown temperature profile (either superadiabatic or subadiabatic depending on T_{CMB} compared with the temperature in the convective region). However, the liquid will be increasingly sulfur rich, changing the thermal conductivity and hence its capacity to transport heat from below. Using Gomi et al. (2013) model for thermal conductivity dependence on impurities in liquid iron, a 2 wt% increase in sulfur content will reduce k by 2.2%, which is slightly more than the excess heat (1.8%) required to operate a critical dynamo. Thus, the growth of the sulfur-rich stratified layer acts to quell dynamo action in the core both by shrinking the dynamo region (increasing $Ra^*_{Q,c}$), and by inhibiting its ability to cool itself (decreasing Q_{adv}).

The calculation for the chosen parameters yield only a marginal suppression of dynamo activity and are uncertain. The main control over whether it arises in Mars is the core's bulk sulfur content: the higher it is, the less extensive stratification will be, but the more likely an inner core will be.

Stratification will be difficult to observe directly. However, Birch's homogeneity test (Birch 1952) of the core wavespeed profile, once it is determined, may yield a verdict on the core's chemical composition. Figure 11 shows a profile of the homogeneity index $\eta_B = dK/dP + g^{-1} d\Phi/dr$ (Dziewonski and Anderson 1981) for a synthetic Mars core wavepseed model based on Rivoldini et al. (2011) (see the "Appendix" section for methods used). $\eta_B \approx 1$ indicates adiabatic compression of a homogeneous substance. The homogeneity index is roughly 1 in the unstratified part of the core but changes substantially through the stratified region, even though the wavespeeds only change slightly, with no major changes in the shape of the travel time curves. If a good radial wavespeed model is available for the core, this test could reveal whether the core is stratified, and therefore unable to fully convect. This would be a strong constraint on Mars' evolution from the timing of the disappearance of its magnetic field.

Table 2 Mars thermophysical properties relevant for dynamo activity

Symbol	Quantity	Value	Units and scale	Source
Ω	Rotational frequency	1.28126	$\times 10^{-5} s^{-1}$	(Turcotte and Schubert 2002)
R_i	Inner core radius	100	km	
R_o	Outer radius of convective region	1100	km	
D	Convective region thickness			$= R_i - R_o$
g_0	Gravitational acceleration at top of convective region	1.99	$m\ s^{-2}$	
α	Thermal expansivity	4.7	$\times 10^{-5} K^{-1}$	(Helffrich and Kaneshima 2004) for Fe-11wt%S
T_o	Temperature at top of convective region	1850	K	Rivoldini et al. (2011) adiabat
T_{CMB}	CMB temperature	1700	K	(Rivoldini et al. 2011)
C_P	Heat capacity	600	$J\ kg^{-1}\ K^{-1}$	(Helffrich and Kaneshima 2004) for Fe-11wt%S
ρ	Density	6390	$kg\ m^{-3}$	(Rivoldini et al. 2011)
ν	Viscosity	1	$\times 10^{-6}\ m^2 s^{-1}$	(Christensen and Aubert 2006)
κ	Thermal diffusivity	2	$\times 10^{-5}\ m^2 s^{-1}$	(Christensen and Aubert 2006)
k	thermal conductivity in convective region	30	$W\ m^{-1} K^{-1}$	(Stacey 1992)
E	Ekman number	8.05	$\times 10^{-14}$	$= \nu(\Omega D^2)^{-1}$
E_k	Thermal Ekman number	1.61	$\times 10^{-12}$	$= \kappa(\Omega D^2)^{-1}$
$Ra^*_{Q,c}$	Modified critical Rayleigh number	9.19	$\times 10^{-9}$	$= Ra_c E E_k$ (Christensen and Aubert 2006); Ra_c from Al-Shamali et al. (2004)

$Ra^*_{Q,c} = \frac{1}{4\pi R_o R_i} \frac{\alpha g_0 Q_{adv}}{\rho C_P \Omega^3 D^2}$; $Q_{ad} = k \frac{\alpha g_0 T}{C_P}$. Total heat flux $Q = Q_{adv} + Q_{ad}$; heat flow excess to operate dynamo $Q_e = Q_{adv}/Q$

Conclusions

This review summarized the knowledge of Mars' interior structure, its inferred composition, and its conjectured seismological properties arising from compositional possibilities. From the fairly detailed liquidus diagram summarizing the existing experimental studies, I discussed the set of processes that could act within Mars' core: an iron "snow" from the core-mantle boundary's surface and a $Fe_{3-x}S_2$ "ground fog" forming at the base of the core. The temperature and compositional range at which these crystallization processes operate can imprint the core with identifiable structural features. From the seismic wavespeed and density profiles, I examined the key raypaths for the likely set of seismic wave arrivals and assessed their ability to diagnose chemical and thermal conditions in Mars' core.

In the near future, after InSight places a seismometer on Mars and records sufficient seismic activity on and within the planet, the resulting data will allow stronger inferences about its composition and evolution to be made. The most important datum to extract will be the radius of Mars' core that will be evident from PcP (and ScS) travel times. This will provide first-order constraints on the planetary Fe/Si ratio. The next most important seismic observation will be the presence or absence of an inner core. If an inner core is detected, temperatures in Mars' core must be much lower than expected (1375–1420 K at the CMB).

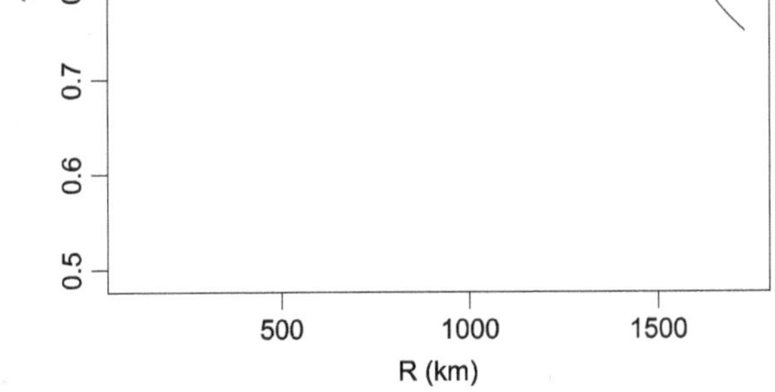

Fig. 11 Plot of homogeneity index for stratified core based on Rivoldini et al. (2011) model. The index η_B is 1 for self-compression of a homogeneous substance. The bulk sulfur content is 11 wt%, and T_{CMB} = 1700 K. This causes the topmost ~ 500 km of the core to be enriched in sulfur compared to the bulk. The index is approximately 1 in the innermost part of the outer core, but uniformly decreases in the topmost 500 km of the outer core where the sulfur concentration increase changes the wavespeed more strongly than the increase in pressure in a homogeneous material does

Wavespeeds in the core, particularly those derived from SKS and SKKS, should be able to constrain the extent of any stratification in the outer core, and, if present, restrict internal conditions to a combination of the bulk sulfur composition and CMB temperature. Finally, provided that sufficiently large seismic events can set the body of the planet oscillating, records of the long-period vibrational eigenfrequencies will yield independent constraints on the planet's density structure. Most planetary probes yield surprises, and there is no reason to expect Mars to fail to astonish.

Appendix

Methods used

Nomenclature for ray paths. A ray path is broken into segments, each described by a letter tabulated below. Each segment either starts at the earthquake focus or at one of the main seismic discontinuities in a planet: (1) the surface; (2) the core-mantle boundary (CMB); (3) the inner core-outer core boundary (ICB). Table 3 is a partial list that omits symbols not discussed in the review. See, for example, Shearer (2009) Chapter 4.9 for an extended discussion with more examples.

Starting from the focus, a wave leaves it as either a P or an S wave. As the wave meets one of the boundaries, it is either transmitted or reflected there and a new symbol describes the next segment in the path. Thus PKIKP is a P wave that crosses the outer core (K) to the inner core (I), where it reaches its maximum depth and turns upward to cross back into the core (K) and the mantle (P). Similarly, SKKS starts as S at the source, enters the core (K), reflects off the bottom of the CMB (K) and leaves the core as an S wave that travels to the surface.

Seismic wave amplitudes. The amplitude of a seismic wave is governed by the intensity of the seismic source, the relative impedance of the material at the source and the receiver, geometric focusing (and defocusing) by shrinkage and expansion of wavefronts by the radial structure of the medium, and by reflections and conversions of seismic waves at major boundaries crossed by the wave front (such as the core-mantle boundary). All except the first and the last of these factors is expressed in the following rela-

Table 3 Symbols used to describe rays

Symbol	Use
P	A downgoing P wave in the mantle
S	A downgoing S wave in the mantle
K	A P wave in the outer core
I	A P wave in the inner core
c	A reflection from the CMB topside
i	A reflection from the ICB topside

tion (Shearer 2009):

$$\frac{\hat{E}(\Delta)}{E_s} = \frac{pv_{src}^2}{4\pi r_{src}^2 r_{rec}^2 \sin\Delta \cos\theta_{src} \cos\theta_{rec}} \left| \frac{dp}{d\Delta} \right| . \quad (1)$$

The subscripts *src* and *rec* designate quantities evaluated at the source and the receiver, respectively. p is the spherical wave slowness, v is the wave speed, r is the radius, Δ is the angular separation between the source and receiver, and θ is the wavefront normal angle with respect to the radially downward direction. Specifically, $\frac{\hat{E}(\Delta)}{E_s}$ is the energy per unit of wavefront relative to the intensity of the source. The seismic wave amplitude is the square root of this quantity.

The effect of reflections and conversions at major boundaries (the core-mantle boundary in Mars' case) is accounted for by using the Zöppritz equations evaluated with appropriate material properties at the boundary and incoming and outgoing wavefront normal angles (Young and Braile 1976). These are combined with the geometric focusing and turned into a ratio of the amplitude of an arrival interacting with the core (PcP, SKKS, etc.) to the direct P or S wave arriving at the same distance.

The source intensity is obtained from Teanby and Wookey (2011) who calculated the seismic acceleration vs distance relation for an isotropic compressional energy source in Mars model at a range of source moments. I use the rule of thumb for crustal earthquakes on Earth (Hardebeck and Shearer 2003) to scale the P source intensity to an S source intensity; those authors found empirical agreement with the theoretical relation of the S intensity being 10 times the P intensity.

Again following Teanby and Wookey (2011), the detection threshold a_{peak} is (Havskov and Alguacil 2004):

$$a_{peak} = \frac{5}{4} n_{rms} [\Delta f]^{1/2} , \quad (2)$$

where Δf is the noise bandwidth, chosen to be (4–0.4) Hz (Teanby and Wookey 2011). Given the reported SP and VBB instrument sensitivities n_{rms} of 10^{-8}m s^{-2} Hz$^{-1/2}$ and 10^{-9}m s^{-2} Hz$^{-1/2}$ (Lognonné et al. 2014), these provide the positions of the lines in Fig. 9. The source moment is converted to a moment magnitude using its defining relation (Hanks and Kanamori 1979).

Seismic wavespeeds in the inner core. Saxena and Ericsson (2015) estimated the thermophysical properties for Fe_3S_2 and Fe_3S at a reference state of 1 bar and 1350 K. I use those reference state values, listed in Table 4, to calculate the seismic wavespeed at core pressures and temperatures. A third order Birch-Murnaghan equation of state is used, as described in Helffrich and Connolly (2009), but with the finite strain-based thermal expansivity pressure dependence as used by Helffrich (2017). The solid equation of state gives both density ρ, isothermal bulk modulus K, and thermal expansivity α at high pressure and temperature. The seismic wavespeed is calculated from $V_P = [(K_s + 4/3\mu)/\rho]^{1/2}$, where the adiabatic bulk modulus $K_s = K(1 + T\alpha\gamma)$. A Poisson solid is assumed in order to estimate the shear modulus μ, yielding $\mu = K_s/3$.

Velocity profiles in a stratified core. Liquid metal seismic wavespeeds in the Fe-S system are calculated using the model of Helffrich and Kaneshima (2004). Deviations of V_P and ρ from a homogenous sulfur concentration in the core are calculated down an adiabat footed at the CMB. The deviations $\delta V_P(r)$ and $\delta\rho(r)$ are expressed as multiplicative factors and applied to the V_P and ρ profiles from Rivoldini et al. (2011) to harmonize the different methods used to calculate wavespeeds in the two studies.

Core liquidus. To approximately model the core's liquidus we account for the melting depression due to the presence of a light element in the core. Because sulfur is cosmochemically abundant, lowers iron's freezing point, and is experimentally investigated, we use data for the Fe-FeS eutectic to account for this effect on the formation of the inner core.

The calculated melting temperature of pure iron (Alfè et al. 2002), calibrated with shock melting results for iron (Nguyen and Holmes 2004) provide the melting temperature for pure iron. I parameterize the liquidus temperature as a function of composition for high iron concentrations using the Fe-FeS eutectic compositions and temperatures investigated as a function of pressure from various sources (Buono and Walker 2015; Chudinovskikh and Boehler 2007; Fei et al. 1997, 2000, Morard et al. 2008). The pressure and compositionally dependent melting temperature is parameterized as

$$T(P, X) = T_{Fe}(P) - cX^2 , \quad (3)$$

where T_{Fe} is the iron melting curve and c is a pressure-dependent coefficient required to make $T(P, X)$ be the eutectic temperature $T_e(P)$ at the eutectic composition $X_e(P)$. The parabolic dependence is in contrast to Dumberry and Rivoldini 2015 who made a linear interpolation between the pure Fe melting point and the eutectic. For Mars's core, P and T is in the experimentally determined range (Chudinovskikh and Boehler 2007; Morard et al.

Table 4 Thermophysical properties

Symbol	Property	Value		Scale and units
		Fe_3S_2	Fe_3S	
V_0	Molar volume	32.848	27.370	cc mol^{-1}
α	Thermal expansivity	1	1	$\times 10^{-5}$K^{-1}
K_0	Isothermal bulk modulus	300	150	GPa
K'	Pressure derivative of bulk modulus	6	3.67	

Reference state 1 bar and 1350 K. $\gamma = 1.4$ for all phases; $\delta_T = K'$ for all phases too

2008). I similarly parameterize the liquidus temperature of $Fe_{(3-x)}S_2$ from Fei et al. (1997, 2000), Stewart et al. (2007) and Morard et al. (2008) to determine the liquidus P-T-X relations for $Fe_{(3-x)}S_2$. Stewart et al. (2007) showed Ni to have an insignificant effect on the melting relations, so its presence is neglected. Due to lack of compositional information, I assume the the incongruent melting $Fe_{(3-x)}S_2$ to FeS + liquid reaches an asymptotic composition of 26 wt% S along Morard et al. (2008) melting curve. In contrast, Dumberry and Rivoldini (2015) assumed that the eutectic composition decreased exponentially with pressure towards a value of 11 wt% S (compare Fig. 3), and did not parameterize the incongruent melting of $Fe_{(3-x)}S_2$ at all.

Abbreviations

CMB: Core-mantle boundary; ICB: Inner core boundary

Acknowledgements

I thank James Dohm for discussions about Mars geology, and him and Shigenori Maruyama for the impetus to write about Mars' internal structure. Matthieu Laneuville and Nick Teanby gave some valuable suggestions to earlier drafts of the manuscript and provided some key preprints.

Funding

Not applicable.

Competing interests

The author declares that he has no competing interest.

References

Acuña MH, Connerney JEP, Ness NF, Lin RP, Mitchell D, Carlson CW, McFadden J, Anderson KA, Rème H, Mazelle C, Vignes D, Wasiliewski P, Cloutier P (1999) Global distribution of crustal magnetization discovered by the Mars Global Surveyor MAG/ER experiment. Science 284:790–793

Al-Shamali FM, Heimpel MH, Aurnou JM (2004) Varying the spherical shell geometry in rotating thermal convection. Geophys Astrophys Fluid Dyn 98:153–169

Alfè D, Price GD, Gillan MJ (2002) Iron under Earth's core conditions: liquid state thermodynamics and high-pressure melting curve from ab-initio calculations. Phys Rev B 65:10–110365165118

Allègre CJ, Poirier J-P, Humler E, Hofmann AW (1995) The chemical composition of the Earth. Earth Planet Sci Lett 134:515–526

Anderson DL, Miller WF, Latham GV, Nakamura Y, Toksöz MN, Dainty AM, Duennebier FK, Lazarewicz AR, Kovach RL, Knight TCDD (1977) Seismology on Mars. J Geophys Res 82:4524–4546

Banerdt W, Smrekar S, Lognonné P, Spohn T, Asmar S, Banfield D, Boschi L, Christensen U, Dehant V, Folkner W, Giardini D, Goetze W, Golombek M, Grott M, Hudson T, Johnson C, Kargl G, Kobayashi N, Maki J, Mimoun D, Mocquet A, Morgan P, Panning M, Pike W, Tromp J, van Zoest T, Weber R, Wieczorek M, Garcia R, Hurst K (2013) InSight: a discovery mission to explore the interior of Mars. In: 44th Lunar and Planetary Science Conference. Lunar and Planetary Inst, Houston. p 1915

Birch F (1952) Elasticity and constitution of the earth's interior. J Geophys Res 57:227–286

Buono AS, Walker D (2015) H, not O or pressure, causes eutectic T depression in the Fe-FeS System to 8 GPa. Meteor Planet Sci 50:547–554

Christensen U. R, Aubert J (2006) Scaling properties of convection-driven dynamos in rotating spherical shells and application to planetary magnetic fields. Geophys J Int 166:97–114

Chudinovskikh L, Boehler R (2007) Eutectic melting in the system Fe-S to 44 GPa. Earth Planet Sci Lett 257:97–103

Dreibus G, Palme H (1996) Cosmochemical constraints on the sulfur content in the Earth's core. Geochim Cosmochim Acta 60:1125–1130

Dreibus G, Wänke H (1985) Mars, a volatile-rich planet. Meteoritics 20:367–381

Dreibus G, Wänke H (1987) Volatiles on Earth and Mars: A comparison. Icarus 71:225–240

Dumberry M, Rivoldini A (2015) Mercury's inner core size and core-crystallization regime. Icarus 248:254–268

Dziewonski A, Anderson DL (1981) Preliminary reference Earth model. Phys Earth Planet Inter 25:297–356

Esposito PB, Banerdt WB, Lindal WL, Sjogren WL, Slade MA, Bills BG, Smith DE, Balmino G (1992) Gravity and topography. In: Kieffer HH, Jakosky BM, Snyder CW, Matthews MS (eds). Mars. U. Ariz. Press, Tucson. pp 209–248

Fei Y, Bertka CM, Finger LW (1997) High-pressure iron-sulfur compound, Fe3S2, and melting relations in the Fe-FeS system. Science 275:1621–1623

Fei Y, Li J, Bertka CM, Prewitt CT (2000) Structure type and bulk modulus of Fe3S, a new iron-sulfur compound. Am Mineral 85:1830–1833

Gomi H, Ohta K, Hirose K, Labrosse S, Caracas R, Verstraete MJ, Hernlund JW (2013) The high conductivity of iron and thermal evolution of the Earth's core. Phys Earth Planet Inter 224:88–103

Hanks TC, Kanamori H (1979) A moment magnitude scale. J Geophys Res 84:2348–2350

Hardebeck JL, Shearer PM (2003) Using S/P amplitude ratios to constrain the focal mechanisms of small earthquakes. Bull Seismol Soc Am 93:2434–2444

Havskov J, Alguacil G (2004) Instrumentation in Earthquake Seismology. Springer, Amsterdam

Helffrich G (2017) A finite strain approach to thermal expansivity's pressure dependence. Am Mineral. (in press)

Helffrich G, Connolly JAD (2009) Physical contradictions and remedies using simple polythermal equations of state. Am Mineral 94:1616–1619

Helffrich G, Kaneshima S (2004) Seismological constraints on core composition from Fe-O-S liquid immiscibility. Science 306:2239–2242

Kilburn MR, Wood BJ (1997) Metal-silicate partitioning and the incompatibility of S and Si during core formation. Earth Planet Sci Lett 152:139–148

Konopliv AS, Asmar SW, Folkner WM, Karatekin O, Nunes DC, Smrekar SE, Yoder CF, Zuber MT (2011) Mars high resolution gravity fields from MRO, Mars seasonal gravity, and other dynamical parameters. Icarus 211:401–428

Konopliv AS, Yoder CF, Standish EM, Yuan D-N, Sjogren WL (2006) A global solution for the Mars static and seasonal gravity, Mars orientation, Phobos and Deimos masses, and Mars ephemeris. Icarus 182:23–50

Lodders K, Fegley B (1997) An oxygen isotope model for the composition of Mars. Icarus 126:373–394

Lognonné P, Banerdt W, Pike T, Giardini D, Christensen U, Banfield D, Mimoun D, Laudet P, de Raucourt S, Bierwirth M, Zweifel P, Calcutt S, Hurst K, Bruce C (2014) SEIS/INSIGHT and Mars seismology: Development status and focus on the Impact detection. Geophys Res Abst 14:12183

Loper DE (2000) A model of the dynamical structure of Earth's outer core. Phys Earth Planet Inter 117:179–196

Lorenz R. D, Nakamura Y (2013) Viking seismometer record: data restoration and dust devil search. In: 44th Lunar and Planetary Science Conference. Lunar and Planetary Institute, Houston. p 1178

Masters G, Gubbins D (2003) On the resolution of density within the Earth. Phys Earth Planet Inter 140:159–167

McDonough WF, Sun S-S (1995) The composition of the Earth. Chem Geol 120:223–253

Monteux J, Amit H, Choblet G, Langlais B, Tobie G (2015) Giant impacts, heterogeneous mantle heating and a past hemispheric dynamo on Mars. Phys Earth Planet Inter 240:114–124

Monteux J, Jellinek AM, Johnson CL (2011) Why might planets and moons have early dynamos? Earth Planet Sci Lett 310:349–359

Morard G, Andrault D, Guignot N, Sanloup C, Mezouar M, Petitgirard S, Fiquet G (2008) In situ determination of Fe-Fe3S phase diagram and liquid structural properties up to 65 GPa. Earth Planet Sci Lett 272:620–626

Nguyen J, Holmes NC (2004) Melting of iron at the physical conditions of the Earth's core. Nature 427:339–342

Nimmo F, Stevenson DJ (2000) Influence of early plate tectonics on the thermal evolution and magnetic field of Mars. J Geophys Res 105(E5):11969–11979

Olson P, Christensen UR (2002) The time-averaged magnetic field in numerical dynamos with non-uniform boundary heat flow. Geophys J Int 151:809–823

Panning MP, Beucler E, Drilleau M, Mocquet A, Lognonné P, Banerdt WB (2015) Verifying single-station seismic approaches using Earth-based data: Preparation for data return from the InSight mission to Mars. Icarus 248:230–242

Panning MP, Lognonné P, Banerdt WB, Garcia R, Golombek M, Kedar S, Knapmeyer-Endrun B, Mocquet A, Teanby NA, Tromp J, Weber R, Beucler E, Blanchette-Guertin J-F, Bozdag E, Drilleau M, Gudkova T, Hempel S, Khan A, Lekić V, Murdoch N, Plesa A-C, Rivoldini A, Schmerr N, Ruan Y, Verhoeven O, Gao C, Christensen U, Clinton J, Dehant V, Giardini D, Mimoun D, Pike WT, Smrekar S, Wieczorek M, Knapmeyer M, Wookey J (2017) Planned Products of the Mars Structure Service for the InSight Mission to Mars. Space Sci Rev:10–10071121401603175. (in press)
Rivoldini A, Hoolst T. V, Verhoeven O, Mocquet A, Dehant V (2011) Geodesy constraints on the interior structure and composition of Mars. Icarus 213:451–472
Rubie DC, Gessmann CK, Frost DJ (2004) Partitioning of oxygen during core formation on the Earth and Mars. Nature 429:58–61
Rückriemen T, Breuer D, Spohn T (2015) The Fe snow regime in Ganymede's core: a deep-seated dynamo below a stable snow zone. JGR-Planets 120:1095–1118
Ruiz J, McGovern PJ, Jiménez-Díaz A, López V, Williams J-P, Hahn BC, Tejero R (2011) The thermal evolution of Mars as constrained by paleo-heat flows. Icarus 216:508–517
Saxena S, Eriksson G (2015) Thermodynamics of Fe-S at ultra-high pressure. CALPHAD 51:202–205
Seidelmann PK, Abalakin VK, Bursa M, Davies ME, de Bergh C, Lieske JH, Oberst J, Simon JL, Standish EM, Stooke P, Thomas PC (2002) Report of the IAU/IAG working group on cartographic coordinates and rotational elements of the planets and satellites: 2000. Celest Mech Dynam Astron 82:83–110
Shearer PM (2009) Introduction to seismology. Cambridge University Press, Cambridge
Sohl F, Spohn T (1997) The interior structure of Mars: implications from SNC meteorites. J Geophys Res 102:1613–1635
Stacey FD (1992) Physics of the Earth. Brookfield Press, Kenmore
Stewart AJ, Schmidt MW, van Westrenen W, Liebske C (2007) Mars: a new core-crystallization regime. Science 316:1323–1325
Takahashi F, Matsushima M (2006) Dipolar and non-dipolar dynamos in a thin shell geometry with implications for the magnetic field of Mercury. Geophys Res Lett 33:10–10292006025792
Taylor J, Teanby N, Wookey J (2013) Estimates of seismic activity in the Cerberus Fossae region of Mars. J Geophys Res-Planets 118:2570–2581
Teanby N, Wookey J (2011) Seismic detection of meteorite impacts on Mars. Phys Earth Planet Inter 186:70–80
Turcotte DL, Schubert G (2002) Geodynamics. Cambridge University Press, Cambridge
Wänke H, Dreibus G (1988) Chemical composition and accretion history of terrestrial planets. Philos Trans R Soc London Ser A 325:545–557
Weiss BP, Vali H, Baudenbacher FJ, Kirschvink JL, Stewart ST, Shuster DL (2002) Records of an ancient Martian magnetic field in ALH84001. Earth Planet Sci Lett 201:449–463
Williams J-P, Nimmo F (2004) Thermal evolution of the Martian core: Implications for an early dynamo. Geology 23:97–100
Wood BJ, Walter MJ, Wade J (2006) Accretion of the Earth and segregation of its core. Nature 441:825–833
Yoder CF, Konopliv AS, Yuan DN, Standish EM, Folkner WM (2003) Fluid core size of Mars from detection of the solar tide. Science 300:299–303
Young GB, Braile LW (1976) Application of Zoeppritz's amplitude and Knott's energy equations. Bull Seismol Soc Am 66:1881–1886

3

Effect of width, amplitude, and position of a core mantle boundary hot spot on core convection and dynamo action

Wieland Dietrich[1*], Johannes Wicht[2] and Kumiko Hori[1,3]

Abstract

Within the fluid iron cores of terrestrial planets, convection and the resulting generation of global magnetic fields are controlled by the overlying rocky mantle. The thermal structure of the lower mantle determines how much heat is allowed to escape the core. Hot lower mantle features, such as the thermal footprint of a giant impact or hot mantle plumes, will locally reduce the heat flux through the core mantle boundary (CMB), thereby weakening core convection and affecting the magnetic field generation process. In this study, we numerically investigate how parametrised hot spots at the CMB with arbitrary sizes, amplitudes, and positions affect core convection and hence the dynamo. The effect of the heat flux anomaly is quantified by changes in global flow symmetry properties, such as the emergence of equatorial antisymmetric, axisymmetric (EAA) zonal flows. For purely hydrodynamic models, the EAA symmetry scales almost linearly with the CMB amplitude and size, whereas self-consistent dynamo simulations typically reveal either suppressed or drastically enhanced EAA symmetry depending mainly on the horizontal extent of the heat flux anomaly. Our results suggest that the length scale of the anomaly should be on the same order as the outer core radius to significantly affect flow and field symmetries. As an implication to Mars and in the range of our model, the study concludes that an ancient core field modified by a CMB heat flux anomaly is not able to heterogeneously magnetise the crust to the present-day level of north–south asymmetry on Mars. The resulting magnetic fields obtained using our model either are not asymmetric enough or, when they are asymmetric enough, show rapid polarity inversions, which are incompatible with thick unidirectional magnetisation.

Keywords: Core convection, Geodynamo, Ancient Martian dynamo, Inhomogeneous CMB heat flux, Numerical simulation

Background

Within our solar system, the three terrestrial planets, Earth, Mercury, and Mars, harbour or once harboured a dynamo process in the liquid part of their iron-rich cores. Vigorous core convection shaped by rapid planetary rotation is responsible for the generation of global magnetic fields. Unlike the dynamo regions of giant planets or the convective zone of the sun, the amount of heat escaping the cores of terrestrial planets is determined by the convection of the overlying mantle. As this vigorous core convection assures efficient mixing and hence a virtually homogeneous temperature T_{core} at the core side of the core mantle boundary (CMB), the temperature gradient at and thus the flux through the CMB are entirely controlled by the lower mantle temperature T_{lm}. The heat flux through the CMB is then

$$q_{\text{cmb}} = k\frac{T_{\text{lm}} - T_{\text{core}}}{\delta_{\text{cmb}}}, \tag{1}$$

where δ_{cmb} is the vertical thickness of the thermal boundary layer on the mantle side and k is the thermal conductivity. Hot mantle features, such as convective upwellings, thermal footprints of giant impacts, or chemical heterogeneities, locally reduce the heat flux through the CMB (e.g., Roberts and Zhong 2006, Roberts et al. 2009). Seismologic investigations of Earth have revealed strong anomalies in the lowermost mantle that can be interpreted in terms of temperature variations (Bloxham 2000). In

*Correspondence: w.dietrich@leeds.ac.uk
[1] Department of Applied Mathematics, University of Leeds, Woodhouse Lane, LS9 2JT Leeds, West Yorkshire, UK
Full list of author information is available at the end of the article

recent years, several authors have investigated the consequences of laterally varying the CMB heat flux. A seismologically based pattern has been used in many studies on the characteristics of Earth (e.g., Olson et al. 2010; Takahashi et al. 2008); this pattern shows a strong effect on the core convection and secular variation of the magnetic field (e.g., Davies et al. 2008). For Mars, low-degree mantle convection or giant impacts may have significantly affected the core convection and the morphology of the induced magnetic field (Harder and Christensen 1996; Roberts and Zhong 2006; Roberts et al. 2009). For example, the strong southern hemispherical preference of the crustal magnetisation can be explained by an ancient dynamo that operated more efficiently in the southern hemisphere. However, this remains a matter of debate because post-dynamo processes may simply have reduced a once more homogeneous crustal magnetisation in the northern hemisphere (Nimmo et al. 2008; Marinova et al. 2008). The asymmetry of Mercury's magnetic field, which is significantly stronger in the north than in the south, could also be partly explained by a non-homogeneous CMB heat flux (e.g., Cao et al. 2014; Wicht and Heyner 2014).

The magnetic field generation process inside Earth's outer core relies on thermal and compositional convection. Thermal convection is driven by secular cooling or the latent heat released upon inner core freezing. Compositional convection arises because the light elements mixed into the outer core alloy are not as easily contained in the inner core. A large fraction is thus released at the inner core front. For planets with no solid inner core, like ancient Mars or early Earth, only the thermal component of the buoyant force can power convection and hence the dynamo process (e.g., Breuer et al. 2010). The buoyancy sources are then not concentrated at the bottom but homogeneously distributed over the core shell. When modelling these processes, secular cooling can be modelled using a buoyancy source equivalent to internal heating, whereas a basal heating source can be used when thermal and/or compositional buoyancy fluxes arise from the inner core boundary. Kutzner and Christensen (2000) and Hori et al. (2012) investigated the dynamic consequences when a core model is driven by either internal or basal heating. In general, the effects of large-scale CMB heat flux anomalies on convection and magnetic field generation are stronger when the system is driven by internal (as in ancient Mars) rather than basal heating, which is more realistic for present-day Earth (Hori et al. 2014). For example, the equatorial symmetry of the flow is more easily broken in the former than the latter case (Wicht and Heyner 2014).

The Mars Global Surveyor (MGS) mission revealed a remarkable equatorial asymmetry in the distribution of magnetised crust (Acuña et al. 1999; Langlais et al. 2004). Interestingly, the crustal topographic dichotomy is well aligned with the pattern of crustal magnetisation (Citron and Zhong 2012). This is the magnetic imprint of an ancient core dynamo driven by thermal convection in the core or tidal dissipation (Arkani-Hamed 2009), which ceased to exist and further magnetise the crust roughly 3.5 Gyrs ago (Lillis et al. 2008). Assuming the hemispherical crustal magnetisation is of internal origin, most numerical models attempting to design a Mars core dynamo model quantified the success of their efforts by comparing the resulting modelled magnetic fields to the actual crustal magnetisation pattern (Stanley et al. 2008; Amit et al. 2011; Dietrich and Wicht 2013; Monteux et al. 2015). Hereafter, the study by Dietrich and Wicht (2013) is abbreviated as DW13. The CMB heat flux in such models is typically modified by a large-scale sinusoidal perturbation increasing the CMB heat flux in the southern hemisphere and reducing it in the northern hemisphere. Because one hemisphere is more efficiently cooled than the other, a strong latitudinal temperature anomaly arises and in turn drives fierce zonal flows via a thermal wind. Such equatorially antisymmetric, axisymmetric (EAA) flows are reported to reach up to 85 % of the total kinetic energy (Amit et al. 2011, DW13) in self-consistent dynamo models. Although such flows are indicative of the induction of hemispherical fields, it remains unclear to what extent their prevalence is due to the (probably unrealistic) large, strong, and axisymmetric forcing patterns. As a consequence of this forcing, the induction process is more concentrated in the southern hemisphere, leading to a hemispherical magnetic field. Even though such hemispherical fields can match the degree of hemisphericity in the crustal magnetisation at the planetary surface, they also show a strong time variability (periodic oscillations). If the typical stable chron epoch is much smaller than the typical crustal buildup time, the system is not able to magnetise the crust to the required intensity (DW13), which requires the relatively homogeneous magnetisation of a layer with a thickness of at least 20 km (Langlais et al. 2004).

In this study, we focus on not only somewhat more complex but also more realistic CMB heat flux variations. The heat flux is reduced in a more localised area of varying position and horizontal extent. Such anomalies may more realistically reflect the effect of mantle plumes or impacts, but may not yield the fundamental equatorial asymmetry in the temperature as efficiently as the simplistic Y_{10} pattern. Further tilting the anomalies away from the axis of rotation may result in the superposition of EAA and equatorially symmetric, non-axisymmetric (ESN) temperature and flow patterns. We also investigate the influence of the shell geometry and the vigour of convection. More generally, we aim to estimate how large, strong, and aligned CMB heat flux anomalies must be to affect core convection and the magnetic field process. It is also of interest to quantify the interaction of flow and field under the control of a laterally variable CMB heat flux. In particular, we address the question of whether the conclusions of DW13 regarding the crustal magnetisation still hold when applied to Mars. Various comparable models have been investigated over the last decade. Those focusing on exploring parameter dependencies, e.g., Amit et al. (2011) and DW13, mainly investigated magnetic cases with a CMB heat flux described by a fundamental spherical harmonic. Those studies featured tilted cases and various amplitudes. The recent study by Monteux et al. (2015) presented simulations featuring anomalies of a smaller horizontal length scale. However, a comprehensive parametrisation of the anomaly width, amplitude, and position has not yet been reported. The rather dramatic results of DW13 may only hold when anomalies of a planetary scale are at work. We therefore also test the robustness of their results with respect to the most common model assumptions.

Methods

We model the liquid outer core of a terrestrial planet as a spherical shell (with inner and outer radii of r_{icb} and r_{cmb}, respectively) containing a viscous, incompressible, and electrically conducting fluid. The core fluid is subject to rapid rotation, vigorous convection, and Lorentz forces due to the induced magnetic fields. The evolution of the fluid flow is thus given by the dimensionless Navier–Stokes equation:

$$\mathrm{E}\left(\frac{\partial \vec{u}}{\partial t}+\vec{u}\cdot\vec{\nabla}\vec{u}\right)=-\vec{\nabla}\Pi+E\nabla^2\vec{u}-2\hat{z}\times\vec{u}+\frac{\mathrm{RaE}}{\mathrm{Pr}}\frac{\vec{r}}{r_{\mathrm{cmb}}}T+\frac{1}{\mathrm{Pm}}(\vec{\nabla}\times\vec{B})\times\vec{B}\,, \tag{2}$$

where $\vec{u}$ is the velocity field, Π is the non-hydrostatic pressure, $\hat{z}$ is the direction of the rotation axis, T is the superadiabatic temperature fluctuation, and $\vec{B}$ is the magnetic field.

The evolution of the thermal energy is affected by temperature diffusion and advection by the flow, such that

$$\frac{\partial T}{\partial t}+\vec{u}\cdot\vec{\nabla}T=\frac{1}{\mathrm{Pr}}\nabla^2 T+\epsilon\,, \tag{3}$$

where ϵ is a uniform heat source density. The generation of magnetic fields is controlled by the induction equation

$$\frac{\partial \vec{B}}{\partial t}=\vec{\nabla}\times(\vec{u}\times\vec{B})+\frac{1}{\mathrm{Pm}}\nabla^2\vec{B}\,. \tag{4}$$

We use the shell thickness $D = r_{\mathrm{cmb}} - r_{\mathrm{icb}}$ as the length scale, the viscous diffusion time D^2/ν as the time scale, and $(\rho\mu\lambda\Omega)^{1/2}$ as the magnetic scale. The mean superadiabatic CMB heat flux density q_0 serves to define the temperature scale $q_0 D/c_p\rho\kappa$. Furthermore, ν is the viscous diffusivity, ρ is the constant background density, μ is the magnetic permeability, λ is the magnetic diffusivity, Ω is the rotation rate, κ is the thermal diffusivity, and c_p is the heat capacity.

Non-penetrative and no-slip velocity boundary conditions are used, and the magnetic fields are matched to the potential field outside the fluid region. For an internally heated Mars-like setup, we fix the heat flux at both boundaries and power the system exclusively by internal heat sources. Because the flux at the inner boundary is set to zero, this leads to a simple balance of heat between the source density ϵ and the mean CMB heat flux q_0. To model the secular core cooling, we fix the mean heat flux at the outer boundary ($q_0 = 1$) and balance the heat source such that

$$\epsilon=\frac{1-\beta}{1-\beta^3}\frac{q_0}{3\mathrm{Pr}}\,, \tag{5}$$

where $\beta = r_{\mathrm{icb}}/r_{\mathrm{cmb}}$ is the aspect ratio of the spherical shell.

The non-dimensional control parameters are the Prandtl number $\mathrm{Pr} = \nu/\kappa$, which is the ratio between the viscous and thermal diffusivities, and the magnetic Prandtl number $\mathrm{Pm} = \nu/\lambda$, which is the ratio of the viscous and magnetic diffusivities. The Ekman number $\mathrm{E} = \nu/\Omega D^2$ relates the viscous and rotational time scales, and the Rayleigh number $\mathrm{Ra} = \alpha g q_o D^4/\nu\kappa^2$ controls the vigour of convection. We fix $\mathrm{Pr} = 1$ and $\mathrm{E} = 10^{-4}$ and use $\mathrm{Pm} = 2$ for the dynamo cases. The Rayleigh number is varied between $\mathrm{Ra} = 2 \times 10^7$ and 1.6×10^8.

Modelling the anomaly

In recent studies focusing on the mantle control of Mars and Earth, it is common for the horizontal variation of the CMB heat flux to be described in terms of spherical harmonics. Especially for Mars, the majority of studies rely on spherical harmonics of degree $l = 1$ and order $m = 0$, i.e., a simple cosine variation (e.g., Stanley et al. 2008; Amit et al. 2011; Aurnou and Aubert 2011, DW13). Notable exceptions are the study by Sreenivasan and Jellinek (2012), in which a localised temperature anomaly was used, and that by Monteux et al. (2015), in which the anomaly pattern was derived from realistic impact models. The former study relies on fixed temperature boundary conditions, basal heating, and strong CMB temperature anomalies and is thus quite different from our systematic approach, which features fixed flux boundary conditions, internal heating, and anomaly amplitudes not exceeding the mean CMB heat flux. Here, we explore more locally confined variations of the CMB heat flux. The thermal CMB anomaly q' is characterised by four parameters: its amplitude q^*, its opening angle Ψ, and the colatitude and longitude (τ, ϕ_0) of its midpoint. The anomaly has the form

$$q' = \begin{cases} \frac{1}{2}\left(\cos\left(2\pi \frac{\alpha}{\Psi}\right) + 1\right) & \text{for } \alpha < \Psi \\ 0 & \text{else} \end{cases} \tag{6}$$

where $\alpha(\theta, \phi)$ is the opening angle between the central vector $\vec{r}(\theta, \phi)$ and the mid-point vector $\vec{r}(\tau, \phi_0)$. Furthermore, the anomaly is normalised, such that the net heat flux through the CMB is constant. The total CMB heat flux is then given by the mean heat flux q_0, the anomaly q', and the normalisation $C(\Psi)$ as

$$q_{\text{cmb}} = q_0 - q^* \left(\frac{q'}{C(\Psi)} - 1 \right), \tag{7}$$

with

$$C(\Psi) = \int_{S_\Psi} q' \sin\theta d\theta d\phi = \frac{1 - \cos\Psi}{4}, \tag{8}$$

where S_Ψ is the area of the anomaly up to its rim, which is given by Ψ. Note that the case of $(\Psi = 180°, \tau = 0)$ is identical to the spherical harmonic Y_{10} mentioned above. Figure 1 shows the CMB heat flux profiles for $\tau = 0$, $q^* = 1$, and various widths Ψ. For such parameters, the heat flux anomaly is axisymmetric and reduces the heat flux at the northern pole to zero.

Numerical model and runs

The magnetohydrodynamics (MHD) equations (Eqs. 2, 3, and 4) were numerically solved using the MagIC3 code in its shared memory version (Wicht 2002; Christensen and Wicht 2007). The numerical resolution is given by 49, 288, and 144 grid points in the radial direction, the azimuthal direction, and along the latitude, respectively. For the higher Rayleigh number cases, the numerical resolution was increased to 61, 320, and 160 points, respectively.

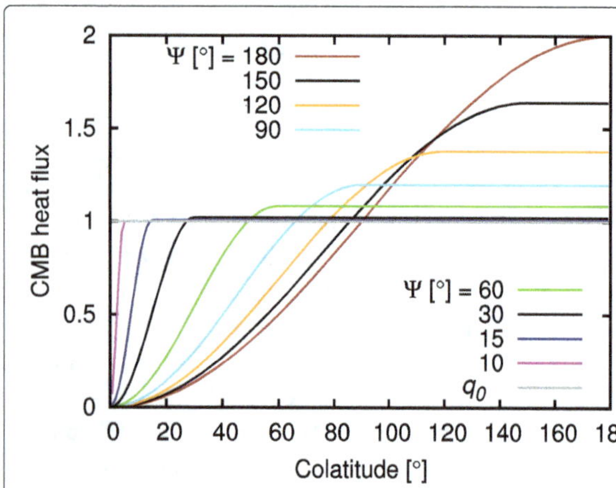

Fig. 1 Latitudinal profile of axisymmetric CMB heat flux. The *lines* show the total CMB heat flux modified by an anomaly with different opening angles Ψ (each value of Ψ is represented by a *different colour*). Note that all profiles are normalised such that there is no net contribution from the anomaly to the mean CMB heat flux ($q_0 = 1$, *grey*). Parameters: $\tau = 0$, $q^* = 1$

We conducted a broad parameter study in which eight anomaly widths between $\Psi = 180°$ (planetary scale) and $\Psi = 10°$ (most concentrated hot spot) were used. Furthermore, we tested four different anomaly amplitudes ranging from $q^* = 0.2$ to $q^* = 1$. The peak position of the anomaly was tilted at six different angles between $\tau = 0°$ (polar anomaly) and $\tau = 90°$ (equatorial anomaly). Because the ancient Martian core was fully liquid at the time the magnetisation was acquired, the thick shell regime is investigated here. Particularly, an inner core with an Earth-like aspect ratio of $\beta = 0.35$ is retained in most of the models to ensure consistency with the existing literature regarding Earth and Mars. However, the influence of the aspect ratio β was tested by varying the aspect ratio between $\beta = 0.35$ and $\beta = 0.15$, which corresponds to the smallest inner core size. The vigour of the convection was varied using four different Rayleigh numbers from $\text{Ra} = 2 \times 10^7$, which is only slightly supercritical, to $\text{Ra} = 1.6 \times 10^8$, which ensures rich turbulent dynamics. We repeated the numerical experiments for simulations including the magnetic field for various amplitudes q^* and tilt angles τ. Together with reference cases with no thermal boundary heterogeneity for each β and Ra, these parameters represent 129 hydrodynamic and 73 self-consistent dynamo simulations (202 simulations in total).

Each case was time integrated until a statistically steady state was reached and then time-averaged over a significant fraction of the viscous (magnetic) diffusion time. The moderate Ekman number allowed each of the hydro cases to be modelled within a computational time of 1 or 2 days when parallelised over 12 cores, whereas dynamo simulations usually require several days.

Previous work and output parameters

Figure 2 illustrates the mean flow and field properties for an unperturbed reference dynamo case with homogeneous boundary heat flux (Fig. 2a) and a commonly studied model with a heterogeneous CMB heat flux (Fig. 2b). In the reference case, ESN convective columns (e.g., Busse 1970) account for 72 % of the total kinetic energy. Table 1 shows the kinetic energy symmetry contributions in the two cases. Axisymmetric flow contributions consist of the zonal flow and the meridional circulation. Both are predominantly equatorially symmetric but have low amplitudes. These equatorially symmetric, axisymmetric (ESA) kinetic energy contributions therefore amount to only 3 % of the total kinetic energy (Table 1). The zonal temperature $\overline{T}$ is also equatorially symmetric, and its colatitudinal gradient is in very good agreement with the z-variation of the zonal flow (last two plots in the first row of Fig. 2a), which proves that this z-variation is caused by thermal

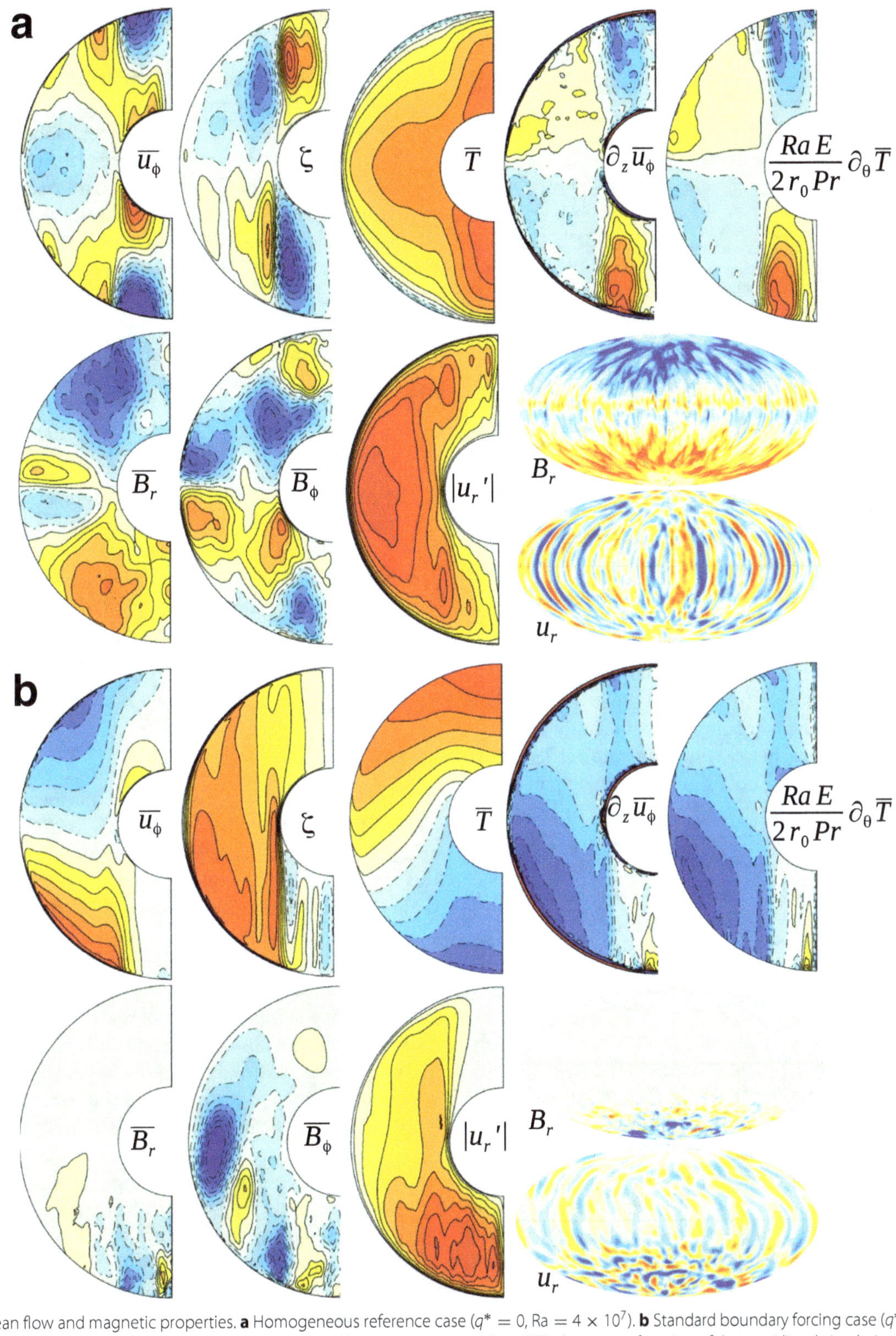

Fig. 2 Mean flow and magnetic properties. **a** Homogeneous reference case ($q^* = 0$, $Ra = 4 \times 10^7$). **b** Standard boundary forcing case ($q^* = 1.0$, $\tau = 0$, $\Psi = 180°$). The first row in each part shows, from left to right, the zonal flow ($\overline{u_\phi}$), the stream function of the meridional circulation (ζ), the zonal temperature ($\overline{T}$), and the two sides of the thermal wind equation ($\partial \overline{u_\phi}/\partial z$ and $RaE/(2r_0 Pr)\, \partial T/\partial \theta$). The second row in each part contains the radial field ($\overline{B_r}$), the mean azimuthal field ($\overline{B_\phi}$), the intensity of non-axisymmetric radial flows ($|u_r'|$), and hammer projections of the radial field at the surface (*top*) and the radial flow at mid-depth (*bottom*). Parameters: $Ra = 4 \times 10^7$, $q^* = 1.0$, $\tau = 0$, $\beta = 0.35$, $Pm = 2$

Table 1 Flow symmetries

q^*	Ψ	C_{ESA}	C_{ESN}	C_{EAA}	C_{EAN}
0	0	0.03	0.72	0.01	0.24
1	180°	0.03	0.09	0.78	0.10
0.5	180°	0.05	0.29	0.34	0.31
0.75	120°	0.05	0.31	0.33	0.30
1	90°	0.05	0.35	0.31	0.24
1	30°	0.04	0.61	0.08	0.27

Time-averaged relative kinetic energy of various flow contributions (C) obeying different symmetries, which are represented by three-letter abbreviations. The first two letters indicate the equatorial symmetry: equatorial symmetric (ES) or equatorial antisymmetric (EA). The last letter indicates the axisymmetry: axisymmetric (A) or non-axisymmetric (N). The first two rows are magnetic runs (Fig. 2), and the other four are hydro runs (Fig. 4). Parameters: Ra $= 4 \times 10^7$

wind. The mean radial and azimuthal magnetic fields (first two plots in the second row of Fig. 2) show the typical equatorial antisymmetry of a dipole-dominated magnetic field. This field is produced by non-axisymmetric flows (intensity $|u_r'|$) in an α^2-mechanism (Olson et al. 1999). The final two plots in the second row of Fig. 2 show the Hammer–Aitoff projections of the radial field at the CMB and the radial flow at mid-depth.

We compare the homogeneous case with the most commonly studied CMB heat flux anomaly: a Y_{10} anomaly with a strong amplitude ($q^* = 1.0$). Such an anomaly cools the southern hemisphere more efficiently than the northern hemisphere; hence, the temperature decreases from the hot north to the cool south, leading to a negative temperature gradient along the colatitude. Such strong temperature anomalies are known to modify the leading order vorticity balance between the pressure and the Coriolis force. The curl of the Navier–Stokes Eq. 2 gives to first order (neglecting viscous and inertia terms):

$$\nabla \times \hat{z} \times \vec{u} = \frac{\mathrm{RaE}}{2\mathrm{Pr}} \frac{1}{r_{\mathrm{cmb}}} \nabla \times (\vec{r}T) . \tag{9}$$

In models with a homogeneous heat flux and relatively small Ekman number, the right-hand side of Eq. 9 is small, and the flow (at least the convective flow) tends to be vertically invariant. Because of the rigid walls applied here, the zonal flow is weak and ageostrophic in the reference case. However, for the boundary-forced system, the right-hand side becomes large. The heterogeneous CMB heat flux mainly cools the southern hemisphere and leaves the northern hemisphere hot. The large-scale temperature asymmetry that develops between the north and south is responsible for driving a significant axisymmetric thermal wind. For the mean azimuthal component of Eq. 9, it is thus found that

$$\frac{\partial \overline{u_\phi}}{\partial z} = \frac{\mathrm{RaE}}{2\mathrm{Pr}} \frac{1}{r_{\mathrm{cmb}}} \frac{\partial \overline{T}}{\partial \theta} . \tag{10}$$

Figure 2 compares the right- and left-hand sides of Eq. 10 and demonstrates that this thermal wind balance is indeed well fulfilled in both the homogeneous and Y_{10} cases. The larger north–south gradient in the latter case drives a zonal wind system with retrograde and prograde jets in the northern and southern hemispheres, respectively. This EAA flow system dominates the kinetic energy once the amplitude of the boundary pattern is sufficiently large (see Table 1). We therefore quantify the influence of the boundary forcing by measuring the relative contribution C_{EAA} of EAA flows by evaluating (in spectral space) the relative kinetic energy of spherical harmonic flow contributions of order $m = 0$ (axisymmetric) and odd degree $l = 2n + 1$ (equatorially antisymmetric)

$$C_{\mathrm{EAA}} = \frac{\int_{r_{\mathrm{icb}}}^{r_{\mathrm{cmb}}} \sum_l E_{2l+1,0}^k r^2 dr}{\int_{r_{\mathrm{icb}}}^{r_{\mathrm{cmb}}} \sum_{l,m} E_{l,m}^k r^2 dr} . \tag{11}$$

The radial flows required to induce radial fields are then concentrated in the southern polar region (Fig. 2, bottom row), where the cooling is more efficient. Hence, the induced radial field is also concentrated in the respective hemisphere. The dominant magnetic field production, however, remains the thermal wind shear, which produces a strong axisymmetric azimuthal field via the Ω-effect. The bottom row of Fig. 2 shows the hemisphere-constrained radial field and the amplified azimuthal toroidal field created by shearing motions around the equator. DW13 reported that the Ω-effect dominates once the Y_{10} amplitude exceeds 60 % of the mean heat flux. The dynamo then switches from α^2- to $\alpha\Omega$-type, initiating periodic polarity reversals that are characteristic of this dynamo type (see also Dietrich et al. 2013). Note that EAA symmetric flows can emerge in dynamo models with a homogeneous heat flux (Landeau and Aubert 2011) when they are internally heated and satisfy a large convective supercriticality. Conversely, the anomalies themselves can also drive sufficiently complex flows to induce magnetic fields in the absence of thermal and compositional buoyancy fluxes (Aurnou and Aubert 2011).

The magnetic field is stronger in the convectively more active hemisphere and shows weaker magnetic flux in the less active hemisphere. Amit et al. (2011) gave an estimate of the mean crustal magnetisation per hemisphere on present-day Mars, which can be compared to the geometry of the model output fields. DW13 defined the magnetic field hemisphericity H_{sur} at the planetary surface as:

$$H_{\mathrm{sur}} = \left| \frac{B_N - B_S}{B_N + B_S} \right| , \tag{12}$$

where B_N and B_S are the radial magnetic fluxes in the northern and southern hemispheres, respectively. Note that the crustal value is 0.55 ± 0.1 (Amit et al. 2011,

DW13). As radial fields are induced by convective flows, an equivalent convective hemisphericity Γ can be defined as

$$\Gamma = \left| \frac{\Gamma_N - \Gamma_S}{\Gamma_N + \Gamma_S} \right| . \tag{13}$$

A formal derivation of Γ_N and Γ_S is given in the 'Results and discussions' section. We further quantify the mean flow amplitude with the hydrodynamic Reynolds number Re for the hydrodynamic simulations and the magnetic Reynolds number Rm for the dynamo simulations. In the latter case, the mean core magnetic field strength is given by the Elsasser number Λ:

$$\mathrm{Re} = \frac{UD}{\nu}, \qquad \mathrm{Rm} = \frac{UD}{\lambda}, \qquad \Lambda = \frac{B^2}{\mu_0 \lambda \rho \Omega} \tag{14}$$

As suggested in the study by Hori et al. (2014) and DW13, heat flux anomalies applied along the equator ($\tau = 90°$) yield special solutions. In this case, the equatorial symmetry of the CMB heat flux is re-established as in the case with homogeneous boundaries, but the azimuthal symmetry and axisymmetry are broken. It has been reported that anomalies of a planetary scale modelled by spherical harmonics of degree and order unity (Y_{11}) lead to flows dominated in spectral space by azimuthal order $m = 1$ resembling the anomaly. In the same manner as Hori et al. (2014), we measure the dominance of $m = 1$ by evaluating

$$E_{m1} = \frac{\int_{r_{icb}}^{r_{cmb}} \sum_l E_{l,1}^k r^2 dr}{\int_{r_{icb}}^{r_{cmb}} \sum_{l,m} E_{l,m}^k r^2 dr} . \tag{15}$$

Results and discussions

Of the 202 simulations performed in this study, selected numerical models are presented in Table 2 to provide an overview of and compare a number of hydrodynamic runs (centre column) and their equivalent dynamo runs (right column). All results in Table 2 were calculated with fixed parameters Pr = 1, Pm = 0/2 (hydrodynamic/dynamo runs), Ra = 4×10^7, E = 10^{-4}, and $\beta = 0.35$ and with variable forcing amplitude q^*, anomaly width Ψ, and tilt angle τ.

Amplitude

As noted above, the amplitude of the CMB heat flux variation is determined by the thermal lower mantle structure. For Earth, the amplitude can exceed the superadiabatic part of the homogeneous flux, indicating that values of $q^* > 1$ may be possible. Here, we restricted ourselves to $q^* \leq 1$, thereby avoiding models with local core heating that may lead to stable stratification.

To isolate the influence of the anomaly amplitude, the tilt angle was fixed at $\tau = 0$, and the amplitude q^* and the width Ψ were varied in small steps. Figure 3a shows the strength of the EAA contribution C_{EAA} with respect to the total kinetic energy. For the largest and strongest anomaly with $\Psi = 180°$ and $q^* = 1$, the highest value of C_{EAA} is found (black line in Fig. 3a). When the width Ψ of the anomaly was reduced and its amplitude was kept fixed, the EAA symmetry contribution reduced accordingly. Hence, the enormous EAA contributions found in DW13 are strongly related to the large scale of the anomaly chosen there. It is thus not (or not only) the breaking of the equatorial symmetry that leads to strong antisymmetric flow contributions. An anomaly with a weaker amplitude ($q^* < 1$) reduces the strength of the EAA contribution C_{EAA} almost linearly. Interestingly, halving the anomaly width has almost the same effect as halving the anomaly amplitude. As an example, for $\Psi = 180°$ and $q^* = 0.5$, the EAA contribution is 0.338, whereas $\Psi = 90°$ and $q^* = 1.0$ yield an EAA contribution of 0.331.

Figure 4 shows the zonally averaged temperatures and azimuthal flows for various combinations of anomaly amplitudes q^* and widths Ψ in four selected models with polar anomalies ($\tau = 0$). Note that the thermal wind balance (Eq. 10) is always satisfied. One could expect that narrower anomalies (smaller Ψ) with stronger horizontal heat flux gradients would have stronger and more concentrated thermal winds, but this does not seem to be the case. The large-scale temperature anomaly between the north and south develops independent of the anomaly width. The first three models in Fig. 4 have similar EAA contribution strengths C_{EAA} (see also Table 1), supporting the quasi-linear increase of EAA symmetry with amplitude and width.

Latitudinal position

Thus far, we have focused on polar anomalies, where the anomaly peak vector is aligned with the rotation axis, which is a special situation. Because those break only the equatorial symmetry and not the azimuthal symmetry, the total CMB heat flux remains colatitude-dependent but axisymmetric. Mantle plumes and giant impacts generally do not sit on or reach the pole. Therefore, we explored several tilt angles τ. The magnetic case with a planetary-scale heat flux anomaly was explored in DW13 across a variety of tilt angles, and it was demonstrated that all tilt angles $\tau < 80°$ lead to a fairly strong EAA contribution. Such behaviour was also observed in a study by Amit et al. (2011).

Figure 3b shows the EAA flow contribution C_{EAA} for four tilt angles τ. The first case, with $\tau = 22°$, was set up such that the anomaly peak vector was located at the colatitude of the tangent cylinder. As will be discussed later, the tangent cylinder and the shell geometry were expected to have a strong influence on the dynamics, but our results show that they have no particular influence on the location of the heat flux anomaly for the thick

Table 2 Selected numerical models with fixed Rayleigh Ra $= 4 \times 10^7$, Ekman E $= 10^{-4}$, and Prandtl Pr $= 1$ numbers and aspect ratio $\beta = 0.35$

			Hydrodynamic				Magnetic						
q^*	Ψ [°]	τ [°]	Re	C_{EAA}	E_{m1}	Γ	Rm	C_{EAA}	E_{m1}	Γ	Λ	H_{sur}	ω
0	–	–	154.7	0.066	0.029	0.057	224.3	0.007	0.064	0.010	7.83	0.005	–
1	180	0	223.3	0.603	0.013	0.686	446.7	0.747	0.015	0.778	1.71	0.573	49.42
		22	220.9	0.589	0.019	–[a]	433.5	0.738	0.022	–[a]	2.50	0.561	46.54
		45	213.3	0.535	0.040	–[a]	406.3	0.691	0.037	–[a]	3.28	0.431	?
		63	198.7	0.375	0.081	–[a]	347.4	0.594	0.071	–[a]	7.94	0.220	–
		90	181.0	0.029	0.239	–[a]	308.7	0.003	0.257	–[a]	2.72	–[a]	–
	120	0	208.0	0.427	0.023	0.432	428.0	0.759	0.002	0.684	3.94	0.415	25.58
		22	208.9	0.419	0.031	–[a]	412.4	0.729	0.024	–[a]	4.12	0.410	22.74
		45	198.1	0.370	0.071	–[a]	360.1	0.654	0.046	–[a]	7.32	0.226	–
		63	184.9	0.229	0.145	–[a]	300.1	0.386	0.097	–[a]	8.68	0.136	–
		90	168.8	0.016	0.247	–[a]	255.6	0.006	0.257	–[a]	7.84	–[a]	–
	90	0	192.5	0.291	0.029	0.265	365.3	0.703	0.003	0.501	7.94	0.318	–
		22	191.4	0.293	0.038	–[a]	369.3	0.698	0.028	–[a]	3.69	0.297	–
		45	184.0	0.243	0.102	–[a]	291.9	0.483	0.055	–[a]	12.87	0.170	–
		63	177.4	0.149	0.164	–[a]	283.9	0.262	0.111	–[a]	6.98	0.118	–
		90	163.2	0.013	0.231	–[a]	239.7	0.006	0.238	–[a]	2.40	–[a]	–
	60	0	172.1	0.169	0.032	0.194	271.8	0.224	0.144	0.194	10.19	0.077	–
		22	172.0	0.178	0.357	–[a]	263.5	0.258	0.111	–[a]	11.8	0.065	–
		45	170.4	0.157	0.093	–[a]	249.1	0.205	0.075	–[a]	11.97	0.060	–
		63	168.9	0.109	0.151	–[a]	249.5	0.109	0.155	–[a]	8.66	0.047	–
		90	158.0	0.015	0.175	–[a]	230.3	0.006	0.237	–[a]	0.99	–[a]	–
	30	0	158.5	0.083	0.031	0.065	232.5	0.010	0.047	0.057	7.12	0.007	–
		22	158.7	0.091	0.027	–[a]	233.6	0.009	0.031	–[a]	6.70	0.007	–
		45	158.1	0.093	0.031	–[a]	239.1	0.017	0.130	–[a]	6.97	0.012	–
		63	157.5	0.087	0.045	–[a]	238.2	0.019	0.167	–[a]	7.54	0.011	–
		90	151.9	0.027	0.062	–[a]	228.7	0.006	0.199	–[a]	8.55	–[a]	–
0.75	180	0	208.4	0.473	0.015	0.532	425.0	0.763	0.019	0.767	3.69	0.413	34.91
	120	0	198.1	0.340	0.025	0.377	368.8	0.716	0.022	0.462	7.80	0.327	–
	90	0	184.5	0.250	0.028	0.245	346.5	0.666	0.023	0.401	7.11	0.284	–
	60	0	168.1	0.154	0.030	0.163	255.2	0.136	0.132	0.043	9.45	0.054	–
	30	0	158.0	0.085	0.027	0.065	231.2	0.008	0.053	0.010	7.09	0.005	–
0.5	180	0	193.9	0.351	0.023	0.383	347.3	0.725	0.031	0.555	10.75	0.258	–
	120	0	181.8	0.257	0.028	0.272	307.2	0.609	0.033	0.379	14.44	0.193	–
	90	0	173.7	0.195	0.027	0.187	269.9	0.321	0.071	0.193	12.36	0.110	–
	60	0	163.5	0.128	0.027	0.115	245.9	0.078	0.116	0.052	8.72	0.035	–
	30	0	157.3	0.078	0.027	0.083	227.7	0.009	0.064	0.038	7.81	0.007	–
0.25	180	0	166.0	0.173	0.028	0.206	239.5	0.183	0.086	0.2	13.20	0.060	–
	120	0	163.5	0.147	0.027	0.163	237.8	0.110	0.117	0.111	11.08	0.052	–
	90	0	161.6	0.122	0.028	0.095	237.7	0.074	0.105	0.048	9.32	0.033	–
	60	0	158.2	0.089	0.026	0.070	231.7	0.009	0.092	0.070	7.59	0.015	–
	30	0	155.3	0.067	0.026	0.043	227.7	0.009	0.068	0.024	7.58	0.008	–

The magnetic Prandtl number was kept constant at Pm $= 2$ throughout all magnetic simulations. If the magnetic field reverses, the oscillation frequency is given in multiples of $2\pi \text{Pm}/\tau_{vis}$. Note that for $q^* = 1$, $\Psi = 180°$, and $\tau = 45°$, the field reverses but with an unclear frequency (?). Non-oscillating dynamos are marked with '-'

[a]The quantities of entries were not calculated. This applies to both convective Γ and magnetic H_{sur} hemisphericity

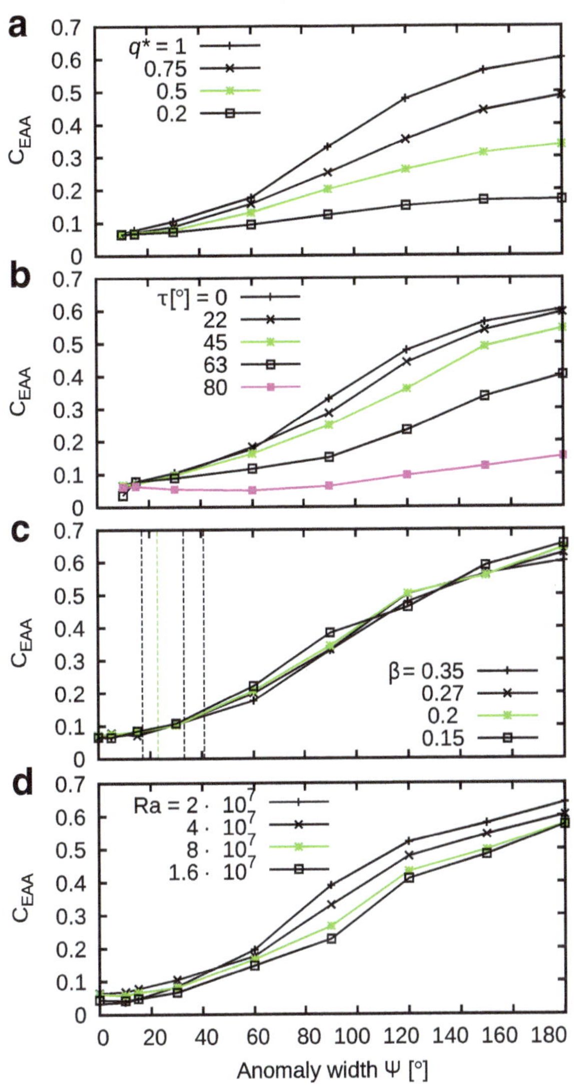

Fig. 3 Influence of various model parameters on EAA contribution. Impact of **a** anomaly amplitude, **b** tilt angle, **c** shell geometry, and **d** vigour of convection on the EAA contribution for varying anomaly width Ψ. The *coloured vertical lines* denote the colatitude of the tangent cylinder. Reference parameters (unless specified otherwise): $Ra = 4 \times 10^7, q^* = 1.0, \tau = 0, \beta = 0.35$

shells studied here. However, tilting the anomaly further from the axis of rotation did not significantly affect the strength of the EAA symmetry for moderate tilt angles ($\tau \leq 45°$). In this case, non-axisymmetric flow contributions are changed little, and the EAA contribution remains surprisingly strong. For $\Psi = 180°$, we can decompose the heat flux anomaly into Y_{10} and Y_{11} contributions. The effective Y_{10} contribution is simply $q^* \cos(\tau)$. The EAA remained significantly stronger than even the effective Y_{10} contribution would suggest. For example, for $\tau = 45°$, we expected $C_{EAA} = C_{EAA}(\tau = 0)/\sqrt{2} = 0.42$, but we obtained $C_{EAA} = 0.54$. The nearly equatorial case showed distinct behaviour. In general, it might be concluded that any anomaly smaller than $\Psi = 60°$ might only weakly influence the core convection. Note that this behaviour is independent of the tilt angle or anomaly amplitude.

Figure 5 illustrates the time-averaged temperature and zonal flow in four different meridional cuts. The sample model here is for $q^* = 1$, $\Psi = 180°$, and $\tau = 45°$. The first plot is positioned at a longitude of $\phi = 0$ and includes the location of the centre of the anomaly (black line). At this value of ϕ, the CMB heat flux at $\theta = 45°$ is exactly zero. The three other plots are taken at intervals of $\phi = 90°$ eastwards. Remarkably, large-scale equatorial asymmetry in the temperature is visible in all cuts; hence, the EAA symmetry driven by the thermal wind is a meaningful measure for the tilted cases as well. Interestingly, not much is visible of the broken azimuthal symmetry.

Aspect ratio

For numerical reasons, we have kept in our model an inner core that is purely driven by internal heat sources. Hori et al. (2010) have shown that such an inner core has little impact on the solution for homogeneous outer boundary conditions. Figure 3c proves that this remains true for the inhomogeneous heat flux explored here. The aspect ratio affects the critical Rayleigh number for the onset of convection. Therefore, the cases compared here have different supercriticality. However, this also seems to have little impact for the limited range of Rayleigh numbers we studied.

Vigour of convection

The Rayleigh number Ra can influence the EAA instability in at least two ways. It directly scales buoyancy forces and thus also scales the thermal wind strength (see Eq. 10). The EAA contribution should therefore grow linearly with Ra. However, Ra also controls the convective vigour and length scale. As Ra grows, the scale decreases while the vigour increases, and both of these changes lead to more effective mixing, which should counteract the impact of the inhomogeneous boundary condition. Figure 3d illustrates that the EAA contribution decreases with growing Ra such that the latter effect seems to dominate. However, for large anomalies, the relative EAA flow symmetry seems to become saturated for all Ra.

Magnetic fields

It has been reported that EAA flows enforced by boundary anomalies dramatically affect the morphology and time dependence of the magnetic field (DW13). A polar planetary-scale anomaly with amplitude $q^* = 1$ and width $\Psi = 180°$ transforms a strong and stationary dipolar-dominated magnetic field into a weaker wave-like hemispherical dynamo that is dominated by an axisymmetric toroidal field (DW13). The altered magnetic field

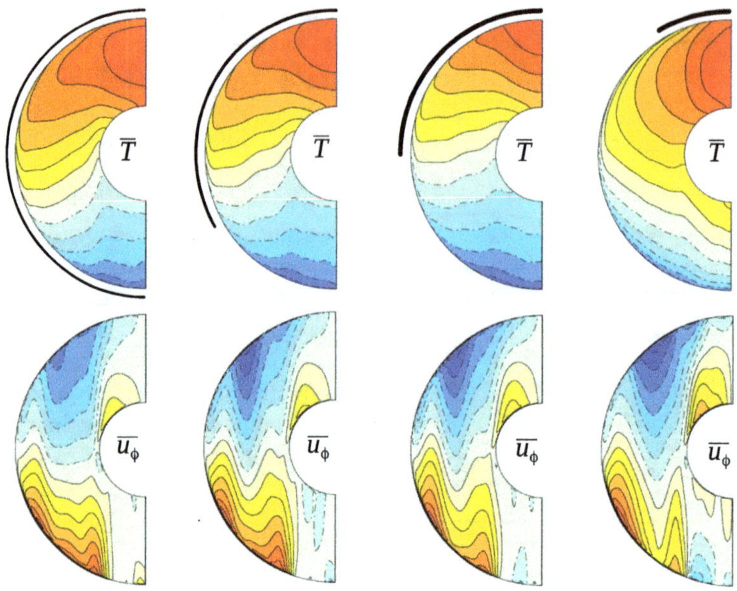

Fig. 4 Zonal temperature and flow. The zonally averaged temperature (*top*) and differential rotation (*bottom*) are shown for four example cases with variable anomaly width Ψ and amplitude q^*. Note that the first three cases have similar EAA symmetric flows. The *black arcs* denote the width of the anomaly and the arc thickness scales with the amplitude. Parameters: Ra $= 4 \times 10^7$, $\tau = 0$

configuration also indicates the strength of the EAA symmetry. DW13 suggested that the magnetic field significantly enhances the EAA contribution. A flow that inhibits a strong EAA symmetry results from a strongly asymmetric temperature anomaly, which is maintained by the thermal wind and hence confines the convective motions into a single hemisphere. This weakens the global amount of convection, potentially inducing magnetic fields. Another possible effect is that the magnetic field may relax the strong rotational constraint on the flow, thus allowing the convection to develop in more three-dimensional rather than columnar structures. This would support the convection in the magnetically active hemisphere and increase the temperature asymmetry and hence the EAA symmetry. Finally, the strong azimuthal toroidal field that emerges in the equatorial region, created by fierce equatorial antisymmetric shear associated with the EAA symmetric differential rotation, can potentially suppress columnar convective flows within this equatorial region (DW13).

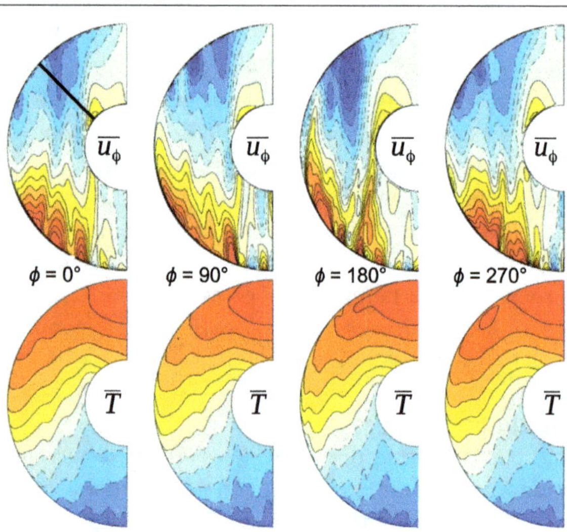

Fig. 5 Meridional cuts of flow and temperature. The temperature (top) and differential rotation (bottom) are shown for four different meridional cuts at ϕ for an anomaly of width $\Psi = 180°$, amplitude $q^* = 1$, and tilt angle $\tau = 45°$. The thick black line denotes the central anomaly peak vector. Parameters: Ra $= 4 \times 10^7$

Figure 6 compares simulations including the magnetic field generation process (solid lines) with the previously discussed hydrodynamic cases (dashed). Figure 6a is restricted to polar anomalies (fixed $\tau = 0$) and takes several anomaly amplitudes q^*, whereas Fig. 6b fixes the amplitude q^* and varies the tilt angle τ; both parts of the figure plot the EAA contribution against the anomaly width Ψ. For smaller or vanishing anomalies, the magnetic field actually suppresses the effect of the heat flux inhomogeneity and reduces the EAA contribution to nearly zero. The magnetic field now further suppresses the equatorially asymmetric contributions that were already relatively weak in the non-magnetic case.

If the anomalies reach a width of $\Psi = 90°$, the magnetic field drastically enhances the EAA symmetry relative

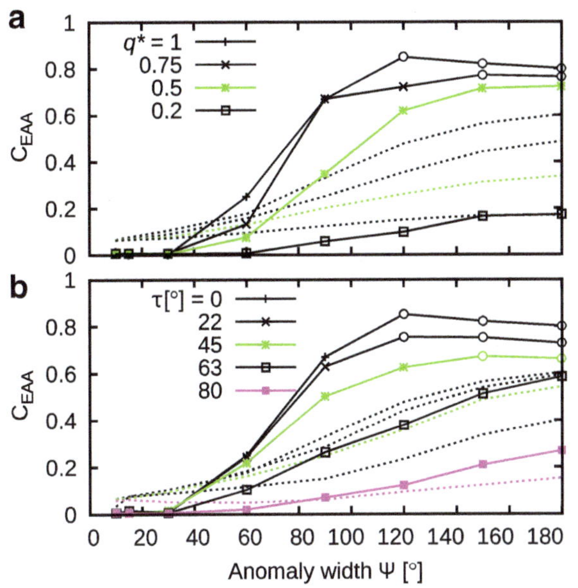

Fig. 6 Effect of magnetic fields on EAA convective mode. Relative EAA kinetic energy contribution for **a** polar anomalies (fixed $\tau = 0$) with various amplitudes q^* and for **b** fixed $q^* = 1.0$ with various tilt angles τ as function of the anomaly width Ψ. For comparison, the hydrodynamic reference cases are included as *dashed curves*. The EAA contributions C_{EAA} for the homogeneously cooled reference cases are 0.01 (dynamo) and 0.06 (hydrodynamic). Oscillatory dynamos are indicated by *empty circles*. Parameters: Ra $= 4 \times 10^7$, Pm $= 2$ (dynamo) or Pm $= 0$ (hydrodynamic)

to the hydrodynamic runs. For $q^* = 0.75$ and $\Psi = 90°$, the magnetic field changed the EAA contribution from 0.25 to 0.67, indicating a stronger convection in the magnetically more active (southern) hemisphere. All dynamos reach a strong magnetic field (Elsasser number $\Lambda \geq 1$), which is when the leading force balance in the momentum equation is between the Coriolis and Lorentz forces. Note that the azimuthal vorticity balance (thermal wind) is still exclusively between the Coriolis force and the buoyancy (see the last two plots in the upper row of the lower set in Fig. 2).

Figure 6b illustrates that the magnetic effect on the EAA contribution remains similar as the tilt angle τ varies. The magnetic field increases the EAA contribution for larger opening angles but decreases it for smaller values of Ψ. Open circles in Fig. 6 show the locations at which oscillatory reversing dynamos have been found. This behaviour is promoted by a strong Ω-effect and therefore requires a large thermal wind shear. Because the main thermal wind contribution is EAA, the respective measure of the EAA symmetry is a good proxy for the Ω-effect (Dietrich et al. 2013). Figure 6 demonstrates that oscillatory dynamos are only found for relatively large EAA contributions. The tilt angle also plays a role in this behaviour. For $\tau = 0$, only models with $C_{EAA} > 0.7$ are oscillatory, whereas for $\tau = 45°$, a non-reversing dynamo still exists at $C_{EAA} = 0.68$. Even though a strong EAA symmetric flow leads to strong shear around the equatorial region, yielding a strong Ω-effect (Dietrich et al. 2013), it can co-exist with non-reversing fields. If, despite the strong shearing, a sufficient fraction of toroidal field is created by non-axisymmetric helical flows, the field remains stable.

For a more detailed investigation of how the magnetic field affects the flow, a measure of the heat transport efficiency was developed. We correlate the convective motion in terms of non-axisymmetric radial flows u'_r with non-axisymmetric temperature perturbations T' over the azimuth and time. Radial integration of this measure gives the mean vertical convective heat transport Γ_c as a function of the colatitude alone:

$$\Gamma_c(\theta) = \int_{r_i}^{r_o} \overline{u'_r T'}(r,\theta)\, r^2\, dr\,, \tag{16}$$

where the overbar denotes the correlation over the azimuthal angle ϕ and time. This measure is closely related to the definition of the Nusselt number proposed by Otero et al. (2002), where another integration along the colatitude is taken. Because of the large temperature variation along the colatitude, we keep Γ_c as a function of the colatitude θ. Figure 7 shows the colatitudinal profiles of Γ_c for a few cases; in each panel, the hydrodynamic (black) and self-consistent dynamo (red) simulations are shown. Of all the simulations, we chose to investigate the reference case with homogeneous heat flux, the case with $\Psi = 90°$ and $q^* = 0.75$, and the most commonly studied case (Y_{10}) given by $\Psi = 180°$ and $q^* = 1.0$ (Fig. 7a, b, and c, respectively). The study case in Fig. 7b was selected because it shows an enormous difference between the hydrodynamic and dynamo simulations and the magnetic field is non-reversing. The case in Fig. 7c features a reversing magnetic field and hence oscillates between weak (hydrodynamic) and strong (dynamo) EAA symmetry. The correlation needed to calculate Γ_c in Fig. 7c is taken over only three time steps, during which the magnetic field remains strong and does not change sign, whereas for the other cases, the correlations are taken over the full magnetic diffusion time with tens of snapshots.

DW13 suggested that the emerging strong axisymmetric toroidal field induced by the EAA shear suppresses the radial, non-axisymmetric convective flows. To test this hypothesis, we calculated the strength of the mean azimuthal magnetic field relative to the total magnetic field. Because some magnetic fields oscillate, we took the time-averaged root mean square azimuthal field rather than the time-averaged field and obtained:

$$\lambda_\phi(\theta) = \frac{1}{\sqrt{B^2}} \int_{r_i}^{r_o} \sqrt{\overline{B_\phi}^2}(r,\theta)\, r^2\, dr\,, \tag{17}$$

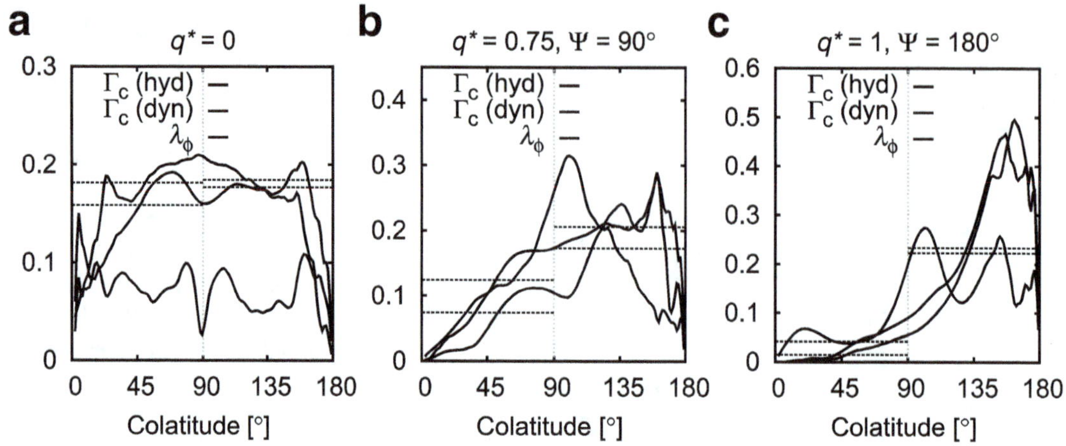

Fig. 7 Radially integrated convective heat transport. Colatitudinal profiles of the vertical convective heat transport Γ_c for hydrodynamic (*black*) and magnetic (*red*) simulations. The *horizontal dashed lines* denote the hemispherical average of each Γ_c. *Blue lines* denote the relative contribution λ_ϕ of the mean azimuthal toroidal field. **a** Reference case with homogeneous boundary heat flux. **b** $\Psi = 90°$ and $q^* = 0.75$. **c** Y_{10} case with a large-scale anomaly: $\Psi = 180°$ and $q^* = 1.0$. Note that whereas **a** and **b** show long time averages, **c** is calculated from a few snapshots because of the reversing magnetic field. Parameters: Ra $= 4 \times 10^7$, $\tau = 0$, Pm $= 2$

which is included in Fig. 7 in blue. As suggested in DW13, we found that λ_ϕ is large when the reduction of convective flows is large (see Fig. 8). The vertical convective heat transport Γ_c is clearly suppressed when the toroidal field indicated by λ_ϕ is large. For the case in Fig. 7c, the enhancement of Γ_c in the magnetically more active hemisphere is visible as well.

Furthermore, we defined the convective hemisphericity Γ, which is equivalent to the magnetic field hemisphericity H_{sur} but is based on the vertical heat transport Γ_c integrated over either the northern or southern hemisphere, such that

$$\Gamma = \left|\frac{\Gamma_N - \Gamma_S}{\Gamma_N + \Gamma_S}\right| \quad , \text{ where } \quad \Gamma_c^{N,S} = \int_{0,\pi/2}^{\pi/2,\pi} \Gamma_c(\theta)\sin\theta d\theta \; . \tag{18}$$

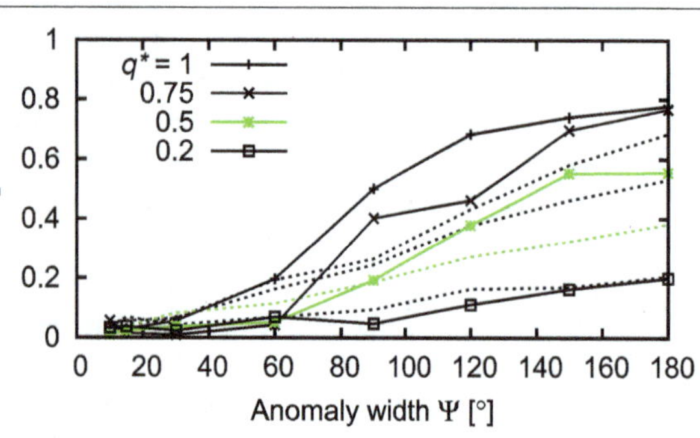

Fig. 8 Hemisphericity of vertical heat transport. Asymmetry of the vertical heat transport in the northern and southern hemispheres. Pure hydrodynamic models (*dashed*) and dynamo simulations (*solid*) are included. Parameters: Ra $= 4 \times 10^7$, $\tau = 0$, Pm $= 2$ (dynamo) or Pm $= 0$ (hydrodynamic)

These values are plotted in Fig. 7 as dashed vertical lines, and the convective hemisphericity Γ is plotted in Fig. 8 for all polar anomalies ($\tau = 0$). According to our results, it can be concluded that the magnetic field (mainly the axisymmetric part of B_ϕ) is responsible for the increased equatorial asymmetry of the convective cooling and thus the boost of EAA symmetric flows (see Fig. 6). For the tilted cases (Fig. 6b), the magnetically driven EAA enhancement also appeared for all tilt angles. Even for $\tau = 63°$, the EAA contribution reached nearly 0.6 for the largest anomaly width Ψ. However, for these tilted cases, the EAA mode increased linearly with Ψ, where for smaller tilt angles, saturation occurred.

Equatorial anomalies

For anomalies tilted towards the equator ($\tau = 90°$), the flows show a strong azimuthal alteration along with the outer boundary heat flux. Hence, significant kinetic energy contributions are expected from flows with an equivalent spectral order of $m = 1$ (see definition of E_{m1}, Eq. 15). Figure 9 gives an overview of the spectral response E_{m1} for the hydrodynamic cases with an anomaly amplitude of $q^* = 1$. Large equatorial anomalies led to an increase from $E_{m1} = 0.026$ in the homogeneous reference case to $E_{m1} = 0.25$ for the largest anomaly width. If the anomaly is tilted away from the equator again, both the equatorial and azimuthal symmetry are broken. Hence, it would be expected that apart from the equatorial case with $\tau = 90°$, smaller tilt angles will also show an enhancement of E_{m1}. Figure 9 also shows various tilt angles between $\tau = 90°$ and 45°, where the E_{m1} amplitude decreases with decreasing tilt angle. Interestingly, the cases with more concentrated anomalies, e.g.,

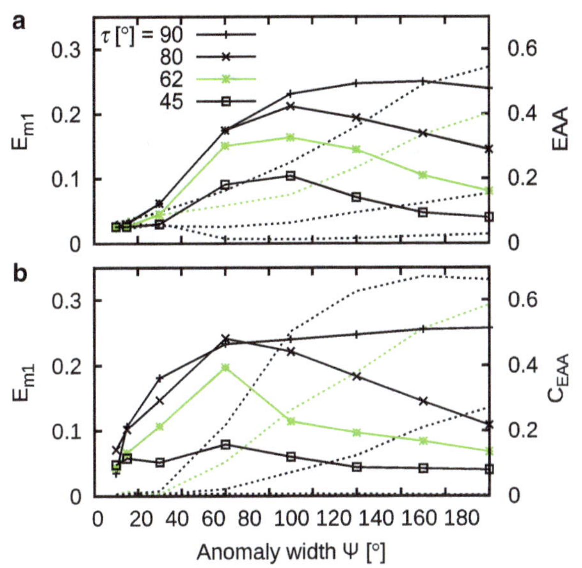

Fig. 9 $m = 1$ dominance for equatorial anomalies. Relative kinetic energy contribution of the $m = 1$ flows (E_{m1}, *solid lines*) and the EAA contribution (*dashed*) for anomalies of various tilt angles (τ) as a function of anomaly width Ψ for **a** hydrodynamic and **b** self-consistent dynamo simulations. Parameters: Ra $= 4 \times 10^7$, $q^* = 1.0$, Pm $= 2$ (dynamo) or Pm $= 0$ (hydrodynamic)

$\Psi = 90°$, yielded higher values of E_{m1} than the planetary-scale anomalies with $\Psi = 180°$. For larger Ψ reaching far enough across the equator, it seems that the equatorial asymmetry takes control, and EAA symmetric flows are established. It then seems reasonable for E_{m1} to decay only for $\tau \neq 90°$. If the magnetic field is included (Fig. 9b), the systematic behaviour of E_{m1} found in the hydrodynamic simulations is rather equivalent. This suggests that the magnetic field is not important for E_{m1} as a measure of the dynamic response to breaking the azimuthal symmetry of the outer boundary heat flux.

Application to Mars

For application to Mars, the time-averaged surface hemisphericity of the radial field is correlated with the EAA symmetry. The two are dynamically linked because a strong EAA kinetic energy contribution relies on strongly equatorially asymmetric convection, which in turn induces a hemispherical field. The magnetic field is extrapolated by a potential field towards the Martian surface with an outer core radius of $r_{cmb} = 1680$ km and a surface radius of $r_{sur} = 3385$ km. The magnetic field hemisphericity H_{sur} at the surface gives a ratio that is a function of the radial field intensities B_N and B_S integrated over the northern and southern hemispheres, as defined in Eq. 12.

Figure 10 shows the correlation between the EAA symmetry and the surface hemisphericity of the radial field. In these simulations, a weak EAA led to weak H_{sur}, as expected. For large EAA symmetry enforced by boundary forcing, the magnetic fields tend to be more hemispherical. The lower bound on the Martian crustal value of H_{sur} is 0.45 (Amit et al. 2011), which is only reached when the anomalies have a large horizontal extent ($\Psi > 120°$) and amplitude ($q^* \geq 0.75$). Even though the EAA contribution can become dominant for smaller and weaker anomalies, it is far more challenging to induce a magnetic field of sufficient surface hemisphericity in this case. Furthermore, all magnetic fields with sufficiently high H_{sur} values are oscillatory, which makes a thick unidirectional magnetisation unlikely (see discussion in DW13).

DW13 explored the parameter dependence of a polar Y_{10} anomaly, changing the Rayleigh, magnetic Prandtl, and Ekman numbers within the numerically accessible limits (see their Fig. 11). Not unexpectedly, decreasing the Ekman number led to smaller hemisphericities because the geostrophic geometry was more severely enforced. This can be counteracted by increasing the anomaly amplitude; however, increasing Ra or Pm also seemed to help, likely because inertia or Lorentz forces more significantly contribute to balancing the Coriolis force. At realistically small Ekman numbers of approximately 3×10^{-15} and appropriate Ra and Pm for Mars, this likely means that unrealistically large heat flux variations would be required to yield the observed hemisphericity. Inertial forces are thought to be small in planetary cores, whereas Lorentz forces should not significantly exceed the strength reached at the smaller Ekman number of 10^5 explored by DW13.

The generally oscillatory nature of high hemisphericities remains a problem at all parameter combinations and geometries explored in DW13 and in the present study. Latitudinal temperature variations paired with their respective gradients in convective efficiency never fail to drive strong thermal winds. These in turn lead to a significant Ω-effect that seems to favour oscillatory dynamos (Dietrich et al. 2013). None of the variations in the general set-up explored in this paper have indicated that this fundamental mechanism is incorrect.

Conclusions

We constructed a suite of 202 numerical models of spherical shell convection and magnetic field generation in which the outer boundary heat flux was perturbed by an anomaly of variable width, amplitude, and position. The convection was driven exclusively by secular cooling, which is an appropriate model for terrestrial planets in the early stages of their evolution, when no inner core is present. The dynamic response of the flow was measured in terms of the expected spectral equivalent of the boundary forcing. For anomalies breaking the equatorial

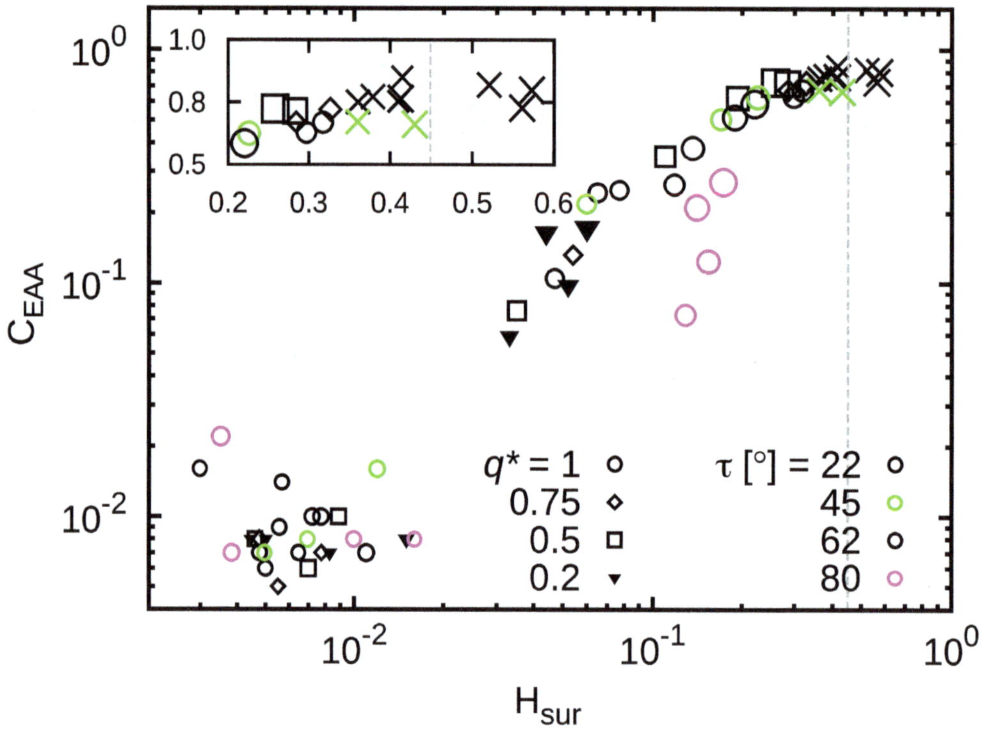

Fig. 10 Magnetic field hemisphericity vs. EAA. Hemisphericity of the radial magnetic field at the Martian surface plotted against the EAA flow contribution. The different *colours*, *symbols*, and *symbol sizes* represent different tilt angles (τ), anomaly amplitudes (q^*), and widths of the anomaly (Ψ), respectively, and the *crosses* indicate reversing dynamos. The specific values of C_{EAA} and further characteristic quantities for the oscillatory dynamos can be found in Fig. 6 and Table 2. The small inset plot (*top left*) includes only points with $H_{sur} \geq 0.2$ and $C_{EAA} \geq 0.5$. The *vertical grey dashed line* indicates the lower limit of the Martian crustal value of H_{sur}. Parameters: The Rayleigh and Ekman numbers are kept constant throughout all simulations in the plot (Ra $= 4 \times 10^7$, E $= 10^{-4}$)

symmetry, the relative strength of EAA kinetic energy was used, and for anomalies breaking the azimuthal symmetry, the relative strength of flows with a spectral order of $m = 1$ was investigated (E_{m1}).

For hydrodynamic models without a magnetic field, the strength of the EAA symmetry was found to increase almost linearly with the amplitude and width of the anomaly. These flows are driven by a large-scale equatorial asymmetry in the axisymmetric temperature field. Hence, a more localised CMB heat flux anomaly does not lead to a stronger or more confined thermal wind, even though the horizontal variation of the heat flux is locally much larger. The simulations also indicated that models perturbed by narrower anomalies or anomalies that are not aligned with the axis of rotation also yield the same fundamental temperature asymmetry. For example, if the anomaly peak is tilted by an angle $\tau \leq 45°$ from the axis of rotation, the EAA symmetry is quite similar to that in the case of the polar anomaly ($\tau = 0$). Furthermore, this suggests that the system is more sensitive to changes in the equatorial symmetry than in the azimuthal symmetry.

For equatorial anomalies ($\tau = 90°$), the spectral response (E_{m1}) reached up to 25 % of the kinetic energy and was only mildly affected by the magnetic field. Interestingly, for tilt angles $45° \leq \tau \leq 80°$, the contribution of E_{m1} could be measured as well and was found to typically be the strongest for moderately sized anomalies $60° \leq \Psi \leq 90°$. Larger anomalies broke the equatorial symmetry as well, thus increasing the EAA symmetry at the cost of the E_{m1} symmetry.

For numerical reasons, the models were run with an inner core present; as such, we further tested the influence of smaller aspect ratios, which proved to be negligible. A similar conclusion was reached by Hori et al. (2010) for a homogeneous outer boundary heat flux and may be extended with this study to boundary-forced models. As the primary purpose of our model is to comprehensively parametrise the various boundary anomalies, out of various other model parameters (Ra, E, Pm, Pr), we only tested the influence of increasing the convective vigour. Our model also indicated that stronger convective stirring does not suppress fundamental temperature asymmetries or the EAA mode. However, it was shown that the model is slightly more sensitive to boundary forcing when convective driving is weaker. DW13 provides a discussion of the possible dependence on the Ekman number (rotation rate) and the magnetic Prandtl number, showing the robustness of a similar model featuring Y_{10}-forcing.

In the presence of dynamo action, the behaviour is far more non-linear. For narrow anomalies, the flow is equatorially symmetrised by the dipole field, whereas anomalies with widths $\Psi \geq 90°$, amplitudes $q^* \geq 0.5$, and tilt angles $\tau \leq 45°$ strongly boost the EAA symmetry. It was shown that the strong azimuthal toroidal field around the equator suppresses the remaining columnar convection and further increases the asymmetry in the temperature and, as a consequence, the antisymmetry in the core flow. Hence, the magnetic field prevents narrow heat flux anomalies from affecting the core convection, where it drastically increases their respective effects when they reach a horizontal extent on the same order as the core radius. This effect is independent of the tilt angle, amplitude, and width of the anomaly when $\tau \leq 45°$, $q^* \geq 0.5$, and $\Psi \geq 90°$. For all models within these boundaries, this in turn also implies that CMB heat flux anomalies can be smaller, weaker, and non-polar while still yielding effects similar to those of the fundamental Y_{10} anomaly. A similar observation was reported recently by Monteux et al. (2015). Hence, our results suggest that the core dynamos of ancient Mars or early Earth are sensitive to CMB heat flux anomalies only if they are strong in amplitude and large in horizontal extent.

Regarding the hemispherical magnetisation of the Martian crust, it seems rather unlikely that the magnetising field is as hemispherical as the crustal pattern suggests; hence, the crustal magnetisation dichotomy seems only realistically explained by additional demagnetisation events of external origins in the northern hemisphere. The results of the numerical models clearly indicate that a sufficiently hemispherical field is possible only if the anomaly is of core scale and significantly affects the CMB heat flux. However, all of these geometrically corresponding hemispherical dynamos show rather frequent polarity reversals and hence would require a crustal rock magnetisation time on the order of the magnetic diffusion time (tens of thousands of years), which might be a rather unrealistic scenario for a thick magnetised layer of at least 20 km. Note that it is possible to create a magnetic field that shows a smaller degree of equatorial asymmetry *and* is stable in time. However, at more realistic model parameters, e.g., a smaller Ekman number, it seems likely that these models remain applicable only when a much stronger forcing is applied to sufficiently break the typical z-invariance of the flow (geostrophy). One common feature consistently found here and in DW13 is that a stronger thermal forcing naturally leads to oscillatory fields.

Thus, the main results obtained in this study are

- Fundamental large-scale equatorial asymmetry in the temperature and hence EAA symmetric flows emerge independent of the width, position, and amplitude of a CMB heat flux anomaly.
- The magnetic field prevents narrow heat flux anomalies from affecting the core convection but drastically increases their respective effects when the anomalies reach horizontal extents on the same order as the core radius.
- At least for the parameters and geometries explored here and in DW13, it is not possible for a hemispherical dynamo to explain the observed dichotomy in Martian crustal magnetisation.

Competing interests
The authors declare that they have no competing interests.

Authors' contributions
WD proposed the topic, designed the study, and carried out the numerical experiments. JW developed the numerical implementation and extended the code to include the variable anomaly. KH and JW collaborated with WD in the construction of the manuscript. All authors read and approved the final manuscript.

Acknowledgements
This work was supported in part by the Science and Technology Facilities Council (STFC). Numerical simulations were undertaken on ARC1, part of the High Performance Computing facilities at the University of Leeds, UK.

Author details
[1]Department of Applied Mathematics, University of Leeds, Woodhouse Lane, LS9 2JT Leeds, West Yorkshire, UK. [2]Max Planck Institute for Solar System Research, Justus-von-Liebig-Weg 3, 37077 Göttingen, Germany. [3]Solar-Terrestrial Environment Laboratory, Nagoya University, Furo-cho, Chikusa-ku, 464-8601 Nagoya, Japan.

References
Acuña MH, Connerney JEP, Ness NF, Lin RP, Mitchell D, Carlson CW, et al. Global distribution of crustal magnetization discovered by the Mars Global Surveyor MAG/ER experiment. Science. 1999;284:790. doi:10.1126/science.284.5415.790.
Amit H, Christensen UR, Langlais B. The influence of degree-1 mantle heterogeneity on the past dynamo of Mars. Phys Earth Planet In. 2011;189: 63–79. doi:10.1016/j.pepi.2011.07.008.
Arkani-Hamed J. Did tidal deformation power the core dynamo of Mars? Icarus. 2009;201:31–43. doi:10.1016/j.icarus.2009.01.005.
Aurnou JM, Aubert J. End-member models of boundary-modulated convective dynamos. Phys Earth Planet In. 2011;187:353–63. doi:10.1016/j.pepi.2011.05.011.
Bloxham J. Sensitivity of the geomagnetic axial dipole to thermal core-mantle interactions. Nature. 2000;405:63–5.
Breuer D, Labrosse S, Spohn T. Thermal evolution and magnetic field generation in terrestrial planets and satellites. Space Sci Rev. 2010;152: 449–500. doi:10.1007/s11214-009-9587-5.
Busse FH. Thermal instabilities in rapidly rotating systems. J Fluid Mech. 1970;44:441–60. doi:10.1017/S0022112070001921.
Cao H, Aurnou JM, Wicht J, Dietrich W, Soderlund KM, Russell CT. A dynamo explanation for Mercury's anomalous magnetic field. Geophys Res Lett. 2014;41:4127–34. doi:10.1002/2014GL060196.
Christensen UR, Wicht J. Numerical dynamo simulations In: Olson P, editor. Treatise on Geophysics. Amsterdam: Elsevier; 2007. p. 245–82.
Citron RI, Zhong S. Constraints on the formation of the Martian crustal dichotomy from remnant crustal magnetism. Phys Earth Planet In. 2012;212:55–63. doi:10.1016/j.pepi.2012.09.008.
Davies CJ, Gubbins D, Willis AP, Jimack PK. Time-averaged paleomagnetic field and secular variation: predictions from dynamo solutions based on lower mantle seismic tomography. Phys Earth Planet In. 2008;169:194–203. doi:10.1016/j.pepi.2008.07.021.
Dietrich W, Schmitt D, Wicht J. Hemispherical Parker waves driven by thermal

shear in planetary dynamos. Europhys Lett. 2013;104:49001. doi:10.1209/0295-5075/104/49001.1402.0343.
Dietrich W, Wicht J. A hemispherical dynamo model: implications for the Martian crustal magnetization. Phys Earth Planet In. 2013;217:10–21. doi:10.1016/j.pepi.2013.01.001.1402.0337.
Harder H, Christensen UR. A one-plume model of Martian mantle convection. Nature. 1996;380:507–9. doi:10.1038/380507a0.
Hori K, Wicht J, Christensen UR. The effect of thermal boundary conditions on dynamos driven by internal heating. Phys Earth Planet In. 2010;182:85–97. doi:10.1016/j.pepi.2010.06.011.
Hori K, Wicht J, Christensen UR. The influence of thermo-compositional boundary conditions on convection and dynamos in a rotating spherical shell. Phys Earth Planet In. 2012;196:32–48. doi:10.1016/j.pepi.2012.02.002.
Hori K, Wicht J, Dietrich W. Ancient dynamos of terrestrial planets more sensitive to core-mantle boundary heat flows. Planet Space Sci. 2014;98: 30–40. doi:10.1016/j.pss.2013.04.007.
Kutzner C, Christensen U. Effects of driving mechanisms in geodynamo models. Geophys Res Lett. 2000;27:29–32. doi:10.1029/1999GL010937.
Langlais B, Purucker ME, Mandea M. Crustal magnetic field of Mars. J Geophys Res. 2004;109:2008. doi:10.1029/2003JE002048.
Landeau M, Aubert J. Equatorially asymmetric convection inducing a hemispherical magnetic field in rotating spheres and implications for the past Martian dynamo. Phys Earth Planet In. 2011;185:61–73. doi:10.1016/j.pepi.2011.01.004.
Lillis RJ, Frey HV, Manga M, Mitchell DL, Lin RP, Acuña MH, et al. An improved crustal magnetic field map of Mars from electron reflectometry: highland volcano magmatic history and the end of the Martian dynamo. Icarus. 2008;194:575–96. doi:10.1016/j.icarus.2007.09.032.
Marinova MM, Aharonson O, Asphaug E. Mega-impact formation of the Mars hemispheric dichotomy. Nature. 2008;453:1216–9. doi:10.1038/nature07070.
Monteux J, Amit H, Choblet G, Langlais B, Tobie G. Giant impacts, heterogeneous mantle heating and a past hemispheric dynamo on Mars. Phys Earth Planet In. 2015;240:114–24. doi:10.1016/j.pepi.2014.12.005.
Nimmo F, Hart SD, Korycansky DG, Agnor CB. Implications of an impact origin for the Martian hemispheric dichotomy. Nature. 2008;453:1220–1223. doi:10.1038/nature07025.
Olson P, Christensen U, Glatzmaier GA. Numerical modeling of the geodynamo: mechanisms of field generation and equilibration. J Geophys Res. 1999;1041:10383–404. doi:10.1029/1999JB900013.
Olson PL, Coe RS, Driscoll PE, Glatzmaier GA, Roberts PH. Geodynamo reversal frequency and heterogeneous core-mantle boundary heat flow. Phys Earth Planet In. 2010;180:66–79. doi:10.1016/j.pepi.2010.02.010.
Otero J, Wittenberg RW, Worthing RA, Doering CR. Bounds on Rayleigh Bénard convection with an imposed heat flux. J Fluid Mech. 2002;473: 191–9. doi:10.1017/S0022112002002410.
Roberts JH, Zhong S. Degree-1 convection in the Martian mantle and the origin of the hemispheric dichotomy. J Geophys Res. 2006;111:6013. doi:10.1029/2005JE002668.
Roberts JH, Lillis RJ, Manga M. Giant impacts on early Mars and the cessation of the Martian dynamo. J Geophys Res. 2009;114:4009. doi:10.1029/2008JE003287.
Sreenivasan B, Jellinek MA. Did the Tharsis plume terminate the Martian dynamo? Earth Planet Sci Lett. 2012;349:209–17. doi:10.1016/j.epsl.2012.07.013.
Stanley S, Elkins-Tanton L, Zuber MT, Parmentier EM. Mars' paleomagnetic field as the result of a single-hemisphere dynamo. Science. 2008;321:1822. doi:10.1126/science.1161119.
Takahashi F, Tsunakawa H, Matsushima M, Mochizuki N, Honkura Y. Effects of thermally heterogeneous structure in the lowermost mantle on the geomagnetic field strength. Earth Planet Sci Lett. 2008;272:738–46. doi:10.1016/j.epsl.2008.06.017.
Wicht J. Inner-core conductivity in numerical dynamo simulations. Phys Earth Planet In. 2002;132:281–302.
Wicht J, Heyner D. Mercury's Magnetic Field in the MESSENGER era In: Jin S, editor. Boca Raton: Taylor & Francis; 2014. p. 223–62.

4

The middle atmospheric circulation of a tidally locked Earth-like planet and the role of the sea surface temperature

Elisavet Proedrou[1,2*†], Klemens Hocke[1,2,3†] and Peter Wurz[1,4]

Abstract

We investigate the influence of the sea surface temperature (SST) changes on the middle atmosphere of a tidally locked Earth-like planet orbiting a G star using the coupled 3D chemistry-climate model CESM1(WACCM). We perform three 90 day simulations. The first simulation is a present-day Earth (PDE) simulation, the second is a simulation of a tidally locked Earth-like planet with a tidally locked aquaplanet sea surface temperature (cold TLE (CLTE)) and the third is a hybrid simulation of a tidally locked Earth-like planet with a present-day Earth sea surface temperature (warm TLE (WTLE)). Our results show that changes in the SST have an influence on the lower stratospheric temperature and the secondary ozone layer. Both atmospheres exhibit a dayside upwelling and a nightside downwelling extending from the surface to the mesosphere. They are also characterised by comparable lower and middle stratospheric horizontal winds and relatively different mesospheric horizontal winds. The temperature of the WTLE atmosphere is altered as a result of the SST changes, compared to the CTLE. Specifically, the WTLE lower tropospheric temperature is increased by 3.7 K on average, due to the absorption of the increased upwelling longwave radiation and the increased sensible and latent heat. The WTLE upper troposphere temperature is decreased by 4 K on average, is adiabatic in nature, and is generated by the increased WTLE upwelling. The WLTE lower stratospheric temperature is increased by 3.8 K on average due to the absorption of the increased upwelling longwave radiation. The lower mesospheric temperature is decreased by 1.13 K on average due to increased mesospheric wave breaking. The upper mesospheric temperature is increased by 4.3 K, and its generation mechanism is currently unknown. Furthermore, the secondary ozone volume mixing ratio is increased by 40.5 %. The occurrence of large-scale vortices and variable jet streams depends, to some extent, on the SST distribution.

Keywords: Exoplanet, Tidally locked Earth, Middle atmosphere, Circulation, Atmospheric dynamics, Absorption lines, M star, Spectral line, SST, Sea surface temperature

Background

In this study, we simulate and analyse how differences in the underlying sea surface temperature (SST) of a tidally locked Earth-like planet orbiting a Sun-like star (TLE) alter its circulation and ozone distribution using a realistic, high-resolution, 3D chemistry-climate model.

To that end, we perform a pure TLE simulation, with the SST of the tidally locked aquaplanet generated by Merlis and Schneider (2010) and a hybrid TLE simulation, which has the sea surface temperature of the present-day Earth. The hybrid simulations is not supposed to be a physical scenario and is only meant to test the importance of SST patterns for slowly rotating planets. We then compare the evolution of the ozone layer towards a steady state and determine their atmospheric structure and circulation. The study aims to determine the level of SST accuracy needed in simulations and observations of future of observed exoplanets.

The importance of the presence of an ozone layer in a planetary atmosphere is its ability to protect lifeforms from the harmful stellar UV radiation. It is, therefore, important for habitability studies, to determine whether

*Correspondence: elisavet.proedrou@iap.unibe.ch
†Equal contributors
[1]Institute of Applied Physics, University of Bern, Sidlerstrasse 5, 3012 Bern, Switzerland
[2]Center for Space and Habitability, University of Bern, Sidlerstrasse 5, 3012 Bern, Switzerland
Full list of author information is available at the end of the article

very different underlying SSTs will significantly alter a planet's ozone layer.

Currently, planets located in the habitable zones of M stars are considered prime targets for the search for terrestrial habitable exoplanets (Tarter et al. 2007). This is due to the abundance of M stars and a good mass and radius star-to-planet ratio. Furthermore, the detection of planets orbiting an M star is easier due to the close proximity of the habitable zone to the M star and leads to a higher probability of the planet becoming tidally locked.

Haberle et al. (1996), Joshi et al. (1997) and Joshi (2003) showed that terrestrial planets located in an M star's habitable zone (0.02–0.2 AU) (Tarter et al. 2007) would be habitable provided that liquid water and the chemical constituents necessary for the emergence of life existed for a sufficiently long period of time. The flaring activity of M stars has been cited as a danger for the habitability of such planets; however, Segura et al. (2010) showed that the flares should not present a threat for the surface life. In our study, we investigate the ozone distribution and, therefore, one condition for the habitability of a tidally locked exoplanet revolving around an M star.

Due to being less massive than the Sun, M stars have weaker UV and visible emission and their spectrum peaks in the infrared. However, the enhanced stellar activity of M stars can result in the generation of flares whose EUV and UV emission is stronger than those of the quiet Sun. The different stellar spectral energy distributions will impact the planet's ozone distribution and the atmospheric temperature structure. Selsis et al. (2000), Segura et al. (2003), Segura et al. (2005), Grenfell et al. (2007), Rugheimer et al. (2013), Rauer et al. (2011) and Godolt et al. (2015) investigated the impact of different stellar spectral energy distributions on the ozone distribution of habitable non-tidally locked terrestrial extrasolar planets.

Selsis (2000) simulated the evolution of the chemical and thermal atmospheric structure of an Earth-like planet's when placed in the orbit of a F9 and K2 star using a 1D atmospheric model. He reported that an increase in the UV/visible light ratio results in the thickening of the planetary ozone layer compared to the Earth.

Segura et al. (2003) calculated the spectra of Earth-like planets orbiting a F2V, a G2V and a K2V star using a 1D coupled radiative-convective photochemical model and determined that the F2V planet would have a lower ozone concentration compared to the G2V or a K2V planet. This contrasts with the results of Selsis et al. (2000).

Segura et al. (2005) calculated the spectra of Earth-like planets orbiting a M star and determined that an ozone layer similar to that of the Earth would develop.

Grenfell et al. (2007) calculated the atmospheric composition changes of an Earth-like planet when subjected to 5–10 % changes of its orbital position around a G2V, a F2V, and a K2V star using a 1D coupled radiative-convective photochemical column model. As the planet moved outwards, the ozone increased by ~10 % due to the decreased stratospheric temperature.

Rugheimer et al. (2013) used a geometrical 1D model global atmosphere model to simulate the spectra of clear and cloudy Earth-like planets orbiting a F, a G, and a K star at the 1 AU equivalent distance. They reported that an increase in either the parent star's UV radiation or its temperature resulted in increased O_3 concentrations and stronger O_3 spectral features.

Rauer et al. (2011) used a plane-parallel, 1D climate model, coupled with a chemistry model to calculate the molecular absorption bands of super-Earth planetary atmospheres orbiting M stars. They reported that the ozone emission of planets orbiting quiet warm M0–M3 dwarfs would be stronger than the Earth's. The ozone emission spectra of planets orbiting cool and quiet M4–M7, on the other hand, stars would be weaker due to increases in the planet's mid-atmospheric temperatures.

Grenfell et al. (2013) used a global-mean radiative-convective-photochemical column model to perform a sensitivity study on the atmosphere of an Earth-like planet. The stellar class and planetary gravity of the planet was varied and the effects of the resulting photochemistry and climate changes investigated. According to their results, the stratospheric ozone generation of an Earth-like planet orbiting an M0 star is still dominated by the Chapman cycle but to a lesser extent compared to the Earth and the smog generation mechanism is strengthened. The Chapman cycle is a set of chemical and photochemical reactions responsible for the formation and destruction of ozone that involves only oxygen reactions. Moving towards cooler stars, the Chapman mechanism is weakened and the smog mechanisms is strengthened. For cool M5-M7 stars, the ozone photochemistry shifted completely from the Chapman production to smog-dominated stratospheric ozone production.

Godolt et al. (2015) studied the influence of F, G and K stars on the stratospheric temperature, climate and potential habitability of Earth-like exoplanets using a state-of-the-art 3D Earth climate model which accounted for local and dynamical processes. They used a fixed Earth-like atmosphere with no atmospheric chemistry and positioned the planets at orbital distances that ensured that the total amount of energy received from the parent stars would equal the solar constant. The results of the study confirmed that when the atmospheric chemistry is maintained unaltered, different stellar spectral energy distributions will result in different ozone heating rates and, therefore, different vertical temperature structures, in accordance with Selsis (2000), Segura et al. (2003), Segura et al. (2005), Grenfell et al. (2007), Rugheimer et al. (2013) and Rauer et al. (2011). Specifically, the stratospheres of

Earth-like planets orbiting stars less massive than the Sun will be characterised by lower ozone heating rates and shallower stratospheric temperature increases compared to the Earth.

The above studies demonstrated that for non-tidally locked, M star orbiting planets, the reduced UV radiation emission would result in a cooler stratosphere and changes in their stratospheric ozone concentration. It is reasonable to assume that tidally locked planets would be equally affected.

The ozone concentration of a tidally locked Earth-like planet orbiting an M star would be altered by the effects of the tidal lock and the altered stellar irradiance spectrum and make the attribution of the ozone distribution changes on either the tidal lock or the altered UV radiation exceedingly challenging. Therefore, we chose to not include the effects of the altered UV radiation in our study and focus only on the impact of the tidal locking on the Earth-like planet orbiting a Sun-like star.

Past studies of tidally locked Earth-like atmospheres limited themselves to the troposphere (Merlis and Schneider 2010; Yang et al. 2013; Grenfell et al. 2014).

Merlis and Schneider (2010) simulated the troposphere of an Earth-like, tidally locked aquaplanet with a rotation period equal to one Earth year using an ideal gas GCM with an active hydrological cycle, a grey radiation scheme and a slab ocean. They reported a maximum subsolar point surface temperature of $\sim$300 K and a minimum antisolar point temperature of 240 K. Since the surface temperature never dropped below 240 K, no atmospheric collapse occurred on the nightside. They also reported the presence of a strong upwelling above the subsolar point on the dayside and a downwelling, centred over the antisolar point, on the nightside. The upwelling was generated by the increased radiative heating, while the downwelling was generated by the increased radiative cooling.

Yang et al. (2013) demonstrated that clouds in the troposphere of a tidally locked aquaplanet, especially when located over the subsolar point, can ensure habitability at almost twice the Earth stellar flux.

Grenfell et al. (2014) used a global-mean, stationary, hydrostatic, atmospheric column model, extending from the surface to $\sim$70 km, to perform a 1D simulation of an Earth-like exoplanet atmosphere. The planet's atmosphere had the starting composition, pressure and temperature of the 1976 Wikipedia US Standard Atmosphere (2016). The UV emission of a cool M7 star was varied, and the resulting climate-photochemical response of the planetary atmosphere, including numerous catalytic processes of O_3 depletion, were calculated. Grenfell et al. (2014) showed that the strongest O_3 emission was generated when the stellar UV radiative flux is ten times that of an M7 star. They also reported that the exoplanet's O_3 9.6 μm spectral line profile was influenced strongly by the 200–350 nm UV output of the parent star.

The influence of the sea surface temperature on the tropospheric and lower stratospheric temperature and dynamics has been investigated by several researchers for the present-day Earth but not for exoplanets.

Braesicke and Pyle (2004) investigated the dynamics and ozone sensitivity to different SSTs using the Met Office Unified Model with simple stratospheric chemistry. They performed multi-annual simulations, each lasting 20 years. The simulations shared the same simplified ozone chemistry but different prescribed SSTs. According to their results, the appearance of extreme events in the winter stratosphere were more strongly correlated with the underlying SSTs rather than with changes of the ozone layer.

Rosenlof and Reid (2008) used data provided by the NOAA/CIRES Climate Diagnostics Center to investigate the long-term tropical lower stratospheric temperature trends over the western Tropical Pacific Ocean, in relation to variations in the SST. Their study revealed the presence of an anticorrelation between stratospheric temperature anomalies and SST anomalies. Rosenlof and Reid (2008) proposed the existence of a fairly direct influence between the underlying ocean and the lower tropical stratosphere and speculated that the anticorrelation is generated by the increased deep tropospheric convection generated by the SST warming.

Decker and Dameris (2008) reported the presence of a lower stratospheric cooling due to higher tropical SSTs. Their study investigated how the strength of the tropical upwelling is influenced by changes in the underlying SST and greenhouse gas concentrations. To that end, they performed two simulations with different SSTs and greenhouse gas concentrations. They concluded that the deep convection was amplified by the higher underlying SST and waves were generated. The waves enhanced the residual meridional circulation and the upwelling of ozone-poor tropospheric air into the lower stratosphere in the tropics.

One study by Hegy et al. (2014) was performed using the Community Earth System Model 1 (Whole Atmosphere Community Climate Model) (CESM1(WACCM)) version 1.0.2 to investigate the initial transient response of the boreal winter stratospheric polar vortex to localised SST warming events. They conducted 20 perpetual winter simulations. According to their results, the polar vortex was weakened by changes in the eddy-driven mean meridional circulation and negative anomalies appeared in the eddy momentum flux convergence. They also demonstrated that the initial state and subsequent internal variation of the extratropical atmosphere is equally important to the type of SST forcing in determining the response of the stratospheric polar vortex. Furthermore, they found that the interactions between the internal variability of the vortex and the SST-driven wave anomalies govern the nature of the polar vortex response to the forcing.

Chen et al. (2010) used an Earth-like aquaplanet general circulation model to study the tropospheric winds and stratospheric Brewer-Dobson circulation sensitivities to SST warmings. The main aim of the study was to investigate the relationship between a warming ocean due to global warming and a change in the Earth's large-scale atmospheric circulation. The authors concluded that the location and sign of the SST perturbations gradient strongly influenced the tropospheric jet and the Hadley cell. They reported an increase in the Brewer-Dobson circulation in the presence of a low-latitude warming and a decrease in the presence of a high-latitude warming extending to the subtropics.

To the extent of our knowledge, the correlation between the SST and the middle atmosphere has not been investigated for a tidally locked Earth-like planet whose circulation largely differs from the Brewer-Dobson circulation of the fast rotating, present-day Earth.

Furthermore, the majority of the exoplanet and tidally locked exoplanet studies to date were performed using 1D and 2D models, and atmospheric transport processes are generally not accurately represented in such models. The few studies performed using a 3D model either focused entirely on the exoplanet's troposphere (Yang et al. 2013; Merlis and Schneider 2010) or studied the stratospheric dynamics, without taking into consideration the stratospheric chemistry and photochemistry (Godolt et al. 2015). None of the past studies investigated the exoplanets' mesosphere. They also did not study how the secondary ozone layer found in the Earth's mesosphere would be affected by the changes in parent star's stellar class, its distance from the star or the planet's tidal locking.

In a tidally locked Earth-like planet, the majority of the stratospheric ozone is located in the lower stratosphere and its distribution depends strongly on the atmospheric transport processes, as shown in Proedrou and Hocke (2016). Our current study uses a 3D realistic chemistry-climate global circulation model, extending from the surface to the top of the mesosphere. The 3D nature of the model and its inclusion of dynamics, chemistry and photochemistry will allow us to correctly calculate the exoplanet's ozone distribution and how a change in the underlying SST will affect it.

The sensitivity of the mesospheric conditions to SST variations is crucial for the observations of the spectral signatures of mesospheric gases, who will be affected by altered temperatures and/or circulation patterns.

Below, we first describe the model used and the simulation setup. Next, we present and analyse the atmospheric evolution to a steady state, the atmospheric circulation, the atmospheric temperature structure and the ozone volume mixing ration distribution ($O_3\,VMR$).

Model description

The simulation is performed using the CESM version 1.04 which consists of five fully coupled geophysical models: atmosphere (ATM), land (LND), ocean (OCN), sea-ice (ICE) and land-ice (GLC). The models have five possible different modes: fully prognostic, data, stub and dead. The fully prognostic mode provides "state-of-the-art climate prediction and analysis tools" (Vertenstein et al. 2012).

The atmospheric model used to perform our study is the Whole Atmosphere Community Climate Model version 4 (CESM1(WACCM)) (Neal et al. 2012). CESM1(WACCM) has been used to simulate the circulation, gravity waves and atmospheric composition changes of the present-day Earth by Pedatella et al. (2014), Pedatella and Liu (2013), Lu et al. (2012), Tan et al. (2012a, 2012b, 2012c), Davis et al. (2013) and Smith et al. (2012). It is fully coupled to the land, ocean and ice models (Vertenstein et al. 2012).

It has a fully compressible horizontal discretisation and a quasi-Lagrangian vertical discretisation approximation which is good for scales larger than 10 km (Neal et al. 2012). It has 66 vertical levels starting on the surface of the planet and up to 5×10^{-6} hPa (2.5–149 km). The model top is located at $\sim$150 km. The vertical coordinate is terrain following below 100 hPa and purely isobaric above 100 hPa. The vertical resolution is 1.1 km in the troposphere, 1.1–1.4 km in the lower stratosphere, 1.75 km at the stratopause and 3.5 km above 65 km (Neal et al. 2012).

CESM1(WACCM) has full tropospheric, stratospheric and mesospheric chemistry, with 57 chemical species (Neal et al. 2012). The gas-phase chemistry is coupled to the Modal Aerosol Model (Neal et al. 2012). The stratospheric distributions of long-lived species are taken from previously performed CESM1(WACCM) simulations.

The longwave and shortwave radiative transfer calculations are provided by the radiation code RRTMG (Iacono et al. 2008; Mlawer et al. 1997). It is capable of efficiently calculating the irradiance and heating rate in broad spectral intervals, while retaining a high level of accuracy relative to measurements and high-resolution line-by-line models. It also distinguishes between the direct and scattered solar radiation.

The shortwave radiative transfer is calculated over 14 bands (0.2 to 12.2 μm) while the longwave radiative transfer is calculated over 16 bands (3.1 to 1000.0 μm). The 16th longwave band includes the infrared contribution from the spectral interval below 3.1 μm. Above 65 km, the model covers the spectrum interval between soft X-rays and extreme ultraviolet irradiances (0.05 nm to Lyman-α (121.6 nm) and the spectrum interval between the Lyman-α (121.6 nm) and 100 μm (Neale et al. 2012).

The total shortwave fluxes are accurate within 1–2 W/m^2 compared to the standard RRTM SW for clear sky conditions and aerosols and 6 W/m^2 in overcast sky conditions. The total longwave fluxes are accurate within 1.0 W/m^2 at all levels. "The longwave radiative transfer is performed over a single angle for one upward and one downward calculation" (Neale et al. 2012). The absorption coefficients for the k-distributions for both the shortwave and the longwave radiation calculations in RRTMG are obtained from the line-by-line radiation model LBLRTM (Clough and Iacono 1995; Clough et al. 2005).

The calculation of the photolysis coefficients is divided into the 120–200 nm and 200–750 nm regions. The total photolytic rate constants are calculated by the model by integrating the product of the wavelength-dependent exo-atmospheric flux, the atmospheric transmission function, the molecular absorption cross section and the quantum yield for each absorbing species. The exo-atmospheric flux is taken from observations and varies over the 11 year solar sunspot cycle. The transmission function is wavelength-dependent and a function of the model abundance of ozone and molecular oxygen. For wavelengths above 200 nm, the molecular absorption cross section and the quantum yield are calculated by the model, while below 200 nm, their values are pre-defined. For NO and O_2 detailed photolysis and parametrisations are included in the model. The impact of clouds on the photolysis rates is parametrised. The impact of tropospheric and stratospheric aerosols on photolysis rates, on the other hand, is not calculated (Neale et al. 2012).

The CESM model uses the Kurucz solar source function, whose radiative transfer calculation is based upon solar measurements. The model assumes a total solar irradiance at the top of the atmosphere equal to TSI = 1368.22 W/m^2. The value is then "scaled in each spectral band through the specification of a time-varying solar spectral irradiance" (Neale et al. 2012). A combination of solar parametrizations is used to specify spectral irradiances over the soft X-ray to Lyman-α and the Lyman-α to 100 μm spectral interval. The first spectral interval fluxes are calculated using the parametrization of Solomon and Qiang (2005). It accepts as input the 10.7 cm solar radio flux (f 10.7), whose daily values are obtained from the NOAA's Space Environment Center (www.sec.noaa.gov) and its 81 day average. The second spectral interval fluxes are calculated using the empirical model of the wavelength-depending sunspot and facular influences (Lean et al. 2000; Wang et al. 2005; Neal et al. 2012).

The modelled sources of radiation absorption and scattering in the shortwave electromagnetic spectrum are H_2O, O_3, CO_2, O_2, CH_4, N_2, clouds, aerosols and Rayleigh scattering. The sources of molecular radiation absorption in the longwave electromagnetic spectrum are H_2O, CO_2, O_3, N_2O, CH_4, O_2, N_2, CFC-11 and CFC-12. The shortwave electromagnetic radiation is calculated only for zenith angles larger than zero (Neale et al. 2012). The zonal mean climatology of the local O_3 concentration, the temperature, the overhead column O_3 and other chemicals is based on satellite and in situ Earth observations (Neal et al. 2012).

The model computes the chemical equilibrium of 36 photochemical species (O_2, O_3, N_2O, NO, NO_2, N_2O_5, HNO_3, NO_3, HO_2NO_2, CH_3OOH, CH_2O, H_2O, H_2O_2, Cl_2, ClO, OClO, Cl_2O_2, HOCl, HCl, $ClONO_2$, BrCl, BrO, HOBr, $BrONO_2$, CH_3Cl, Ccl_4, CH_3CCl_3, CFC11, CFC12, CFC113, HCFC22, CH_3Br, CF_3Br, CF_2ClBr, CO_2, CH_4). The model atmospheric tracers are O_x, NO_x, HO_x, ClO_x and BrO_x chemical families, excluding O_2, along with CH_4 and its degradation products. The main photochemical and chemical interactions responsible for changing the atmospheric ozone concentration are described by the reaction rate of ozone $\frac{d(O_3)}{dt} + J_{O_3}(O_3) + k_3(O)(O_3) + a_2(H)(O_3) + a_6(OH)(O_3) + a_{6b}(HO_2)(O_3) + b_4(NO)(O_3) + b_9(NO_2)(O_3) + d_2(Cl)(O_3) + e_2(Br)(O_3) = k_2(M)(O_2)(O)$. Where $\frac{d(O_3)}{dt}$ is the production rate of O_3, J_{O_3} is the photolysis rate of the ozone and $k_3, a_2, a_6, a_{6b}, b_4, b_9, d_2, e_2, k_2(M)$ are the chemical rate constants of the reactions and are taken from JPL06-2 (Sander et al. 2006).

A polar stratospheric cloud parametrisation scheme is incorporated in the model. It is activated when the cosine of the solar zenith angle is larger than zero at stratospheric altitudes and the stratospheric temperature drops below 195 K (Neal et al. 2012).

The momentum, sensible heat flux, latent heat flux, land surface albedos and upward longwave electromagnetic radiation are used to calculate the atmospheric radiation. The upward longwave radiation is given by the difference of the incident and the absorbed fluxes. The incident flux values are, in turn, determined from the daily values of the solar radio flux (F 10.7) which are provided by the National Oceanic and Atmospheric Administration's (NOAA) Space Environment Center (2016).

The starting values of the zonal mean climatology of the local O_3 concentration, the temperature, the overhead column O_3 and other chemicals used by the model are provided by satellite and in situ Earth observations (Neal et al. 2012). A detailed model description can be found in Neal et al. (2012).

Methods

Simulation setup

We perform three 90 day simulations. Two for a TLE planet, with different SST distributions, and one for the present-day Earth (Fig. 1).

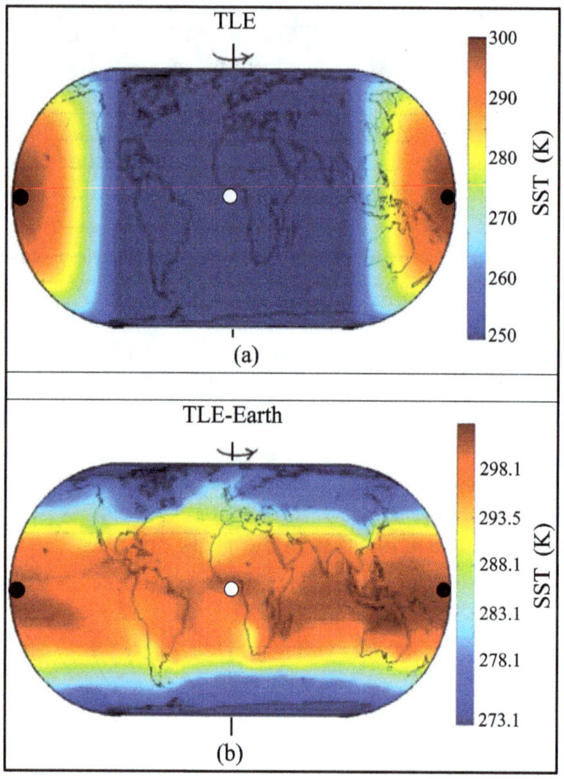

Fig. 1 CTLE and WTLE sea surface temperature setups. **a** CTLE. *Shades of red* indicate high temperatures and define the dayside of the planet. *Shades of blue* indicate low temperature and define the nightside of the planet. **b** WTLE. *Shades of red* indicate high temperatures. *Shades of blue* indicate low temperature. The subsolar point is indicated with a *black point* on either side of the figure. The antisolar point is indicated with a *white point* in the centre of the figure for both simulations

The present-day Earth simulation is an unaltered simulation of the present-day Earth, which starts on the Spring Equinox (21.03.2000) and ends 90 days later (18.06.2000).

The TLE simulations simulate a tidally locked Earth-like planet orbiting a Sun-like star. The tidally locked planets are located at a distance of 1 AU from the parent star and have the same eccentricity as the present day Earth, a stellar irradiance at the top of the model $1366.96 < S_o < 1368.60$ and a rotation rate that is 1/365th of the Earth's rotation rate. Both of their subsolar points are permanently located (0.17° N, −178.17° E), which is the position of the Earth's subsolar point at 00:00 UT on the day of the Spring Equinox (21.03.2000).

The parameters altered to achieve the tidal lock and a perpetual equinox are:

- The number of seconds in a siderial day: altered from 86, 164 s/day to 3.15×10^7 s/day
- The planet's rotational velocity: set to 2.31×10^{-12} rad/s
- The planet's new rotation rate: set to 1° every 87,600 s (the Earth's rotation rate is 1° every 240 s)
- The position of the subsolar point: set to (0.17° N, −178.17° E) for the duration of the simulations
- The SST: changed to resemble the SST reported by Merlis and Schneider (2010)
- The solar zenith angle (SZA): set to a constant value by means of the present-day Earth (PDE) Julian date.

The SZA is the main parameter used for the assessment of the incoming solar radiation in the CESM model. By setting the Julian date to a constant value we stop the periodic 24 h variation of the SZA at all grid points. Figure 2 shows the shortwave solar radiation flux at the top of the model for the WTLE and CTLE simulations. The solar

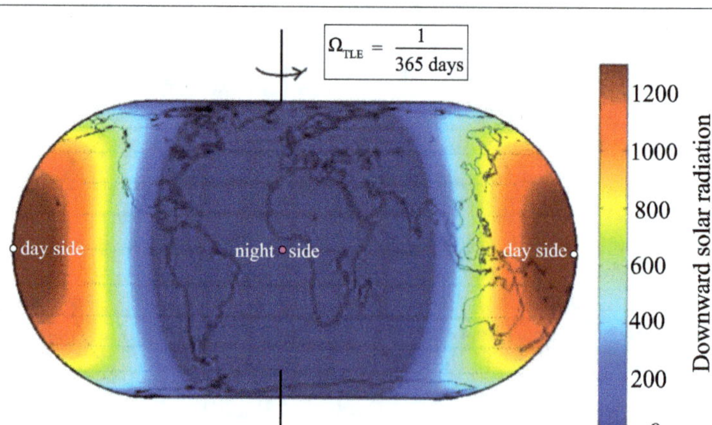

Fig. 2 Incident shortwave radiation at the top of the model for both the CTLE and the WTLE. *Shades of red* indicate high incoming solar radiation and define the dayside of the planet. *Shades of blue* indicate low incoming solar radiation. The *dark blue* visible in the centre of the plot indicates a lack of incoming solar radiation and defines the nightside of the planet. The subsolar point is indicated with a white *point* on either side of the figure. It is centred approximately at (0°, 180°) over the Pacific Ocean. The antisolar point is indicated with a magenta point in the centre of the figure and is centred approximately at (0°, 0°)

point is depicted by a white dot located on either side of the figure, while the antisolar point is depicted by a magenta point located in the centre of the figure.

As mentioned in the Introduction, the tidally locked planets are identical except for their sea surface temperature distributions. The pure TLE SST distribution resembles the SST reported by Merlis and Schneider (2010) (Fig. 1b). It was characterised by a uniform nightside SST of 250 K and a monotonically and isotropically increasing dayside SST that reached a maximum of 300 K within a 30 degrees radius centred on the subsolar point. On account of the overall cooler SST, the simulation will be called cold TLE (CTLE).

The hybrid TLE SST distribution is that of the present-day Earth. It is characterised by temperatures of 290–300 K in the tropics and 270–280 K in the polar regions. Due to the overall warm SST, the simulation using this SST will be hereafter called warm TLE (WTLE). This WTLE simulation is not a realistic physical scenario but is rather used to test the importance of SST patterns for slowly rotating planets. Both SSTs can be seen in Fig. 1.

The CESM1(WACCM) requires the presence of an active land model and, therefore, has the general topography and continents of the Earth. The atmospheric and surface initialisation data are identical for all simulations and all start with same 3D ozone and wind fields.

In the next subsection, we present the middle atmospheric circulation (Ω, HW) and temperature (T) and ozone distribution (TOC, O_3VMR) of the two simulations and we discuss the degree by which they are influenced by the two extreme SSTs. We first discuss the atmospheric adjustment time of the two tidally locked Earths (TLEs) and compare it to the present-day Earth. Then, we present and discuss their atmospheric circulation, atmospheric temperature structure and latitude-longitude O_3VMR cross section.

It should be noted that while the possibility of tidal locking of an Earth-like planet within the habitable zone of a Sun-like star has a low probability, the similarity of the solar spectrum and atmospheric composition to the PDE fosters the intercomparison and interpretation of the differences in the model runs. Furthermore, since several past studies of tidally locked exoplanets have used models with solar spectrum and atmospheric composition similar to that of the PDE, this setup fosters the intercomparison and interpretation of the changes with past models."

Data analysis

The zonal wind (U), the meridional wind (V), the vertical wind (Ω), the atmospheric temperature (T), the air number density (M) and the O_3VMR are extracted from the output datasets of the simulation run. The above parameters are then interpolated for the altitudes of 1 to 140 km. Next, the zonally averaged zonal, meridional and vertical wind are calculated from 5 to 110 km. The O_3 concentration is calculated for the altitudes of 1 to 110 km by multiplying the ozone volume mixing ratio with the air number density ($[O_3] = O_{3VMR} \times [air]$). As a next step, the total ozone column density (TOC) is calculated by interpolating the O_3 concentration between the altitudes of 5 and 110 km ($TOC = \int_5^{110} [O_3]\, dz$). The altitude limits are imposed in order to avoid the data gaps generated by the presence of several high mountains and mountain ranges at low altitudes (e.g Himalayas, Kilimanjaro) and the air density data gaps above 110 km altitude. The TOC, HW, Ω and T global means are determined by calculating their zonal means as a function of latitude and then weighting them with the surface area of the latitude belts (surface area preserving mean). The change in percent between the CTLE and WTLE parameters is calculated using the following equation:

$$\Delta = \frac{\text{parameter}_{\text{CTLE}} - \text{parameter}_{\text{WTLE}}}{\text{parameter}_{\text{WTLE}}} * 100 \qquad (1)$$

where the term parameter is used to refer to any of the TOC, Ω, HW, T and O_3VMR.

Terminology

The dayside of a tidally locked planet is the permanently illuminated hemisphere that is always facing the parent star, while the nightside is the hemisphere that is never illuminated by the parent star. The subsolar point is defined as the position of maximum solar flux, located at the centre of the dayside and has a SZA equal to zero (SZA = 0°). The antisolar point is defined as the centre of the nightside (SZA = 180°).

Results and discussion

Atmospheric steady state

According to our results, the middle atmospheres of the CTLE and WTLE adjust to the new radiative and dynamical conditions, reaching a steady-state condition within 80 days from the start of the simulation. Figure 3 shows the adjustment times of the CTLE and WTLE total ozone column (TOC), and the global-mean stratospheric horizontal wind (HW), global-mean stratospheric vertical wind (Ω) and global-mean stratospheric temperature (T) for the 30–40 km altitude range. The temporal evolutions of the PDE TOC, HW, Ω and T are also provided.

The CTLE and the WTLE TOC, HW and Ω adjustment times and values are not significantly altered by the SST change, as can be seen in Fig. 3a–c. The percent change in their adjustment times is shown in Table 1. The CTLE and WTLE T adjustment times are shown in Fig. 3d. As one can see, their adjustment times are comparable despite their altered slope and the CTLE values are decreased compared to the WTLE. This difference is also visible

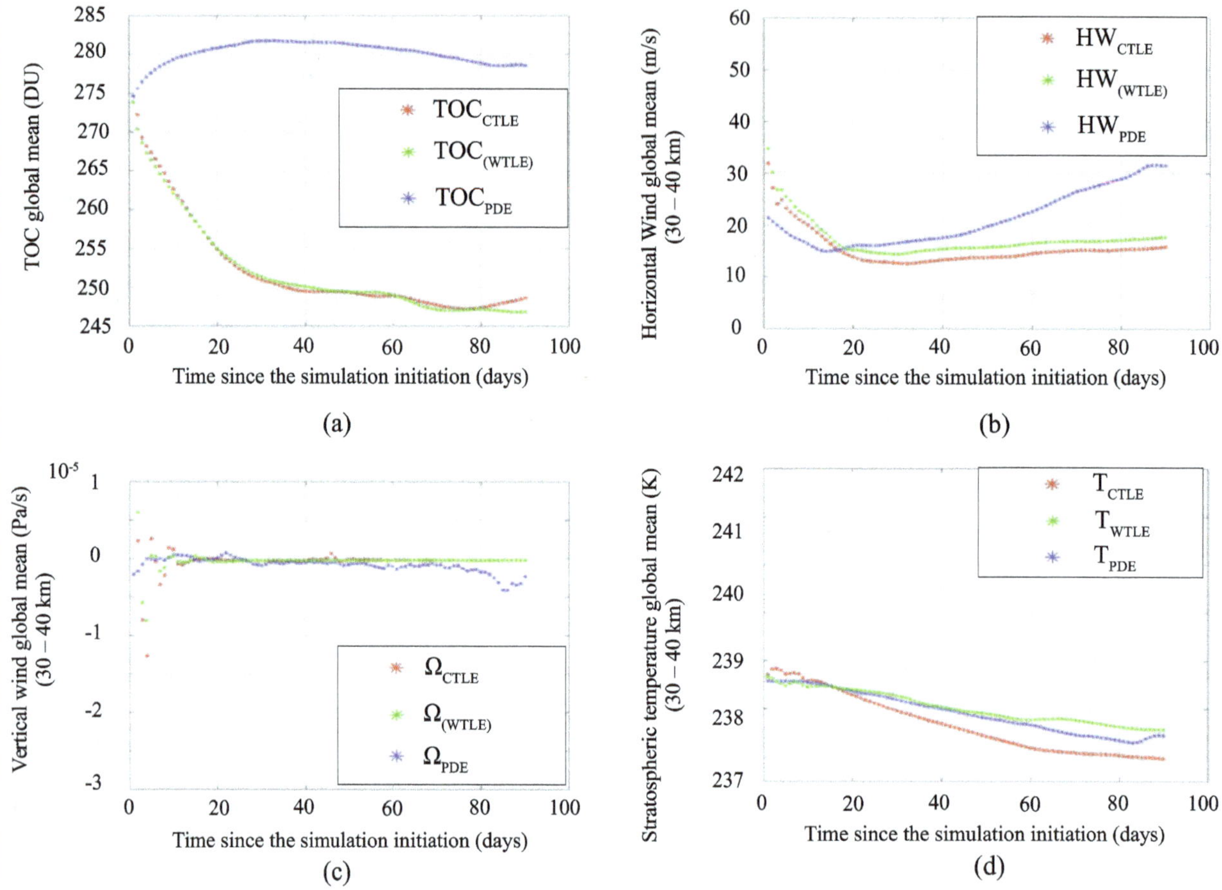

Fig. 3 CTLE, WTLE and PDE atmospheric temporal evolution. **a** TOC. **b** HW. **c** Ω. **d** *T*. The CTLE and WTLE TOC, HW, and T are exponentially decaying to their steady-state value with an e-folding time of 30, 20, and 40 days, respectively. The CTLE and WTLE vertical winds are exponentially increasing to their steady-state value with an e-folding time of 15 days

in the CTLE and WTLE T global average which will be presented in the "Vertical wind" section.

The CTLE and WTLE e-folding times of the TOC, HW, Ω and T along with their PDE equivalents and standard deviations (σ) are shown in Table 1. They are 30, 20, 15 and 40 days, respectively. Therefore, their adjustment times are 60, 40, 30 and 80 days, respectively. A comparison between the CTLE, the WTLE and the PDE σ in Table 1 reveals that the PDE has higher standard deviations (σ_{PDE}) compared to the CTLE and WTLE standard deviations (σ_{CTLE} and σ_{WTLE}, respectively).

Atmospheric circulation

The tidal lock results in a breakdown of the Brewer-Dobson circulation (Fig. 4) in both the CTLE and the WTLE. As can be seen in Figs. 5, 6, 7, 8 and 9, it is replaced by a very different atmospheric circulation. Despite the atmospheric changes in the circulation, tro-

Table 1 Comparison of the PDE, CTLE and WTLE *TOC* and the middle stratospheric *HW*, Ω and *T* e-folding times, steady-state mean values and their standard deviations

	PDE	σ_{PDE}	CTLE	σ_{CTLE}	WTLE	σ_{WTLE}
TOC (DU)	280.1	1.44	248.7	1.00	248.64	1.42
HW (m/s)	21.19	5.21	14.21	1.00	16.10	2.22
Ω (Pa/s)	-8.6×10^{-7}	9.6×10^{-7}	-1.7×10^{-7}	1.3×10^{-7}	-2.0×10^{-7}	-9.7×10^{-8}
T_{strat} (*K*)	239.9	0.85	238.4	0.43	239.6	0.27

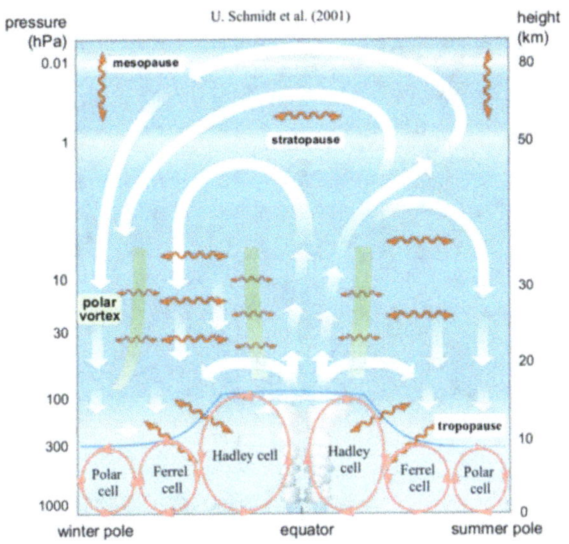

Fig. 4 The Earth's Brewer-Dobson circulation. The Hadley, Ferrel and Polar cells are visible in the figure for both hemispheres. The *thick white arrows* indicate the transformed Eulerian-mean mass stream function. The *wavy orange arrows* indicate two-way mixing processes. The *thick green lines* indicate the presence of stratospheric transport and mixing barriers. (Adapted from Bonisch et al. 2011)

posphere, stratosphere, and mesosphere are present in both simulations. Below, we present a description of the atmospheric circulation of the two simulations.

Vertical wind

Snapshots of the last model step of the PDE, CTLE, and WTLE vertical wind are shown in Figs. 5, 6 and 7, respectively. The PDE vertical circulation is characterised by many upwelling and downwelling regions in both the daytime and the nighttime hemispheres. In the tidally locked planets on the other hand, the dayside air rises as a result of being radiatively heated by the constant shortwave radiation flux it receives from the parent star. On the nightside, the lack of incoming radiation results in the radiatively cooling of the air which subsequently sinks towards the surface. This results in the appearance of a dayside upwelling and a nightside downwelling extending from the surface to the mesosphere. Merlis and Schneider (2010) also predicted the presence of such a vertical wind setup for the troposphere of a tidally locked Earth-like aquaplanet orbiting a Sun-like star.

A comparison between Figs. 6 and 7 reveals that the vertical winds of the two simulations are very similar. They are characterised by different small-scale variability vortices and jets of different magnitudes and locations in the CTLE and WLTE simulations. The WTLE atmosphere is characterised by a greater small-scale variability compared to the CTLE. We can, therefore, conclude that the SST change has only a limited effect on the vertical wind.

Horizontal wind

Snapshots of the last model step of the CTLE and WTLE horizontal wind at stratospheric and mesospheric altitudes are shown in Figs. 8 and 9. The CTLE horizontal wind at 24 km is shown in Fig. 8a, while the WTLE horizontal wind at 24 km is shown in Fig. 9a. A comparison between the two figures reveals that in both simulations, the horizontal wind is characterised by the presence of a global eastward zonal jet stream.

The CTLE and WTLE stratospheric horizontal wind at 36 km is shown in Figs. 8b and 9b. In both simulations,

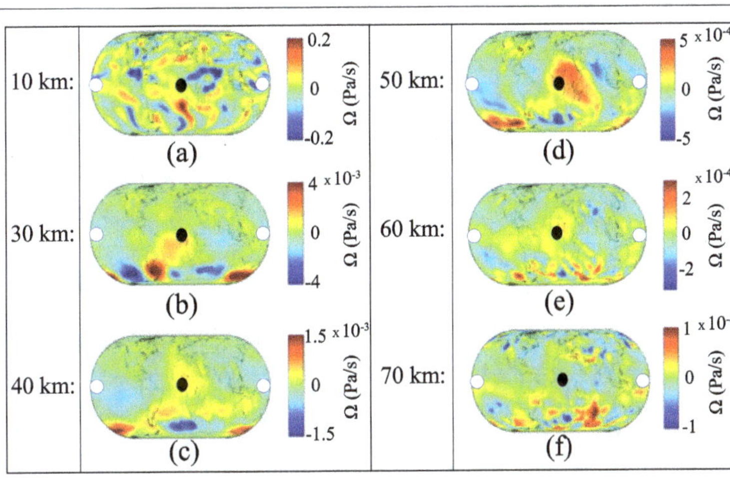

Fig. 5 PDE longitude-latitude vertical wind circulation. **a** Vertical wind at 10 km. **b** Vertical wind at 30 km. **c** Vertical wind at 40 km. **d** Vertical wind at 50 km. **e** Vertical wind at 60 km. **f** Vertical wind at 70 km. *Warm colours* (positive values) indicate a downwelling wind. *Cold colours* (negative values) indicate an upwelling wind. Each figure is centred on the antisolar point which is indicated with a *black dot*. The subsolar point is indicated with a *white dot* on either side of each figure

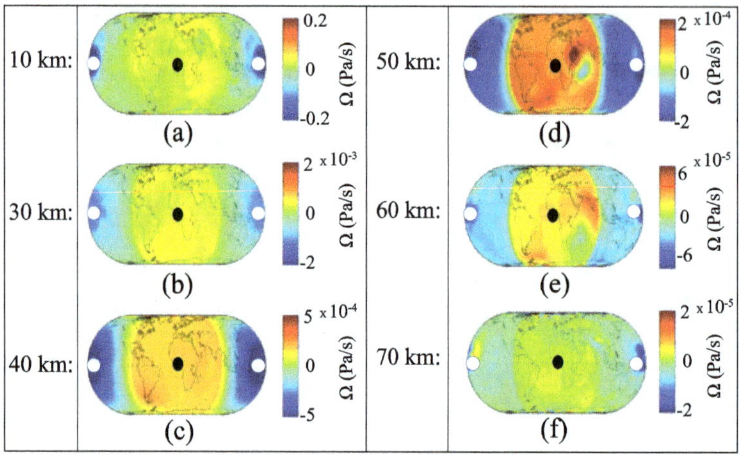

Fig. 6 CTLE longitude-latitude vertical wind circulation. **a** Vertical wind at 10 km. **b** Vertical wind at 30 km. **c** Vertical wind at 40 km. **d** Vertical wind at 50 km. **e** Vertical wind at 60 km. **f** Vertical wind at 70 km. *Warm colours* (positive values) indicate a downwelling wind. *Cold colours* (negative values) indicate an upwelling wind. Each figure is centred on the antisolar point which is indicated with a *black dot*. The subsolar point is indicated with a *white dot* on either side of each figure

a westward global zonal jet stream with an accompanying vortex located at polar latitudes is present. The vortex can be seen on the left hand side of the Southern and Northern Hemispheres in Figs. 8b and 9b. The blue coloured regions on the right hand side of the Southern and Northern hemispheres, on the other hand, are regions of low HW speeds. The WTLE horizontal wind is slightly weaker and has a wider jet stream compared to the CTLE.

The situation is different at mesospheric altitudes. At 60 km altitude, the zonal jet stream is replaced by large-scale vortices in both the CTLE and the WTLE, located at different geographical locations.

The mesospheric wave-breaking mechanism may explain this phenomenon. Gravity waves are generated by tropospheric convection followed by the release of latent heat. They then propagate upwards and break when they reach the mesosphere. There, they deposit their energy and, in the process, decelerate the wind flow (Nappo 2013). As a result of the different SST heat distributions in the WTLE and CTLE, the convectively induced gravity waves are generated at different locations with different energy budgets. This leads to different wave-breaking patterns and different gravity wave energy dissipation rates and gives rise to different horizontal wind circulations patterns in the two simulations.

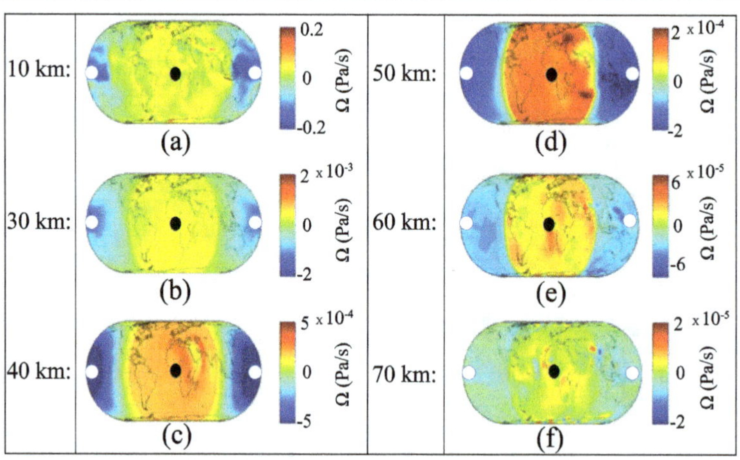

Fig. 7 WTLE longitude-latitude vertical wind circulation. **a** Vertical wind at 10 km. **b** Vertical wind at 30 km. **c** Vertical wind at 40 km. **d** Vertical wind at 50 km. **e** Vertical wind at 60 km. **f** Vertical wind at 70 km. *Warm colours* (positive values) indicate a downwelling wind. *Cold colours* (negative values) indicate an upwelling wind. Each figure is centred on the antisolar point which is indicated with a *black dot*. The subsolar point is indicated with a *white dot* on either side of each figure

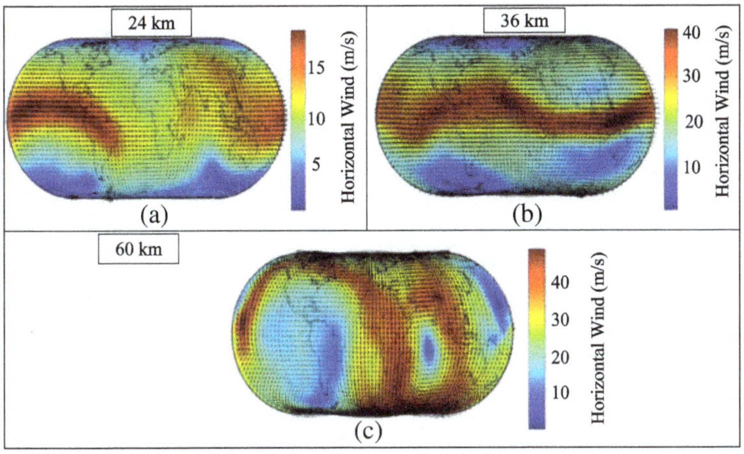

Fig. 8 CTLE horizontal circulation at select altitudes. **a** Horizontal wind at 24 km. **b** Horizontal wind at 36 km. **c** Horizontal wind at 60 km. The *black arrows* indicate the wind vector while the *colour shading* shows the wind speed

Gravity waves are also generated through interaction between the planet's surface orography and the lower tropospheric wind field (Nappo 2013). The encounter between the tropospheric wind and a hill, mountain or mountain range vertically displaces the stably stratified wind flow upwards. This leads to the generation of upwards propagating gravity waves which transport energy and mean-flow momentum towards the middle and upper atmosphere. There, it is deposited through the wave-breaking mechanism (Nappo 2013). In the case of the WTLE and CTLE, both simulations have the same Earth-like orography but different SSTs. As a result different sea and land breezes are generated, which lead to different lower tropospheric horizontal wind distributions (1.4 m/s in average) and therefore different terrain-generated gravity wave distributions. Consequently, the mesospheric energy deposition and wind deceleration are different for the two simulations. Therefore, the reported changes in the mesospheric horizontal wind are possibly induced by differences in the energy and momentum fluxes of the convectively and orographically generated gravity waves of the two simulations.

As can be seen in Fig. 3b, the middle atmospheric horizontal wind speeds of the CTLE and the WTLE are slower compared to the PDE. They are also characterised by stronger and wider jet streams and the presence of large scale vortices at various geographical locations. The slower wind speeds could be attributed to the smaller CTLE and WTLE Coriolis force, compared to the PDE.

We can, consequently, conclude that the mesospheric circulation is more sensitive to the underlying SST compared to the stratospheric circulation, as is made evident by the regional differences in the vertical and horizontal wind maps. The discrepancy between the stratospheric

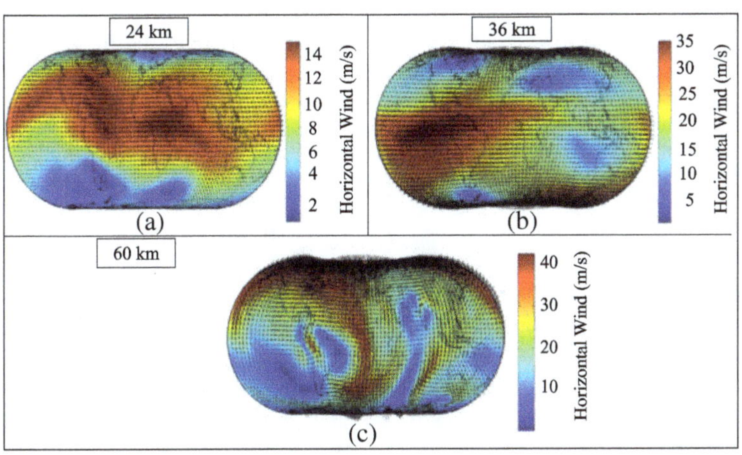

Fig. 9 WTLE horizontal circulation at select altitudes. **a** Horizontal wind at 24 km. **b** Horizontal wind at 36 km. **c** Horizontal wind at 60 km. The *black arrows* indicate the wind vector while the *colour shading* shows the wind speed

and mesospheric circulation can be explained by the fact that the stratospheric dynamics are mainly driven by the insolation of stratospheric ozone.

Atmospheric temperature structure

Next, we present the temperature structure of the CTLE and WTLE atmospheres, which can be seen in Fig. 10a. An inspection of Fig. 10a, b, which depicts the snapshots of the last model step, shows that the WTLE's lower troposphere ($z < 3$ km) is warmer than its CTLE equivalent by a factor of 3.7 K on average. This temperature is generated by the increased upwelling longwave radiation and sensible and latent heat from the underlying WTLE SST which is warmer by approximately 23.7 K compared to the CTLE.

In the 10–15 km region, the WTLE temperature is cooler by 4 K on average compared to the CTLE, as can be seen in the purple highlighted region of Fig. 10b. The adiabatic cooling is generated by the increased upwelling of the subsolar region air, which is radiatively and convectively heated by the warm WTLE SST. This result is in agreement with the studies performed by Rosenlof and Reid (2008), Breasicke and Pyle (2004), Decher and Dameris (2008), who reported the generation of an upper tropospheric cooling by an SST warming, due to the increased deep tropospheric convection generated by the SST warming.

Between 18 and 30 km, the increased upwelling longwave radiation emitted from the warm WTLE SST is absorbed by ozone and radiatively heats the WTLE stratosphere, increasing its temperature by 3.8 on average. This warming is shown in the green highlighted region of Fig. 10b and is in agreement with expectations of radiative coupling between the surface infrared emission and the lower stratosphere.

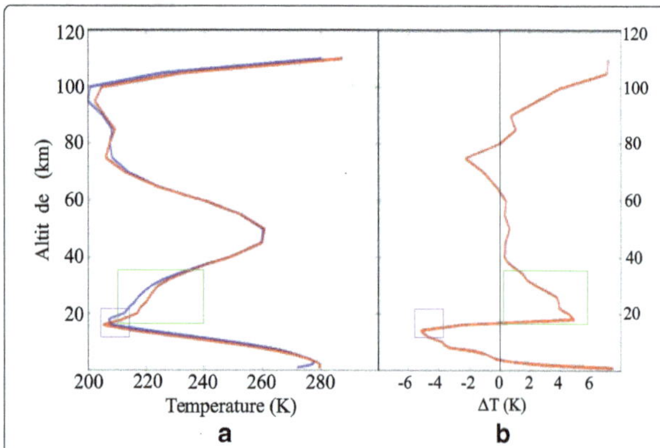

Fig. 10 CTLE and WTLE temperature global mean on day 90. **a** Comparison between the temperature global means of the two planets. The CTLE temperature is depicted using the *blue colour line*. The WTLE is depicted using the *red colour line*. The altitude range is 0–120 km. **b** The difference $\Delta T_{altitude}$ of the two planets

To investigate this coupling, we correlate the atmospheric temperature differences with the SST differences obtained from the two simulations. They are calculated by the following equations:

$$\Delta T_{\text{altitude}} = T_{\text{WTLE}} - T_{\text{CTLE}} \quad (2)$$

gives the difference between the WTLE and the CTLE temperatures at a given altitude and geographic location, while

$$\Delta \text{SST} = \text{SST}_{\text{WTLE}} - \text{SST}_{\text{CTLE}} \quad (3)$$

gives the difference between the WTLE and the CTLE SSTs at a given geographic location.

Figure 11 shows the relationship between the ΔSST and the ΔT at 1 km altitude ($\Delta T_{1\text{km}}$). The figure indicates that there is a linear correlation between the two quantities and that the SST changes have a direct radiative effect on the lower tropospheric temperature. Using the least squares method, we determine that the correlation between the ΔSST and the $\Delta T_{1\text{km}}$ is given by:

$$\Delta T_{24\text{km}}(\Delta \text{SST}) = (0.2768 \pm 0.0131)\Delta \text{SST} + (1.678 \pm 0.394) \quad (4)$$

A 30 K change of the SST corresponds to a 8.3 K change of the lower stratospheric temperature.

Figure 12 shows the relationship between the ΔSST and the ΔT at 24 km altitude ($\Delta T_{24\text{km}}$). As can be seen, a linear correlation exists between the ΔSST and $\Delta T_{24\text{km}}$, which suggests that SST changes have a direct radiative effect on the lower stratospheric temperature. Using the least square method, we determine that the correlation between the $\Delta T_{1\text{km}}$ and the $\Delta T_{24\text{km}}$ is given by

$$\Delta T_{24\text{km}}(\Delta \text{SST}) = (0.00168 \pm 0.000495)\Delta \text{SST} + (3.82 \pm 0.015) \quad (5)$$

A 30 K change of the SST corresponds to a 0.05 K change of the lower stratospheric temperature. The inclination of the regression line is rather small. This may be due to the strong mixing and quasi-isothermal behaviour of the lower stratosphere of tidally locked Earth-like planets.

The above results suggest that the increased thermal radiation emitted by the WTLE's surface together with the upwelling sensible and latent heat are absorbed by the lower tropospheric greenhouse gases and then increases the WTLE's temperature. The longwave radiation not absorbed by the lower troposphere reaches the lower stratosphere. There, it is absorbed by the ozone 9.6 μm absorption band, a band expected to be present in the spectra of exoplanets capable of sustaining an ozone layer. Therefore, a stronger absorption line could be expected for warmer SSTs.

The lower mesospheric temperature (70–80 km) on the other hand is cooler by 1.13 K on average compared to

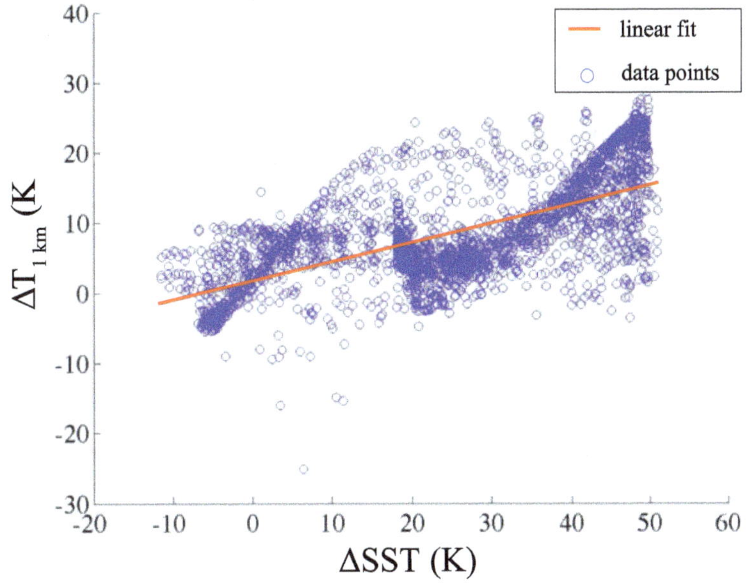

Fig. 11 Scatter plot of ΔSST and ΔT_{1km}. The plot shows the correlation between the surface Δ SST and the ΔT at 1 km altitude, together with a linear regression to the data

the CTLE. This temperature difference can be explained using the mesospheric wave-breaking mechanism mentioned above. The higher number and more energetic gravity waves generated by the warm WTLE SST lead to higher wave momentum deposition in the lower mesosphere and, therefore, to a lower mesospheric adiabatic cooling.

The picture reverses in the upper mesosphere (90–110 km), where the WTLE temperature is higher by 4.3 K on average, compared to the CTLE. This temperature difference may be induced by circulation changes generated by different gravity wave fluxes from below, but the details of the underlying processes remain an open question.

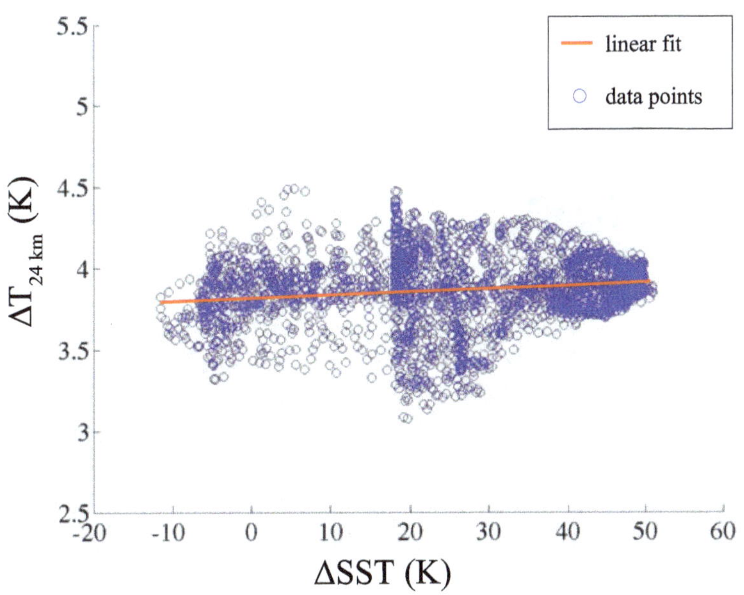

Fig. 12 Scatter plot of ΔSST and ΔT_{24km}. The plot shows the correlation between the surface Δ SST and the ΔT at 24 km altitude, together with a linear regression to the data

We can therefore conclude that increased SST leads to enhanced temperatures in the lower stratosphere of a tidally locked Earth-like exoplanet. Our results indicate that this lower stratospheric heating could be due to the enhanced upwelling of infrared radiation from the warm SST. Generally, the difference between the WLTE and the CLTE simulation is smaller than we expected.

Ozone volume mixing ratio distribution

Figure 13 shows the snapshots of the last model step of the CTLE and WTLE dayside and nightside ozone volume mixing ratio O_3VMR distribution. It depicts the cross section of the CTLE and WTLE O_3VMR distribution along the 1st and the 180th meridian, respectively.

A comparison between Figs. 13b and d reveals that the primary ozone layer (30–40 km altitude) is not significantly affected. The increased WTLE middle atmospheric temperature (Fig. 10) results in the decrease of the WTLE dayside O_3VMR by 1.8 % and a nightside decrease of 1.27 %.

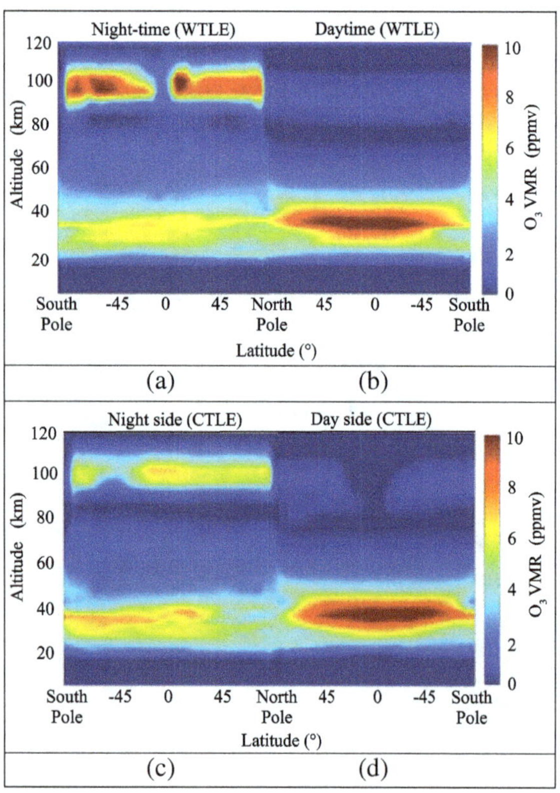

Fig. 13 WTLE and CTLE height-latitude cross section of O_3VMR along the meridian through the subsolar point and the antisolar point. **a** Vertical cross section of the WTLE nighttime hemisphere. **b** Vertical cross section of the WTLE daytime hemisphere. **c** Vertical cross section of the CTLE nightside hemisphere. **d** Vertical cross section of the CTLE dayside hemisphere

On the other hand, the nighttime secondary ozone layer, located in both cases in the 85–110 km altitude range, is significantly affected. On Earth, the secondary ozone layer appears only during nighttime. On the daytime hemisphere, the strong UV radiation destroys any ozone molecules that might form shortly after their creation. On the nightside, ozone is generated and accumulated on the nightside by recombination of atomic and molecular oxygen.

The WTLE O_3VMR is enhanced by 40.5 % compared to the CTLE. It is possible that the mesospheric circulation changes, generated by the different upwelling gravity wave fluxes, are responsible for the changes in the secondary ozone layer, but the details of the underlying processes remain an open question.

We can, therefore, conclude that the use of an inaccurate SST in a tidally locked Earth-like exoplanet may significantly affect the secondary ozone layer, while the primary ozone layer remains almost unaffected.

The gap visible at mesospheric altitudes over the WTLE equator in Fig. 13a appears as a result of local upwelling of ozone-poor air from lower altitudes and is located in the regions close to the Atlas mountains and the Atlantic Ocean. It is possibly generated by upwelling gravity waves, not present in the CTLE.

Conclusions

We studied the sensitivity of a tidally locked Earth-like atmosphere to its underlying SST by simulating the atmospheric circulation, temperature, and ozone distribution changes generated by two different SST distributions. The first SST distribution represented the well-observed SST of the fast-rotating, present-day Earth, and the simulation using it was named WLTE. The second SST distribution was the SST of a tidally locked Earth-like aquaplanet. It was provided by Merlin and Schneider (2010) and the simulation using it was named CLTE.

The main conclusion of our study is that for slow-rotating planets, the drastic SST changes do not lead to similarly strong changes in the middle atmosphere of the planet, as the vertical and horizontal winds, the primary ozone layer and the atmospheric thermal structure are relatively insensitive to the changes in the underlying SST.

The underlying SST changes result in small-scale variability changes of the vertical and horizontal winds. The strongest effect appears at mesospheric altitudes, possibly due to differences in the upwelling gravity wave flux generated by the different SST distributions. Our results are of interest for both exoplanets and for Earth climate trend studies.

The primary ozone layer O_3VMR is altered by 1.8 % on the dayside and by 1.27 % on the nightside. The secondary ozone layer, on the other hand, is altered by 40.5 %. The generation mechanism is currently an open question and under investigation.

The lower tropospheric temperature global average is increased, on average, by 3.7 K compared to the CTLE due to the warmer WTLE SST. The WTLE tropospheric warming causes the hot air to ascend and cool adiabatically, generating a cooling of 4 K on average in the altitude regions between 10 and 15 km, compared to the CTLE. This result is in agreement with past studies on the effects of SST warmings on the Earth's upper troposphere. Between 18 and 30 km, a stratospheric heating of 3.8 K on average, is present in the WTLE compared to the CTLE, generated by the absorption of the upwelling longwave radiation by ozone through the 9.6 μm line. The enhanced, upwelling longwave radiation is generated by the warmer WTLE SST. The lower mesospheric cooling (1.13 K) on average is also generated by the warmer WTLE, due to increased wave generation and subsequent wave-breaking in the mesosphere. An upper mesospheric warming (4.3 K) is also present in the WTLE compared to the CTLE, but its generation mechanism is currently an open question and under investigation.

Our results show that the majority of the changes in the middle atmospheric circulation appear due to the tidal locking which results in the breakdown of the Brewer-Dobson circulation. The vertical circulation is replaced by a dayside downwelling, centred on the subsolar point and a nightside upwelling, centred on the antisolar point. The tidal locking also alters the horizontal wind at all altitudes, resulting in lower wind speeds, wider and stronger jet streams and the appearance of large scale vortices at various locations compared to the PDE, possibly due to the reduced Coriolis force of the TLE.

A future study would aim to explain the physical processes responsible for the increase of the upper mesospheric temperature and strengthening of the secondary ozone layer accompanying the increase in the underlying SST.

Abbreviations

ATM, atmospheric geophysical model; CESM, community earth system model; CICE, sea-ice data model; CLM, community land model; CPL, coupler of the CESM; docn, prescribed data ocean model; GLC, land-ice geophysical model; ICE, sea-ice geophysical model; LND, land geophysical model; MODIS, moderate-resolution imaging spectroradiometer; NOAA, National oceanic and atmospheric administration; OCN, ocean geophysical model; SZA, solar zenith angle; TLE, tidally locked earth; WTLE, warm tidally locked earth; CTLE, cold tidally locked earth; TOC, total O_3 Content; WACCM, whole atmosphere community climate model; HW, horizontal wind; Ω, vertical wind; E_{kin}, integrated atmospheric kinetic energy; T, temperature; O_3VMR, ozone volume mixing ratio

Acknowledgements

We would like to thank the Center for Space and Habitability (University of Bern) for the PhD fellowship and the Institute of applied Sciences (University of Bern) for the funding that made this study possible. We would like to thank the WACCM forum for the invaluable information it provided. We would also like to thank Ansgar Schanz and Martin Leiner for the computer technical support they provided. Finally, we would like to thank Niklaus Kämpfer and Helmut Lammer for the valuable discussions and advises.

Authors' contributions

EP performed the simulations and the data analysis. KH contributed to the data analysis and the interpretation. PW contributed to the revising of the paper. All authors read and approved the final manuscript.

Competing interests

The authors declare that they have no competing interests.

Author details

[1]Institute of Applied Physics, University of Bern, Sidlerstrasse 5, 3012 Bern, Switzerland. [2]Center for Space and Habitability, University of Bern, Sidlerstrasse 5, 3012 Bern, Switzerland. [3]Oeschger Centre for Climate Change Research, University of Bern, Falkenplatz 16, 3012 Bern, Switzerland. [4]Physics Institute, University of Bern, Sidlerstrasse 5, 3012 Bern, Switzerland.

References

Braesicke P, Pyle JA (2004) Sensitivity of dynamics and ozone to different representations of SSTs in the unified model. Q J R Meteorol Soc 130:2033–2045. doi:10.1256/qj.03.183

Chen G, Plumb RA, Lu J (2010) Sensitivities of zonal mean atmospheric circulation to SST warming in an aquaplanet model. Geophys Res Lett 37:12701. doi:10.1029/2010GL043473

Clough S, Iacono M (1995) Line-by-line calculation of atmospheric fluxes and cooling rates 2. Application to carbon dioxide, ozone, methane, nitrous oxide and the halocarbons. J Geophys Res 100:16519–16535. doi:10.1029/95JD01386

Clough SA, Shephard MW, Mlawer EH, Delamere JS, Iacono MJ, Cady-Pereira K, Boukabara S, Brown PD (2005) Atmospheric radiative transfer modeling: a summary of the AER codes. J Quant Spectrosc Radiat Transfer 91:233–244. doi:10.1016/j.jqsrt.2004.05.058

Davis RN, Du J, Smith AK, Ward WE, Mitchell NJ (2013) The diurnal and semidiurnal tides over Ascension Island (° S, 14° W) and their interaction with the stratospheric quasi-biennial oscillation: studies with meteor radar, eCMAM and WACCM. Atmos Chem Phys 13(18):9543–9564

Deckert R, Dameris M (2008) Higher tropical SSTs strengthen the tropical upwelling via deep convection. Geophys Res Lett 35:10813. doi:10.1029/2008GL033719

Godolt M, Grenfell JL, Hamann-Reinus A, Kitzmann D, Kunze M, Langematz U, von Paris P, Patzer ABC, Rauer H, Stracke B (2015) 3D climate modeling of Earth-like extrasolar planets orbiting different types of host stars. Planetary Space Sci 111:62–76. doi:10.1016/j.pss.2015.03.01

Grenfell JL, Stracke B, von Paris P, Patzer B, Titz R, Segura A, Rauer H (2007) The response of atmospheric chemistry on earthlike planets around F, G and K Stars to small variations in orbital distance. Planetary Space Sci 55:661–671. doi:10.1016/j.pss.2006.09.00

Grenfell JL, Gebauer S, Godolt M, Palczynski K, Rauer H, Stock J, von Paris P, Lehmann R, Selsis F (2013) Potential biosignatures in super-earth atmospheres II. Photochemical responses. Astrobiology 13:415–438. doi:10.1089/ast.2012.0926

Grenfell JL, Gebauer S, Paris P. v, Godolt M, Rauer H (2014) Sensitivity of biosignatures on Earth-like planets orbiting in the habitable zone of cool M-dwarf stars to varying stellar UV radiation and surface biomass emissions. Planetary Space Sci 98:66–76. doi:10.1016/j.pss.2013.10.00

Haberle RM, McKay CP, Tyler D, Reynolds RT, Doyle LR (1996) Circumstellar Habitable Zones. In: Doyle LR (ed). Proceedings of The First International Conference. Travis House Publications, Menlo Park, CA. p 29

Hegyi BM, Deng Y, Black RX, Zhou R (2014) Initial transient response of the winter polar stratospheric vortex to idealized equatorial pacific sea surface temperature anomalies in the ncar waccm. J Clim 27:2699–2713. doi:10.1175/JCLI-D-13-00289.1

Iacono M, Delamere J, Mlawer M, Shephard E, Clough S, Collins W (2008) Radiative forcing by long-lived greenhouse gases: calculations with the AER radiative transfer models. J Geophys Res Atmos:113. doi:10.1029/2008JD009944

Joshi MM, Haberle RM, Reynolds RT (1997) Simulations of the atmospheres of synchronously rotating terrestrial planets orbiting M dwarfs Conditions for atmospheric collapse and the implications for habitability. Icarus 129:450–465

Joshi M (2003) Climate model studies of synchronously rotating planets. Astrobiology 3(2):415–427. doi:10.1089/153110703769016488

Lean J (2000) Evolution of the sun's spectral irradiance since the maunder minimum. Geophys Res Lett 27(16):2425–2428. doi:10.1029/2000GL000043

Lu X, Liu H-L, Liu AZ, Yue J, McInerney JM, Li Z (2012) Momentum budget of the migrating diurnal tide in the Whole Atmosphere Community Climate Model at vernal equinox. J Geophys Res Atmos:17. http://dx.doi.org/10.1002/2013JA019421

Merlis TM, Schneider T (2010) Atmospheric dynamics of Earth-like tidally locked aquaplanets. J Adv Model Earth Syst 2:13. doi:10.3894/JAMES.2010.2.1

Mlawer E, Taubman S, Brown P, Iacono M, Clough S (1997) Radiative transfer for inhomogeneous atmospheres: RRTM, a validated correlated-k model for the longwave. J Geophys Res 102:16663–16682. doi:10.1029/97JD00237

Nappo CJ (2013) An introduction to atmospheric gravity waves. Academic Press

National Oceanic and Atmospheric Administration's Space Environment Center (2016). wikipedia: http://www.swpc.noaa.gov/

Neale RB, Gettelman A, Park S, Chen C, Lauritzen PH, Williamson DK, Conley AJ, Kinnison D, Marsh D, Smith AK, Vitt F, Garcia R, Lamarque JF, Mills M, Tilmes S, Morrison H, Cameron-Smith W, Collins WD, Iacono MT, Easter RC, Liu X, Ghan SJ, Rasch PJ, Taylor MA (2012) Description of the NCAR Community Atmosphere Model (CAM 5.0). CESM database. http://www.cesm.ucar.edu/models/cesm1.0/cam/docs/description/cam5_desc.pdf

Pedatella NM, Liu H-L (2013) Influence of the El Niño Southern Oscillation on the middle and upper atmosphere. J Geophys Res Space Phys 118:2744–2755. http://dx.doi.org/10.1002/2013JA019421

Pedatella NM, Fuller-Rowell T, Wang H, Jin H, Miyoshi Y, Fujiwara H, Shinagawa H, Liu H-L, Sassi F, Schmidt H, Matthias V, Goncharenko L (2014) The neutral dynamics during the 2009 sudden stratosphere warming simulated by different whole atmosphere models. J Geophys Res Space Phys. http://dx.doi.org/10.1002/2013JA019421

Proedrou E, Hocke K (2016) Characterising the three-dimensional ozone distribution of a tidally locked earth-like planet. Earth Planets Space 68(1):1–20. doi:10.1186/s40623-016-0461-x

Rauer H, Gebauer S, Paris PV, Cabrera J, Godolt M, Grenfell JL, Belu A, Selsis F, Hedelt P, Schreier F (2011) Potential biosignatures in super-Earth atmospheres I. Spectral appearance of super-Earths around M dwarfs. Astron Astrophy 529:8. doi:10.1051/0004-6361/20101436

Rosenlof KH, Reid GC (2008) Trends in the temperature and water vapor content of the tropical lower stratosphere: sea surface connection. J Geophys Res 113:06107. doi:10.1029/2007JD009109

Rugheimer S, Kaltenegger L, Zsom A, Segura A, Sasselov D (2013) Spectral fingerprints of Earth-like planets around FGK stars. Astrobiology 13:251–269. doi:10.1089/ast.2012.088

Sander SP, Friedl RR, Ravishankara AR, Golden DM, Kolb CE, Kurylo MJ, Molina MJ, Moortgat GK, Keller-Rudek H, Finlayson-Pitts BJ, Wine PH, Huie RE, Orkin VL (2006) Chemical kinetics and photochemical data for use in atmospheric studies. Evaluation Number 15. National Aeronautics and Space Administration, Jet Propulsion Laboratory, Pasadena, CA

Segura A, Krelove K, Kasting JF, Sommerlatt D, Meadows V, Crisp D, Cohen M, Mlawer E (2003) Ozone concentrations and ultraviolet fluxes on Earth-like planets around other stars. Astrobiology 3:689–708. doi:10.1089/153110703322736024

Segura A, Kasting JF, Meadows V, Cohen M, Scalo J, Crisp D, Butler RAH, Tinetti G (2005) Biosignatures from Earth-like planets around M-dwarfs. Astrobiology 5:706–725. doi:10.1089/ast.2005.5.70

Segura A, Walkowicz LM, Meadows V, Kasting J, Hawley S (2010) The Effect of a strong stellar flare on the atmospheric chemistry of an Earth-like planet orbiting an M Dwarf. Astrobiology 10:751–776. doi:10.1089/ast.2009.037

Selsis F (2000) Darwin and Astronomy the Infrared Space Interferometer. In: Schü"rmann B (ed). ESA Special Publication, vol 451. p 133

Smith KL, Polvani LM, Marsh DR (2012) Mitigation of 21st century Antarctic sea ice loss by stratospheric ozone recovery. Geophys Res Lett 39(20):L20701. doi:10.1029/2012GL053325, http://dx.doi.org/10.1029/2012GL053325

Solomon S, Qiang L (2005) Solar extreme-ultraviolet irradiance for general circulation models. J Geophys Res 110:10306. doi:10.1029/2005JA011160

Tan B, Chu X, Liu H-L, Yamashita C, Russell JM (2012a) Atmospheric semidiurnal lunar tide climatology simulated by the whole atmosphere community climate model. J Geophys Res 117:1–11. doi:10.1029/2012JA017792

Tan, B, Chu, X, Liu H-L, Yamashita C, Russell JM (2012b) Zonal-mean global teleconnection from 15 to 110 km derived from SABER and waccm. J Geophys Res Atmos 117:1–11. doi:10.1029/2011JD016750

Tan B, Chu X, Liu H-L, Yamashita C, Russell JM (2012c) Parameterization of the inertial gravity waves and generation of the quasi-biennial oscillation. J Geophys Res Atmos:117. doi:10.1029/2011JD016778

Tarter JC, Backus PR, Mancinelli RL, Aurnou JM, Backman DE, Basri GS, Boss AP, Clarke A, Deming D, Doyle LR, Feigelson ED, Freund F, Grinspoon DH, Haberle RM, Hauck SAI, Heath MJ, Henry TJ, Hollingsworth JL, Joshi MM, Kilston S, Liu MC, Meikle E, Reid IN, Rothschild LJ, Scalo J, Segura A, Tang CM, Tiedje JM, Turnbull MC, Walkowicz LM, Weber AL, Young RE (2007) A reappraisal of the habitability of planets around M dwarf stars. Astrobiology 7:30–65. doi:10.1089/ast.2006.012

Vertenstein M, Craig T, Middleton A, Feddema D, Fisher C (2012). CESM 1.0.4 user's guide. available from: http://www.cesm.ucar.edu/models/cesm1.0/cesm/cesm_doc_1_0_4/book1.html

Wang Y-M, Lean JL, Sheeley NRJ (2005) Modeling the sun's magnetic field and irradiance since 1713. Astrophys J 625(1):522

Wikipedia U.S Standard Atmosphere (2016). https://en.wikipedia.org/wiki/U.S._Standard_Atmosphere

Yang J, Cowan NB, Abbot DS (2013) Stabilizing cloud feedback dramatically expands the habitable zone of tidally locked planets. Astrophys J Lett 771:45. doi:10.1088/2041-8205/771/2/L4

5

Feasibility of a magma ocean dynamo on Mars

George Helffrich

Abstract

Crustal magnetization of rocks in regions of Mars surface testifies to an era of dynamo activity. I examine the possibility that the dynamo that operated then, in the first 400–600 Ma after Mars formed, was powered by a crystallizing subsurface magma ocean. Of the ways that a magma ocean dynamo could operate, on Mars only turbulent and magnetostrophic dynamos seem feasible; geostrophic dynamos do not seem so unless very high heat flows, 100–1000 times present, are invoked. Given the anticipated information from the future InSight lander mission, it will be difficult to assess where the dynamo originated unless an inner core is discovered, rendering the dynamo likely to have operated in the core.

Keywords: Mars, Dynamo, InSight, Magma ocean

Introduction

Mars' era of dynamo activity was discovered when the Mars Global Surveyor (MGS) spacecraft dropped to a low orbit during an aerobraking manoeuver and recorded quite high levels of crustal magnetization (Acuña et al. 1999). The heavily cratered, older southern highlands exhibit the strongest signals, restricted to the crust; there is no presently operating dynamo due to the lack of a planetary field detectable from high orbit. The much lower levels of crustal magnetism in the younger, northern planetary hemisphere (100 times smaller according to Whaler and Purucker (2005)), led the discoverers to infer that a dynamo that operated early in Mars' history up until the early Noachian epoch (for $\sim$0.5 Gyr). Some debate on the validity of this inference ensued until the magnetization of individual crystals in SNC meteorites was measured and the same crystals were dated, showing magnetic ages of 3.9–4.1Ga (Weiss et al. 2002). More recent estimates based on crater counts overlap with the age range 4.1 $\pm$ 0.2 Ga (Chassefière et al. 2007) further supporting this conclusion.

Dynamos require a power source to operate. They are driven by fluid motions in conductive materials, commonly assumed to be the planet's core. If there is no motion, the field will decay on a time scale given by the magnetic diffusivity. For the Earth, this decay time is around 10,000 years (Stevenson 2003) and, given Mars' smaller core size, corresponding shorter by the square of the radius ratio, or 2500 years.

There are three viable sources for powering dynamos. The first is heat from cooling of the core or a thermal dynamo. This may be accretion heat (essentially released gravitational potential energy) or radiogenic heating of the core itself. Most radiogenic elements with long lifetimes (U, Th, K) are lithophile (McDonough and Sun 1995) and do not significantly concentrate themselves in the core. Thus radioactivity is probably not a significant factor in powering thermal dynamos.

The second is by crystallization of a solid inner core out of the liquid outer core, a significant contribution to the power for the Earth's dynamo now (Labrosse 2003). The crystallization not only releases latent heat, it also enriches the alloying elements in the core liquid by excluding them from the solid. The exclusion makes the remaining core liquid less dense and thus causes it to rise buoyantly from the crystallizing region, essentially recovering gravitational potential energy. This fluid motion is able to drive the dynamo quite efficiently and for a long time. For example, scaling laws for the growth of the Earth's inner core (Breuer et al. 2010) suggest that it requires another 10–20 Gyr to crystallize the whole core and thus extinguish dynamo activity. This dynamo may be

Correspondence: george@elsi.jp
Earth-Life Science Institute, Tokyo Institute of Technology, 2-12-1-IE-1 Ookayama, Meguro-ku, Tokyo 152-8550, Japan

described as a compositional type because it is the changing composition of the crystallizing fluid that primarily contributes to its running.

The final way to drive a dynamo is by precession of the solid body surrounding the core forced by tidal coupling to the planet's rotational bulge. Though this seems energetically efficient, conditions for its operation have not been explored as thoroughly as the other two (Wu and Roberts 2009). In the case of Mars, with no present-day field, and yet tidal interactions with its moons Diemos and Phobos, it clearly does not operate now and, by inference, probably never did.

Now, consider what type of dynamo Mars' early one might have been. If it was compositional, then it should still be running now, or Mars' core must be solid. The latter possibility is ruled out by the elastic tidal Love number k_2, whose value is 0.159 ± 0.009, which indicates that the core is at least partly liquid (Konopliv et al. 2006). It is possible that the core is not completely frozen, and the annulus of remaining liquid is too thin to support dynamo activity (Stevenson 2001). This state must have been achieved by ~500 Myr, which is about the limit age of magnetized areas of the Martian crust. Again, using the scaling laws for inner core growth (Breuer et al. 2010), the growth time requires that the inner core nucleated quite early in Mars' accretion history; 100 year after core formation. Given that accretion times are on the order of 10 Myr (Brasser 2013), the scaling places the inner core's (and thus its magnetic field's) onset time before accretion's end and requires inner core growth during accretion and during growth of the core itself. If accretion is violently cataclysmic, physical disruption of the core renders this scenario implausible.

It is also possible that compositional dynamos might be run by crystallization elsewhere in the core rather than its center. Breuer et al. (2010), summarizing earlier suggestions (Stevenson 1983), note that saturation of metal in light elements might occur at the lower pressures of a magma ocean that might later exsolve from the metal due to oversaturation at higher pressures after the metal segregates to form the core. Flow of exsolved material upwards in the core could drive a dynamo, which Hirose et al. (2017) show that it is quite efficient and possibly operates the present-day Earth's. No computationally simulated dynamos have yet been constructed to operate in this regime, though Hori et al. (2012) note that volumetric buoyancy sources, which could characterize exsolution-driven dynamos, yield more multipolar fields and hence less spatial regularity to the field, perhaps mimicking Mars' surface magnetism. If, however, the Earth's dynamo operates this way (Hirose et al. 2017), its strong dipolarity contradicts the volumetric source field morphology inference.

Consequently, thermal dynamos seem apposite for Mars. Thermal dynamos have a characteristic property that if conditions for running them become unfavorable anywhere in a quasi-adiabatic, convecting core, they are unfavorable everywhere in the quasi-adiabatic region (Stevenson 2001). Thus, thermal dynamos essentially are either on or off, controlled by the flux of heating down the adiabatic gradient to the surface of the core from its interior. This provides a way to end dynamo activity simply by regulating the heat drawn from the core into the mantle. If Mars had an early plate tectonic era (Anguita et al. 2006) that evolved into a stagnant lid form of convection (O'Neill et al. 2007), heat flowing out of the core would wane with the change in tectonic style.

A few Mars development scenarios are based on this phenomenon. Williams and Nimmo (2004) suggest that Mars' core formed quite rapidly and was ~150 K hotter than the base of the mantle. With that extra heat, a dynamo could operate by simple core cooling by convective withdrawal of heat by the mantle from the core-mantle boundary, with the mantle operating in a stagnant lid regime (i.e., no plate tectonics). After ~0.5 Gyr, the core would be cooled sufficiently that dynamo action would stop. Even added radiogenic ^{40}K in the core is not able to provide enough extra heat early in Mars' history to run a dynamo. In their view, Mars' core is still entirely liquid.

Reese and Solomatov (2010) invoke a giant impact origin for the Martian dynamo. The impact causes superheating of Mars' core and thus is a more elaborate form of the previous model. They simulated impact of a planetesimal with Mars and the transfer of the impact-derived heat of the differentiated impactor's metal to the previously formed core, stratifying the hotter metal onto the core. This model attributes the southern highland crust to new crust formed after recrystallization after the impact, inverting the usual relative age relations obtained from crater counts (Frey 2006a, b). One further insight that this study provides is that the core may develop in a state where its exterior is hotter than its interior due to the heat gained by the accumulated metal at the base of a magma ocean descending in the gravitational field of the planet. This temperature profile in the core suppresses convection and thus any dynamo activity. Hence, one also needs to examine whether dynamo activity is possible in the mantle while it is still wholly or partly liquid in the magma ocean phase of development, one of the motivations for this work.

Elkins-Tanton et al. (2005) use the evolution of convection in the early Mars mantle to cause its core to be thermally perturbed into convective motion. The progressive crystallization of an early magma ocean led to the enrichment of denser, iron-rich material at its top in its final stages. This is a convectively unstable configuration that eventually overturns, bringing cold material into contact with the core and starting convection in it due to high heat flow across the core-mantle boundary. The convection is transient, lasting 15–150 Myr, ending because the mantle warms and evolves to an adiabatic state that no longer draws sufficient heat from the core to cool it and run its dynamo.

A final intriguing possibility is that the core was not involved in early dynamo activity on Mars at all. Rather, the dynamo was in the mantle as it cooled through the magma ocean phase after having developed an early crust (Stevenson 2001). Stegman et al. (2003) investigated this possibility for running a lunar dynamo, and Ziegler and Stegman (2013) similarly sought an explanation for Earth's early magnetic field by this mechanism. Here, we apply the same methods to investigate whether a mantle dynamo is feasible for Mars.

Planetary dynamo styles

Dynamos arise from the interplay of convection in a conductive material and electromotive forces arising from the convective motions that generally act to resist motion. They can be characterized by the dominant balance of forces that maintain steady state convection—steady in the sense that though the motion may reorganize itself and adopt different flow patterns, it does not stop. From them, governing relations for characteristic convective speeds may be obtained (Christensen 2010). A brief introduction to the styles arising in magma ocean dynamos follows.

In the style of magma ocean envisaged by Solomatov (2007), temperatures are high, viscosities low, and motion rapid. This leads to a highly turbulent flow regime which is indifferent to the planet's rotational state. Under these conditions, the buoyancy forces due to rising plumes of hot material and sinking plumes of cold material of possibly different composition (an additional source of buoyancy) are resisted by inertia of the fluid itself: packets of fluid need to move out of the way to allow the buoyant bits to flow (Starchenko and Jones 2002). In contrast, Starchenko and Jones (2002) think this force balance to be unrealistic given that rapid flows lead to correspondingly large magnetic forces that resist the flow and create electrical currents that dissipate energy through the electrical resistance of the fluid. They infer that the Lorenz force (the combination of electrical attraction/repulsion and an electrical charge's interaction with a magnetic field) must play a leading role in resisting motion. Moreover, the effects of rotation (giving rise to the Coriolis force) and pressure buildup in a fluid due to its being influenced by the Coriolis force also serve to counteract buoyancy-driven fluid movement. This leads to what they call a MAC force balance (Magnetic-Archimedian (buoyancy)-Coriolis).

A third convective style was motivated from observations of the patterns of convection in a rotating spherical container conducted by Aubert et al. (2001). They found that two main factors cooperated to balance the buoyancy forces: the fluid's inertia and the Coriolis force. This leads to a flow pattern organized into rotating columns that run parallel to the rotation axis through the height of the container. They call this a CIA force balance (Coriolis-Inertia-Archimedian (buoyancy)). In contrast to the previous style, here, the Lorenz force does not play a role.

Which of these convective styles characterizes an early Mars magma ocean? It is not at all clear. Hence the implications of each will be assessed, and their likelihood for driving an early Mars dynamo will be evaluated.

Methods/Experimental

The controlling factor for a dynamo is its magnetic Reynolds number, R_m. It is defined through

$$R_m = vL\mu_0\sigma \quad , \tag{1}$$

where μ_0 is a physical constant, σ is electrical conductivity, L is the length scale of the fluid flow (the mantle depth), and v is a characteristic fluid velocity. We need R_m to be greater than ~40 for dynamo activity to ensue (Olson and Christensen 2006). Silicate melt conductivity data is somewhat sparse, so I use 1 atm values for basaltic melt from Pommier et al. (2010) evaluated at a temperature of 1700 K, which, according to Sohl and Spohn (1997), is a characteristic temperature in Mars' mid-mantle. This indicates a value for σ of 8.9–22.9 S m^{-1}. A characteristic speed is harder to specify because it depends on the governing process. The three processes identified by Ziegler and Stegman (2013) operating in rapidly rotating fluid systems are turbulence, the buoyancy-Lorenz force (magnetostrophic) balance, and the buoyancy-inertia-coriolis (geostrophic) force balance. These are expressed as follows:

$$\begin{aligned} v_T &= 14\left[\frac{Q_m \alpha g}{\rho C_P}\,\frac{1}{\Omega}\right]^{1/2} && \text{turbulence} \\ v_{\mathrm{MAC}} &= \left[\frac{Q_m \alpha g}{\rho C_P}\,\frac{1}{\Omega}\right]^{1/2} && \text{buoyancy} - \text{Lorenz} \\ v_{\mathrm{CIA}} &= \left[\frac{Q_m \alpha g}{\rho C_P}\right]^{2/5}\left[\frac{L}{\Omega}\right]^{1/5} && \text{buoyancy} - \text{inertia} - \text{coriolis} \end{aligned} \tag{2}$$

These in turn depend on Q_m, the heat flux through the convective region, the rotation rate Ω (2π/day length in seconds), and other properties listed in Table 1. I evaluate R_m for the various possibilities using Eqs. (1) and (2).

Table 1 Thermophysical properties for Mars mantle relevant to dynamo activity

Property	Symbol	Value	Scale and dimensions
Magnetic permeability	μ_0	4π	$\times 10^{-7}$ H m^{-1} (definition)
Thermal expansivity	α	2.3	$\times 10^{-5}$ K^{-1} [a]
Heat capacity	C_P	1200	J kg^{-1} K^{-1} [b]
Gravity	g	3	m s^{-2} [a]
Density	ρ	4000	kg m^{-3} [a]
Rotation rate	Ω	71	$\times 10^{-6}$s^{-1} [c]
Heat flux	Q_m	519	GW [a,c]

[a] Sohl and Spohn (1997) evaluated in the transition zone
[b] Nimmo and Stevenson (2000)
[c] Present day value

Results and Discussion

The sensitivity to these parameters is shown in Fig. 1. It shows that turbulent flow speeds (ν_T) and magnetostrophic flow speeds (ν_{MAC}) are capable of operating a dynamo in Mars' mantle ($R_m > 40$), provided that it is liquid. It appears more difficult to do this with geostrophic flows (ν_{CIA}) because the required heat flows are too large: 100–1000 times the present-day estimates. Convection models for the evolution of heat flow of Mars (Breuer et al. 2010) show that heat flows high enough to run a magnetostrophic mantle dynamo existed between 0–1 Gyr after accretion ended, which would provide the required heat flow increase in the time frame for the dynamo's existence. It therefore seems possible that the earliest-surviving crust in the magma ocean phase of planetary evolution could record a magnetic field generated by a mantle dynamo, independent of whether a core dynamo was also operating. As the magma ocean internally crystallized, its depth L lessens, eventually dropping R_m below the critical value (Eq. (1)). Thus, the last formed magmatic output would be emplaced without any internal field.

Takahashi and Matsushima (2006) showed that some thin annular dynamos are non-dipolar, in contrast to thicker shelled dynamos that, except where driven by internal heat generation or are at high Rossby number $Ro = \nu(\Omega L)^{-1}$, are dipolar (Olson and Christensen 2006). The magma ocean dynamos considered here have $Ro \leq 1 \times 10^{-5}$, decidedly low, so their behavior will be controlled by their annular depth and will be non-dipolar. Consequently, as the magma ocean crystallizes, any dynamo operating in it will become weaker and multipolar. This may account for the lack of coherently magnetized domains in the Martian highlands Connernery et al. (1999). The Martian dynamo may have been reversing, an inference from the initial discovery of the pattern of crustal magnetization (Connerney et al. 1999) that showed stripes of alternately polarized magnetic domains evocative of seafloor magnetism on the Earth. This concept involves crustal age variations recording epochs of different polarity, whose analogous mechanism on Mars is not obvious. Rather, later interpretations invoke non-specific mechanisms involving crustal recycling to explain the polarity variation (Whaler and Purucker 2005), yet they still assume age differences for differently magnetized areas of the crust. Cratering records seem to bear this out (Chassefière et al. 2007), as they do for the weak-to-absent crustal magnetization of the Northern hemispheric dichotomy and the Argyre and Hellas impact basins (Acuña et al. 1999; Whaler and Purucker 2005).

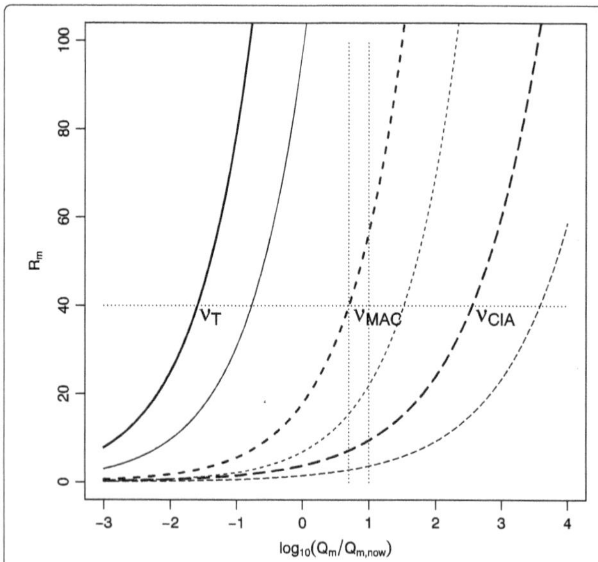

Fig. 1 Feasibility of running a dynamo in a liquid silicate mantle in Mars as a function of mid-mantle heat flux Q_m. The heat flux is scaled relative to the present-day heat flux $Q_{m,now}$ as estimated by Sohl and Spohn (1997). ν_T, ν_{MAC}, and ν_{CIA} show three families of potential dynamos driven by different types of flows (see text). *Thick and thin lines* of each pair denote low and high estimates of silicate electrical conductivity, σ (8.9–22.9 S m^{-1}). R_m values greater than 40 (*horizontal dotted line*) indicate feasible range for dipolar dynamos. Turbulent flow speeds are capable of running dynamos at any mantle heat flow provided that the mantle is liquid. Magnetostropic dynamos are also feasible in the high heat flux epoch early in Mars evolution (*vertical dotted lines*; 0–1 Gyr according to Breuer et al. (2010)). Geostrophic dynamos do not seem feasible under any realistic heat flow scenario

Nimmo (2000) proposes that the origin of the magnetic lineations is dike emplacement in the crust, a process that requires crustal extension and moving intrusion loci in space and time. No matter what the dynamo origin, the crustal magnetization pattern apparently necessitates contemporaneous tectonism, which the geological evidence elsewhere supports (Anguita et al. 2006).

Conclusions

In the near future, after the InSight lander mission (NASA 2013) places a seismometer on Mars and records sufficient seismic activity around and within the planet (Taylor et al. 2013), it should be feasible to use the resulting data to draw stronger inferences about Mars' composition and evolution. First-order constraints on the planetary Fe/Si ratio (Rivoldini et al. 2011; Sohl and Spohn 1997) will be provided by the radius of Mars' core, which will be evident from core-reflected P and S (PcP and ScS) and core penetrating (SKS, SKKS) travel times. The next most important seismic observation relevant for dynamo activity will be the presence or absence of an inner core. If an inner core is detected, the dynamo had a compositional power source available and the straightforward inference is to assume that the dynamo originated in the core. Absence of an inner core will still not discriminate between a mantle and a core dynamo, nor whether any core dynamo was thermal or compositional, unless there is some seismic evidence of reaction between an exsolved core component and the mantle or in the core's wave speed structure itself.

Acknowledgements
I thank John Hernlund and Hiroaki Matsui for discussions on the subject of dynamos and James Dohm for insights into Martian geology.

Funding
This work was not funded by any agency whatsoever; the work arose from the circumstances of day-to-day activities at my home institute.

Authors' contributions
All authors read and approved the final manuscript.

Competing interests
The author declares that he has no competing interests.

References
Acuña MH, Connerney JEP, Ness NF, Lin RP, Mitchell D, Carlson CW, McFadden J, Anderson KA, Rème H, Mazelle C, Vignes D, Wasiliewski P, Cloutier P (1999) Global distribution of crustal magnetization discovered by the Mars Global Surveyor MAG/ER experiment. Science 284:790–793

Anguita F, Fernandez C, Cordero G, Carrasquilla S, Anguita J, Nunez A, Rodriguez S, Garcia J (2006) Evidences for a Noachian-Hesperian orogeny in Mars. Icarus 185:331–357

Aubert J, Brito D, Nataf H-C, Cardin P, Masson J-P (2001) A systematic experimental study of rapidly rotating spherical convection in water and liquid gallium. Phys Earth Planet Inter 128:51–74

Brasser R (2013) The formation of Mars: building blocks and accretion time scale. Space Sci Rev 174:11–25

Breuer D, Labrosse S, Spohn T (2010) Thermal evolution and magnetic field generation in terrestrial planets and satellites. Space Sci Rev 152:449–500

Chassefière E, Leblanc F, Langlais B (2007) The combined effects of escape and magnetic field histories at Mars. Planet Space Sci 55:343–357

Christensen UR (2010) Dynamo scaling laws and applications to the planets. Space Sci Rev 152:565–590

Connerney JEP, Acuña MH, Wasiliewski P, Ness NF, Rème H, Mazelle C, Vignes D, Lin RP, Mitchell D, Cloutier P (1999) Magnetic lineations in the ancient crust of Mars. Science 284:794–798

Elkins-Tanton LT, Zaranek SE, Parmentier EM, Hess PC (2005) Early magnetic field and magmatic activity on Mars from magma ocean cumulate overturn. Earth Planet Sci Lett 236:1–12

Frey HV (2006a) Impact constraints on, and a chronology for, major events in early Mars history. J Geophys Res 111:10–10292005002449

Frey HV (2006b) Impact constraints on the age and origin of the lowlands of Mars. Geophys Res Lett 33:10–10292005024484

Hirose K, Morard G, Sinmyo R, Umemoto K, Hernlund J, Helffrich G, Labrosse S (2017) SiO_2 crystallization and compositional evolution of the Earth's core. Nature 543:99–102

Hori K, Wicht J, Christensen UR (2012) The influence of thermo-compositional boundary conditions on convection and dynamos in a rotating spherical shell. Phys Earth Planet Inter 196:32–48

Konopliv AS, Yoder CF, Standish EM, Yuan D-N, Sjogren WL (2006) A global solution for the Mars static and seasonal gravity, Mars orientation, Phobos and Deimos masses, and Mars ephemeris. Icarus 182:23–50

Labrosse S (2003) Thermal and magnetic evolution of the Earth's core. Phys Earth Planet Inter 140:127–143

McDonough WF, Sun S-s (1995) The composition of the Earth. Chem Geol 120:223–253

NASA NASA (2013) NASAfacts - InSight. (Publ. no. JPL 400-1513, Rev. 2 6/13)

Nimmo F (2000) Dike intrusion as a possible cause of linear Martian magnetic anomalies. Geology 28:391–394

Nimmo F, Stevenson DJ (2000) Influence of early plate tectonics on the thermal evolution and magnetic field of Mars. J Geophys Res 105:11969–11979

Olson P, Christensen U. R (2006) Dipole moment scaling for convection-driven planetary dynamos. Earth Planet Sci Lett 250:561–571

O'Neill C, Jellinek AM, Lenardic A (2007) Conditions for the onset of plate tectonics on terrestrial planets and moons. Earth Planet Sci Lett 261:20–32

Pommier A, Gaillard F, Malki M, Pichavant M (2010) Methodological re-evaluation of the electrical conductivity of silicate melts. Am Mineral 95:284–291

Reese CC, Solomatov VS (2010) Early martian dynamo generation due to giant impacts. Icarus 207:82–97

Rivoldini A, Hoolst TV, Verhoeven O, Mocquet A, Dehant V (2011) Geodesy constraints on the interior structure and composition of Mars. Icarus 213:451–472

Sohl F, Spohn T (1997) The interior structure of Mars: implications from SNC meteorites. J Geophys Res 102:1613–1635

Solomatov V (2007) Magma oceans and primordial mantle differentiation. In: Schubert G (ed). Treatise on Geophysics (2nd Ed.) Vol. 9. pp 81–104

Starchenko SV, Jones CA (2002) Typical velocities and magnetic field strengths in planetary interiors. Icarus 157:426–435

Stegman DR, Jellinek AM, Zatman SA, Baumgardner JR, Richards MA (2003) An early lunar dynamo driven by thermochemical mantle convection. Nature 421:143–146

Stevenson D (1983) Planetary magnetic fields. Rep Prog Phys 46:555–620

Stevenson D (2001) Mars' core and magnetism. Nature 412:214–219

Stevenson D (2003) Planetary magnetic fields. Earth Planet Sci Lett 208:1–11

Takahashi F, Matsushima M (2006) Dipolar and non-dipolar dynamos in a thin shell geometry with implications for the magnetic field of Mercury. Geophys Res Lett 33:10–10292006025792

Taylor J, Teanby NA, Wookey J (2013) Estimates of seismic activity in the Cerberus Fossae region of Mars. J Geophys Res 118:2570–2581

Weiss BP, Vali H, Baudenbacher FJ, Kirschvink JL, Stewart ST, Shuster DL (2002) Records of an ancient Martian magnetic field in ALH84001. Earth Planet Sci Lett 201:449–463

Whaler KA, Purucker ME (2005) A spatially continuous magnetization model for Mars. J Geophys Res 110:10–10292004002393

Williams J-P, Nimmo F (2004) Thermal evolution of the Martian core: implications for an early dynamo. Geology 23:97–100

Wu C-C, Roberts PH (2009) On a dynamo driven by topographic precession. Geophys Astrophys Fluid Dyn 103:467–501

Ziegler LB, Stegman DR (2013) Implications of a long-lived basal magma ocean in generating Earth's ancient magnetic field. Geochem Geophys Geosyst 14:4735–4742

6

Towards more realistic core-mantle boundary heat flux patterns: a source of diversity in planetary dynamos

Hagay Amit[1*], Gaël Choblet[1], Peter Olson[2], Julien Monteux[3], Frédéric Deschamps[4], Benoit Langlais[1] and Gabriel Tobie[1]

Abstract

Mantle control on planetary dynamos is often studied by imposing heterogeneous core-mantle boundary (CMB) heat flux patterns on the outer boundary of numerical dynamo simulations. These patterns typically enter two main categories: Either they are proportional to seismic tomography models of Earth's lowermost mantle to simulate realistic conditions, or they are represented by single spherical harmonics for fundamental physical understanding. However, in reality the dynamics in the lower mantle is much more complicated and these CMB heat flux models are most likely oversimplified. Here we term *alternative* any CMB heat flux pattern imposed on numerical dynamos that does not fall into these two categories, and instead attempts to account for additional complexity in the lower mantle. We review papers that attempted to explain various dynamo-related observations by imposing alternative CMB heat flux patterns on their dynamo models. For present-day Earth, the alternative patterns reflect non-thermal contributions to seismic anomalies or sharp features not resolved by global tomography models. Time-dependent mantle convection is invoked for capturing past conditions on Earth's CMB. For Mars, alternative patterns account for localized heating by a giant impact or a mantle plume. Recovered geodynamo-related observations include persistent morphological features of present-day core convection and the geomagnetic field as well as the variability in the geomagnetic reversal frequency over the past several hundred Myr. On Mars the models aim at explaining the demise of the paleodynamo or the hemispheric crustal magnetic dichotomy. We report the main results of these studies, discuss their geophysical implications, and speculate on some future prospects.

Keywords: Magnetic field; Dynamo; Core-mantle boundary; Heat flux

Introduction

General

Various geophysical observations have been recovered by imposing heterogeneous heat flux patterns on the outer boundary of numerical dynamo simulations. For the Earth, the most popular prescribed heat flux patterns are either proportional to seismic velocity anomalies obtained from tomography models of the lowermost mantle to mimic realistic conditions, or spherical harmonic degree and order 2 which is the dominant term in these tomography models (Aubert et al. 2007, 2008; Bloxham 2002; Christensen and Olson 2003; Davies et al. 2008; Glatzmaier et al. 1999; Gubbins et al. 2007, 2011; Heimpel and Evans 2013; Kutzner and Christensen 2004; Olson and Amit 2014; Olson and Christensen 2002; Olson et al. 2010; Sreenivasan 2009; Sreenivasan and Gubbins 2011; Takahashi et al. 2008). For other planets, various single harmonics are often used, for example degree-1 for Mars (Amit et al. 2011a; Dietrich and Wicht 2013; Stanley et al. 2008) and degree-2 for Mercury and Saturn (Cao et al. 2014; Stanley 2010, respectively), reflecting the low-resolution information on planetary deep interiors. Various single harmonics were imposed for the geodynamo as well in order to obtain fundamental physical understanding of the complex dynamo simulations (e.g., Glatzmaier et al. 1999).

*Correspondence: Hagay.Amit@univ-nantes.fr
[1]CNRS UMR 6112, Université de Nantes, Laboratoire de Planétologie et de Géodynamique, 2 rue de la Houssinière, Nantes, F-44000, France
Full list of author information is available at the end of the article

Inferring core-mantle boundary (CMB) heat flux from tomography models relies on the assumption that seismic and thermal anomalies in the lowermost mantle are highly correlated. However, this idealized linear relation may not hold in reality. Observational and numerical studies of the lower mantle suggest that the structures and dynamics there are far more complex. First, non-thermal contributions (e.g., compositional, phase changes) perturb the idealized linear seismic-thermal mapping (Bull et al. 2009; Deschamps et al. 2012; Lay et al. 2008; Nakagawa and Tackley 2008; Tackley 2011; Trampert et al. 2004). Second, while the spatial resolution inherent to tomographic models in the lowermost mantle (e.g., Lekic et al. 2012) exceeds the resolution of the observed core field, strong lateral gradients of shear wave velocities are not captured. As revealed by detailed waveform analyses at several locations, boundaries of large low shear-wave velocity provinces (LLSVPs) might however be very sharp (e.g., Ni et al. 2005; To et al. 2005). Similarly, synthetic single harmonic patterns for planetary mantle heterogeneity, for example for the CMB conditions that prevailed during the paleo dynamo of Mars, are oversimplified. Even if the Martian anomaly was indeed large-scale (Elkins-Tanton et al. 2005; Harder and Christensen 1996; Ke and Solomatov 2006; Roberts and Zhong 2006; Roberts et al. 2009), it was likely more complex than a single harmonic pattern. Note finally that simulating the geodynamo at earlier times requires information about the time-dependent mantle convection pattern, which is not witnessed by present-day tomography (Yoshida and Santosh 2011; Zhang and Zhong 2011; Zhang et al. 2010; Zhong et al. 2007).

Here we term any CMB heat flux pattern that is neither tomographic nor single harmonic as *alternative*. For the Earth, the alternative patterns include those that account for non-thermal effects or non-resolved small scales in various ways (Amit and Choblet 2009, 2012; Amit et al. in press) and those that model the mantle heterogeneity in the past hundreds of Myr in various ways (Amit and Olson 2015; Olson et al. 2013). For Mars, the alternative patterns are obtained by modeling localized mantle heating induced by a mantle plume (Sreenivasan and Jellinek 2012) or a giant impact (Kuang et al. 2014; Monteux et al. 2015).

Figure 1a illustrates some proposed dynamical scenarios of Earth's lower mantle that result in alternative CMB heat flux patterns (see also, e.g., Lay et al. 2008). Subducting slabs bring cold material to the CMB and locally increase the heat flux. Time-dependent plate motions therefore change the location of these large CMB heat flux regions. Further complications arise from the presence of post-perovskite phase transition (or even its double crossing) in the slab reservoirs of the lowermost mantle. LLSVPs may represent thermochemical piles of a not only hot but also compositionally different material (e.g., Trampert et al. 2004). Growth and collapse of these thermochemical piles, as observed in analog experiments (Davaille 1999), may yield temporal changes in the total heat loss through the CMB and the amplitude of the lateral heterogeneity. Ultra low velocity zones (ULVZs) at the edges of LLSVPs (McNamara et al. 2010) form hot regions that are too narrow to be captured by global large-scale tomography models.

Figure 1b illustrates the effects of localized mantle heating on Mars' CMB heat flux. Giant impacts heat the mantle and reduce the CMB heat flux right below the impact site. Likewise, mantle plumes that originate at the lowermost mantle and may have produced extensive volcanism such as Tharsis are also potential zones of reduced CMB heat flux. It may also be naively expected that core heating by impacts would yield a CMB heat flux increase. However, core impact heating leads to the formation of a hot layer at the top of the core (within 10 kyr) which prevents the core from cooling (Arkani-Hamed 2012; Arkani-Hamed and Olson 2010; Roberts and Arkani-Hamed 2014). Thus, core impact heating could inhibit the dynamo generation during a timescale that is governed by the efficiency of the surrounding mantle to extract the impact induced core thermal anomaly (from several kyr up to 100 Myr).

Numerical simulations of mantle convection may shed light on the relation between the observed seismic anomalies and the CMB heat flux heterogeneity. Nakagawa and Tackley (2008) assumed a linear equation of state to relate their models' thermal, mineral, and chemical anomalies with seismic anomalies. Their purely thermal case gives nearly perfect linear relation between seismic and thermal anomalies, whereas adding post-perovskite effects breaks the linearity but maintains a unique relation (Fig. 2a). The situation is more complicated when chemical anomalies are accounted for; when the density contrast is large enough, a parallel branch appears and the thermal-seismic relation becomes non-unique (Fig. 2b). Implementation of the results of Nakagawa and Tackley (2008) might be problematic because due to limited vertical resolution global tomographic models do not sample precisely the CMB but rather a vertically averaged region above it.

The outline of this review paper is as follows. In the "Alternative models of core-mantle boundary heat flux" subsection, we introduce motivations for alternative CMB heat flux models. Related dynamo properties inferred from observations are listed in subsection "Relevant observed planetary dynamo properties". In the section "Methods", we recall the governing equations and control parameters of numerical dynamo simulations. We describe the incorporation of mantle control effects by imposing an outer thermal boundary condition. In the section "Review", we describe the results of papers that

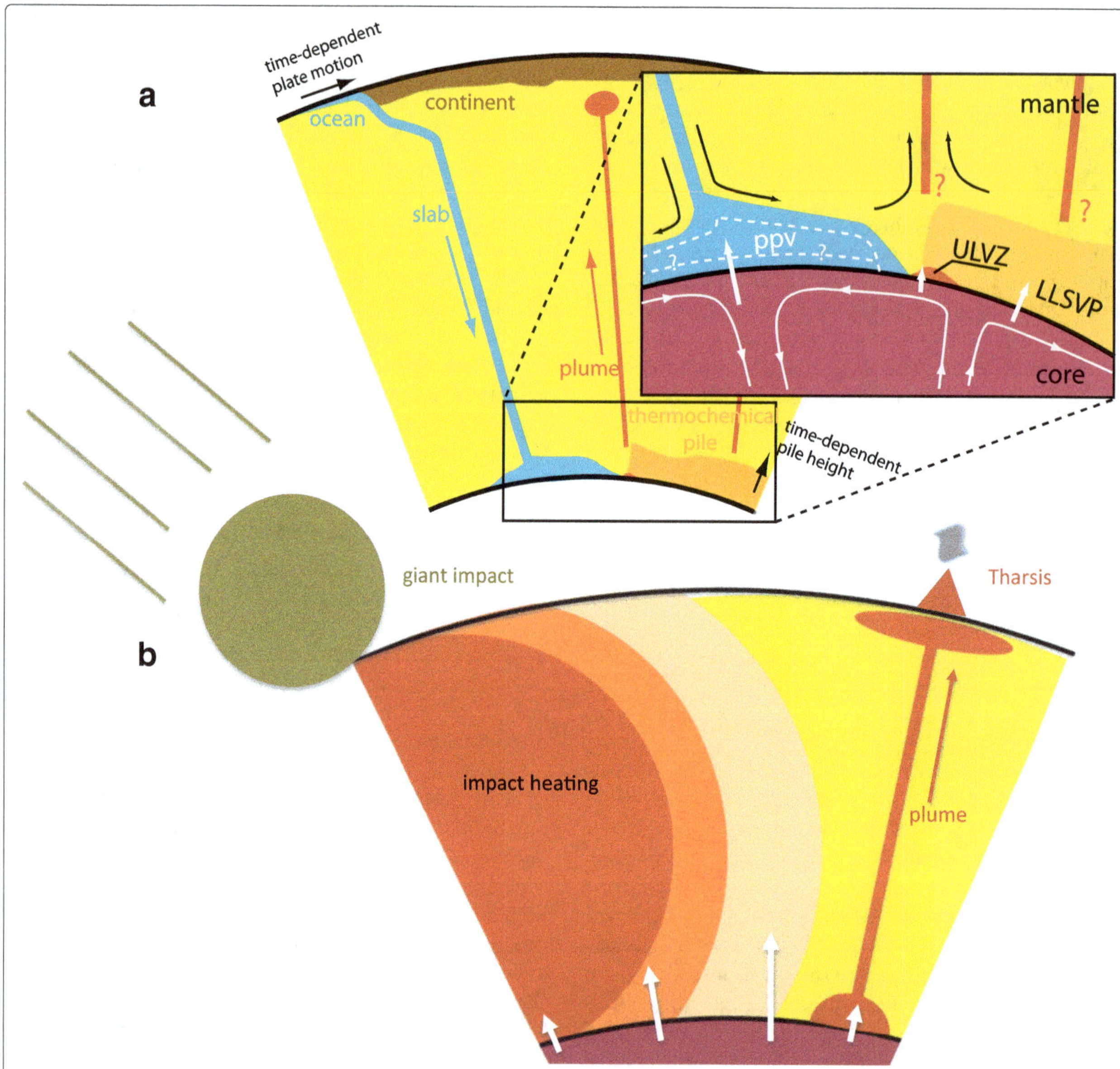

Fig. 1 Cartoons showing various possible dynamical complexities in the Earth's **a** and Mars' **b** lowermost mantles that may motivate modeling of alternative CMB heat flux models

imposed alternative outer boundary conditions on numerical dynamos. For each of these papers, we specify the procedure used to construct the alternative CMB heat flux model and we report how the perturbation from the conventional reference pattern affected the resulting planetary dynamo features. We discuss the results in the subsection "Discussion", and we propose some future prospects in subsection "Future prospects".

Alternative models of core-mantle boundary heat flux

The existence of the post-perovskite phase transition (Murakami et al. 2004; Oganov and Ono 2004) was invoked to explain the seismic heterogeneity of the lowermost mantle (Lay et al. 2006). Post-perovskite is needed to explain core-reflected (Hernlund et al. 2005) and diffracted (Cobden et al. 2012) waves. It may also be a good candidate for the D" discontinuity, at least in regions where P- and S-waves have opposite signs (Cobden and Thomas 2013). In contrast, in probabilistic tomography models the thermal distribution is similar with and without post-perovskite (Mosca et al. 2012). Mantle convection simulations that incorporate post-perovskite effects indicate that its presence affects mantle dynamics (Cizkova et al. 2010; Nakagawa and Tackley 2011) and distorts the thermal-seismic relation from its linearity (Nakagawa and Tackley 2008).

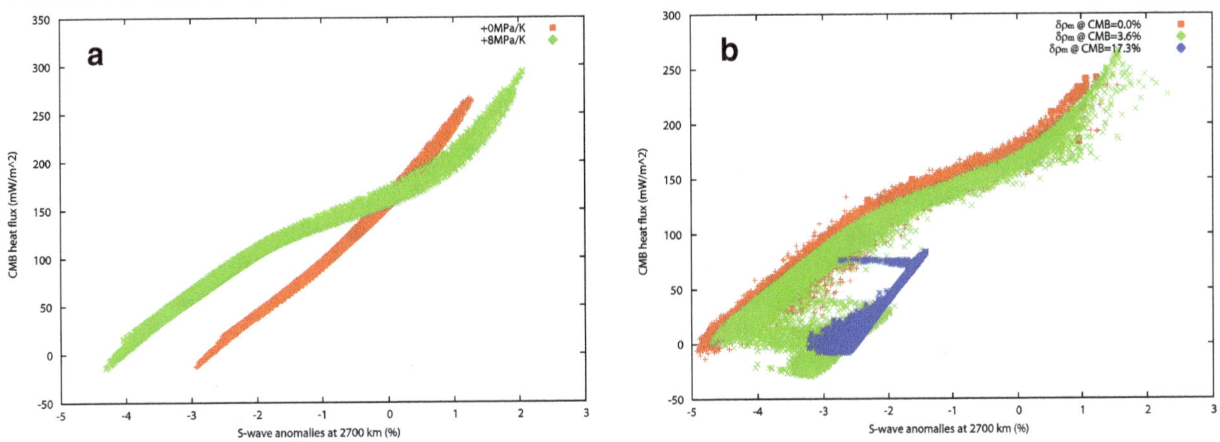

Fig. 2 Scatter plots of CMB heat flux anomalies vs. seismic shear velocity anomalies at the lowermost mantle based on mantle convection simulations. **a** Isochemical cases without (*red*) and with (*green*) post-perovskite. **b** Thermochemical cases with various density contrasts $\delta\rho_m$ at the CMB. From Nakagawa and Tackley (2008)

Sharp margins of LLSVPs were suggested by local studies of the lowermost mantle (Ni et al. 2002; To et al. 2005; Wang and Wen 2007). Dynamical models indicate that ULVZs likely correspond to the hottest regions of LLSVPs (McNamara et al. 2010). Thin hot narrow regions on the edges of the LLSVPs were identified in two end-member models of lower mantle dynamics (Bull et al. 2009). The low spatial resolution of global tomography models excludes such short wavelength features (e.g., Ritsema et al. 2007).

LLSVPs are better explained if a compositional component is also present (Deschamps et al. 2012; Mosca et al. 2012; Trampert et al. 2004). These structures could be primitive or may consist of recycled crustal material, although the seismic sensitivity of mantle material indicates that the latter is less likely (Deschamps et al. 2012). By taking into account seismic normal modes, Ishii and Tromp (1999) provided an independent constraint on mantle density distributions, which can be used to resolve thermal and compositional contributions to seismic anomalies, as done by probabilistic tomography studies (Mosca et al. 2012; Trampert et al. 2004).

One obvious motivation to consider alternative CMB heat flux patterns is the time-dependence of mantle dynamics. Global plate motion models indicate that the distribution of continents has evolved from a spherical harmonic degree-1 heterogeneity before the breakup of supercontinent Pangea to its present-day degree-2 structure (Seton et al. 2012). Some mantle convection models constrained by reconstructed time-dependent plate motions suggest that the supercontinent formation and breakup was accompanied by substantial change in mantle convection pattern (Zhang and Zhong 2011; Zhang et al. 2010; Zhong et al. 2007). According to these dynamical models, the low seismic velocity structures below the Pacific and Africa predate and postdate, respectively, the Pangea breakup (McNamara and Zhong 2005; Zhang et al. 2010). These time-dependent mantle convection patterns would inevitably correspond to time-dependent CMB heat flux with patterns distinctively different from that inferred from present-day tomography models.

In contrast, the pattern of LLSVPs has been interpreted as a root mantle structure that prevailed for several hundreds of Myr (Burke 2011; Burke et al. 2008; Dziewonski et al. 2010). Evidence for long-lived mantle structure comes from hotspot reconstructions (Torsvik et al. 2006) and paleomagnetic pole locations that indicate substantial true polar wander in the Mesozoic (Courtillot and Besse 1987). The dynamical origin of this (stationary) degree-2 pattern may be two mantle superplumes (Romanowicz and Gung 2002; Torsvik et al. 2010), plume clusters (Schubert et al. 2004) or dense chemical piles above the CMB (McNamara and Zhong 2005; Tackley 2002; Tan and Gurnis 2007). Permanent or not, the thermochemical piles are clearly dynamic and respond to changes in the global mantle circulation, through variations in their footprints above the CMB and their height in the lower mantle. Variations in either or both of these would affect the mean CMB heat flux and the amplitude of its heterogeneity.

Alternative CMB heat flux models were also constructed for planets other than the Earth. It has been argued that the emergence of the Tharsis volcanic complex coincided with the termination of the Martian dynamo (Jellinek et al. 2008; Johnson and Phillips 2005; Lillis et al. 2008). Mars' mantle may contain a single huge plume (Harder and Christensen 1996; Srámek and Zhong 2010). If Tharsis emerged from such a deep mantle plume, then it may indeed have affected the thermal heterogeneity on Mars' CMB.

Another possible scenario is localized mantle heating by giant impacts. This process enhances the iron/silicate separation and modifies the heat repartition between the mantle and the core. Thus, large impacts potentially affected the efficiency of core cooling and dynamo activity during (Monteux et al. 2011, 2012) or after terrestrial core formation (Glenn Sterenborg and Crowley 2013; Ke and Solomatov 2009; Monteux and Arkani-Hamed 2014; Monteux et al. 2013; Reese and Solomatov 2010). This large-scale heating also strongly influences mantle dynamics (Reese et al. 2002; Roberts and Arkani-Hamed 2012; Roberts and Barnouin 2012; Watters et al. 2009) and hence the heat repartition in the deepest part of terrestrial bodies. The onset or demise of terrestrial dynamos were modeled by giant impacts (Arkani-Hamed and Olson 2010; Kuang et al. 2014; Monteux and Arkani-Hamed 2014; Monteux et al. 2013; Roberts et al. 2009). Large impacts heated Mars' mantle below the impact site, with the heated volume depending mostly on the impactor's size and velocity (Pierazzo et al. 1997; Tonks and Melosh 1993). The impact-driven heating may be modeled by a uniform spherical temperature increase with rapid temperature decrease away from the isothermal sphere (Monteux et al. 2007, 2013; Senshu et al. 2002). Hence, for large impacts, the CMB heat flux may be affected.

Giant impacts may also have affected the morphology of the Martian paleomagnetic field. These events were invoked to explain the Martian and lunar topographic hemispheric dichotomies (Jutzi and Asphaug 2011; Marinova et al. 2008; Nimmo et al. 2008). Numerical dynamos with a synthetic Y_1^0 heat flux pattern lead to magnetic fields concentrated at the large heat flux hemisphere (Amit et al. 2011a; Dietrich and Wicht 2013; Stanley et al. 2008). Giant impacts may produce a similar effect with a more realistic CMB heat flux pattern (Monteux et al. 2015).

Relevant observed planetary dynamo properties

Alternative (or any) heterogeneous CMB heat flux is expected to influence the dynamo on very long timescales (Olson and Christensen 2002), but some statistical preference can already be detected on shorter timescales (Amit et al. 2010a; Bloxham 2002). Stronger heat flux heterogeneity amplitudes increase the probability of a high correlation between a dynamo property and its mantle-driven long-term average (Aubert et al. 2007). If the heterogeneity amplitude is very large compared to the mean superadiabatic CMB heat flux, locking may occur and a mantle signature may appear in a snapshot (Willis et al. 2007).

Because neither the dynamo equations nor the geometry of planetary outer cores contain any special longitude, any persistent deviation of a dynamo property from axisymmetry can only be explained by boundary heterogeneity. In contrast there are special latitudes, obviously the geographic equator and poles, but also the latitude of the tangent cylinder (Aurnou et al. 2003). However, there is no preferred hemisphere (neither in the dynamo equations nor in the geometry of planetary outer cores), so any persistent deviation of a dynamo property from equatorial symmetry may also require boundary heterogeneity. Below we list the main dynamo-related observations that motivated imposing heterogeneous CMB heat flux, in particular alternative patterns.

The primary dynamo observation is the observed magnetic field at the surface, from which a truncated (or large-scale) part can be downward continued to model the radial magnetic field on the CMB. Many studies attempted to reproduce the locations of high-latitude intense geomagnetic flux patches. The existence of two of these prominent geomagnetic field features at each hemisphere (Jackson et al. 2000) is consistent with the Y_2^2 dominance of the lowermost mantle seismic heterogeneity (e.g., Masters et al. 2000). The longitudes of the flux patches and the large seismic shear velocity structures seem to be correlated (Gubbins 2003), although some patches exhibit significant mobility (Amit et al. 2011b; Korte and Holme 2010). The time-average radial field of the dynamo model of Aubert et al. (2008) decently reproduces the longitudes of the flux patches in the paleomagnetic field (Kelly and Gubbins 1997). However, a shift between the locations of the large heat flux structures and the patches was observed, with the shift value depending on the dynamo internal parameters (Aubert et al. 2007; Olson and Christensen 2002; Takahashi et al. 2008). Alternatively, some geomagnetic (or paleomagnetic) spectral information, e.g., the number of patches at each hemisphere (corresponding to the dominant order), may be used as an observational constraint (Amit and Choblet 2009).

While high-latitude intense flux patches near the edge of the tangent cylinder (Aurnou et al. 2003) are a robust feature of dipole-dominated numerical dynamos (Christensen et al. 1998), the observed intense geomagnetic flux patches at low-latitude (Jackson 2003) are less characteristic of these models. The morphological criteria for Earth-like dynamos proposed by Christensen et al. (2010) identify well magnetic field models that resemble the geomagnetic field, but these criteria contain no information about the latitudinal distribution of magnetic flux. Alternative CMB heat flux heterogeneity offers a ready explanation for localized low-latitude intense flux patches (Amit and Choblet 2012; Amit et al. in press). However, such flux patches are not necessarily connected to boundary heterogeneity. Mobile low-latitude flux patches are seen in numerical dynamos without boundary heterogeneity when a strong enough influence of rotation on outer core convection is present (Wicht et al. 2011).

Other geodynamo properties apart from the magnetic field may also be used as observational constraints to dynamo models with heterogeneous CMB heat flux. For example, core flow models inverted from the geomagnetic secular variation (Holme 2007) show active Atlantic vs. quiet Pacific hemispheres (Amit and Olson 2006; Aubert et al. 2013; Gillet et al. 2009; Holme and Olsen 2006; Hulot et al. 2002). In addition, the seismic velocity, anisotropy, and attenuation of the inner core exhibit an east-west hemispheric dichotomy (Tanaka and Hamaguchi 1997), although these three seismic properties may sample different parts of the inner core (Deuss 2014; Irving and Deuss 2011). Aubert et al. (2008) recovered the observed patterns of the paleomagnetic field and the inner core hemispheric dichotomy in the corresponding time-average quantities of their dynamo model. The inner core hemispheric dichotomy may indeed be mantle-driven (Aubert et al. 2008; Gubbins et al. 2011), or alternatively independent inner core dynamics (Alboussiére et al. 2010; Monnereau et al. 2010) may produce anomalous bouyancy flux that would affect the outer core dynamics and the properties of the CMB (Aubert et al. 2013; Olson and Deguen 2012).

On much longer timescales of hundreds of Myr, the enigmatic variability of the geomagnetic reversal frequency may constrain changes in mantle conditions. In particular, the existence of long superchrons of tens of Myr in which no reversal was found in the paleomagnetic record in contrast to periods of hyper-reversing behavior (Gradstein et al. 2012; Merrill et al. 1998) is difficult to reconcile without time-dependent CMB conditions (Amit and Olson 2015; Biggin et al. 2012; Olson et al. 2013). In mantle convection simulations constrained by reconstructed plate motions the timing of the Kiaman Reversed Superchron (KRS) at around 270–315 Ma coincides with the low mean and equatorial CMB heat flux (Zhang and Zhong 2011). However, fast plate motions during the most recent Cretaceous Normal Superchron (CNS) may result in large CMB heat flux and frequent reversals (Olson et al. 2013).

Strong crustal magnetic field measured by Mars Global Surveyor is suggestive of a core dynamo that ceased at some time during the Noachian (Acuña et al. 2001; Langlais and Purucker 2007; Lillis et al. 2008; Milbury et al. 2012). Localized mantle heating may have affected the efficiency of the dynamo generation on Mars. A strong mantle plume rooted at the CMB could have terminated the Martian dynamo (Sreenivasan and Jellinek 2012). Alternatively, if the Martian dynamo was subcritical (Kuang et al. 2008), a moderate CMB heat flux anomaly induced by a giant impact could have stopped the dynamo (Kuang et al. 2014).

It has been proposed that the hemispheric dichotomy in Mars' crustal magnetic field was caused by heterogeneous CMB heat flux when the Martian paleo dynamo operated (Stanley et al. 2008). The magnitude of this dichotomy (Amit et al. 2011a) is recovered only with strongly convecting reversing dynamo models (Dietrich and Wicht 2013). Thus, in order to record hemispheric magnetization, the Martian crust should have cooled over a shorter period than the typical chron time (Lebrun et al. 2013), which poses a constraint on the feasibility of the hemispherical dynamo scenario for explaining the hemispheric dichotomy in Mars' crustal magnetic field (Dietrich and Wicht 2013; Monteux et al. 2015).

Methods

The following set of self-consistent non-dimensional Boussinesq (i.e., incompressible) magnetohydrodynamics equations for dynamo action due to thermochemical convection of an electrically conducting fluid in a rotating spherical shell were solved (e.g., Olson et al. 1999):

$$E\left(\frac{\partial \vec{u}}{\partial t} + \vec{u}\cdot\nabla\vec{u} - \nabla^2\vec{u}\right) + 2\hat{z}\times\vec{u} + \nabla P = Ra\frac{\vec{r}}{r_o}T + \frac{1}{Pm}(\nabla\times\vec{B})\times\vec{B} \quad (1)$$

$$\frac{\partial \vec{B}}{\partial t} = \nabla\times(\vec{u}\times\vec{B}) + \frac{1}{Pm}\nabla^2\vec{B} \quad (2)$$

$$\frac{\partial T}{\partial t} + \vec{u}\cdot\nabla T = \frac{1}{Pr}\nabla^2 T + \epsilon \quad (3)$$

$$\nabla\cdot\vec{u} = 0 \quad (4)$$

$$\nabla\cdot\vec{B} = 0 \quad (5)$$

where $\vec{u}$ is the velocity, $\vec{B}$ is the magnetic field, T is co-density (Braginsky and Roberts 1995), t is time, $\hat{z}$ is a unit vector in the direction of the rotation axis, P is modified pressure which contains centrifugal effects, $\vec{r}$ is the position vector, r_o is the core radius, and ϵ is co-density source or sink. Four internal non-dimensional parameters in (1)–(5) control the dynamo action. The Ekman number represents the ratio of viscous and Coriolis forces

$$E = \frac{\nu}{\Omega D^2} \quad (6)$$

where ν is kinematic viscosity. The Prandtl number is the ratio of kinematic viscosity to thermal diffusivity κ

$$Pr = \frac{\nu}{\kappa} \quad (7)$$

and the magnetic Prandtl number is the ratio of kinematic viscosity to magnetic diffusivity λ

$$Pm = \frac{\nu}{\lambda} \quad (8)$$

The strength of buoyancy force driving the convection relative to retarding forces is represented by the

Rayleigh number. When the CMB heat flux is strongly superadiabatic, Ra can be defined based on the mean CMB heat flux q_0 as (Olson and Christensen 2002)

$$Ra = \frac{\alpha g_0 q_0 D^4}{k\kappa\nu} \quad (9)$$

where α is thermal expansivity, g_0 is gravitational acceleration on the outer boundary at radius r_o, D is shell thickness, and k is thermal conductivity. The heat flux based Ra (9) was used with either generic thermochemical convection (Amit and Choblet 2009; Amit et al. in press; Amit and Choblet 2012) or purely thermal convection (Monteux et al. 2015). Alternatively, when convection is dominantly chemical, Ra can be defined based on the rate of light elements release into the outer core due to inner core freezing as (Olson 2007a)

$$Ra = \frac{\beta g_0 D^5 \dot{\chi}}{\kappa\nu^2} \quad (10)$$

where $\dot{\chi}$ is the time rate of change of the light element concentration (mixing ratio) in the outer core due to inner core growth. The mixing ratio based Ra (10) was invoked with dominantly (or purely) chemical convection (Amit and Olson 2015; Olson et al. 2013). The final internal control parameter is ϵ, the sink (or source) term that appears in the co-density equation (Christensen and Wicht 2007), which parameterizes the effects of mixing of light elements in the outer core due to inner core freezing, secular cooling of the outer core, and radioactivel heat sources.

Various alternative CMB heat flux models were imposed on the outer boundary of the dynamo models. In the "Review" section below, we describe in details the procedures for their constructions. The non-dimensional heat flux anomaly amplitude is classically defined in numerous studies by the peak-to-peak difference normalized by twice the mean value (e.g., Olson and Christensen 2002) as

$$q^* = \frac{q_{\max} - q_{\min}}{2q_0} \quad (11)$$

When the CMB heat flux is localized (11) appears to be inappropriate, hence various alternative measures were proposed. Sreenivasan and Jellinek (2012) proposed their control parameter f as the peak-to-background difference normalized by the background heat flux H_B

$$f = \frac{q_B - q_{\min}}{q_B} \quad (12)$$

They also defined the ratio of the average heat flux, either over the plume footprint or over the entire CMB, to the background heat flux, i.e. $\bar{q}_{\text{plume}}/q_B$ or $\bar{q}_{CMB}/q_B$, respectively (Sreenivasan and Jellinek 2012). Kuang et al. (2014) defined their control parameter ϵ_l as the relative post-impact heterogeneity amplitude, as in (12). In contrast, according to the model of mantle heating by impact of Monteux et al. (2015), in the vicinity of the impact, the lower mantle is heated to the core temperature and the CMB heat flux becomes zero, whereas away from it the lower mantle temperature is unchanged. Therefore, the control parameter q_0^r of Monteux et al. (2015) is based on the portion of the CMB heated by the impact

$$q_0^r = \frac{q_0^h - q_0}{q_0^h} \quad (13)$$

where q_0^h and q_0 are the pre- and post-impact mean CMB heat flux values, respectively.

To dimensionalize the dynamo models output, D, D^2/ν, and $\sqrt{\rho\mu_0\lambda\Omega}$ were used to scale length, time, and magnetic field, respectively, where ρ is the fluid density and μ_0 the permeability of free space. Co-density is scaled either by Dq_0/k when (9) is used or by $\rho\beta D^2\dot{\chi}/\nu$ when Ra is defined by (10). For studying the present-day geodynamo an Earth-like inner-outer core radii ratio of $r_i/r_o = 0.35$ was applied (Amit and Choblet 2009, 2012; Amit et al. in press). A smaller ratio of $r_i/r_o = 0.2$ was applied for Mars (Monteux et al. 2015) to approach no inner core conditions while avoiding numerical instability. The past geodynamo was modeled with a variable aspect ratio based on thermal history modeling (Olson et al. 2013). Numerical dynamos with rigid insulating boundary conditions were analyzed. Overall, the models differ in the imposed outer boundary heat flux pattern, the amplitude of its variation and the convection vigor and style—all of these choices depend on the particular geophysical application.

Parametric studies of numerical dynamos found two main types of solutions: non-reversing dipole-dominated and reversing multipolar (Christensen and Aubert 2006; Kutzner and Christensen 2002; Olson and Christensen 2006). Careful choice of control parameters may yield some "Earth-like" overlap of dipole-dominated reversing dynamos (Olson 2007a; Wicht et al. 2009). In this review, studies that focus on morphological features of the present-day field had dipole-dominated non-reversing models as a reference state, whereas in studies of reversal frequency either reversing dynamos or models of both types were analyzed. Both types of solutions were examined in the context of the paleo dynamo of Mars. Table 1 summarizes the control parameters, convection styles, inner to outer core ratios, and dynamo types in the papers reviewed here.

Review

To our best knowledge, only eight papers have examined the effects of alternative (i.e., neither tomographic nor single harmonic) CMB heat flux on planetary dynamos (Amit and Choblet 2009, 2012; Amit and Olson 2015; Amit et al. in press; Kuang et al. 2014; Monteux et al. 2015; Olson et al. 2013; Sreenivasan and Jellinek

Table 1 Summary of dynamo models control parameters, geometry, convection styles and reversibility in the following papers: AC09 (Amit and Choblet 2009), AC12 (Amit and Choblet 2012), ADC15 (Amit et al. in press), ODHZ13 (Olson et al. 2013), AO15 (Amit and Olson 2015), SJ12 (Sreenivasan and Jellinek 2012), KJRF14 (Kuang et al. 2014) and MACLT15 (Monteux et al. 2015). In all papers $Pr = 1$. Control parameters correspond to the definitions in this paper (see "Methods" section), so the values may differ from the original papers if different definitions were used (in particular for the Ekman number). Alternative definitions of CMB heat flux heterogeneity amplitudes q^*_{eff} (termed here *effective*) include f of SJ12 (12), ϵ_l of KJRF14 (also 12) and q^r_0 of MACLT15 (13). The Rayleigh number for SJ12 and KJRF14 is given in terms of the critical value for the onset of convection Ra_c. The time-dependent inner core size in ODHZ13 is determined based on a core evolution model

Paper	*E*	*Ra*	*Pm*	q^*	q^*_{eff}	r_i/r_o	*Conv*	*Rev?*
AC09	$3 \cdot 10^{-4}$	10^6	3	0.5	–	0.35	TC	No
AC12	$3 \cdot 10^{-4}$	$(0.5-1) \cdot 10^6$	$3-7$	$0.5-0.8$	–	0.35	TC	No
ADC15	$3 \cdot 10^{-4}$	$(1-3) \cdot 10^6$	$3-10$	$0.7-1.34$	–	0.35	TC	No
ODHZ13	$5.75 \cdot 10^{-3}$	$(1.5-4) \cdot 10^4$	20	$0.67-1.1$	–	Variable	C	Both
AO15	$(1-6) \cdot 10^{-3}$	$(0.6-18) \cdot 10^5$	20	$0.65-2.22$	–	0.35	C	Both
SJ12	$5 \cdot 10^{-5}$	$12Ra_c$	1	–	$4-29$	0.35	TC	No
KJRF14	$2.6 \cdot 10^{-5}$	$\sim Ra_c$	1	–	$0.05-0.3$	0.31	T	Yes
MACLT15	$(1-3) \cdot 10^{-4}$	$(0.735-4) \cdot 10^7$	3	$0.3-0.8$	$0.04-0.26$	0.2	T	Both

Conv convection type, *T* thermal, *TC* thermochemical, *C* chemical, *Rev?* reversibility

2012). Here we describe for each of these papers the CMB heat flux models that were imposed as outer boundary conditions on the dynamo simulations. Then, for each paper, we compare the results with those obtained with a reference heat flux model (Table 2) and we report the consequences for the resulting dynamo properties.

Lower mantle thermal-chemical-phase heterogeneity and geodynamo morphological features

Here we review the papers by Amit and Choblet (2009), Amit and Choblet (2012), and Amit et al. (in press). As in previous studies, a reference CMB heat flux anomaly of purely thermal origin δq_T was assumed to be linearly proportional to the lowermost mantle seismic shear velocity anomaly δv_s:

$$\frac{\delta q_T}{q_T} \propto \frac{\delta v_s}{v_s} \tag{14}$$

Figure 3a shows the CMB heat flux based on the tomography model of Masters et al. (2000) and (14). Nakagawa and Tackley (2008) found that when incorporating post-perovskite phase transition in their mantle convection models, the δq to δv_s relation becomes an inclined step function (Fig. 2a), with the spreading effect well approximated by a tanh function. Following equation (B8) of Nakagawa and Tackley (2008), Amit and Choblet (2009) attempted to parameterize the thermal-phase interpretation δq_{Tp} as

$$\frac{\delta q_{Tp}}{q_{Tp}} = \frac{\delta q_T}{q_T} + A\left[0.5\left(\tanh\left(-\frac{1}{B}\frac{\delta q_T}{q_T}\right) - 1\right)\right] + D \tag{15}$$

Table 2 Summary of dynamo models imposed outer boundary heat flux patterns in AC09 (Amit and Choblet 2009), AC12 (Amit and Choblet 2012), ADC15 (Amit et al. in press), ODHZ13 (Olson et al. 2013), AO15 (Amit and Olson 2015), SJ12 (Sreenivasan and Jellinek 2012), and MACLT15 (Monteux et al. 2015). For each paper, the reference and alternative patterns are given. Tomographic always refers to present-day, HF1 is the time-dependent model of Zhang and Zhong (2011). For more details, see text

Paper	Reference	Alternative
AC09	Tomographic	Thermal + post-perovskite
AC12	Y_2^2	Y_2^2 + ridges
ADC15	Tomographic	Probabilistic tomography
ODHZ13	Tomographic	HF1
AO15	Homogeneous	Rotated Y_2^0
SJ12	Homogeneous	Plume-induced
KJRF14	Homogeneous	Impact-induced
MACLT15	Degree-1	Impact-induced

The non-dimensional parameters in (15) are the amplitude A, the tanh spreading factor B, and the offset D. For a given amplitude A, the spreading factor B is constrained by reaching the tanh asymptote while maintaining an inclined step function (rather than Z-like curve).

Figure 3b shows a heat flux anomaly map that accounts for post-perovskite according to (15). The patterns in Fig. 3a, b are similar, but the relative magnitudes of the positive/negative heat flux structures vary. When the post-perovskite effect is included, the positive structures weaken and spread over larger areas, whereas the negative structures remain strong and more localized (Fig. 3b). This is due to the particular pattern of the tomographic model that contains larger amplitude narrower regions of negative seismic velocity anomalies ($\delta v_s, \delta q_T < 0$) as opposed to broader regions of weaker positive anomalies (Fig. 3a). The tanh function strongly reduces intermediate δv_s while it reduces extreme δv_s more mildly, resulting in a spreading effect acting stronger on the positive δq_T.

Following Aubert et al. (2008), Amit and Choblet (2009) compared three long-term time-average properties of their dynamo models with relevant observations: the radial magnetic field on the outer boundary with the time-average paleomagnetic field over the past 5 Myr (Kelly and Gubbins 1997), the flow at the top of the shell with the core flow model of Amit and Christensen (2008) obtained from inversions of the historical geomagnetic secular variation (Jackson et al. 2000) and averaged over the period 1840–1990, and the inner boundary buoyancy flux with seismic anomalies of the upper inner core (Tanaka and Hamaguchi 1997). As an example, we highlight here the latter comparison (Fig. 4). The inner boundary buoyancy flux is characterized by two isolated positive low-latitude structures both in the thermal case (Fig. 4a, denoted as T) as well as when accounting for post-perovskite (Fig. 4b, denoted as Tp1). In addition, however, the ICB buoyancy fluxes contain other orders. More vigorous convection leads to stronger filtering of the CMB order 2 content and the ICB pattern remains dominantly order 1 (Aubert et al. 2008), whereas slightly supercritical convection leaves the ICB pattern with nearly non-distorted order 2 (Gubbins et al. 2011). Amit and Choblet (2009) found that the ratio of order 1 to order 2 in Fig. 4b is significantly larger than that in 4a. This suggests that accounting for post-perovskite improves the agreement of the dynamo models with the order 1 dominance in seismic properties of the upper inner core (Tanaka and Hamaguchi 1997), although the phase may vary substantially from one dynamo model to another. Amit and Choblet (2009) argued that accounting for post-perovskite also improves the recovery of the Atlantic/Pacific hemispheric dichotomy in core flow activity (Amit and Christensen 2008; Gillet et al. 2009) and the single intense

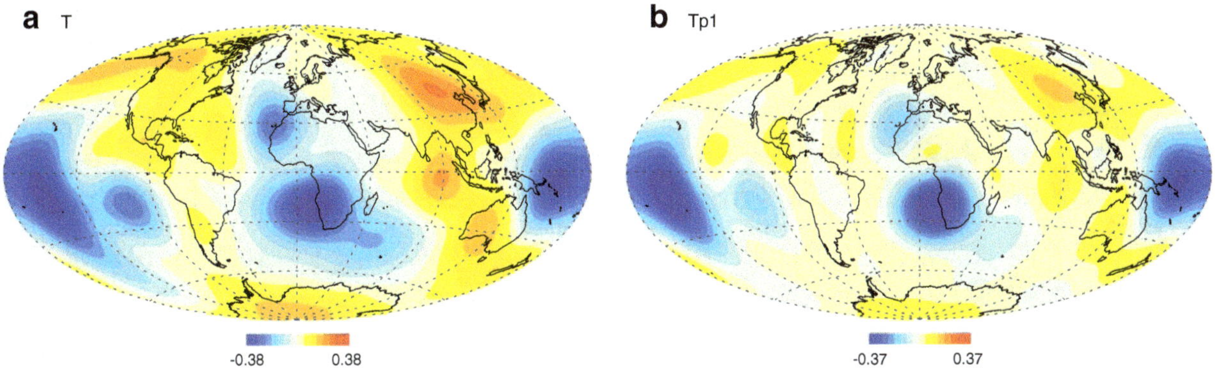

Fig. 3 Non-dimensional CMB heat flux patterns based on the mantle tomography model of Masters et al. (2000): **a** Thermal linear interpretation T and **b** interpretation including a post-perovskite effect Tp1. From Amit and Choblet (2009)

paleomagnetic flux patch in the southern hemisphere (Kelly and Gubbins 1997).

Next, narrow features that are not captured by global tomography models were considered. Lower mantle tomography models are dominated by a Y_2^2 component (e.g., Masters et al. 2000), which is classically interpreted as the largest scale temperature anomaly. Such a pattern was imposed by many authors as heat flux on the outer boundary of their numerical dynamo models (e.g., Aubert et al. 2007; Bloxham 2002; Olson and Christensen 2002; Sreenivasan 2009). Amit and Choblet (2012) used the Y_2^2 pattern as a reference case and superimposed on it low heat flux ridges at the four margins of hot and cold large-scale structures (Table 2). The ridges were modeled using a Gaussian geometry with an amplitude equal to that of the hot large-scale features (Fig. 5).

Figure 6 presents two close-ups of the time-average dynamo model velocity and magnetic field, one on the center of Fig. 5 (i.e., center of a ridge with a positive heat flux anomaly to its east), the other centered on a ridge with a positive heat flux anomaly to its west. In places where a positive heat flux anomaly appears east of the ridge (Fig. 6 left), the poloidal flow dominates and the fluid diverges from the hot ridge. Stronger field appears east of the ridge where cold fluid downwelling concentrates magnetic flux. In contrast, in places where a positive heat flux anomaly appears west of the ridge (Fig. 6 right), the upwelling associated with this ridge is broken at the equator, westward toroidal flow dominates, and the magnetic field exhibits patches of comparable intensity on both sides of the ridge. Note that the fluid upwellings appear very close to the longitudes of the hot CMB heat flux ridges, in contrast to previous studies of numerical dynamos with a Y_2^2 CMB heat flux pattern in which a phase shift was found between the heat flux and the flow structures (Aubert et al. 2007; Olson and Christensen 2002; Takahashi et al. 2008).

Amit and Choblet (2012) interpreted the low-latitude dynamics around the ridge locations by a combined effect of westward drift in the time-average homogeneous dynamo (Aubert 2005) and the steady boundary-driven flow that roughly goes from hot to cold regions (Gubbins 2003). When cold material is east of the ridge the two flows counteract and the ridge acts as a flow barrier, whereas when cold material is west of the ridge both flows are westward and the barrier effect is diminished. Applying these inferences to the tomography model of

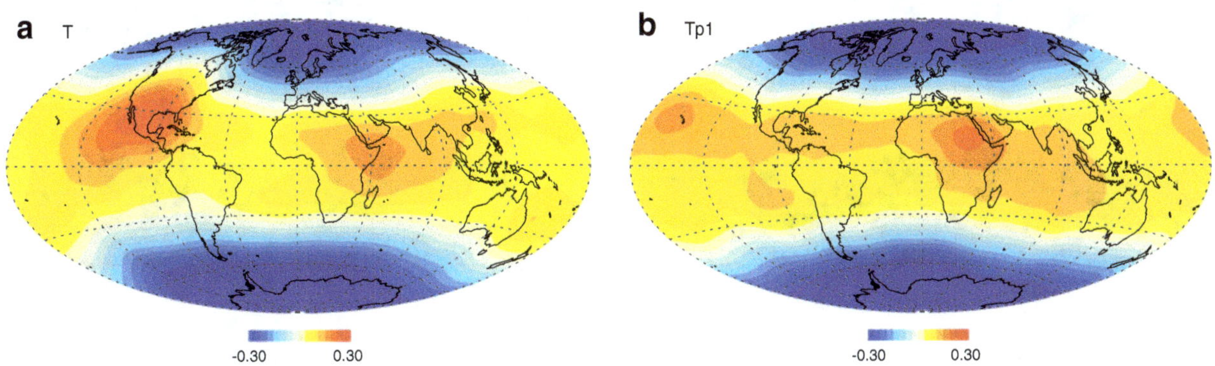

Fig. 4 Time-average inner-core boundary buoyancy flux anomalies normalized by the mean buoyancy flux: **a** Model T and **b** Model Tp1. From Amit and Choblet (2009)

Fig. 5 Non-dimensional CMB heat flux based on a Y_2^2 pattern with narrow hot ridges superimposed on the boundaries between large-scale hot and cold structures. From Amit and Choblet (2012)

Masters et al. (2000), Amit and Choblet (2012) predicted (in a statistical sense) fluid upwelling at the top of the core and a barrier for azimuthal flow below central Asia and the Indian Ocean and below the American Pacific coast. Below east Asia and Oceania and below the Americas they predicted persistent intense low-latitude geomagnetic flux patches. These predictions are in agreement with preferred longitudes of virtual geomagnetic poles during reversals both in paleomagnetic data (Hoffman 1996; Love 1998; Merrill and McFadden 1999) and in tomographic reversing dynamo models (Coe et al. 2000; Kutzner and Christensen 2004). It also resonates with the persistent production of low-latitude geomagnetic flux patches below Indonesia during the historical era (Jackson et al. 2000).

Finally, probabilistic tomography may be used to isolate the thermal component of the observed seismic anomalies. A limitation of the probabilistic tomography models published so far is that only even spherical harmonic degrees are included (Mosca et al. 2012; Trampert et al. 2004). To overcome this problem, Amit et al. (in press) mapped the ratio between shear-wave velocity δlnV_s^e (even degrees) and thermal anomalies δT^e (even degrees) in probabilistic tomography and applied this mapping to a conventional lower mantle seismic tomography model δlnV_s^C (even and odd degrees) to obtain a rescaled thermal anomaly distribution δT^P with both even and odd degrees. Mathematically they wrote

$$\delta T^P = \frac{\delta T^e}{\delta lnV_s^e}\delta lnV_s^C \tag{16}$$

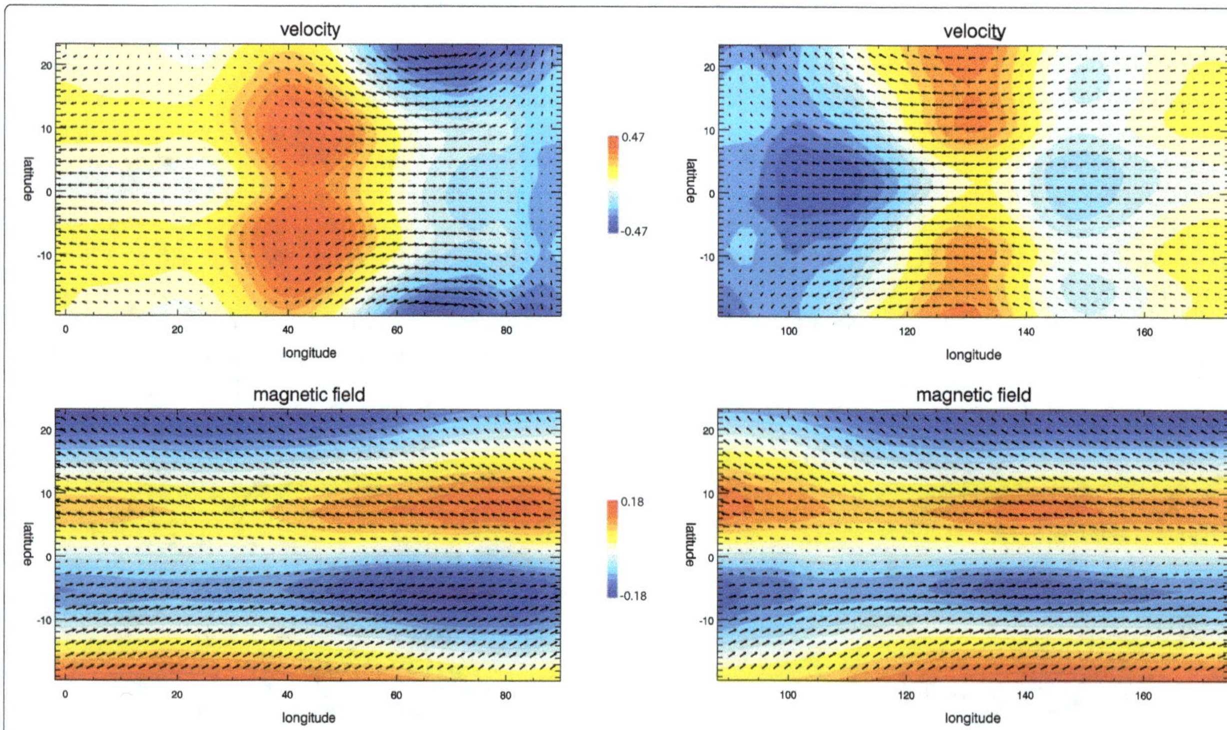

Fig. 6 Zoom into two ridges at low-latitude of the time-average of a dynamo case with imposed CMB heat flux as in Fig. 5, with a positive heat flux anomaly to the east (*left*, where Fig. 5 is centered) and with a positive heat flux anomaly to the west (*right*). Non-dimensional velocity at the *top* of the free stream is shown in the *top* panel, non-dimensional magnetic field in the *bottom* panel. Radial components are in *colors*, tangential components in *arrows*. From Amit and Choblet (2012)

To avoid numerical instabilities at regions where δlnV_s^e is low, at each node of the probabilistic tomography model δT^e and δlnV_s^e were sampled according to their full probability distribution functions (PDFs) rather than their mean values, resulting in a full PDF of the ratio $\delta T^e/\delta lnV_s^e$. Amit et al. (in press) applied the above procedure on the conventional seismic tomography model of Masters et al. (2000) and the probabilistic tomography model of Mosca et al. (2012) for the seismic and thermal anomalies. The resulting CMB heat flux pattern inferred from probabilistic tomography (Fig. 7) exhibits positive structures mostly concentrated at low latitude.

The dynamo models with an imposed probabilistic tomography heat flux pattern are characterized by stronger magnetic and convective activity at low-latitude than the dynamo models with conventional tomography heat flux. The timeseries of the ratio of RMS low-latitude magnetic flux to RMS high-latitude magnetic flux as well as the ratio of maximum low-latitude magnetic flux to maximum high-latitude magnetic flux (Fig. 8) show larger mean values and much larger temporal fluctuations in the probabilistic dynamo models (Fig. 8b) than those obtained with conventional tomography (Fig. 8a). For comparison, Amit et al. (in press) found that the historical geomagnetic field model *gufm1* (Jackson et al. 2000) exhibits for the period 1840–1990 average RMS and maxima ratios of 0.675 and 0.725, respectively. The higher resolution model *CHAOS-4* (Olsen et al. 2014) for the period 2000–2010 is characterized by somewhat larger average ratios of 0.797 and 0.882, and the average ratios in the lower resolution archeomagnetic field model *Cals3k.3* (Korte et al. 2009) for the past three millennia are 0.470 and 0.648. Accounting for the temporal fluctuations in the dynamo models with the CMB heat flux pattern inferred from probabilistic tomography, any of these models may occasionally recover the observed low- to high-latitude geomagnetic flux ratios.

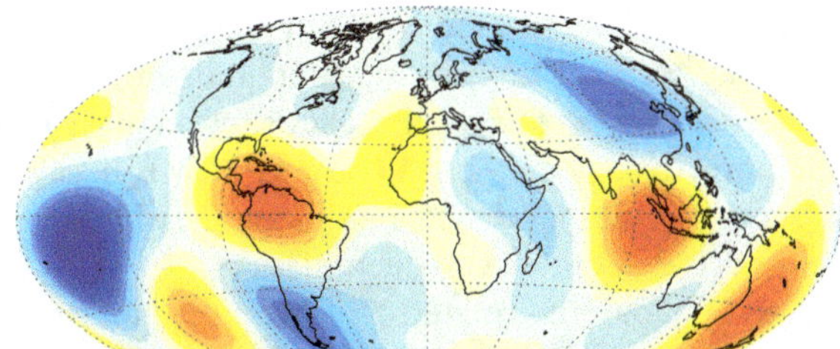

Fig. 7 Non-dimensional normalized CMB heat flux pattern inferred from probabilistic tomography. The same pattern was used for different q^* values (see Table 1), hence the absence of a *color bar*. *Red/blue* denotes positive/negative heat flux anomalies, respectively. From Amit et al. (in press)

Lower mantle history and geomagnetic reversal frequency

Here we review the papers by Olson et al. (2013) and Amit and Olson (2015). Zhang and Zhong (2011) imposed a reconstructed plate tectonics model for the period 0–120 Ma (Lithgow-Bertelloni and Richards 1998) and Atlantic hemisphere continent reconstructions for older periods (Scotese 2001) as a time-dependent mechanical upper boundary condition on their mantle thermochemical convection model to infer the time-dependent CMB heat flux. According to one of the CMB heat flux models of Zhang and Zhong (2011), termed HF1, relatively large/low mean CMB heat flux corresponds to fast/slow plate reconstructions (Lithgow-Bertelloni and Richards 1998; Scotese 2001), respectively. In addition, HF1 includes temporal changes in the heat flux pattern (Fig. 9) and amplitude (Table 1), most notably a transition from degree-1 dominance during Pangea to degree-2 dominance at present.

Olson et al. (2013) imposed snapshots of HF1 (Fig. 9) on chemical convection dominated numerical dynamos. In addition, they accounted for the time-dependent core evolution that is controlled by the mean CMB heat flux. Using thermal history modeling (Labrosse 2003), Olson et al. (2013) determined the time-dependent CMB heat flux and consequently the time-dependent inner core size and inner core boundary buoyancy flux. According to these models, a few tens of Myrs after inner core nucleation outer core convection indeed becomes dominantly chemical. Very high core thermal conductivity (e.g., Pozzo et al. 2012) implies that higher CMB heat flux, radiogenic heating, or some presently unknown form of chemical differentiation in the core would be needed to power the geodynamo prior to inner core nucleation. However, the debate on core thermal conductivity is still ongoing (Zhang et al. 2015). Finally, Olson et al. (2013) also accounted for the observed nearly linear increase in length of day (Williams 2000) to derive a time-dependent rotation rate. Overall, they established the relative temporal evolution of the Rayleigh and Ekman numbers for the period corresponding to the HF1 model.

Olson et al. (2013) compared the time-dependent reversal frequency of two types of dynamo models, one with time-dependent and another with fixed present-day CMB heat flux (both cases include time-dependent core evolution). This comparison allows distinguishing between the direct effects of the CMB heterogeneity and the effects of the core evolution. Figure 10 shows comparison of the reversal frequency of the dynamo models of Olson et al. (2013) with the observed paleomagnetic reversal frequency record (Gradstein et al. 2012). The dynamo model with fixed CMB heat flux is characterized by weaker temporal variability in reversal frequency. In contrast, the dynamo model with time-dependent CMB heat

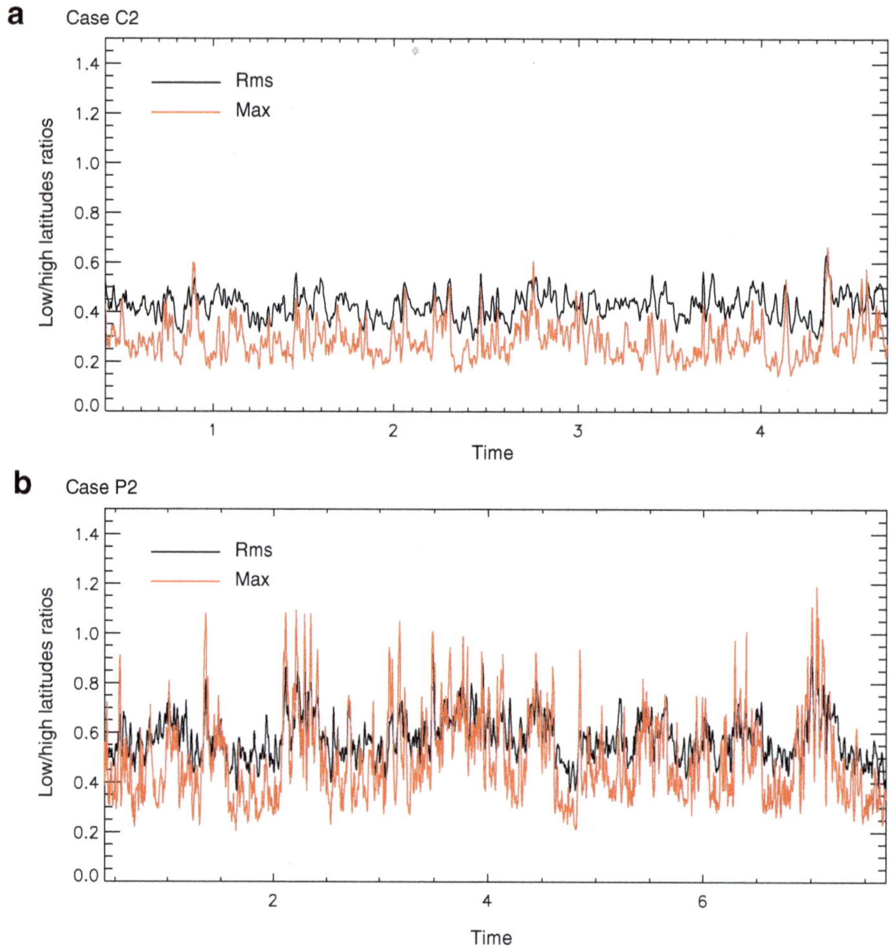

Fig. 8 Timeseries of low/high-latitude magnetic flux ratios for dynamo models with the same control parameters but different outer boundary heat flux patterns, one with a pattern from conventional tomography (**a**, case C2 of Amit et al. in press) and the other with a pattern inferred from probabilistic tomography (**b**, case P2 of Amit et al. in press). *Black/red curves* are RMS/maxima ratios, respectively. From Amit et al. (in press)

flux shows larger reversal frequency fluctuations including superchrons and rather hyper-reversing periods. In particular, at 275 Ma when Pangea was at its height and prior to it, the dynamo model with time-dependent CMB heat flux recovers the low reversal frequency, whereas that is not the case for the dynamo model with fixed CMB heat flux. Olson et al. (2013) argued that the transition from low to high reversal frequency is driven by the supercontinent cycle. Both dynamo models fail to reproduce the robust Cretaceous Normal Superchron at 120 Ma. However, Olson et al. (2013) demonstrated that the reversal frequency is strongly sensitive to mild changes in the CMB heat flux (mean, pattern, and its amplitude).

In contrast, the lower mantle may exhibit a quasi-stationary pattern. Amit and Olson (2015) proposed that the extreme variability in the recorded geomagnetic reversal frequency, from hyper-reversing periods to superchrons, may be explained by lower mantle piles with a fixed lateral pattern but time-dependent height. Amit and Olson (2015) proposed a simple model of the D" layer with fixed temperature difference, conserved volume with time, and a conductive equilibrium. Based on these assumptions, they derived analytically the dependence of the mean CMB heat flux and the amplitude of its lateral heterogeneity on the pile height. They showed that superplume growth increases both the mean CMB heat flux and the amplitude of its lateral heterogeneity, whereas pile collapse decreases these two quantities.

Figure 11 shows the rotated Y_2^0 CMB heat flux pattern based on the model of lower mantle seismic heterogeneity by Dziewonski et al. (2010). The mean heat flux was assumed to be close to the core adiabat, i.e., the dynamo models are dominated by chemical convection. The heterogeneity pattern was generated using spherical harmonics Y_2^0, Y_2^2, and Y_1^1, with relative amplitudes of 10 : 10 : 1, respectively. Amit and Olson (2015) determined the

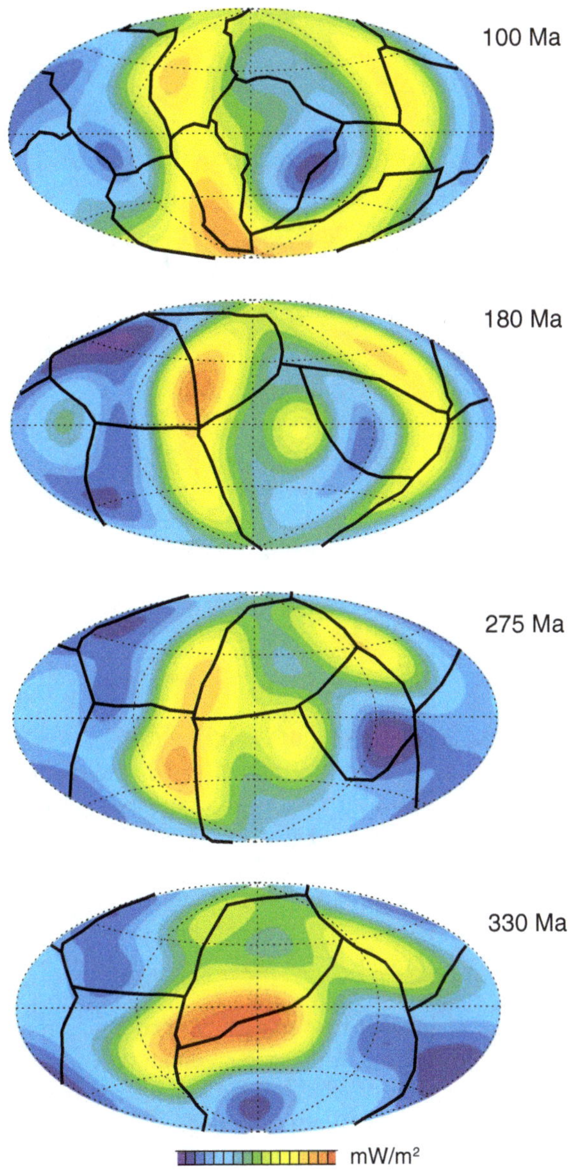

Fig. 9 Snapshots of CMB heat flux (*colors*) and plate boundaries (*solid black lines*) from the mantle history model HF1 of Zhang and Zhong (2011). From Olson et al. (2013)

amplitudes of mean heat flux and heterogeneity in their dynamo models based on their analytical piles model as functions of pile height. In addition, they set the Rayleigh number based on the mean heat flux from their analytical piles model.

Figure 12 shows the non-dimensional reversal frequency as functions of pile height and mean outer boundary heat flux for the rotated Y_2^0 (Fig. 11) dynamos. The error bars correspond to $\sqrt{N}/\tau_d$ (where N is the reversal frequency and τ_d is the dipole free decay time), consistent with a Poisson distribution (Lhuillier et al. 2013). The reversal frequency increases approximately linearly beyond onset. There is a fairly small window separating the pile height that would produce magnetic superchrons and the pile height that would produce hyper-reversing dynamos. A relatively moderate increase of ~35% in CMB heat flux corresponds to a transition from dynamo models exhibiting superchrons to those in a hyper-reversing state. Amit and Olson (2015) concluded that pile growth may stimulate polarity reversals, whereas pile collapse may inhibit reversals.

Localized mantle heating and the paleo dynamo of Mars

Here we review the papers by Sreenivasan and Jellinek (2012), Kuang et al. (2014), and Monteux et al. (2015). Sreenivasan and Jellinek (2012) simulated the localized CMB anomaly induced by the mantle plume that produced Tharsis by

$$q = q_{\min} exp\left(-\left[(\theta-\theta_p)^2/\delta_\theta^2 + (\phi-\phi_p)^2/\delta_\phi^2\right]\right) \tag{17}$$

where θ and ϕ are latitude and longitude, θ_p and ϕ_p are the corresponding coordinates of the plume center, and δ_θ^2 and δ_ϕ^2 are the corresponding exponential decay coefficients, set to be 15° at the equator. The localized CMB plume-induced heat flux heterogeneity is visualized in the inlet of Fig. 13a.

The dynamo models of Sreenivasan and Jellinek (2012) are strongly supercritical (Table 1), requiring large heterogeneity amplitudes to terminate the dynamo (Fig. 13a). Increasing heterogeneity amplitude attenuates the field and eventually leads to the dynamo failure (Fig. 13b). In addition, it is easier to kill the dynamo if the mantle plume is centered at low-latitude (Fig. 13a).

The underlying mechanism for the demise of the dynamo due to the localized CMB heating is visualized in Fig. 14. The boundary heterogeneity weakens the axial flow and reverses its direction. As a result, the kinetic helicity also weakens significantly. The weakening of helicity reduces the field generation efficiency (Moffatt 1978) and leads to the demise of the dynamo (Sreenivasan and Jellinek 2012).

Kuang et al. (2014) also studied the possibility of Martian dynamo cessation due to localized CMB heating, but by a giant impact. The impact-induced thermal heterogeneity is modeled by shock wave heating emanating from the impact site (Pierazzo et al. 1997; Watters et al. 2009). The temperature rapidly decays from the impact center. In contrast to Sreenivasan and Jellinek (2012), the dynamo models of Kuang et al. (2014) are subcritical with moderate heterogeneity amplitudes of $\epsilon_l \le 0.3$ (see Table 1). These subcritical dynamos were found to be hemispheric (Kuang et al. 2008), in contrast to other studies that argued that under subcritical conditions helicity generation is

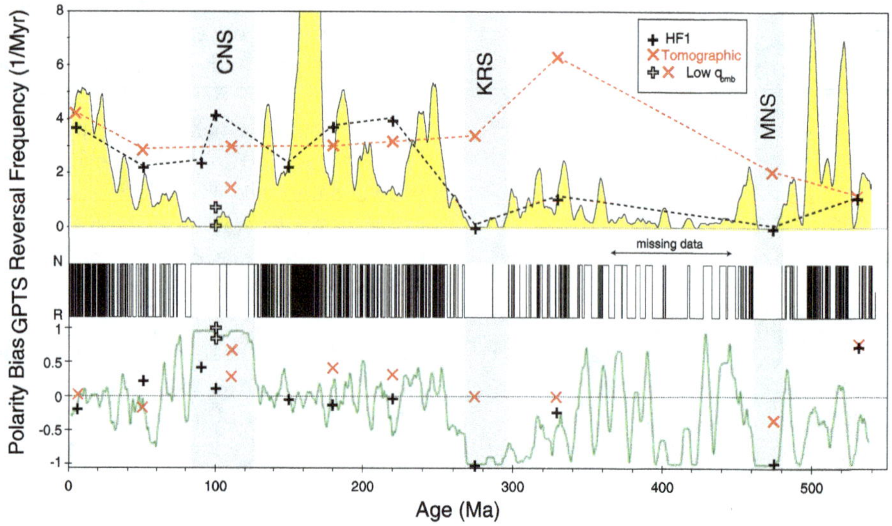

Fig. 10 *Top*: Geomagnetic reversal frequency (*yellow* and *barcode below*) based on the Geomagnetic Polarity Time Scale (GPTS) database of Gradstein et al. (2012) vs. reversal frequency in dynamo models with time-dependent CMB heat flux (*black pluses*) and fixed present-day tomographic pattern (*red crosses*). The Cretaceous Normal Superchron (CNS), Kiaman Reversed Superchron (KRS), and Moyero Normal Superchron (MNS) are *grey shaded*. *Bottom*: Polarity bias, i.e., the difference between time spent in normal and reversed polarities normalized by their sum. From Olson et al. (2013)

more efficient when the field is dipole-dominated (Hori and Wicht 2013; Sreenivasan and Jones 2011).

Kuang et al. (2014) also found that a giant impact at low-latitude and with a large heterogeneity amplitude is more likely to terminate the dynamo. Figure 15 shows that when the impacts where placed at latitude $60°N$ and higher the dynamos survived. The dynamos failed if the CMB heat flux heterogeneity amplitude was larger than a critical value. This critical value decreases with increasing latitude of the impact site. The dynamo failure occurs when the helicity is reduced in the hemisphere where most of the field is expected to be induced.

Localized mantle shock heating by a giant impact could have also altered the morphology of the Martian paleo dynamo. Monteux et al. (2015) uniquely determined the CMB heat flux lateral variations as well as the reduction in the mean CMB heat flux by the size and geographic location of the impactor. Figure 16 shows the CMB heat flux patterns resulting from an impactor of 800 km radius falling on the north pole (left) or the equator (right) of Mars' surface. Because it is thought that there is no inner core in Mars (Breuer et al. 2010; Schubert and Spohn 1990), purely thermal convection was modeled, with a volumetric homogeneous heat source ϵ that compensated for the loss of heat through the outer boundary (Amit et al. 2011a; Dietrich and Wicht 2013). A small inner core corresponding to $r_i/r_o = 0.2$ was retained to avoid numerical instabilities, but the inner boundary was set to be convectively passive (Table 1). Both non-reversing and reversing dynamos were examined.

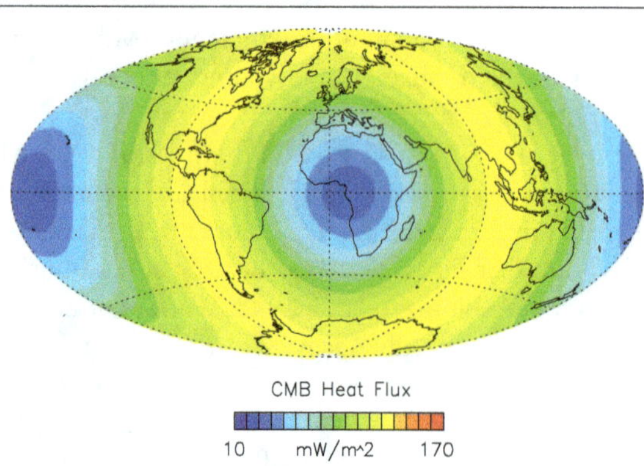

Fig. 11 CMB heat flux pattern in the dual superplume configuration. The mean heat flux is 100 mW/m^2 and the peak-to-peak variation is 60 mW/m^2. From Amit and Olson (2015)

Figure 17 shows the time-average RMS radial field at the CMB and the time-average field intensity at the surface of Mars for a dynamo model with an impactor of radius 800 km falling on the north geographic pole (Fig. 16 left). The impactor yields an hemispheric field similar to those obtained with synthetic Y_1^0 CMB heat flux patterns (Amit et al. 2011a; Dietrich and Wicht 2013; Stanley et al. 2008). Monteux et al. (2015) found that a polar impactor leads to a north-south hemispheric magnetic dichotomy that is stronger than an east-west dichotomy created by an equatorial impactor (in agreement with Amit et al. 2011a, Dietrich and Wicht 2013).

Monteux et al. (2015) found that an impactor radius of about 1000 km could have recovered the magnetic dichotomy observed in the Martian crustal field.

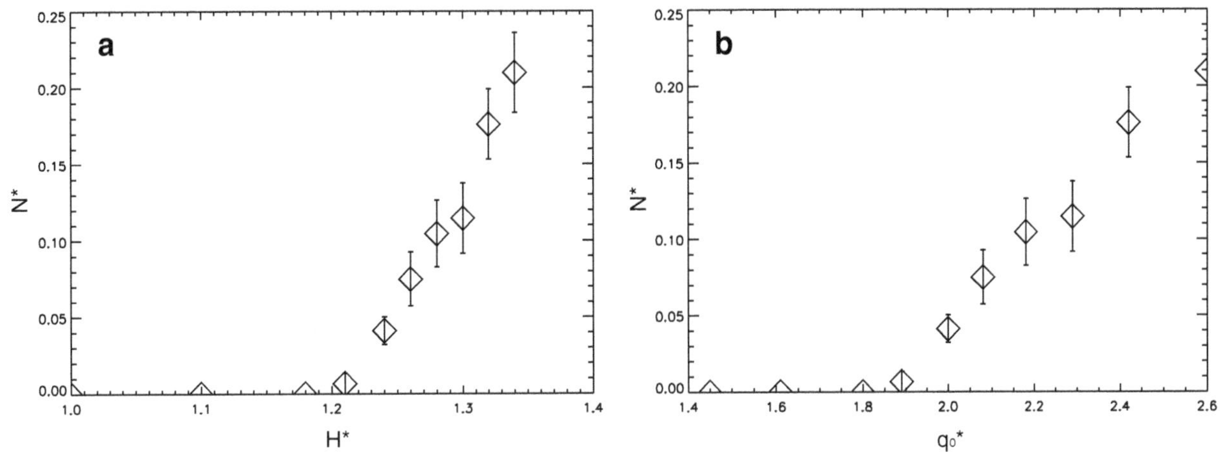

Fig. 12 Non-dimensional reversal frequency vs. non-dimensional pile height **a** and vs. non-dimensional mean CMB heat flux **b** in the set of dual pile (rotated Y_2^0) dynamo models. From Amit and Olson (2015)

However, strong enough magnetic dichotomies were obtained only for frequently reversing dynamos (as in Dietrich and Wicht 2013). Therefore, such dynamos may account for the observed field dichotomy only if very rapid post-impact magma cooling took place (Lebrun et al. 2013).

Conclusions

Discussion

The study of mantle control on planetary dynamos may be classified into three categories. In the first category, numerical dynamos with homogeneous boundary conditions have already reproduced successfully the most fundamental observed geodynamo properties, in particular the dipole dominance of the field (Glatzmaier and Roberts 1997) and its reversibility (Glatzmaier and Roberts 1995). Parametric studies of numerical dynamos show that increasing convection vigor (which is controlled by the mean CMB heat flux) causes a transition from dipole dominated non-reversing to multipolar reversing dynamos (Aubert et al. 2009; Christensen and Aubert 2006; Kutzner and Christensen 2002). The very different magnetic field geometries and amplitudes observed for the various planets and moons (Stevenson 1983) may represent different locations of their dynamos in parameter space with respect to this transition (Olson and Christensen 2006). In the second category, heterogeneous but simplified (i.e., not alternative) boundary conditions may help explain more refined dynamo properties, in particular for the Earth where detailed observations are available. In a third category, numerical dynamos with imposed alternative CMB heat flux can be considered as the next step to approach the recovery of more refined observations using more realistic CMB conditions.

Persistent morphological features of the present-day geodynamo were explained using alternative CMB heat flux models (Amit and Choblet 2009, 2012; Amit et al. in press). The modifications of the CMB heat flux patterns due to post-perovskite and thermal ridges are mild, and therefore their impact on the dynamo properties are also mild. Thus, in order to detect the differences between the reference dynamo models and those modified by the post-perovskite and thermal ridges effects (Table 2), long-term time-averaging is required (Amit and Choblet 2009, 2012). In contrast, the CMB heat flux pattern inferred from probabilistic tomography is dramatically distinctive from its reference (Amit et al. in press), and therefore provides a genuinely new insight into the possible control of the mantle heterogeneity on the geodynamo.

Dynamo models with imposed CMB heat flux inferred from both thermal ridges and probabilistic tomography were invoked to explain the presence of intense geomagnetic flux patches at low latitude (Amit and Choblet 2012; Amit et al. in press). In the latter models, enhanced low-latitude convective and magnetic activity is found, in particular during some occasional bursts during which the models exceed the observed relative low-latitude geomagnetic flux (Amit et al. in press). These robust low-latitude geomagnetic features (Jackson 2003) are more difficult to reproduce in dynamo models than high-latitude flux patches (Christensen et al. 1998), possibly due to flux dispersion by upwelling (Wicht et al. 2009) that reconnects with the downwellings near the tangent cylinder in the dynamo's meridional circulation (Aubert 2005). Intense low-latitude geomagnetic flux patches, which are indeed easier to recover with alternative CMB heat flux, have so far not been included explicitly as a morphological constraint for Earth-like dynamo models (Christensen et al. 2010).

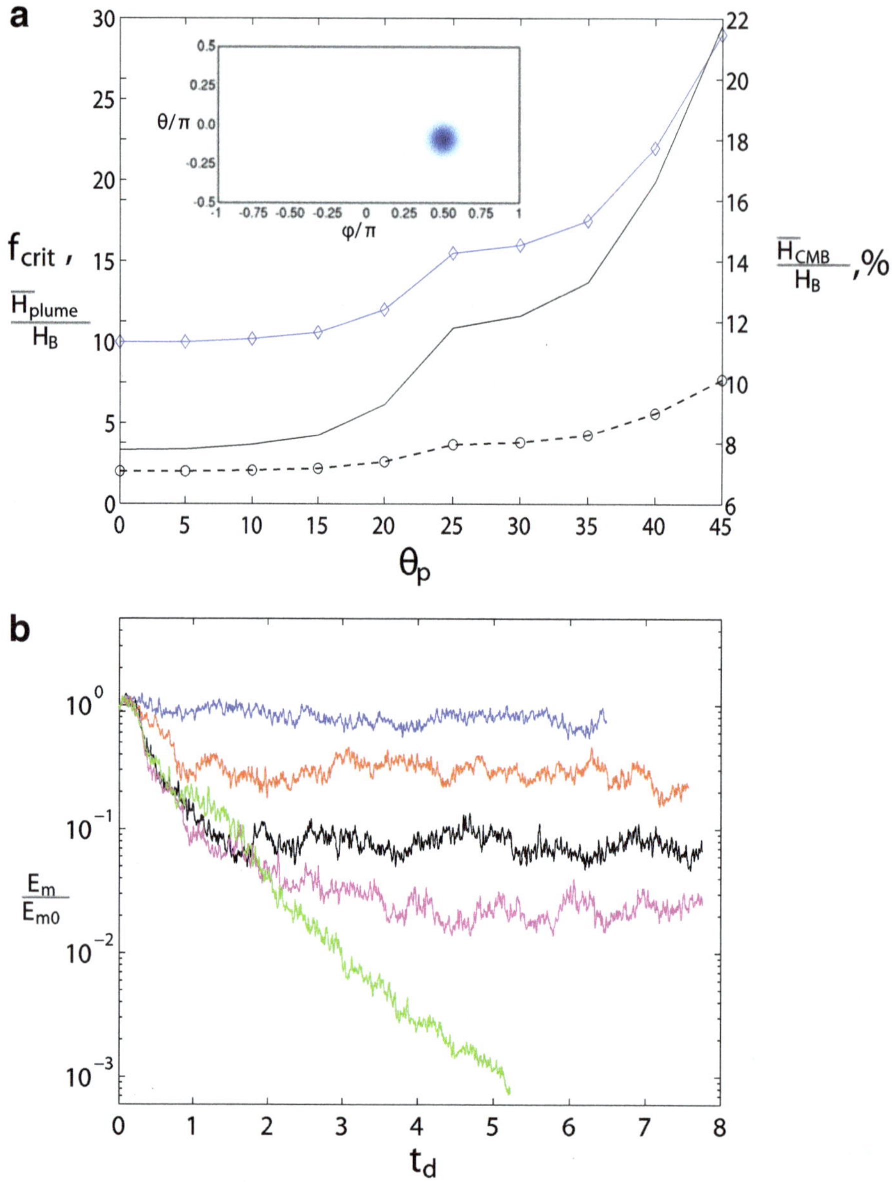

Fig. 13 a Critical heat flux heterogeneity amplitudes required for dynamo failure vs. plume latitude for a mantle plume centered at 15°S (see CMB heat flux pattern in inlet). The three curves are based on three definitions of the heterogeneity amplitude: f (*blue*, Eq. (12)), $\bar{H}_{CMB}/H_B$ (*solid black*), and $\bar{H}_{plume}/H_B$ (*dashed black*). **b** Timeseries of normalized magnetic energy for a mantle plume centered at 45°S with $f = 0$ (*blue*), $f = 4$ (*red*), $f = 10$ (*black*), $f = 25$ (*magenta*), and $f = 29$ (*green*). From Sreenivasan and Jellinek (2012)

Due to the very low viscosity and very large scale of planetary outer cores, dynamo models operate in a parametric regime very far from realistic conditions (e.g., Christensen and Wicht 2007; Glatzmaier 2002). The situation is more severe if long-term time-averaging is required to detect the impact of alternative or any heterogeneous mantle control, because long simulations are needed for adequate statistical convergence of the time-average dynamo properties. An especially extreme case concerns the study of reversal frequency which requires very long simulations to obtain significant statistics of rare events. Indeed the Ekman and magnetic Prandtl numbers in the studies by Olson et al. (2013) and Amit and Olson (2015) are very large (Table 1); the Ekman number is more than 10 orders of magnitude larger than its geophysical estimate (e.g., Olson 2007b). Amit and Olson (2015) explored some simulations with a lower Ekman number. They found qualitative agreement with their larger Ekman number results, but computational limitations prevented them from providing a confident quantitative analysis, in this case of the reversal frequency vs. pile height.

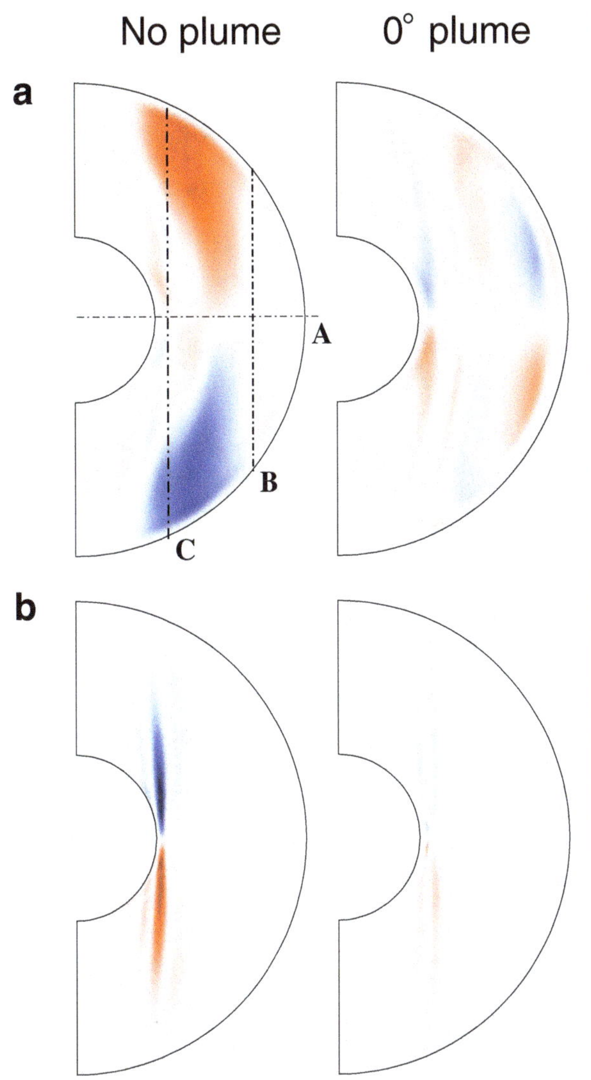

Fig. 14 Zonal profiles of non-dimensional axial velocity **a** and non-dimensional axial kinetic helicity **b** in dynamo models without (*left*) and with (*right*) a plume. The minimum and maximum axial velocity values are [−40.4, 34.4] (*left*) and [−19.3, 18.3] (*right*); the minimum and maximum axial kinetic helicity values are [−174.6, 137.8] (*left*) and [−23.6, 24.8] (*right*). From Sreenivasan and Jellinek (2012)

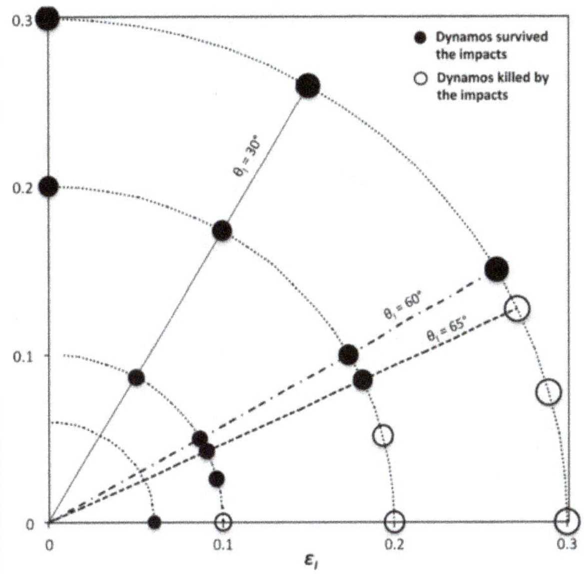

Fig. 15 Survivability of dynamos as a function of CMB heat flux heterogeneity amplitude ϵ_l (radial distance) and the co-latitude of the impact θ_l (angle from vertical axis). From Kuang et al. (2014)

The variability in the paleomagnetic reversal frequency was explained by alternative time-dependent CMB heat flux models (Amit and Olson 2015; Olson et al. 2013). The larger temporal variability in reversal frequency in the dynamo models with time-dependent CMB heat flux compared to those with a fixed tomographic pattern, in particular the presence of superchrons in the former, supports a time-dependent mantle convection structure over the past ~500 Myr (Olson et al. 2013). It also shows that core evolution by its own cannot account for the large variability in the paleomagnetic reversal frequency. The HF1 model (Zhang and Zhong 2011) exhibits temporal variations in the mean CMB heat flux, its pattern and the amplitude of lateral heterogeneity (Fig. 9 and Table 1), all of which may affect reversal frequency (see Olson and Amit 2014, and references therein).

According to the HF1 model, the CMB heat flux pattern evolved from degree-1 dominance during Pangea to degree-2 at present. This transition is reflected in the time-average magnetic field of the dynamo models with prescribed heat flux pattern inferred from HF1. Olson et al. (2013) showed that at present, the dipole is very axial, whereas with increasing age the dipole axis substantially deviated from the geographic pole. The symmetry of the early time-average magnetic field highly correlates with the symmetry of the CMB heat flux heterogeneity, exhibiting magnetic field concentrations aligned with the axis of the large CMB heat flux.

Dziewonski et al. (2010) proposed that the CMB heat flux is quasi-stationary with two stable states that may each persist for hundreds of Myr. When the plates are organized in one supercontinent at Earth's surface, the CMB heat flux may be characterized by a rotated Y_1^0 pattern; when two continents are present, the CMB heat flux is expressed by a rotated Y_2^0 pattern. Amit and Olson (2015) showed that in both scenarios increasing pile height increases the mean CMB heat flux, the amplitude of its heterogeneity, and the resulting reversal frequency. For a given pile height, the rotated Y_1^0 pattern yields larger mean heat flux, heterogeneity amplitude and reversal frequency than the rotated Y_2^0 pattern (Amit and Olson

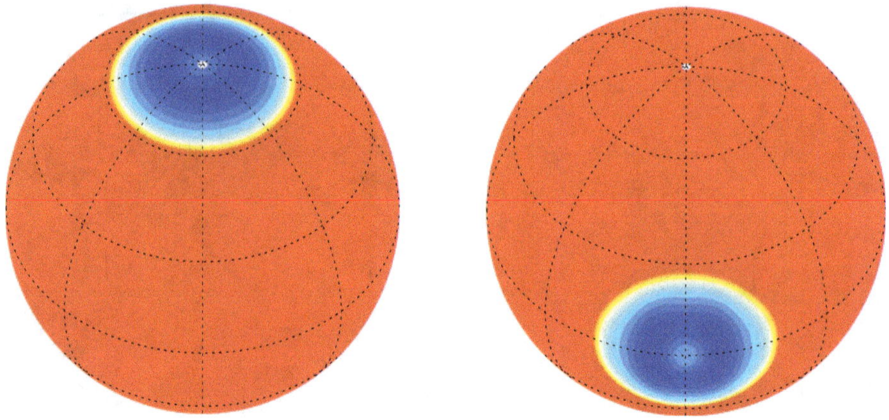

Fig. 16 Non-dimensional CMB heat flux resulting from an impactor of radius 800 km falling on the north pole (*left*) or the equator (*right*). The same pattern was used for different q^* values (see Table 1), hence the absence of a color bar. *Red/blue* denotes positive/negative heat flux anomalies, respectively. From Monteux et al. (2015)

2015), suggesting higher sensitivity of the paleomagnetic reversal frequency to changing mantle conditions prior to the KRS.

The dependence of the reversal frequency on the heterogeneity amplitude is not trivial. In particular, the rotated Y_2^0 case contains polar cooling that according to most studies may suppress reversals (Glatzmaier et al. 1999; Kutzner and Christensen 2004). Polar cooling results in axial-dipole dominance due to flux concentration at high-latitudes (Amit et al. 2010b) and therefore indeed suppresses reversals (a state termed geographic control by Olson and Amit 2014). However, when the dynamo is far from the onset of reversals, Olson and Amit (2014) found that the opposite occurs and polar cooling increases reversal frequency by producing more turbulent conditions (which they termed inertial control). In the piles model of Amit and Olson (2015), the effects of variable mean CMB heat flux and heterogeneity amplitude are non-separable. However, if indeed the rotated Y_2^0 pattern works in unison with the mean CMB heat flux to trigger reversals, it is suggestive of inertial rather than geographic control (Olson and Amit 2014).

The demise of the Martian dynamo was explained by a localized CMB heating (Kuang et al. 2014; Sreenivasan and Jellinek 2012). Sreenivasan and Jellinek (2012) imposed plume-induced thermal anomalies with large heterogeneity amplitudes on highly supercritical numerical dynamos, whereas Kuang et al. (2014) imposed impact-induced thermal anomalies with moderate heterogeneity amplitudes on subcritical dynamos. Under supercritical conditions, the CMB anomaly causes the dynamo to decay relatively slow—the field drops by one order of

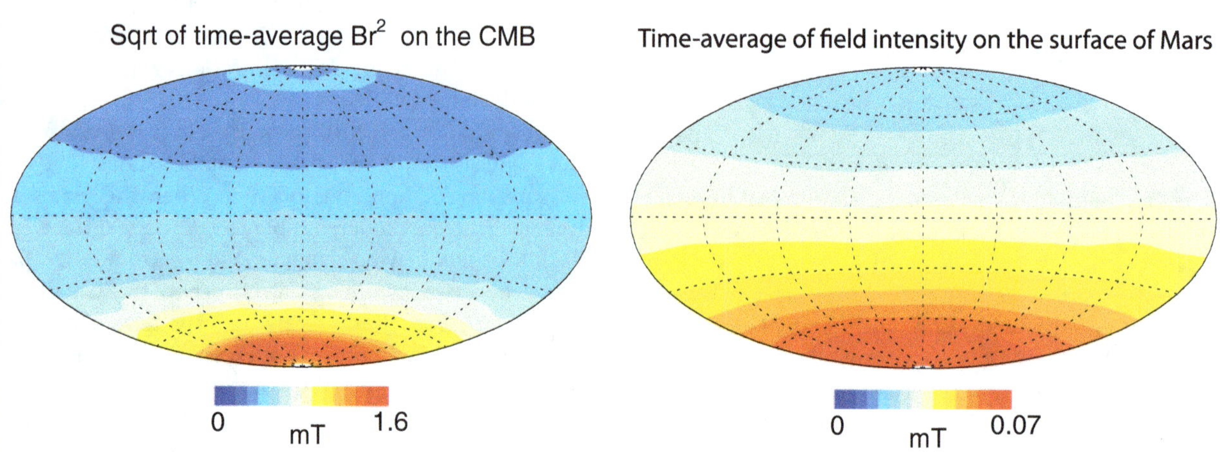

Fig. 17 Time-average magnetic field properties on the CMB (*left*) and on the surface of Mars (*right*) in a dynamo model with imposed outer boundary heat flux due to a polar impactor as in Fig. 16. From Monteux et al. (2015)

magnitude in ~1.7 magnetic diffusion times (Sreenivasan and Jellinek 2012), whereas under subcritical conditions the termination is more sudden—the field decays ~4 times faster (Kuang et al. 2014). The large heterogeneity amplitudes in the dynamo models of Sreenivasan and Jellinek (2012) imply that below the mantle plume the heat is flowing unrealistically into the core. Another problem is that under vigorous convection the dynamo is expected to restart once the mantle anomaly vanishes. Note that reducing dynamo efficiency by heterogeneous CMB heat flux is not trivial; another pattern may have the opposite effect, as was shown for Y_2^2 heterogeneity which enhances helical core flow and hence improves dynamo efficiency (Sreenivasan 2009).

In both cases (Kuang et al. 2014; Sreenivasan and Jellinek 2012) it was found that low-latitude localized heating is favorable for terminating the dynamo, despite the distinctive role of convection in these two types of dynamo models. Sreenivasan and Jellinek (2012) argued that the critical latitude for plume location in order to kill the dynamo depends on the lowest latitude where axial columns driven by rapid rotation (Busse 1970) impinge the CMB (see Fig. 14a). Overall, these results may pose upper bounds on the amount of possible planet reorientation in the history of the Martian paleodynamo (Kuang et al. 2014; Sreenivasan and Jellinek 2012). The possibility of a major true polar wander (TPW) event on Mars is under debate. Significant TPW has been advocated for Early Noachian from studies of the crustal geoid associated to dynamical models for the formation of the crustal dichotomy (Roberts and Zhong 2007). TPW was also proposed for the subsequent 4 Gyr from studies of paleomagnetic poles (Hood et al. 2005) and paleoshorelines (Perron et al. 2007) or from theoretical models (Matsuyama et al. 2006). However, to our best knowledge, observed geological evidence for such processes (e.g., Grimm and Solomon 1986) are still lacking. Furthermore, models of lithospheric shell rotation associated with the apparent migration of Tharsis (e.g., Srámek and Zhong 2012) could lead to tectonic observations identical to that caused by TPW, although the two underlying physical processes are very different.

Some of the results obtained by Kuang et al. (2014) seem somewhat counter-intuitive, in particular the field morphology (see their Fig. 3). Their homogeneous CMB heat flux subcritical dynamo model indeed produces a hemispheric magnetic field (as previously reported by Kuang et al. 2008), with a peak at a specific latitude (~30°S). However, the field may reverse and this peak can migrate to different latitudes with time. It is therefore not clear why would the peak magnetic field of the heterogeneous dynamos almost coincide with that of the homogeneous case. It is also not intuitive that the field becomes spatially less heterogeneous with increasing heterogeneity amplitude. In addition, there is no suggestion for a depleted field at the location of the impact, as may be expected. Overall, it seems that the effects of mantle control on subcritical dynamos are distinctive from those in supercritical dynamos.

Finally, Monteux et al. (2015) imposed a localized impact-induced CMB heat flux pattern on numerical dynamos to explain the observed hemispheric dichotomy in the Martian crustal field (Acuña et al. 1998; Langlais et al. 2004). The amplitude of the hemispheric magnetic dichotomy is mostly controlled by the horizontal Rayleigh number which represents the vigor of the convection driven by the lateral variations of the CMB heat flux. Monteux et al. 2015 showed that, for a given horizontal Rayleigh number, an impact-induced CMB heat flux heterogeneity is more efficient than a synthetic degree-1 heterogeneity in generating strong hemispheric magnetic dichotomies. However, even with the more efficient impact-induced CMB heat flux pattern, several important conditions should be met in order to accept the impact-induced hemispheric dynamo scenario as the origin of the crustal field dichotomy, in particular the need for a very rapid crustal cooling rate (Dietrich and Wicht 2013). Dynamo models that recover the observed magnetic hemisphericity have a short average chron time of about 18 kyr (Monteux et al. 2015). However, in the case of molten material induced by a giant impact a major part of the atmosphere can be eroded and crustal formation may occur within only 1000 years (Lebrun et al. 2013; Solomatov 2007), possibly enabling the recording of the magnetic hemisphericity in the crust.

Future prospects

Because the CMB heat flux is an input to the dynamo models, understanding the dynamics of the lower mantle is a fundamental key to improve core dynamics models. In addition, progress in dynamo modeling is obviously relevant for all core-related problems, including the influence of heterogeneous (alternative or not) boundary conditions. Improved observations, in particular higher quality planetary magnetic field models, may provide tighter constraints on the dynamo models. Here we list some future prospects concerning alternative CMB heat flux and its influence on planetary dynamos. We also mention alternative scenarios that may also explain the non-trivial dynamo-related observations.

Knowledge of planetary magnetic fields will continue to grow in the coming years thanks to several new satellite missions that are currently flying or will soon be launched. Geomagnetic field and secular variation models will benefit from the successful Swarm ESA mission (Olsen et al. 2013). Three identical and magnetically dedicated satellites were launched simultaneously during the end of 2013 at two different altitudes to monitor with unprecedented quality the Earth's magnetic field and its secular variation. Measurements of the NASA MESSENGER mission around Mercury revealed a weak, dipole-dominated, and hemispheric magnetic field (Anderson et al. 2012). However, these measurements have not yet been fully exploited, and it is anticipated that alternative modeling techniques (Oliveira et al. 2015) and/or the very low-altitude measurements at the end of the mission will bring new insights on the field morphology of Mercury. The ESA/JAXA BepiColombo mission that will be launched in 2017 will complete the geographical coverage of low altitudes of the planet and will lead to the first global view of the magnetic field of Mercury (Benkhoff et al. 2010). The description of the past Martian magnetic field and its interpretation in terms of past dynamo processes will be improved by the recently inserted-into-orbit MAVEN mission. These measurements will be completed by the upcoming InSight mission, which will study the internal structure of Mars and bring constraints on the size of the liquid core. The CASSINI mission to Saturn (Dougherty et al. 2004) is extended until 2016, and the forthcoming Juno (around Jupiter) and later JUICE (around Jupiter moons) missions (Grasset et al. 2013) will also bring new measurements around the giant gas planets and their moons. These new planetary magnetic field models will provide more reliable observational constraints on the processes generating the fields in the interior of the planets and moons.

The identification of the post-perovskite phase transition (Murakami et al. 2004) provided better constraints on the Earth's deepest mantle. Seismic observations with improved tomographic images as well as sophisticated analyses of seismic waves were obtained. A better description of lower mantle materials with laboratory experiments reaching the appropriate high pressure range as well as time-consuming ab initio quantum chemistry computations became possible. As can be expected, the emerging picture bears a much greater complexity which includes the probable role of compositional variations and possible contributions of mantle melting and spin transitions of iron. While the last decade opened new questions such as the complex effect of composition on the post-perovskite phase transition, the precise nature of the melting curve and the points of dissent that remain between laboratory experiments and computational mineralogy, the establishment over the coming decade of an integrative framework (Shim and Lay 2014) would offer a much clearer interpretation of seismic observations and allow to better extract the purely thermal contribution from tomographic models.

Progress in understanding the dynamics of the Earth mantle in the last hundreds of Myr ultimately requires a better description of plate motions in the framework of mantle convection. Promising research paths include conceptual developments such as the understanding of lithospheric damage (Bercovici and Ricard 2014) as well as an improved treatment of plate-like behavior in spherical simulations of mantle dynamics (Bello et al. 2014; Rolf et al. 2012, 2014). In the meantime, improved reconstructions of plate velocities (Seton et al. 2012) can be prescribed to mantle convection models to investigate the role of specific ingredients such as mantle rheology, phase transitions and the nature of a possible dense basal layer. A more thorough investigation of mantle convection models than the pioneering attempt of Zhang and Zhong (2011) may establish more robust CMB heat flux evolution models than HF1.

One direct consequence of the results of the reversal frequency studies is the possibility to invert the record of paleomagnetic reversal frequency (Gradstein et al. 2012) for time-dependent core energetics. If the relation between pile height and CMB heat flux (mean and heterogeneity amplitude) is indeed approximately linear (Amit and Olson 2015; Olson and Amit 2014), the observed reversal frequency may be transformed into a relative CMB heat flux. This may provide a glimpse into the history of core convection.

As for the distant past of the Earth, prospects for other planetary objects mostly rely on modeling progress. Specific planetary scenarios that involve large impacts will most probably remain attractive in the next decade, concerning the ancient Martian magnetic field as well as other planets. Better descriptions of such events than the simplified approaches mentioned above can be simulated (Marinova et al. 2008, 2011). In terms of CMB heat flux, the geometry of impact heating with both a mantle contribution and a possible core contribution (Roberts and Arkani-Hamed 2014) could be described by such numerical tools. As mentioned in Monteux et al. (2015), the presence of impact-induced molten material is key in recording the impact-influenced magnetic field, and the interplay between the cooling time of this material and the transient dynamo properties will ultimately dictate the nature of remnant magnetism. In this regard, impact simulations could also supply patterns and volumes of ejecta deposits on the planetary surface. Describing the dynamics of the bulk molten material could involve on longer timescale phenomena such as isostatic adjustment of the retained melt region and lateral spreading as a gravity current (Reese et al. 2011).

Applications to other planets (apart from the Earth and Mars) may also be considered by invoking alternative CMB heat flux models. A Y_2^0 CMB heat flux pattern combined with internal heating may trigger an unstable odd convective mode which may explain the hemispheric field of Mercury (Cao et al. 2014; Wicht and Heyner 2014) as observed by MESSENGER (Anderson et al. 2012). While their dynamical scenario is convincing, the justification for a Y_2^0 thermal heterogeneity at the mantle of Mercury is somewhat vague. Examination of gravity and topographic heterogeneities in Mercury may provide a more realistic (alternative) model for the CMB heat flux which can be invoked to recover MESSENGER's magnetic observations.

Due to computational limitations, dynamo models operate in a parametric regime very far from that of planetary core conditions (Glatzmaier 2002). This is especially restrictive for long simulations, as is the case when the long-term impact of an heterogeneous mantle is investigated. One of the most challenging problems in terms of computer power is that of modeling strongly time-variable magnetic reversal frequency, because adequate statistics require extremely long simulations. For example, the Ekman numbers used by Olson et al. (2013) and Amit and Olson (2015) are about four orders of magnitudes off the lowest values that were reported for numerical dynamos (Sakuraba and Roberts 2009). Improved numerical techniques which will allow running simulations with more challenging parameters, in particular lower Ekman and magnetic Prandtl numbers, are therefore of great interest for studies of mantle control on planetary dynamos. More efficient, massively parallelized dynamo codes are currently under development (http://geodynamics.org/cig/working-groups/geodynamo/).

More accurate incorporation of physical phenomena in dynamo models will also improve their predictions. In particular, correct implementation of thermochemical convection is necessary. Currently, most studies use the co-density formulation (Braginsky and Roberts 1995). Apart from neglecting double-diffusive convection, this approach fails to accurately prescribe the inner boundary condition. Some models have already incorporated thermal and chemical convection with appropriate boundary conditions in numerical dynamos (Manglik et al. 2010; Takahashi 2014; Trümper et al. 2012). Tracer methods which are commonly used in mantle convection simulations (e.g., Tackley and King 2003) may be applied to chemical convection in dynamo codes for more accurate numerical schemes. Other physical complications such as stable stratification in some regions of the outer core (Gubbins and Davies 2013) and compressibility may also be incorporated into dynamo models.

Observed dynamo properties that require boundary heterogeneity may alternatively be explained by lateral anomalies at the inner core boundary. Olson and Deguen (2012) showed that the dipole eccentricity in archeomagnetic field models can be reproduced by dynamo simulations with heterogeneous Y_1^1 inner boundary bouyancy flux pattern, corresponding to the east-west seismic anomalies at the upper inner core (Tanaka and Hamaguchi 1997). Aubert et al. (2013) showed that the localization of the westward drift of the geomagnetic field at the Atlantic hemisphere is controlled by the same heterogeneous inner boundary buoyancy flux. In these studies, the inner boundary heterogeneity amplitude was set to be significantly larger than that of the outer boundary, therefore the role of the CMB heterogeneity becomes secondary. It would be interesting to explore dynamo models with comparable amplitude heterogeneities on both boundaries to investigate competing effects.

Progress in understanding mantle dynamics will lead to better description of boundary conditions for the dynamo and thus improved knowledge of core dynamics. On the other hand, magnetic observations may be used to constrain mantle convection scenarios. Overall, a multidisciplinary approach with interactive exchange of information among the different envelopes may advance the study of planetary deep interiors.

Abbreviations

CMB: core-mantle boundary; LLSVPs: large low shear-wave velocity provinces; ULVZs: ultra low velocity zones; KRS: Kiaman reversed superchron; CNS: cretaceous normal superchron; ESA: European Space Agency; NASA: National Aeronautics and Space Administration; MESSENGER: MErcury surface: Space ENvironment: GEochemistry: and ranging; JAXA: Japan Aerospace eXploration Agency; MAVEN: Mars atmosphere and volatile evolution; InSight: Interior exploration using seismic investigations: Geodesy and Heat Transport; JUICE: JUpiter ICy moon explorer.

Competing interests

The authors declare that they have no competing interests.

Authors' contributions

The idea to this paper was in fact conceived following an invitation of HA by Takashi Nakagawa to give a talk on the topic of this paper in SEDI 2014. HA and GC wrote the first draft. GC, FD, and JM designed the illustrative Fig. 1. All authors discussed and improved the draft. All authors read and approved the final manuscript.

Acknowledgements

We thank Takashi Nakagawa, Binod Sreenivasan, and Weijia Kuang for providing material from their papers and for helpful discussions. H. Amit, G. Choblet, J. Monteux, and B. Langlais were partly supported by ANR project MARMITE (ANR-13-BS05-0012). J. Monteux was partly funded by ANR project Oxydeep (ANR-13-BS06-0008). F. Deschamps was supported by Academia Sinica grant 102-CDA-M02. We are grateful to James Roberts and Johannes Wicht for detailed reviews that significantly improved this review paper.

Author details

[1] CNRS UMR 6112, Université de Nantes, Laboratoire de Planétologie et de Géodynamique, 2 rue de la Houssinière, Nantes, F-44000, France. [2] Department of Earth and Planetary Sciences, Johns Hopkins University, 21218 Baltimore, MD, USA. [3] Laboratoire Magmas et Volcans, Université Blaise Pascal - CNRS - IRD, OPGC, 5 rue Kessler, 63038 Clermont Ferrand, France. [4] Institute of Earth Sciences, Academia Sinica, 128 Academia Road Sec. 2, Nangang, Taipei 11529, Taiwan.

References

Acuña M, Connerney J, Wasilewski P, Lin R, Mitchell D, Anderson K, Carlson C, McFadden J, Remè H, Mazelle C, Vignes D, Bauer S, Cloutier P, Ness N (2001) Magnetic field of mars: Summary of results from the aerobraking and mapping orbits. J Gephys Res 106:23403–23417

Acuña MH, Connerney JEP, Wasilewski P, Lin RP, Anderson KA, Carlson CW, McFadden J, Curtis DW, Mitchell D, Reme H, Mazelle C, Sauvaud JA, d'Uston C, Cros A, Medale JL, Bauer SJ, Cloutier P, Mayhew M, Winterhalter

D, Ness NF (1998) Magnetic field and plasma observations at Mars: Initial results of the Mars global surveyor mission. Science 279:1676–1680
Alboussiére T, Deguen R, Melzani M (2010) Melting-induced stratification above the Earth's inner core due to convective translation. Nature 466:744–747
Amit H, Aubert J, Hulot G (2010a) Stationary, oscillating or drifting mantle-driven geomagnetic flux patches J Geophys Res B07108. doi:10.1029/2009JB006542
Amit H, Choblet G (2009) Mantle-driven geodynamo features - effects of post-perovskite phase transition. Earth Planets Space 61:1255–1268
Amit H, Choblet G (2012) Mantle-driven geodynamo features - effects of compositional and narrow d" anomalies. Phys Earth Planet Inter 190-191:34–43
Amit H, Christensen U (2008) Accounting for magnetic diffusion in core flow inversions from geomagnetic secular variation. Geophys J Int 175:913–924
Amit H, Christensen U, Langlais B (2011a) The influence of degree-1 mantle heterogeneity on the past dynamo of Mars. Phys Earth Planet Inter 189:63–79
Amit H, Deschamps F, Choblet G (in press) Numerical dynamos with outer boundary heat flux inferred from probabilistic tomography – consequences for latitudinal distribution of magnetic flux. Geophys J Int. doi:10.1093/gji/ggv332
Amit H, Korte M, Aubert J, Constable C, Hulot G (2011b) The time-dependence of intense archeomagnetic flux patches. J Geophys Res 116:B12106. doi:10.1029/2011JB008538
Amit H, Leonhardt R, Wicht J (2010b) Polarity reversals from paleomagnetic observations and numerical dynamo simulations. Space Sci Rev 155:293–335
Amit H, Olson P (2006) Time-average and time-dependent parts of core flow. Phys Earth Planet Inter 155:120–139
Amit, H, Olson P (2015) Lower mantle superplume growth excites geomagnetic reversals. Earth Planet Sci Lett 414:68–76
Anderson BJ, Johnson CL, Korth H, Purucker ME, Winslow RM, Slavin JA, Solomon SC, McNutt RL, Raines JM, Zurbuchen TH (2012) The global magnetic field of Mercury from MESSENGER orbital observations. Science 333:1859–1862
Arkani-Hamed J (2012) Life of the Martian dynamo. Phys Earth Planet Inter 196:83–96
Arkani-Hamed J, Olson P (2010) Giant impact stratification of the Martian core. Geophys Res Lett 37:L02201. doi:10.1029/2009GL041417
Aubert J (2005) Steady zonal flows in spherical shell fluid dynamos. J Fluid Mech 542:53–67
Aubert J, Amit H, Hulot G (2007) Detecting thermal boundary control in surface flows from numerical dynamos. Phys Earth Planet Inter 160:143–156
Aubert J, Amit H, Hulot G, Olson P (2008) Thermo-chemical wind flows couple Earth's inner core growth to mantle heterogeneity. Nature 454:758–761
Aubert J, Finlay CC, Fournier A (2013) Bottom-up control of geomagnetic secular variation by the Earth's inner core. Nature 502:219–223
Aubert J, Labrosse S, Poitou C (2009) Modelling the paleo-evolution of the geodynamo. Geophys J Int 179:1414–1428
Aurnou J, Andreadis S, Zhu L, Olson P (2003) Experiments on convection in Earth's core tangent cylinder. Earth Planet Sci Lett 212:119–134
Bello L, Coltice N, Rolf T, Tackley PJ (2014) On the predictability limit of convection models of the Earth's mantle. Geochem Geophys Geosys 15. doi:10.1002/2014GC005254
Benkhoff J, van Casteren J, Hayakawa H, Fujimoto M, Laakso H, Novara M, Ferri P, Middleton HR, Ziethe R (2010) BepiColombo - Comprehensive exploration of Mercury: Mission overview and science goals. Planet Space Sci58. doi:10.1016/j.pss.2009.09.020
Bercovici D, Ricard Y (2014) Plate tectonics, damage and inheritance. Nature 508:513–516
Biggin AJ, Steinberger B, Aubert J, Suttie N, Holme R, Torsvik TH, van der Meer DG, van Hinsbergen DJJ (2012) Possible links between long-term geomagnetic variations and whole-mantle convection processes. Nature Geosci 5:526–533
Bloxham J (2002) Time-independent and time-dependent behaviour of high-latitude flux bundles at the core-mantle boundary. Geophys Res Lett 29. doi:10.1029/2001gl014543
Braginsky SI, Roberts PH (1995) Equations governing convection in Earth's core and the geodynamo. Geophys Astrophys Fluid Dyn 79:1–97
Breuer D, Labrosse S, Spohn T (2010) Thermal evolution and magnetic field generation in terrestrial planets and satellites. Space Sci Rev 152:449–500
Bull AL, McNamara AK, Ritsema J (2009) Synthetic tomography of plume clusters and thermochemical piles. Earth Planet Sci Lett 278:152–162
Burke K (2011) Plate tectonics, the Wilson cycle, and mantle plumes: geodynamics from the top. Annu Rev Earth Planet Sci 39:1–29
Burke K, Steinberger B, Torsvik TH, Smethhurst MA (2008) Plume generation zones at the margins of large low shear velocity provinces on the core-mantle boundary. Earth Planet Sci Lett 265:49–60
Busse FH (1970) Thermal instabilities in rapidly rotating systems. J Fluid Mech 44:441–460
Cao H, Aurnou JM, Wicht J, Dietrich W, Soderlund KM, Russell CT (2014) A dynamo explanation for Mercury's anomalous magnetic field. Geophys Res Lett 41. doi:10.1002/2014GL060196
Christensen U, Aubert J (2006) Scaling properties of convection-driven dynamos in rotating spherical shells and application to planetary magnetic fields. Geophys J Int 166:97–114
Christensen U, Aubert J, Hulot G (2010) Conditions for Earth-like geodynamo models. Earth Planet Sci Lett 296:487–496
Christensen U, Olson P (2003) Secular variation in numerical geodynamo models with lateral variations of boundary heat flow. Phys Earth Planet Inter 138:39–54
Christensen U, Olson P, Glatzmaier G (1998) A dynamo model interpretation of geomagnetic field structures. Geophys Res Lett 25:1565–1568
Christensen U, Wicht J (2007) Numerical dynamo simulations. In: Olson P (ed). Treatise on Geophysics Vol 8. Elsevier Science, Amsterdam
Cizkova H, Cadek O, Matyska C, Yuen D (2010) Implications of post-perovskite properties for core-mantle dynamics. Phys Earth Planet Inter 180:235–243
Cobden L, Mosca I, Trampert JJR (2012) On the likelihood of post-perovskite near the core-mantle boundary: a statistical interpretation of seismic observations. Phys Earth Planet Inter 210-211:21–35
Cobden L, Thomas C (2013) The origin of d" reflections a systematic study of seismic array data sets. Geophys J Int 194:1091–1118
Coe RS, Hongre L, Glatzmaier GA (2000) An examination of simulated geomagnetic reversals from a palaeomagnetic perspective. Phil Trans R Soc Lond 358:1141–1170
Courtillot V, Besse J (1987) Magnetic field reversals, polar wander, and core-mantle coupling. Science 237:1140–1147
Davaille A (1999) Simultaneous generation of hotspots and superswells by convection in a heterogeneous planetary mantle. Nature 402:756–760
Davies CJ, Gubbins D, Willis AP, Jimack PK (2008) Time-averaged paleomagnetic field and secular variation: Predictions from dynamo solutions based on lower mantle seismic tomography. Phys Earth Planet Inter 169:194–203
Deschamps F, Cobden L, Tackley PJ (2012) The primitive nature of large low shear-wave velocity provinces. Earth Planet Sci Lett 349-350:198–208
Deuss A (2014) Heterogeneity and anisotropy of Earth's inner core. Ann Rev Earth Planet Sci 42:103–126
Dietrich W, Wicht J (2013) A hemispherical dynamo model: Implications for the Martian crustal magnetization. Phys Earth Planet Inter 217:10–21
Dougherty MK, Kellock S, Southwood DJ, Balogh A, Smith EJ, Tsurutani BT, Gerlach B, Glassmeier K-H, Gleim F, Russell CT, Erdos G, Neubauer FM, Cowley SWH (2004) The Cassini magnetic field investigation. Space Sc Rev 114. doi:10.1007/s11214-004-1432-2
Dziewonski AM, Lekic V, Romanowicz BA (2010) Mantle anchor structure: an argument for bottom up tectonics. Earth Planet Sci Lett 299:69–79
Elkins-Tanton LT, Zaranek SE, Parmentier EM, Hess PC (2005) Early magnetic field and magmatic activity on Mars from magma ocean cumulate overturn. Earth Planet Sci Lett 236:1–12
Gillet N, Pais MA, Jault D (2009) Ensemble inversion of time-dependent core flow models. Geochem Geophys Geosyst 10. doi:10.1029/2008GC002290
Glatzmaier GA (2002) Geodynamo simulations: how realistic are they Annu Rev Earth Planet Sci Lett 30:237–257
Glatzmaier GA, Coe R, Hongre L, Roberts PH (1999) The role of the earth's mantle in controlling the frequency of geomagnetic reversals. Nature 401:885–890
Glatzmaier GA, Roberts PH (1995) A three-dimensional self-consistent computer simulation of a geomagnetic field reversal. Nature 377:203–209
Glatzmaier, GA, Roberts PH (1997) Simulating the geodynamo. Contemp Phys 38:269–288

Glenn Sterenborg M, Crowley JW (2013) Thermal evolution of early solar system planetesimals and the possibility of sustained dynamos. Phys Earth Planet Inter 214:53–73

Gradstein F, Ogg J, Schmitz M, Ogg G (2012) The Geologic Time Scale 2012. Elsevier Science, Amsterdam

Grasset O, Dougherty MK, Coustenis A, Bunce EJ, Erd C, Titov D, Blanc M, Coates A, Drossart P, Fletcher LN, Hussmann H, Jaumann R, Krupp N, Lebreton J-P, Prieto-Ballesteros O, Tortora P, Tosi F, Van Hoolst T (2013) JUpiter ICy moons Explorer (JUICE): An ESA mission to orbit Ganymede and to characterise the Jupiter system. Planet Space Sci 78. doi:10.1016/j.pss.2012.12.002

Grimm RE, Solomon SC (1986) Tectonic tests of proposed polar wander paths for Mars and the Moon. Icarus 65:110–121

Gubbins D (2003) Thermal core-mantle interactions: theory and observations. In: Dehant V, Creager K, Karato S, Zatman S (eds). Earth's Core: dynamics, structure and rotation. AGU Geodynamics Series American Geophysical Union, Washington DC

Gubbins D, Davies CJ (2013) The stratified layer at the core-mantle boundary caused by barodiffusion of oxygen, sulphur and silicon. Phys Earth Planet Inter 215:21–28

Gubbins D, Sreenivasan B, Mound J, Rost S (2011) Melting of the Earth's inner core. Nature 473:361–363

Gubbins D, Willis PW, Sreenivasan B (2007) Correlation of Earth's magnetic field with lower mantle thermal and seismic structure. Phys Earth Planet Inter 162:256–260

Harder H, Christensen UR (1996) A one-plume model of Martian mantle convection. Nature 380:507–509

Heimpel MH, Evans ME (2013) Testing the geomagnetic dipole and reversing dynamo models over Earth's cooling history. Phys Earth Planet Inter 224:124–131

Hernlund J, Thomas C, Tackley PJ (2005) A doubling of the post-perovskite phase boundary and structure of the Earth's lowermost mantle. Nature 434:882–886

Hoffman KA (1996) Transitional paleomagnetic field behavior: Preferred paths or patches? Surv Geophys 17:207–211

Holme R (2007) Large-scale flow in the core. In: Olson P (ed). Treatise on Geophysics, Vol 8. Elsevier Science, Amsterdam

Holme R, Olsen N (2006) Core surface ow modelling from high-resolution secular variation. Geophys J Int 166:518–528

Hood LL, Young CN, Richmond NC, Harrison KP (2005) Modeling of major Martian magnetic anomalies: Further evidence for polar reorientations during the Noachian. Icarus 177. doi:10.1016/j.icarus.2005.02.008

Hori K, Wicht J (2013) Subcritical dynamos in the early Mars' core: Implications for cessation of the past Martian dynamo. Phys Earth Planet Inter 219:21–33

Hulot G, Eymin C, Langlais B, Mandea M, Olsen N (2002) Small-scale structure of the geodynamo inferred from Oersted and Magsat satellite data. Nature 416:620–623

Irving JCE, Deuss A (2011) Hemispherical structure in inner core velocity anisotropy. J Geophys Res 116. doi:10.1029/2010JB007942

Ishii M, Tromp J (1999) Normal-mode and free-air gravity constraints on lateral variations in velocity and density of Earth's mantle. Science 285:1231–1236

Jackson A (2003) Intense equatorial flux spots on the surface of the Earth's core. Nature 424:760–763

Jackson A, Jonkers ART, Walker MR (2000) Four centuries of geomagnetic secular variation from historical records. Phil Trans R Soc Lond A 358:957–990

Jellinek AM, Johnson CL, Schubert G (2008) Constraints on the elastic thickness, heat flow, and melt production at early Tharsis from topography and magnetic field observations. J Geophys Res 113:E09004

Johnson CL, Phillips RJ (2005) Evolution of the tharsis region of Mars: insights from magnetic field observations. Earth Planet Sci Lett 230:241–254

Jutzi M, Asphaug E (2011) Forming the lunar farside highlands by accretion of a companion moon. Nature 476:69–72

Ke Y, Solomatov VS (2006) Early transient superplumes and the origin of the Martian crustal dichotomy. J Geophys Res 111:10001

Ke Y, Solomatov VS (2009) Coupled core-mantle thermal evolution of early Mars. J Geophys Res 114:1–12

Kelly P, Gubbins D (1997) The geomagnetic field over the past 5 million years. Geophys J Int 128:315–330

Korte M, Donadini F, Constable C (2009) The geomagnetic field for 0-3ka: 2. a new series of time-varying global models. J Geophys Res 10:Q06008. doi:10.1029/2008GC002297

Korte M, Holme R (2010) On the persistence of geomagnetic flux lobes in global Holocene field models. Phys Earth Planet Inter 182:179–186

Kuang W, Jiang W, Roberts J, Frey HV (2014) Could giant basin-forming impacts have killed martian dynamo Geophys Res Lett 41:8006–8012

Kuang W, Jiang W, Wang T (2008) Sudden termination of Martian dynamo? implications from subcritical dynamo simulations. Geophys Res Lett 35:L14204

Kutzner C, Christensen UR (2002) From stable dipolar towards reversing numerical dynamos. Phys Earth Planet Inter 131:29–45

Kutzner C, Christensen UR (2004) Simulated geomagnetic reversals and preferred virtual geomagnetic pole paths. Geophys J Int 157: 1105–1118

Labrosse S (2003) Thermal and magnetic evolution of the earth's core. Phys Earth Planet Inter 140:127–143

Langlais B, Purucker M (2007) A polar magnetic paleopole associated with Apoolinaris Patera Mars. Planet Space Sci 55:270–279

Langlais B, Purucker M, Mandea M (2004) Crustal magnetic field of Mars. J Geophys Res 109:E02008

Lay T, Herlund J, Buffett B (2008) Core-mantle boundary heat flow. Nature Geosci 1:25–32

Lay T, Hernlund J, Garnero EJ, Thorne MS (2006) A post-perovskite lens and D" heat flux beneath the central Pacific. Science 314:1272–1276

Lebrun T, Massol H, Chasseflére E, Davaille A, Marcq E, Sarda P, Leblanc F, Branedeis G (2013) Thermal evolution of an early magma ocean in interaction with the atmosphere. J Geophys Res 118:1155–1176

Lekic V, Cottaar S, Dziewonski A, Romanowicz B (2012) Cluster analysis of global lower mantle tomography: A new class of structure and implications for chemical heterogeneity. Earth Planet Sci Lett 357-358:68–77

Lhuillier F, Hulot G, Gallet Y (2013) Statistical properties of reversals and chrons in numerical dynamos and implications for the geodynamo. Phys Earth Planet Inter 220:19–36

Lillis RJ, Frey HV, Manga M (2008) Rapid decrease in Martian crustal magnetisation in the Noachian era: implications for the dynamo and climate of early mars. Geophys Res Lett 35:L14203

Lithgow-Bertelloni C, Richards MA (1998) Dynamics of cenozoic and mesozoic plate motion. Rev Geophys 36:27–78

Love JJ (1998). Paleomagnetic volcanic data and geometric regularity of reversals and excursions 103:12435–12452

Manglik A, Wicht J, Christensen UR (2010) A dynamo model with double diffusive convection for Mercury's core. Earth Planet Sci Lett 289: 619–628

Marinova MM, Aharonson O, Asphaug E (2008) Mega-impact formation of the Mars hemispheric dichotomy. Nature 453:1216–219

Marinova MM, Aharonson O, Asphaug E (2011) Geophysical consequences of planetary-scale impacts into a Mars-like planet. Icarus 211:960–985

Masters G, Laske G, Bolton H, Dziewonski A (2000) The relative behavior of shear velocity, bulk sound velocity, and compressional velocity in the mantle: Implications for chemical and thermal structure. In: Karato S, Forte A, Liebermann R, Masters G, Stixrude L (eds). Earths deep interior, Vol 117. AGU Monograph, Washington D.C

Matsuyama I, Mitrovica JX, Manga M, Perron JT, Richards MA (2006) Rotational stability of dynamic planets with elastic lithospheres. J Geophys Res 111. doi:10.1029/2005JE002447

McNamara A, Garnero E, Rost S (2010) Tracking deep mantle reservoirs with ultra-low velocity zones. Earth Planet Sci Lett 299:1–9

McNamara A, Zhong S (2005) Thermochemical structures beneath Africa and the Pacific Ocean. Nature 437:1136–1139

Merrill RT, McElhinny MW, McFadden PL (1998) The Magnetic Field of the Earth: Paleomagnetism, the Core, and the Deep Mantle. Academic Press, San Diego, California, USA

Merrill RT, McFadden PL (1999) Geomagnetic polarity transitions 37:201–226

Milbury C, Schubert G, Raymond CA, Smrekar SE, Langlais B (2012) The history of Mars' dynamo as revealed by modeling magnetic anomalies near Tyrrhenus Mons and Syrtis Major. J Geophys Res 117. doi:10.1029/2012JE004099

Moffatt HK (1978) Magnetic Field Generation in Electrically Conducting Fluids. Cambridge University Press, Cambridge, UK

Monnereau M, Calvet M, Margerin L, Souriau A (2010) Lopsided growth of Earth's inner core. Science 238:1014–1017

Monteux J, Amit H, Choblet G, Langlais B, Tobie G (2015) Giant impacts, heterogeneous mantle heating and a past hemispheric dynamo on Mars. Phys Earth Planet Inter 240:114–124

Monteux J, Arkani-Hamed J (2014) Consequences of giant impacts in early Mars: Core merging and Martian dynamo evolution. J Geophys Res 119:480–505

Monteux J, Coltice N, Dubuffet F, Ricard Y (2007) Thermo-mechanical adjustment after impacts during planetary growth. Geophys Res Lett 34:24201–24205

Monteux J, Jellinek AM, Johnson CL (2011) Why might planets and moons have early dynamos Earth Planet Sci Lett 310:349–359

Monteux J, Jellinek AM, Johnson CL (2013) Dynamics of core merging after a mega-impact with applications to Mars' early dynamo. Icarus 226:20–32

Monteux J, Schaeffer N, Amit H, Cardin P (2012) Can a sinking metallic diapir generate a dynamo J Geophys Res 117:E10005. doi:10.1029/2012JE004075

Mosca I, Cobden L, Deuss A, Ritsema J, Trampert J (2012) Seismic and mineralogical structures of the lower mantle from probabilistic tomography. J Geophys Res 117. doi:10.1029/2011JB008851

Murakami M, Hirose K, Sata N, Ohishi Y, Kawamura K (2004) Post-perovskite phase transition in MgSio3. Science 304:855–858

Nakagawa T, Tackley PJ (2008) Lateral variations in cmb heat flux and deep mantle seismic velocity caused by a thermal-chemical-phase boundary layer in 3d spherical convection. Earth Planet Sci Lett 271:348–358

Nakagawa T, Tackley PJ (2011) Effects of low-viscosity post-perovskite on thermo-chemical mantle convection in a 3-D spherical shell. Geophys Res Lett 38:L04309

Ni S, Helmberger DV, Tromp J (2005) Three-dimensional structure of the African superplume from waveform modelling. Geophys J Int 161:283–294

Ni S, Tan E, Gurnis M, Helmberger D (2002) Sharp sides to the African superplume. Science 296:1850–1852

Nimmo F, Hart SD, Horycansky DG, Agnor CB (2008) Implications of an impact origin for the Martian hemispheric dichotomy. Nature 453:1220–1223

Oganov A, Ono S (2004) Theoretical and experimental evidence for a post-perovskite phase of MgSio3 in Earth's D" layer. Nature 430:445–448

Oliveira JS, Langlais B, Pais MA, Amit H (2015) A new method to model partially distributed magnetic field measurements, with application to Mercury. J Geophys Res. 120. doi:10.1002/2014JE004734

Olsen N, Luehr H, Finlay CC, Sabaka TJ, Michaelis I, Rauberg J, Tøffner-Clausen L (2014) The CHAOS-4 geomagnetic field model. Geophys J Int 197:815–827

Olson P (2007a) Gravitational dynamos and the low frequency geomagnetic secular variation. Proc Nat Acad Sci 104:20159–20166

Olson, P (2007b) Overview. In: Olson P (ed). Treatise on Geophysics, Vol 8. Elsevier Science, Amsterdam

Olsen N, Friis-Christensen E, Floberghagen R, Alken P, Beggan CD, Chulliat A, Doornbos E, da Encarnação JT, Hamilton B, Hulot G, van den IJssel J, Kuvshinov A, Lesur V, Lühr H, Macmillan S, Maus S, Noja M, Olsen PEH, Park J, Plank G, Püthe C, Rauberg J, Ritter P, Rother M, Sabaka TJ, Schachtschneider R, Sirol O, Stolle C, Thébault E, Thomson AWP, et al (2013) The Swarm Satellite Constellation Application and Research Facility (SCARF) and Swarm data products. Earth, Planets and Space 65(11):1189–1200. doi:10.5047/eps.2013.07.001

Olson P, Amit H (2014) Magnetic reversal frequency scaling in dynamos with thermochemical convection. Phys Earth Planet Inter 229:122–133

Olson P, Christensen U (2002) The time averaged magnetic field in numerical dynamos with nonuniform boundary heat flow. Geophys J Int 151:809–823

Olson P, Christensen U (2006) Dipole moment scaling for convection-driven planetary dynamos. Earth Planet Sci Lett 250:561–571

Olson P, Christensen UR, Glatzmaier GA (1999) Numerical modeling of the geodynamo: Mechanisms of field generation and equilibration. J Geophys Res 104:10383–110404

Olson P, Coe RS, Driscoll PE, Glatzmaier GA, Roberts PH (2010) Geodynamo reversal frequency and heterogeneous core-mantle boundary heat flow. Phys Earth Planet Inter 180:66–79

Olson P, Deguen R (2012) Lopsided inner core growth and eccentricity of the geomagnetic dipole. Nat Geosci 5(8):565–569

Olson P, Deguen R, Hinnov LA, Zhong S (2013) Controls on geomagnetic reversals and core evolution by mantle convection in the Phanerozoic. Phys Earth Planet Inter 214:87–103

Perron JT, Mitrovica JX, Manga M, Matsuyama I, Richards MA (2007) Evidence for an ancient Martian ocean in the topography of deformed shorelines. Nature 447:840–843

Pierazzo E, Vickery AM, Melosh HJ (1997) A reevaluation of impact melt production. Icarus 127:408–423

Pozzo M, Davies C, Gubbins D, Alfè D (2012) Thermal and electrical conductivity of iron at Earth's core conditions. Nature 485:355–358

Reese CC, Orth CP, Solomatov VS (2011) Impact megadomes and the origin of the martian crustal dichotomy. Icarus 213:433–442

Reese CC, Solomatov VS (2010) Early martian dynamo generation due to giant impacts. Icarus 207:82–97

Reese CC, Solomatov VS, Baumgardner JR (2002) Survival of impact-induced thermal anomalies in the Martian mantle. J Geophys Res 107:1–12

Ritsema J, McNamara A, Bull A (2007) Tomographic filtering of geodynamic models: Implications for model interpretation and large-scale mantle structure. J Geophys Res 112:B01303

Roberts JH, Arkani-Hamed J (2012) Impact-induced mantle dynamics on Mars. Icarus 218:278–289

Roberts JH, Arkani-Hamed J (2014) Impact heating and coupled core cooling and mantle dynamics on Mars. J Geophys Res 119:729–744

Roberts JH, Barnouin OS (2012) The effect of the Caloris impact on the mantle dynamics and volcanism of Mercury. J Geophys Res 117:E02007

Roberts JH, Lillis RJ, Manga M (2009) Giant impacts on early Mars and the cessation of the Martian dynamo. J Geophys Res 114:E04009. doi:10.1029/2008JE003287

Roberts JH, Zhong S (2006) Degree-1 convection in the Martian mantle and the origin of the hemispheric dichotomy. J Geophys Res 111:E06013

Roberts JH, Zhong S (2007) The cause for the north-south orientation of the crustal dichotomy and the equatorial location of Tharsis on Mars. Icarus 190(1):24–31

Rolf T, Coltice N, Tackley PJ (2012) Linking continental drift, plate tectonics and the thermal state of the Earth's mantle. Earth Planet Sci Lett. 351–352:134–146

Rolf T, Coltice N, Tackley PJ (2014) Statistical cyclicity of the supercontinent cycle. Geophys Res Lett 41. doi:10.1002/2014GL059595

Romanowicz B, Gung Y (2002) Superplumes from the core-mantle boundary to the lithosphere: Implications for heat flux. Science 296:513–516

Sakuraba A, Roberts PH (2009) Generation of a strong magnetic field using uniform heat flux at the surface of the core. Nat Geosci 2:802–805

Schubert G, Masters G, Olson P, Tackley P (2004) Superplumes or plume clusters Phys Earth Planet Inter 146:147–162

Schubert G, Spohn T (1990) Thermal history of Mars and the sulfur content of its core. J Geophys Res 95:14095–14104

Scotese CR (2001) Atlas of Earth history. In: PALEOMAP Progress Rep. 90-0497. Dep. of Geol, Univ. of Tex. at Arlington

Senshu H, Kuramoto K, Matsui T (2002) Thermal evolution of a growing mars. J Geophys Res 107:1–13

Seton M, Müller RD, Zahirovic S, Gaina C, Torsvik T, Shephard G, Talsma A, Gurnis M, Turner M, Maus S, Chandler M (2012) Global continental and ocean basin reconstructions since 200 Ma. Earth-Sci Rev 113: 212–270

Shim S-H, Lay T (2014) Post-perovskite at ten. Nat Geosci 7:621–623

Solomatov VS (2007) Magma oceans and primordial mantle differentiation. In: Schubert G (ed). Treatise on Geophysics, Vol 9. Elsevier Science, Amsterdam

Srámek O, Zhong S (2010) Long-wavelength stagnant lid convection with hemispheric variation in lithospheric thickness: link between Martian crustal dichotomy and Tharsis J Geophys Res 115:E09010

Srámek O, Zhong S (2012) Martian crustal dichotomy and Tharsis formation by partial melting coupled to early plume migration. J Geophys Res 117:E01005

Sreenivasan B (2009) On dynamo action produced by boundary thermal coupling. Phys Earth Planet Inter 177:130–138

Sreenivasan B, Gubbins D (2011) On mantle-induced heat flow variations at the inner core boundary. Phys Earth Planet Inter 187:336–341

Sreenivasan B, Jellinek AM (2012) Did the tharsis plume terminate the Martian dynamo? Earth Planet Sci Lett 209–217

Sreenivasan B, Jones CA (2011) Helicity generation and subcritical behaviour in rapidly rotating dynamos. J Fluid Mech 688:5–30

Stanley S (2010) A dynamo model for axisymmetrizing Saturn's magnetic field. Geophys Res Lett 37:5201

Stanley S, Elkins-Tanton L, Zuber MT, Parmentier EM (2008) Mars' paleomagnetic field as the result of a single-hemisphere dynamo. Science 321:1822–1825

Stevenson D (1983) Planetary magnetic fields. Rep Prog Phys 46:555–620

Tackley P (2002) The strong heterogeneity caused by deep mantle layering. Geophys Geochem Geosyst 3. doi:10.1029/2001GC000167

Tackley PJ (2011) Living dead slabs in 3-D: The dynamics of compositionally-stratified slabs entering a 'slab graveyard' above the core-mantle boundary. Phys Earth Planet Inter 188:150–162
Tackley PJ, King SD (2003) Testing the tracer ratio method for modeling active compositional fields in mantle convection simulations. Geochem Geophys Geosyst 4. doi:10.1029/2001GC000214
Takahashi F (2014) Double diffusive convection in the Earth's core and the morphology of the geomagnetic field. Phys Earth Planet Inter 226:83–87
Takahashi F, Tsunakawa H, Matsushima M, Mochizuki N, Honkura Y (2008) Effects of thermally heterogeneous structure in the lowermost mantle on the geomagnetic field strength. Earth Planet Sci Lett 272:738–746
Tan E, Gurnis M (2007) Compressible thermochemical convection and application to lower mantle structures. J Geophys Res 112. doi:10.1029/2006JB004505
Tanaka S, Hamaguchi H (1997) Degree one heterogeneity and hemispherical variation of anisotropy in the inner core from PKP(BC)-PKP(DF) times. J Geophys Res 102:2925–2938
To A, Romanowicz B, Capdeville Y, Takeuchi N (2005) 3D effects of sharp boundaries at the borders of the African and Pacific superplumes: Observation and modeling. Earth Planet Sci Lett 233:137–153
Tonks WB, Melosh HJ (1993) Magma ocean formation due to giant impacts. J Geophys Res 98:5319–5333
Torsvik TH, Burke K, Steinberger B, Webb SJ, Ashwel LD (2010) Diamonds sampled by plumes from the core-mantle boundary. Nature 466:352–355
Torsvik TH, Smethhurst MA, Burke K, Steinberger B (2006) Large igneous provinces generated from the margins of the large low-velocity provinces in the deep mantle. Geophys J Int 167:1447–1460
Trampert J, Deschamps F, Resovsky J, Yuen D (2004) Probabilistic tomography maps chemical heterogeneities throughout the lower mantle. Science 306:853–856
Trümper T, Breuer M, Hansen U (2012) Numerical study on double-diffusive convection in the Earth's core. Phys Earth Planet Inter 194–195:55–63
Wang Y, Wen L (2007) Geometry and P and S velocity structure of the "African Anomaly". J Geophys Res 112:B05313
Watters WA, Zuber MT, Hager BH (2009) Thermal perturbations caused by large impacts and consequences for mantle convection. J Geophys Res 114:E02001
Wicht J, Heyner D (2014) Mercury's magnetic field in the messenger era. In: Shuanggen J (ed). Planetray Geodesy and Remote Sensing. CRC Press, London
Wicht J, Stellmach S, Harder H (2009) Numerical models of the geodynamo: From fundamental Cartesian models to 3D simulations of field reversals. In: Glassmeier H, Soffel H, Negendank J (eds). Geomagnetic Field Variations - Space-time structure, processes, and effects on system Earth. Springer, Berlin
Wicht, J, Stellmach S, Harder H (2011) Numerical dynamo simulations: From basic concepts to realistic models. In: Freeden W, Nashed M, Sonar T (eds). Handbook of Geomathematics. Springer, Berlin - Heidelberg - New York
Williams GE (2000) Geological constraints on the Precambrian history of Earth's rotation and the Moon's orbit. Rev Geophys 38:37–59
Willis PW, Sreenivasan B, Gubbins D (2007) Thermal core-mantle interaction: Exploring regimes for 'locked' dynamo action. Phys Earth Planet Inter 165:83–92
Yoshida M, Santosh M (2011) Supercontinents, mantle dynamics and plate tectonics: A perspective based on conceptual vs. numerical models. Earth Sci Rev 105:1–24
Zhang N, Zhong S (2011) Heat fluxes at the Earth's surface and core-mantle boundary since Pangea formation and their implications for the geomagnetic superchrons. Earth Planet Sci Lett 306:205–2016
Zhang N, Zhong SJ, Leng W, Li ZX (2010) A model for the evolution of the Earth's mantle structure since the early Paleozoic. J Geophys Res 115:B06401
Zhang P, Cohen RE, Haule K (2015) Effects of electron correlations on transport properties of iron at Earth's core conditions. Nature 517:605–607
Zhong S, Zhang N, Li Z-X, Roberts JH (2007) Supercontinent cycles, true polar wander, and very long-wavelength mantle convection. Earth Planet Sci Lett 261:551–564

7

Distribution, morphology, and morphometry of circular mounds in the elongated basin of northern Terra Sirenum, Mars

Ryodo Hemmi and Hideaki Miyamoto*

Abstract

An elongated, flat-floored basin, located in the northern part of Terra Sirenum on Mars, holds numerous enigmatic mounds (100 m wide) on the surface of its floor. We investigated their geological context, spatial distribution, morphological characteristics, and morphometric parameters by analyzing a variety of current remote sensing data sets of Mars. Over 700 mounds are identified; mapping of the mounds shows the spatial density of about 21 per 100 km^2 and appearances of several clusters, coalescence, and/or alignment. Most of the mounds have smoother surface textures in contrast to the rugged surrounding terrain. Some of the mounds display depressions on their summits, meter-sized boulders on their flanks, and distinct lobate features. We also perform high-resolution topographic analysis on 50 isolated mounds, which reveals that their heights range from 6 to 43 m with a mean of 18 m and average flank slopes of most mounds are below 10°. These characteristics are consistent with the deposition and extension of mud slurries with mud breccia and gases extruded from subsurface, almost equivalent to terrestrial mud volcanism. If so, both abundance of groundwater and abrupt increase in pore fluid pressure are necessary for triggering mud eruption. Absolute crater retention ages suggest that the floor of the basin located among middle Noachian-aged highland terrains has been resurfaced during the Late Hesperian Epoch. Because of cross-cutting relationships with the basin and the mounds superposed on the basin floor, the faults and fissures (part of Memnonia Fossae) are thought to have formed during and/or after the period of mound formation. Compressional stress fields which likely formed Memnonia Fossae and Mangala Valles, expected from the dike emplacement model of Wilson and Head (JGR 107:1-1–1-24, 2002), may have facilitated undercompaction or overpressurization of subsurface fluids, focused pore fluids beneath the basin, and opened conduits along faults for upwelling voluminous sediments and fluids.

Keywords: Mars, Terra Sirenum, Remote sensing, Mud volcanism

Introduction

Across the southern highlands of Mars, potential groundwater and/or hydrothermal circulation during the Noachian to Hesperian or possibly Amazonian ages has been repeatedly suggested, as evidenced by the morphology of valley networks, regional fissures and associated fluvial channels (e.g., Tanaka and Chapman 1990; Wilson and Head 2002), and distinct spectral signatures of phyllosilicates, chlorides, and sulfates (e.g., Glotch et al. 2010; Wray et al. 2011; Michalski et al. 2017). However, fluxes of groundwater and surface water, the interaction between groundwater activity and heat flux, the presence of a purported cryosphere (Clifford and Parker 2001) or subsurface clathrate of carbon dioxide and methane, and the timing of a halt of possible groundwater activity or whether liquid water is still extant, remain uncertain.

On early Mars (Noachian to Hesperian Periods with approximate absolute age ranging from 4.1 to 3.1 Gyr (Tanaka et al. 2014)), groundwater/hydrothermal systems, estimated from the presence of clay minerals on a global scale (Ehlmann et al. 2013), may have occurred, with diverse evidence including a number of putative mud volcanoes being reported in both the northern

* Correspondence: hm@sys.t.u-tokyo.ac.jp
Department of Systems Innovation, School of Engineering, The University of Tokyo, 7-3-1 Hongo, Bunkyo-ku, Tokyo 113-8656, Japan

lowlands (e.g, Tanaka et al. 2005) and relatively old, southern highlands (e.g., Pondrelli et al. 2011; Okubo 2016). However, the spatial extent of potential mud volcanism on Mars has yet to be fully understood. Moreover, criteria for distinguishing between mud volcanoes and other morphologically similar landforms (e.g., cinder cones, tuff cones, rootless cones, pingos) have yet to be established.

Our study area, an unnamed NNE-SSW elongated depression (25–28°S, 203–205°W; ~ 150 km long, < ~ 50 km wide), is located on the northern part of Terra Sirenum in the southern highlands of Mars (Fig. 1). This basin was previously mapped as the middle or early Noachian-aged highland units (mNh and eNh units; Tanaka et al. (2014)) which are the same geological units as the surrounding cratered terrain. Based on the analysis of high-resolution hyperspectral data, Wray et al. (2011) suggested that these surrounding impact basins, especially Columbus and Cross craters, contain light-toned layered deposits including sulfates and phyllosilicate minerals, which imply the deep paleolakes, fed by upwelling groundwater, may have existed and formed these deposits during the late Noachian period.

The southern part of Memnonia Fossae, one of the largest graben systems that radially extended from the Tharsis rise, traverses the basin from east to west, where dike intrusions and associated melting of cryosphere may have occurred along these faults (Wilson and Head 2002),

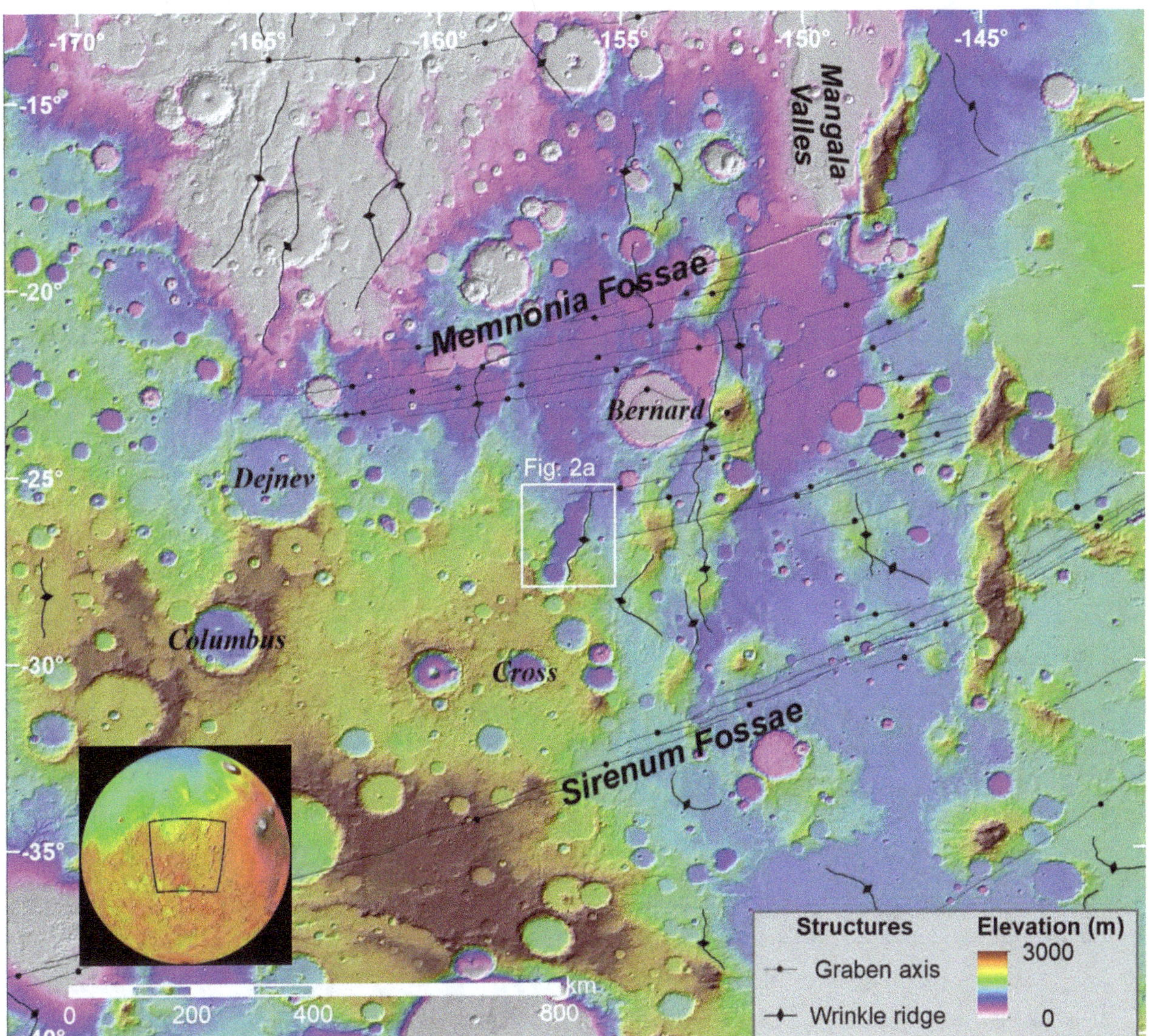

Fig. 1 Color-coded topographic map of the northern part of Terra Sirenum. Base map is the gridded MOLA topography data (512 pixels/degree) overlying the corresponding MOLA-shaded relief map (Smith et al. 2001) in simple cylindrical projection. Regional structures, interpreted by Tanaka et al. (2014), are drawn as solid black lines with symbols. A white rectangle indicates the location of Fig. 2. The inset shows the location of the study area on MOLA global topography map rendered with the Google Earth

which is supported by the presence of Mangala Valles, the major outflow channel sourced from the northern Memnonia Fossae. Cross-cutting relationship between Memnonia Fossae and the basin shows the activity of Memnonia Fossae followed the basin formation. Wrinkle ridges, with N-S trending axis observed around this region, indicate the circum-Tharsis W-E compressional fields (Head et al. 2002), possibly associated with the Tharsis rise formation on global scale.

The shape of the elongated basin appears to be a chain of several circular sub-basins with flat floors (Fig. 2). This may indicate the integration of multiple impact craters caused by erosional processes with subsequent crater-filling depositions. Profiles of global topography model (Fig. 2c) show that inside of the basin has a mostly flat-floored surface with a slight variation in elevation (between 800 and 900 m in altitude) surrounded by the mountain ranges with altitudes of more than 1200 m (except the northern part of the basin), while the terrain outside of the basin has an inclined topography on a regional scale: the south of the basin has relatively high altitudes (~ 2000 m) than the north (~ 1000 m). No obvious fluvial channel into this basin exists.

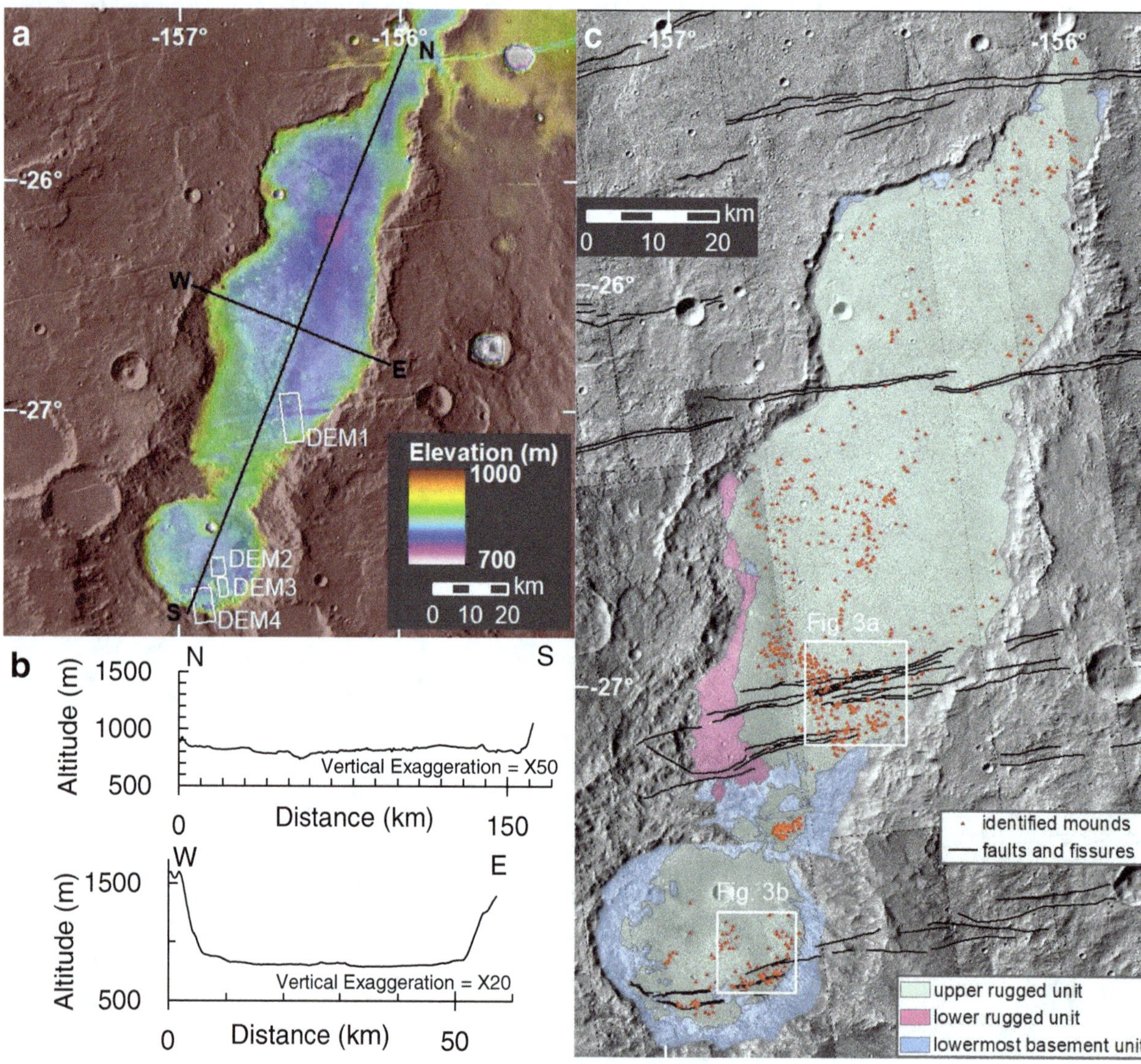

Fig. 2 Close-up view of the study area, the floor of an elongated basin. **a** Color-coded MOLA topography data (463 m/pixel) of the study area, underlain by THEMIS daytime infrared image (100 m/pixel). The elevation range is stretched to enhance the variation in elevation change inside the basin floor. Black solid lines (N–S and W–E) represent the locations of topographic profiles in panel **b**. White rectangles indicate the footprints of four DEMs generated in this study (DEM1 to DEM4; Fig. 7). **b** Cross-sectional topographic profiles of the basin, extracted from the MOLA topography model in panel **a**. **c** Geology map of the study area. The surface of this area has been classified into three units. Identified faults and fissures constitute a part of Memnonia Fossae. The base image, a mosaic of CTX images, was utilized for the geological mapping

Numerous well-preserved circular mounds are distributed inside this basin floor (Fig. 3), which contrasts to other neighboring regions on Terra Sirenum. Recently acquired high-resolution images enable us to characterize their stratigraphy, morphology, and morphometry. Thus, this study focuses on the characterization of the circular mounds on the elongated basin to reveal their formational mechanisms and their relation to the occurrence of the basin and regional geological structures. We propose that development of faults facilitated the migration and overpressurization of subsurface fluids and triggered mud extrusions. This would lead to better understanding of the evolution of hydrology and structural geology across Terra Sirenum and the assessment of its astrobiological potential.

Methods/Experimental

Data sets and processing tools

The following data products are used in this study (mostly as Geographic Information System (GIS) data sets provided by the NASA Planetary Data System, Arizona State University, University of Arizona and the United States Geological Survey (USGS) database): the Mars Odyssey's Thermal Emission Imaging System (THEMIS; Christensen et al. 2004) daytime infrared image mosaic for geological mapping and crater counting at 100 m/pixel, the Mars Global Surveyor (MGS) Mars Orbiter Laser Altimeter (MOLA; Smith et al. 2001) high-accuracy altimetry profile data (or Precision Experiment Data Records (PEDRs)) for the alignment of digital elevation models (DEMs) to Mars global topography, MOLA global topography models at 128 pixels per degree (also known as Mission Experiment Gridded Data Records, or MEGDRs) for examining the regional topography (Fig. 1), and the Mars Reconnaissance Orbiter's (MRO) Context Camera (CTX; Malin et al. (2007)) images at ~ 5–6 m/pixel for geological mapping and the generation of DEMs, and MRO High Resolution Imaging Science Experiment (HiRISE; McEwen et al. (2007)) images at up to ~ 25 cm/pixel

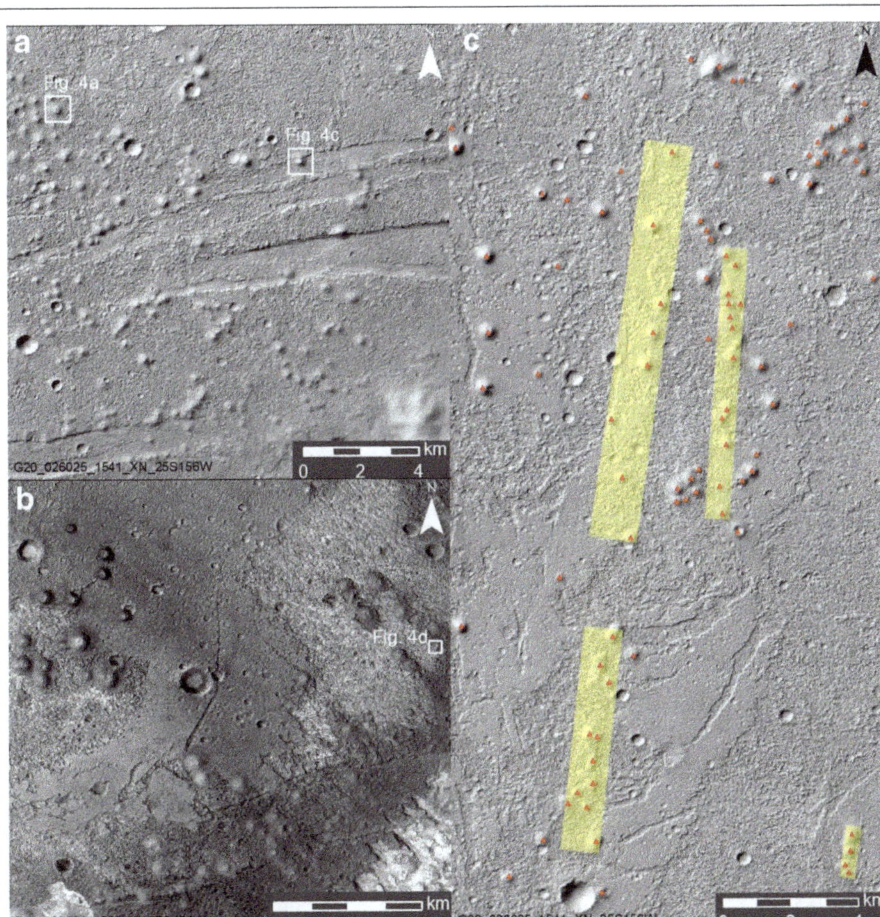

Fig. 3 Example of the mound fields. **a** Mounds located in the middle part of the upper rugged unit. Illumination from the upper left. Part of CTX image. **b** Mounds located in the southern part of the upper rugged unit. Illumination from the upper left. Part of CTX image. **c** Interpreted areas of the alignments (yellow rectangles) of the mounds (red triangles)

for the identification of small-scale morphologies and the generation of DEMs. The Java Mission-planning and Analysis for Remote Sensing (JMARS; https://jmars.asu.edu/) software developed by Arizona State University, the USGS Integrated Software for Imagers and Spectrometers (ISIS; https://isis.astrogeology.usgs.gov//) and the ESRI ArcGIS 10.2, and the Geospatial Data Abstraction Library (GDAL) were mainly utilized for browsing the datasets, processing raw data products, and analyzing the image data on the GIS platform.

Generation of Digital Elevation Models

The workflow of creating high-resolution DEMs and orthoimages is twofold: (1) generating CTX-derived DEMs (18 m post-spacing) tied to MOLA PEDRs and a cartographically controlled mosaic of THEMIS daytime infrared images (Fergason et al. 2013) and (2) making HiRISE-derived DEMs (1 or 2 m post-spacing) tied to the CTX-derived DEMs. In either step, photometric calibration, bundle adjustment based on the ground control points collected from the THEMIS mosaic or CTX-derived DEMs with orthoimages, and map projection of stereo images were performed using ISIS's applications, and then, stereophotogrammetric triangulation, orthorectification, and the alignment of the CTX-derived DEMs to MOLA PEDR were performed with the NASA Ames Stereo Pipeline (ASP) (Broxton and Edwards 2008; Moratto et al. 2010; Beyer et al. 2014; Shean et al. 2016). In this study, a total of two CTX-derived DEMs (used only for co-registration of HiRISE-derived DEMs) and four HiRISE-derived DEMs were extracted (Table 1). Vertical accuracy (or expected precision) of each DEM is in the range of 0.19 to 0.90 m, which has been calculated from the product of 0.2 pixel (stereo matching error) and maximum image resolution between the two images, divided by the parallax-height ratio (e.g., Kirk et al. 2008; Becker et al., 2015; Beddingfield et al. 2015).

Measurements of geometric parameters of mounds

Morphometric parameters (basal diameters and maximum mound heights) of the 50 (sub-)circular mounds were measured using four HiRISE-derived DEMs. Each of the mounds, confirmed in the orthoimages overlaid with the DEMs, was manually delineated using ArcGIS, and then, the mean elevation values along the outlines and the areas and the maximum elevation values inside each outline were measured. Basal diameter and maximum mound height were defined as the diameter of the circle area equivalent to the measured area and the difference between the maximum and mean elevation values, respectively. Highly degraded mounds, irregularly shaped mounds, coalesced mounds, mounds superposed on highly inclined substrata, mounds less than a hundred meters diameter (equivalent to only 5 or 10 DEM pixels), and mounds outside of the valid DEMs were excluded for the measurements due to the difficulty of identifying their outlines and increasing the uncertainty in morphometric parameters. Considering the horizontal resolution and vertical accuracy of each DEM and the variability of manual outlining, errors in basal diameters and maximum mound heights are estimated to be at most ± 8 m and ± 2.0 m, respectively.

Moreover, statistics of slope angles of each mound (mean, minimum, maximum, standard deviation) were also examined using the slope maps (maximum gradient values per pixel scale, in degrees) directly converted from HiRISE-derived DEMs (Table 2). For some mounds, we also extracted the topographic profiles of some mounds in cross-sections as well using the ArcGIS 3D analyst tool.

Estimates of surface formation ages

We followed the commonly accepted techniques of statistical analysis of impact craters on planetary bodies, crater size-frequency distribution (CSFD) (Crater Analysis Techniques Working Group 1979), to estimate the absolute model ages of the formation of the target areas. For mapping and measuring the impact craters on the targeted areas in THEMIS and CTX images, we used the CraterTools software (Kneissl et al. 2011) on the

Table 1 List of the details of four HiRISE-derived DEMs and two CTX-derived DEMs generated in this study

DEM ID	Spatial resolution (m/post)	Vertical accuracy (m)	Stereo pairs used for the generation of DEMs				
			Image 1		Image 2		Parallax-height ratio
			Image ID	Resolution (m/pixel)	Image ID	Resolution (m/pixel)	
DEM1	2.0	0.24	ESP_047862_1525	0.53	ESP_048152_1525	0.52	0.435
DEM2	2.0	0.90	ESP_039370_1520	0.51	ESP_039581_1520	0.25	0.085
DEM3	1.0	0.55	ESP_039581_1520	0.25	ESP_047150_1520	0.26	0.093
DEM4	2.0	0.19	ESP_040293_1520	0.55	ESP_047506_1520	0.26	0.416
CTX_DEM1	18	2.17	F10_039726_1522_XI_27S156W	5.56	F11_040003_1522_XI_27S156W	5.31	0.485
CTX_DEM2	18	12.02	F09_039370_1524_XI_27S156W	5.13	F10_039581_1523_XI_27S156W	5.09	0.084

Table 2 Geometric parameters of 50 mounds measured in this study

DEM used for measurement	Mound ID	Center latitude (°N)	Center longitude (°E)	Basal diameter (m)	Maximum height (m)	Height-width ratio	Slope angles (°)			
							Maximum	Minimum	Median	SD
DEM1	Mound1_1	− 26.9533	− 156.5379	284	21.5	0.076	35.8	0.2	9.8	3.3
DEM1	Mound1_2	− 26.9606	− 156.5390	165	8.6	0.052	35.0	0.3	6.6	2.3
DEM1	Mound1_3	− 26.9645	− 156.5376	234	13.4	0.057	58.2	0.3	8.2	3.7
DEM1	Mound1_4	− 26.9673	− 156.5329	395	22.2	0.056	51.8	0.1	9.9	4.7
DEM1	Mound1_5	− 26.9768	− 156.5351	420	35.5	0.084	58.1	0.1	11.6	4.5
DEM1	Mound1_6	− 26.9508	− 156.5083	160	7.3	0.046	14.5	0.1	5.7	2.4
DEM1	Mound1_7	− 26.9594	− 156.5057	442	42.9	0.097	48.6	0.1	13.0	5.0
DEM1	Mound1_8	− 27.0061	− 156.5324	324	26.5	0.082	47.9	0.1	11.2	4.0
DEM1	Mound1_9	− 27.0232	− 156.4857	205	12.7	0.062	33.9	0.1	7.9	3.1
DEM1	Mound1_10	− 27.0236	− 156.4804	367	26.6	0.072	66.0	0.1	10.0	4.5
DEM1	Mound1_11	− 27.0424	− 156.5225	260	17.5	0.067	17.9	0.1	8.9	3.0
DEM1	Mound1_12	− 27.0478	− 156.5259	205	10.6	0.052	18.2	0.2	6.9	2.8
DEM1	Mound1_13	− 27.0521	− 156.5224	200	11.6	0.058	35.8	0.1	6.9	3.3
DEM1	Mound1_14	− 27.0504	− 156.5158	217	11.1	0.051	17.2	0.0	7.8	3.7
DEM1	Mound1_15	− 27.0428	−156.5069	203	12.6	0.062	67.0	0.2	8.1	2.9
DEM1	Mound1_16	−27.0316	−156.4589	404	28.6	0.071	49.4	0.0	10.0	4.2
DEM1	Mound1_17	−27.0554	−156.4517	271	18.8	0.070	22.9	0.0	8.8	3.2
DEM1	Mound1_18	−27.0659	−156.4866	367	31.6	0.086	54.4	0.0	10.6	4.0
DEM1	Mound1_19	−27.0683	−156.4712	343	21.1	0.062	34.0	0.0	7.3	3.4
DEM1	Mound1_20	−27.0747	−156.4829	238	17.0	0.071	18.7	0.3	9.1	2.5
DEM1	Mound1_21	−27.0810	− 156.4814	331	28.7	0.087	41.8	0.2	10.8	3.6
DEM1	Mound1_22	− 27.0836	− 156.4749	345	25.2	0.073	33.6	0.1	9.4	3.5
DEM1	Mound1_23	− 27.0843	− 156.4679	249	15.3	0.062	24.4	0.0	8.3	3.1
DEM1	Mound1_24	− 27.0878	− 156.4687	184	11.7	0.063	27.8	0.2	7.5	3.0
DEM1	Mound1_25	− 27.0851	− 156.4642	184	11.9	0.064	24.7	0.1	8.5	3.3
DEM1	Mound1_26	− 27.0812	− 156.4620	328	24.5	0.075	35.3	0.0	9.3	3.0
DEM1	Mound1_27	− 27.0741	− 156.4570	212	15.3	0.072	20.2	0.1	8.3	3.1
DEM1	Mound1_28	− 27.0893	− 156.4569	391	30.1	0.077	36.2	0.2	9.4	3.4
DEM1	Mound1_29	− 27.0903	− 156.4467	237	13.1	0.055	68.0	0.0	8.2	4.3
DEM1	Mound1_30	− 27.0867	− 156.5048	230	7.8	0.034	43.2	0.1	5.5	3.1
DEM1	Mound1_31	− 27.0994	− 156.5155	336	23.6	0.070	58.2	0.1	9.2	3.2
DEM1	Mound1_32	− 27.0982	− 156.5081	214	9.1	0.043	23.9	0.1	5.9	2.5
DEM1	Mound1_33	− 27.1091	− 156.5083	209	12.6	0.060	17.3	0.3	7.9	2.5
DEM1	Mound1_34	− 27.1069	− 156.4880	270	17.4	0.065	18.7	0.1	8.6	2.7
DEM1	Mound1_35	− 27.0951	− 156.4716	286	17.6	0.062	26.6	0.2	7.9	2.7
DEM1	Mound1_36	− 27.0999	− 156.4687	276	16.2	0.059	31.2	0.1	7.9	3.4
DEM1	Mound1_37	− 27.1021	− 156.4605	190	8.5	0.045	17.3	0.1	5.7	2.3
DEM2	Mound2_1	− 27.6464	− 156.8530	392	28.8	0.073	66.0	0.1	12.6	6.4
DEM2	Mound2_2	− 27.6443	− 156.8323	301	20.3	0.067	51.1	0.0	10.0	4.9
DEM2	Mound2_3	− 27.7031	− 156.8175	253	19.0	0.075	59.0	0.1	11.1	6.4
DEM3	Mound3_1	− 27.7394	− 156.8234	171	8.6	0.050	67.3	0.2	17.4	10.3
DEM3	Mound3_2	− 27.7409	− 156.8065	276	27.6	0.100	75.7	0.0	19.1	10.7
DEM3	Mound3_3	− 27.7376	− 156.7896	205	16.6	0.081	75.0	0.1	19.8	11.0

Table 2 Geometric parameters of 50 mounds measured in this study *(Continued)*

DEM3	Mound3_4	− 27.7535	− 156.7985	186	10.6	0.057	76.6	0.1	19.4	11.3
DEM4	Mound4_1	− 27.7796	− 156.9253	265	16.6	0.062	23.1	0.0	8.0	3.0
DEM4	Mound4_2	− 27.7784	− 156.9155	242	17.0	0.070	31.8	0.1	8.2	3.2
DEM4	Mound4_3	− 27.7954	− 156.9167	203	11.5	0.057	17.6	0.1	7.0	2.5
DEM4	Mound4_4	− 27.8110	− 156.9188	207	9.5	0.046	16.8	0.2	6.2	2.3
DEM4	Mound4_5	− 27.8032	− 156.8946	206	13.5	0.065	20.2	0.3	8.0	2.5
DEM4	Mound4_6	− 27.8081	− 156.8855	147	6.0	0.041	13.8	0.1	5.4	2.0

ArcGIS ArcMap platform, due to its advantage of reducing the measurement errors dependent on different map projections. We then utilized the CraterStats II software (Michael and Neukum 2010) to plot the cumulative form of the CSFD. Absolute model ages are based on the crater production function of (Hartmann and Neukum 2001) with the crater chronology function from (Ivanov 2001). Note that we excluded the craters less than 100 m in diameter for these plots to decrease the biases influenced by possible overlooking of numerous smaller craters. In the event that small-diameter craters appeared overprinted by resurfacing events (e.g., erosion, mantling), suggested by a distinct "kink" in the CSFD, the resurfacing correction (Michael and Neukum 2010) was also applied to reconstruct the slopes of the CSFD consistent with the isochrons.

Results and discussion

Geologic classification of the basin floor

Based on the surface textures and stratigraphic relationships identifiable in HiRISE and CTX images, we classified the basin floor into three geological units (Fig. 2c): (1) upper rugged unit (surface area = ~ 3500 km^2) which has a rugged surface with superposed impact craters and the target mounds, (2) lower rugged unit which has a rugged surface with impact craters much more abundant than the overlying upper rugged unit (surface area = ~ 180 km^2), (3) lowermost basement unit (surface area = ~ 530 km^2) which has rocky hillocks or degraded bedrock interbedded with strata of light-toned layered rocks (Wray et al. 2011) in places. This study focuses on the upper rugged unit which includes all of the targeted mounds.

Distribution of the mounds

Using the CTX and HiRISE images, we identified a total of 744 mounds from the upper rugged unit (Fig. 2c), giving the mean of their areal density of ~ 21 per 100 km^2. The resultant map of the identified mound fields shows the coalescence of adjacent mounds (Fig. 3a), the mounds superimposed just on the faults (Figs. 3a, 4c), and the relatively high spatial density of mounds in certain areas (e.g., 61 mounds in a 13 km^2 area). Interestingly, in the middle part of the basin floor (Fig. 3c), the four or more individual mounds spacing several hundreds of meters are distributed within the NNE-SSW trending elongated areas, sub-parallel to the long axis of the basin. This characteristic is analogous to the "alignments" of igneous and mud volcanoes on Earth (e.g., Bonini 2012).

Morphology of the mounds

Observations of HiRISE and CTX images show a variety of morphologies of the mounds. One of the most common characteristics of the mounds is a smooth surface texture relative to the surrounding rugged surface covered with blocky materials and dunes (Fig. 4), whereas brightness of the mound surfaces in visible grayscale images vary in places (Fig. 3). For instance, on the one hand, Fig. 3a reveals that the surface brightness of the mounds are almost uniform and similar to that of the surrounding areas. On the other hand, the mounds at the upper left and bottom of Fig. 3b (only several kilometers apart) show relatively dark and bright surfaces, respectively. These suggest that the materials covering the mound surfaces have different albedos on kilometer scale which may have been influenced by different grain sizes, mineral compositions, states of erosion, and etc. In addition, every mound is superposed on the surrounding terrain, supported by the appearance of smooth mound materials embaying the topographic lows of the surrounding rough surface. Some of the mounds have the circular features (mostly depressions) on their summit (Fig. 4b). Along the rims of these summit pits, clusters of meter-scale boulders can be observed.

Several mounds overlying the fissures (Fig. 4c) show the obvious morphology of deformation or degradation. This may indicate that the formation of these fractures followed the formation of the mounds at least in some locations.

A few mounds exhibit distinct flow-like features along their periphery (e.g., Fig. 4d). This is completely different from the morphology of lobate flow-like extension associated with mass movement because of the absence of typical scarps, ridges, or hummocky surfaces.

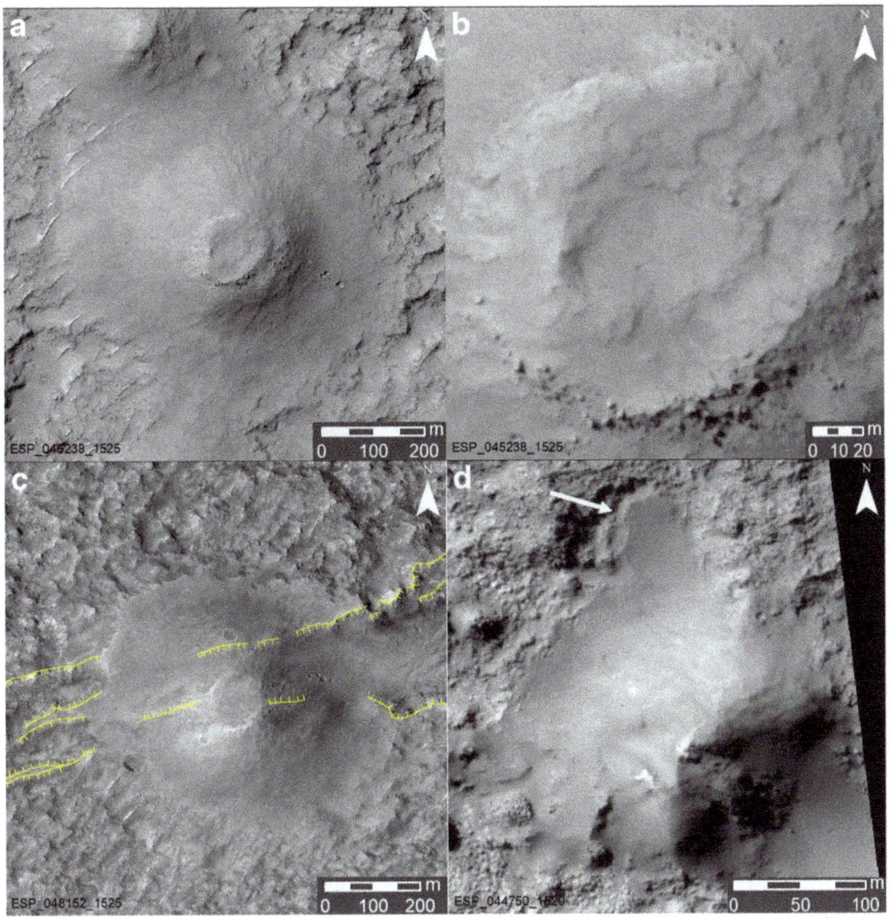

Fig. 4 General morphologic characteristics of the mounds. **a** Smooth surface texture relative to the surrounding terrain. A circular depression observed on the summit of a mound. Part of HiRISE image with illumination from the upper left. **b** Close-up of the summit depression in panel **a**. Clusters of boulders (up to 6 m wide) densely distributed along its rim. **c** A mound deformed by the underlying fissures (yellow lines) determined from the grayscale image and DEM4 (Fig. 5d). Part of HiRISE image with illumination from the upper left. **d** Distinct lobate feature (~ 70 m long) along the periphery of a mound (pointed by arrow). Part of HiRISE image with illumination from the upper left

Morphometry and geometry of the mounds

Morphometric parameters of 50 mounds (~ 6.7% of all the mounds) were measured using the four HiRISE-derived DEMs (Fig. 5). We obtained the basal diameters from 147 to 441 m (mean = 265 m, median = 245 m, SD = 77) and the maximum mound heights from 6.0 to 42.9 m (mean = 17.8, median = 16.6, SD = 8.2) with the height-to-diameter ratio of 0.034 to 0.10 (see Fig. 6 and Table 2).

Cross-sectional topographic profiles of mounds show that their summits have a wide variety of shapes: cratered summit (e.g., Fig. 7a, b), truncated summit (e.g., Fig. 7c, d), and conical summit (e.g., Fig. 7e, f). Summit depressions on some mounds have depths up to several meters, which bottoms are higher in elevation than the surrounding plains. Flanks of each mound are gently sloping, some of which are characterized by upward convex geometry (Fig. 7b, d). Slope values of each mound suggest that most of the mounds have mean slope values lower than ~ 10° (Table 2), except for locations of impact craters and DEM artifacts.

Possible origin of the mounds

Circular mounds of several hundred meters in diameter sharing similar surface textures as such mounds on Mars can be found on Earth. Proposed terrestrial analog features include impact craters, scoria cones, rootless cones, tuff cones/rings, pingos, and mud volcanoes (e.g., Farrand et al. 2005; Meresse et al. 2008; Burr et al. 2009; de Pablo and Komatsu 2009; Oehler and Allen 2010; Brož et al. 2015).

Impact craters are the sites of past meteor impacts, generally found on planetary surfaces, which is a bowl-shaped depression surrounded by raised rims and radially emanating ejecta deposits. Most of the mounds exist in the absence of these characteristics. Though some of

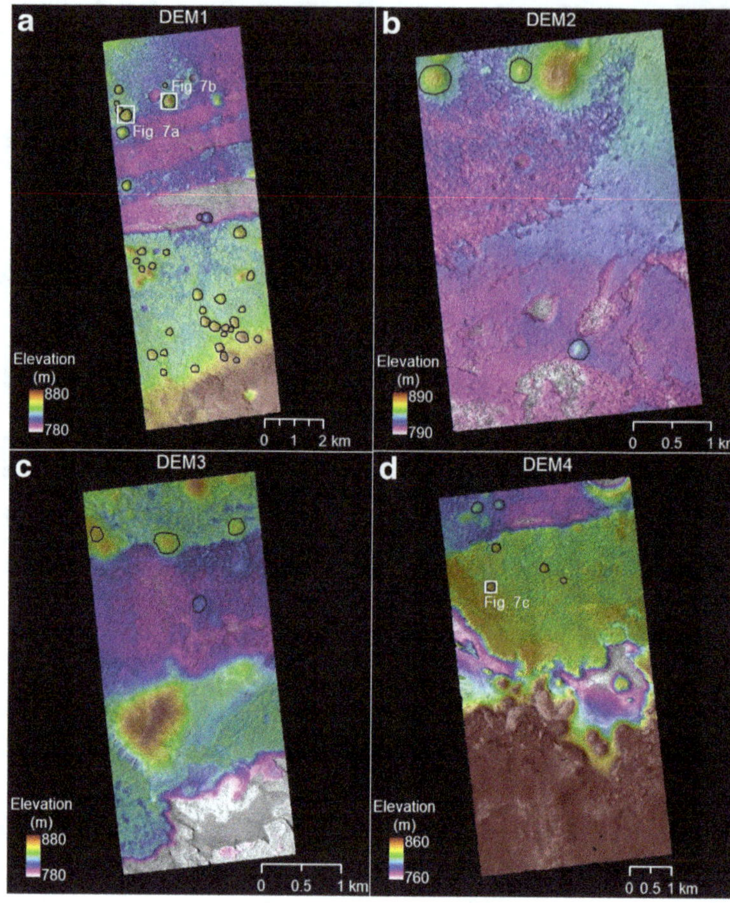

Fig. 5 Four colorized DEMs draped over grayscale orthoimages. **a** DEM1. **b** DEM2. **c** DEM3. **d** DEM4. Measured mounds are marked by black outlines. See Fig. 2 and Table 1 for their locations and details, respectively

the mounds show the smooth materials overlying the surrounding terrains and distinct crater on the summit with raised rims having ejecta-like boulders (e.g. Fig. 4) similar to the morphology of typical impact craters, the depths of summit depression are too shallow to reach the elevations of the surrounding plains, which is difficult to explain the pitted mounds by the impact origin. We thus concluded that the impact origin is unlikely.

Scoria cones (also known as cinder cones) are the conical volcanic edifices formed by explosive eruptions and associated deposition of pyroclastics around individual vents. Some of the mounds show similar characteristics of typical scoria cones; however, the results of this study are also inconsistent with the scoria cone scenario. Morphology of circular terrestrial scoria cones have the mean basal diameters of 800 m and mean height of 250 m (Wood 1979), twice or more larger values than those of the mounds measured in this study (Fig. 8). Moreover, steep or flat-floored summit morphology (Fig. 7) hardly occur on the summit of scoria cones on Earth and Mars. Associated lava flows found on terrestrial and Martian scoria cones (e.g., Meresse et al. 2008) are also lacking in the mound fields of this study. However, there remains the possibility that later erosional or depositional events deformed the original topography of scoria cones and produced the current morphology and topography, which cannot be conclusively determined without accurate in situ observations.

Rootless cones (or pseudo-craters) are conical or ring-like structures formed by the emplacement of lava flows over wet ground and the subsequent explosions (Greeley and Fagents 2001). Although some of the terrestrial rootless cones have the diameter range comparable to the mounds measured in this study (e.g., Noguchi and Kurita 2015; Fig. 8), there is a critical difference between them: whether mound is made of the same material as that of the surrounding plains or not. Rootless cones are constructed by the pyroclasts sourced from the underlying lava flows, so that if they form on the surface of Mars, little clear contrast in surface albedo between the cones and the surrounding plains is expected to be

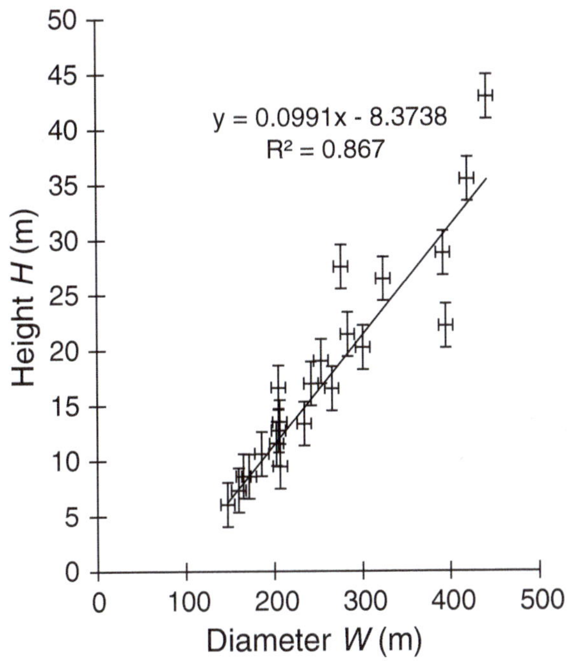

Fig. 6 Plot showing the maximum mound height versus basal diameter for 50 mounds (see also Table 2). The solid line represents the best linear least squares fit

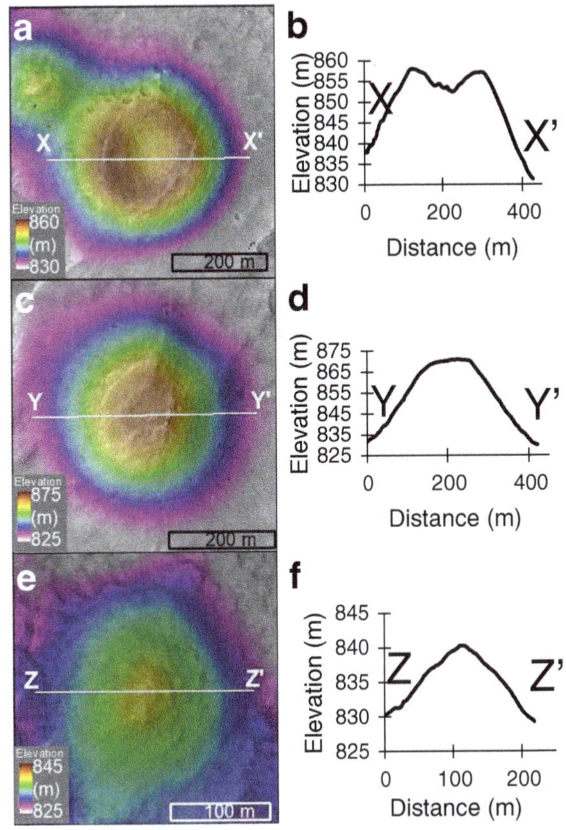

Fig. 7 Topographic profiles of type mounds. **a** Close-up of a mound in DEM4 (white rectangle in Fig. 5d; mound 4_4 in Table 2). The solid line XX′ corresponds to the profile in panel **b**. **b** Cross-sectional elevation profile of the mound in panel **a**. Vertical exaggeration is 10. **c** Close-up of a mound in DEM1 (white rectangle in Fig. 5b; mound 1_4 in Table 2). The solid line YY′ corresponds to the two profiles in panel **d**. **d** Cross-sectional elevation profiles of the mound in panel **c**. Vertical exaggeration is 10. **e** Close-up of a mound in DEM1 (white rectangle in Fig. 5b; mound 1_7 in Table 2). The solid line ZZ′ corresponds to the two profiles in panel **f**. **f** Cross-sectional elevation profiles of the mound in panel **e**. Vertical exaggeration is 5

formed due to their similar basaltic compositions. This contradicts that the surface brightness of some mounds differ to that of the surrounding surface in a single observation (e.g., Fig. 3b). Furthermore, terrestrial rootless cones are morphologically composed of central depressions and associated raised rims, which previously proposed rootless cones on Mars exhibit (e.g., Greeley and Fagents 2001); however, some of the studied mounds do not have any summit depression but rather conical summits (e.g., Fig. 7e, f). Thus, it is not reasonable to interpret all of the target mounds as rootless cones.

Pingos are periglacial landforms formed by uplift of the ground surface due to freezing of groundwater, development of subsurface ice lenses, and subsequent increase in bulk volume (Mackay 1979). The morphometry and summit morphology of pingos, as suggested by Jones et al. (2012), are consistent with these of the mounds in our study. However, these processes of pingo formation are expected to produce domical features with same surface textures as the surrounding plains, which is contrary to the difference in surface texture between the mounds and the basin floor (Fig. 3b). In addition, evident fractured surfaces, which pingo candidates on Mars (Dundas and McEwen 2010) exhibit, were not observed in any of the mounds. Martian equivalent also have 175 m diameter and ~ 10 to ~ 20 m height (Dundas and McEwen 2010), which is slightly higher than the observed mounds with similar diameters in this study. We therefore propose that pingo origin is a less plausible explanation for the formation of the mounds.

Tuff rings, tuff cones, and maars are volcanic craters/cones with topographic rims, which result from phreatomagmatic explosions (i.e., the interaction between groundwater/standing water and magmatic intrusions). These terrestrial cones/rings and Martian tuff rings/cones (Brož and Hauber 2013) have basal diameters one or more orders of magnitude larger than the mounds measured in this study (Fig. 8). Furthermore, large summit craters with high crater-to-base ratios (Pike and Clow 1981) are also not common characteristics in the mounds. Therefore, we concluded the phreatomagmatic

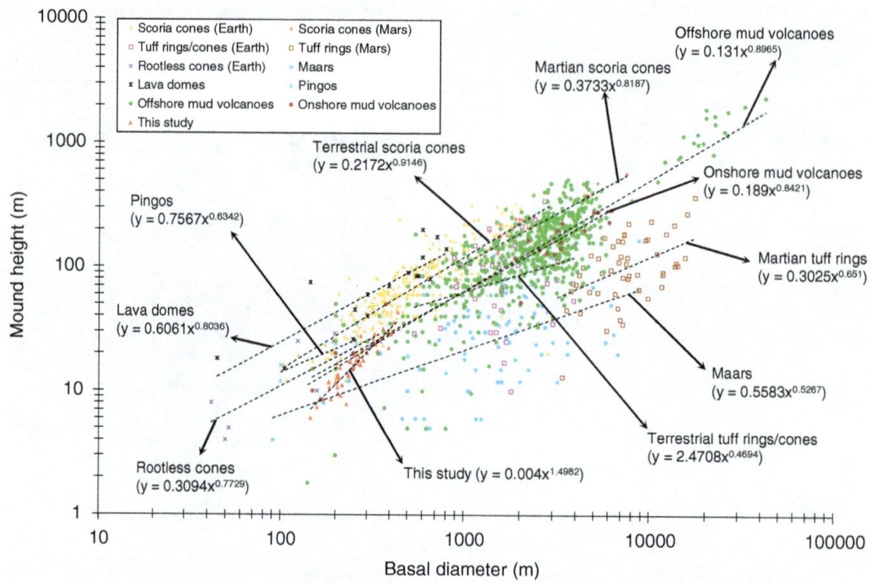

Fig. 8 Morphometric comparison between the measured mounds and other terrestrial analogs. Log-log plots showing mound height versus basal diameter. Dashed lines represent the best linear least squares fits of parameters of each feature. Terrestrial analogs with height and diameter values used for morphometric comparison are offshore mud volcanoes on Earth (Kioka and Ashi 2015; Kirkham et al. 2017), onshore mud volcanoes on Earth (Chigira and Tanaka 1997; Brož and Hauber 2013), potential tuff rings and cones on Mars (Brož and Hauber 2013), potential scoria cones on Mars (Brož et al. 2015), terrestrial maars (Pike 1978), terrestrial lava domes (Pike 1978), terrestrial cinder/scoria cones (Pike 1978; Favalli et al. 2009; Rodríguez et al. 2010), terrestrial pingos (Cabrol et al. 2000), terrestrial tuff rings/cones (Pike 1978), and terrestrial rootless cones (Pike 1978). Morphometric data of elongated (not circular) mounds is not used. Mound maximum height is preferentially used as mound height if available

explosions creating volcanic edifices are less likely origins of the mounds in the basin.

Mud volcanoes are the surface manifestation of mixtures of sediments (e.g., silts, fine sands, brecciated materials) and fluids (i.e., liquid water and gas) migrated upwards from source reservoirs up to several kilometers deep, which can produce a variety of morphologies (e.g., conical mounds, calderas, pools, diapiric intrusions, mud flows) along terrestrial regional compressional fields, mostly subduction zones on a global scale (e.g., Milkov 2000; Kopf 2002). The observations of mound depressions, boulders on the rims, and flow-like features can be interpreted as vent-like structures of the subsurface conduits, mud breccias, and mud flows of typical mud volcanoes on Earth, respectively. The various morphology of cone summits on mud volcanoes in cross-section, reported in Yusifov and Rabinowitz (2004), well correspond to the observations of cone morphologies in Fig. 7.

As summarized in Fig. 8, offshore (submarine) mud volcanoes have the wide range of the morphometric parameters (100 m to 10 km in diameter and several meters to kilometers in height), and these are well consistent with those of onshore mud volcanoes. This suggests that the condition of ambient fluid (i.e., air or liquid water) during mound formation do not affect the final size and topography of the mounds critically. In fact, Murton and Biggs (2003) argued that mud flow spreading across seafloor is driven by the "reduced gravity" (g') which is proportional to the product of surface gravity and the density contrast between seawater and mud. They theoretically established that the mound radius and height are proportional to the eighth root of g' and inversely proportional to the fourth root of g', respectively, so that the surface gravity and density (or pressure) of ambient materials are minor contributor relative to the other conditions (e.g., rheology of mud, flow rate, eruption duration). Therefore, if mud volcanoes formed on the free air or underwater conditions of Mars by the eruption of mud having the same flux as the terrestrial ones, they are expected to show similar morphometry to those of both subaqueous and subaerial mud volcanoes on Earth. Importantly, the smaller range of both onshore and offshore mud volcanoes (~ 100 to ~ 500 m diameters) have the heights of 1 to 50 m, which is consistent with the heights of the investigated mounds of this study (Fig. 8). Thus, if they are of a mud volcanic origin, both subaerial and subaqueous conditions are possible to explain the origin of the investigated mounds on Mars and difficult to distinguish which condition existed during the periods of mound formation.

The inconsistency between the studied mounds and the other analog features are quite difficult to accurately interpret due to different formation ages, different erosion rates, different properties of terrain underlying the mounds (e.g., thickness, bulk soil density), and different surface gravity; however, at least Martian mud volcanoes would not be expected to plot closely with Martian scoria cones, Martian tuff rings/cones, or Martian maars (if existed) in Fig. 8. This is because under low gravity and low atmospheric pressures on Mars surface, larger cone diameters and lower cone heights than those on Earth are expected due to the wider ranges of scoria distribution by far-reaching ballistic emplacement (Wilson and Head 1994; Brož et al. 2015). Actually, the morphometries of candidates of Martian scoria cones (Brož et al. 2015) and Martian tuff rings/cones (Brož and Hauber 2013) are plotted in the larger diameters than the terrestrial equivalents, and they are plotted in different way from that of the observed mounds in this study (Fig. 8). Although the plots in Fig. 8 do not thoroughly compiled all of the previously obtained data, most of the ranges and the trends would not be expected to change significantly. Though exact locations of potential Martian maars have not been reported ever before, they would be expected to be plotted in the larger diameter ranges if they formed on Mars because of the similar mechanism of spatter emplacement to that of scoria cones under low gravity and thinner atmosphere than those of Earth. Data of terrestrial lava domes, terrestrial rootless cones, and terrestrial pingos are not completely compiled in Fig. 8; however, their morphology are completely different from the that of the mud volcanoes and the targeted mounds of this study, so that they are readily discernable even if they have morphometric parameters to those of the studied mounds.

Based on the above discussions, we propose the mud volcanism best explains the origins of the circular mounds in the study area. If such mounds are formed through mud volcanisms, the large amount of clay minerals, aquifers maintained perhaps under moderate heat fluxes, tectonic stresses, and generation of hydrocarbon gases may have occurred within the subsurface below this basin floor for at least limited periods, which would be of great significance in the context of astrobiology. Hence, this basin and mounds can be one of the best candidate sites for future landing exploration.

Surface model age determination

The cumulative plots of the CSFD, based on the crater counts on the sample region of the southern highlands including the basin (Fig. 9a, b), show the surface model age of 4.0 Ga, well consistent with previously described geological units of early Noachian southern highlands (Platz et al. 2013; Tanaka et al. 2014). In contrast, the model age estimates of the upper rugged unit of the basin floor (Fig. 9c, d) is relatively young. A chronology function fitted to the population of impact craters (more than 1.2 km across) provides the estimated value of 3.5 Ga. This age is comparable to the values of some areas of the surrounding highland plains (Platz et al. 2013). In addition, the application of resurfacing correction to the CSFDs of two smaller diameter ranges, 0.6–1.0 km and 150–500 m, provides the age estimates of 2.9 Ga (Hesperian/Amazonian boundary (Hartmann 2005)) and 1.2 Ga (early Amazonian), respectively (Fig. 9d).

Formational history of the basin environments forming mud volcanoes

Based on the combinations of model ages and the cross-cutting and stratigraphic relationship of geological units and structures, we provide a simple consistent explanation for the formational scenario of Terra Sirenum, the elongated basin, the basin floor, the circular mounds, and the fractures of Memnonia Fossae: (1) heavy bombardment events cratered the northern part of past Terra Sirenum area and formed the chains of several basin structure around ~ 4.0 Ga (pre-elongated basin), (2) several cratered surfaces were integrated into an elongated trough-like structure at ~ 3.5 Ga by degradational processes (formation of lowermost basement unit), (3) subsequent infilling of the fluid-laden sediments into the elongated basin, possibly sourced from faults along Memnonia Fossae via upwelling groundwater melted by dike intrusion (Wilson and Head 2002), (4) sedimentary layers with interbedded bright materials (Wray et al. 2011) formed by evaporation (formation of lower rugged unit), (5) local east-west compressional stresses with long-term sediment loading caused overpressurization of groundwater at depth, opened conduits for triggering mud extrusions onto the surface, and deformed some of the mud volcanoes during Hesperian/Amazonian periods, (6) aeolian activities partially buried and excavated the basin interior from Amazonian to date (formation of upper rugged unit).

Additional high-resolution spectral data coverage obtained by the MRO spacecraft in the future will improve mineralogical characterization of the mounds and further discussions of possible diagenetic processes and/or hydrothermal alteration.

Conclusions

We examined the distribution, stratigraphic relationship, morphologies, and morphometric parameters of the circular mounds in the elongated basin of the northern Terra Sirenum. Though an eroded volcanic cone origin cannot

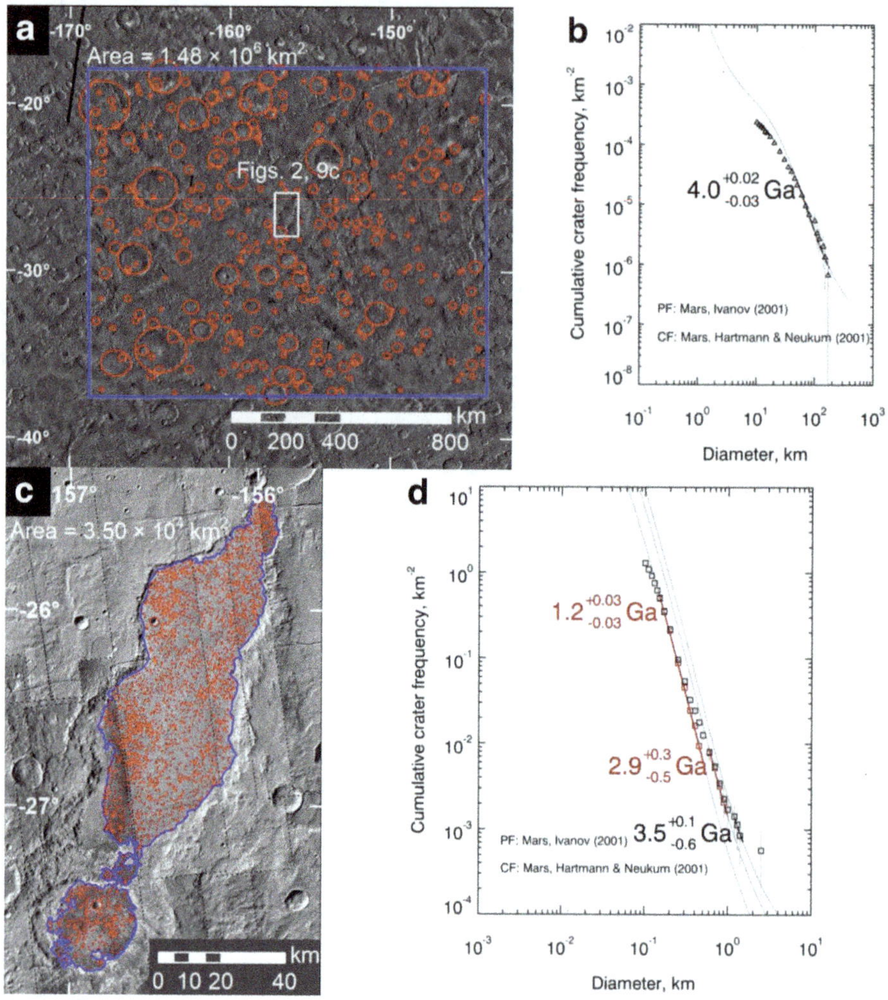

Fig. 9 Crater count-based model age determination. **a, b** Part of northern Terra Sirenum (Fig. 1) (**c, d**), the upper rugged unit (Fig. 2c). **a** Mapped impact craters with diameters > 10 km (red circles) on a sample region (blue rectangle) surrounding the study area (white rectangle). Base map is from the daytime THEMIS infrared image mosaic (100 m/pixel) which was used to mark the craters. Solar illumination is from the left of the scene. **b** Cumulative size-frequency plot of the impact craters of > 100 m diameter ($N = 348$). Production function curve is fitted to the diameter range of 50 to 175 km. **c** Identified impact craters > 100 m diameters (red circles, $N = 4549$) on the upper rugged unit (blue outlines). The CTX image mosaic (base map) was used to map the craters. **d** Cumulative crater size-frequency distribution and absolute model ages. Production functions fitted to the three ranges of the crater diameters: 1.2–1.5 km (black), 0.6–1.0 km (red), and 150–500 m (red). Red plots indicate the resurfacing correction (Michael and Neukum 2010)

be ruled out, the combination of their distribution and meter-scale morphology with their morphometry favors a mud volcano origin. If the mud volcano hypothesis is true, their relatively young surfaces suggest that the formation of source reservoirs and conduit openings along regional fissures for erupting mud and groundwater might have occurred during more recent times than Noachian age. This is consistent with other post-Noachian features in the region such as outflow channels which are linked to faults.

Abbreviations

ASP: Ames Stereo Pipeline; CTX: Context Camera; DEM(s): Digital elevation model(s); GDAL: Geospatial Data Abstraction Library; GIS: Geographic Information System; HiRISE: High Resolution Imaging Science Experiment; ISIS: Integrated Software for Imagers and Spectrometers; JMARS: Java Mission-planning and Analysis for Remote Sensing; MEGDRs: Mission Experiment Gridded Data Records; MGS: Mars Global Surveyor; MOLA: Mars Orbiter Laser Altimeter; MRO: Mars Reconnaissance Orbiter; NASA: National Aeronautics and Space Administration; PEDRs: Precision Experiment Data Records; THEMIS: Thermal Emission Imaging System; USGS: United States Geological Survey

Acknowledgements

We thank HiRISE and CTX science teams for producing a number of publicly available, high-resolution image datasets. We also appreciate the developers of USGS's ISIS and Arizona State University's JMARS for enabling us to handle Mars image datasets. We are grateful to NASA Ames Stereo Pipeline developers for creating an open-source software used to generate DEMs. We acknowledge Trent Hare, Elphia Howington-Kraus, Ross Beyer, Oleg Alexandrov, and Zachary Morrato for their helpful comments on DEM extraction and GIS-based processing, and James M. Dohm for preliminary discussions. The detailed reviews by Chris Okubo and an anonymous reviewer improved the clarity of the manuscript and are greatly appreciated. This work has been supported by JSPS KAKENHI Grant Numbers 23340126, 25120006, 15K21718, 16K13890, and 17H02953. We appreciate the Space Museum TeNQ/TOKYO DOME CORPORATION for their great support of the branch of Space Exploration, Education, and Discovery (SEED), the University Museum, the University of Tokyo.

Funding
Not applicable.

Authors' contributions
RH proposed the topic, conceived and designed the study, analyzed the data, and interpreted the results. HM collaborated with RH in the discussion and construction of the manuscript. All authors read and approved the final manuscript.

Competing interests
The authors declare that they have no competing interest.

References
Becker KJ, Archinal BA, Hare TH, Kirk RL, Howington-Kraus E, Robinson MS, Rosiek MR (2015) Criteria for automated identification of stereo image pairs. In: Abstracts of the 46th Lunar and Planetary Science Conference, 2015, The Woodlands, Texas, p 2703.

Beddingfield CB, Burr DM, Emery JP (2015) Fault geometries on Uranus' satellite Miranda: implications for internal structure and heat flow. Icarus 247:35–52. doi:10.1016/j.icarus.2014.09.048

Beyer RA, Alexandrov O, Moratto ZM (2014) Aligning terrain model and laser altimeter point clouds with the Ames Stereo Pipeline. In: Abstracts of the 45th Lunar and Planetary Science Conference, 2014, The Woodlands, Texas, p 2902

Bonini M (2012) Mud volcanoes: indicators of stress orientation and tectonic controls. Earth Sci Rev 115:121–152. doi:10.1016/j.earscirev.2012.09.002

Broxton MJ, Edwards LJ (2008) The Ames Stereo Pipeline: automated 3D surface reconstruction from orbital imagery. In: Abstracts of the 39th Lunar and Planetary Science Conference, 2008, League City, Texas, p 2419

Brož P, Hauber E (2013) Hydrovolcanic tuff rings and cones as indicators for phreatomagmatic explosive eruptions on Mars. Journal of Geophysical Research: Planets 118:1656–1675. doi:10.1002/jgre.20120

Brož P, Čadek O, Hauber E, Rossi AP (2015) Scoria cones on Mars: Detailed investigation of morphometry based on high-resolution digital elevation models. Journal of Geophysical Research: Planets 120:1512–1527. doi:10.1002/2015JE004873

Burr DM, Bruno BC, Lanagan PD, Glaze LS, Jaeger WL, Soare RJ, Tseung J-MWB, Skinner JA, Baloga SM (2009) Mesoscale raised rim depressions (MRRDs) on Earth: a review of the characteristics, processes, and spatial distributions of analogs for Mars. Planet Space Sci 57:579–596. doi:10.1016/j.pss.2008.11.011

Cabrol NA, Grin EA, Pollard WH (2000) Possible frost mounds in an ancient Martian lake bed. Icarus 145:91–107. doi:10.1006/icar.1999.6326

Chigira M, Tanaka K (1997) Structural features and the history of mud volcanoes in southern Hokkaido, northern Japan. J Geol Soc Jpn 103:781–791. doi:10.5575/geosoc.103.781

Christensen PR, Jakosky BM, Kieffer HH, Malin MC, McSween HY, Nealson K, Mehall GL, Silverman SH, Ferry S, Caplinger M (2004) The Thermal Emission Imaging System (THEMIS) for the Mars 2001 Odyssey Mission. Space Sci Rev 110:85–130. doi:10.1023/B:SPAC.0000021008.16305.94

Clifford SM, Parker TJ (2001) The evolution of the Martian hydrosphere: implications for the fate of a primordial ocean and the current state of the northern plains. Icarus 154:40–79. doi:10.1006/icar.2001.6671

de Pablo MÁ, Komatsu G (2009) Possible pingo fields in the Utopia basin, Mars: geological and climatical implications. Icarus 199:49–74. doi10.1016/j.icarus.2008.09.007

Dundas CM, McEwen AS (2010) An assessment of evidence for pingos on Mars using HiRISE. Icarus 205:244–258. doi:10.1016/j.icarus.2009.02.020

Ehlmann BL, Berger G, Mangold N, Michalski JR, Catling DC, Ruff SW, Chassefière E, Niles PB, Chevrier V, Poulet F (2013) Geochemical consequences of widespread clay mineral formation in Mars' ancient crust. Space Sci Rev 174: 329–364. doi:10.1007/s11214-012-9930-0

Farrand WH, Gaddis LR, Keszthelyi L (2005) Pitted cones and domes on Mars: observations in Acidalia Planitia and Cydonia Mensae using MOC, THEMIS, and TES data. Journal of Geophysical Research: Planets 110:E05005. doi:10.1029/2004JE002297

Favalli M, Karátson D, Mazzarini F, Pareschi MT, Boschi E (2009) Morphometry of scoria cones located on a volcano flank: a case study from Mt. Etna (Italy), based on high-resolution LiDAR data. J Volcanol Geotherm Res 186:320–330. doi:10.1016/j.jvolgeores.2009.07.011

Fergason RL, Lee EM, Weller L (2013) THEMIS geodetically controlled mosaics of Mars. In: Abstracts of the 44th Lunar and Planetary Science Conference, 2013, The Woodlands, Texas, p 1642

Glotch TD, Bandfield JL, Tornabene LL, Jensen HB, Seelos FP (2010) Distribution and formation of chlorides and phyllosilicates in Terra Sirenum, Mars. Geophys Res Lett 37:L16202. doi:10.1029/2010GL044557

Greeley R, Fagents SA (2001) Icelandic pseudocraters as analogs to some volcanic cones on Mars. Journal of Geophysical Research: Planets 106:20527–20546. doi:10.1029/2000JE001378

Crater Analysis Techniques Working Group (1979) Standard techniques for presentation and analysis of crater size-frequency data. Icarus 37:467–474. doi:10.1016/0019-1035(79)90009-5

Hartmann WK (2005) Martian cratering 8: isochron refinement and the chronology of Mars. Icarus 174:294–320. doi:10.1016/j.icarus.2004.11.023

Hartmann WK, Neukum G (2001) Cratering chronology and the evolution of Mars. Space Sci Rev 96:165–194. https://doi.org/10.1023/A:1011945222010

Head JW, Kreslavsky MA, Pratt S (2002) Northern lowlands of Mars: evidence for widespread volcanic flooding and tectonic deformation in the Hesperian Period. Journal of Geophysical Research: Planets 107:3-1-3-29. doi:10.1029/2000JE001445

Ivanov BA (2001) Mars/moon cratering rate ratio estimates. Space Sci Rev 96:87–104. doi:10.1023/A:1011941121102

Jones BM, Grosse G, Hinkel KM, Arp CD, Walker S, Beck RA, Galloway JP (2012) Assessment of pingo distribution and morphometry using an IfSAR derived digital surface model, western Arctic Coastal Plain, Northern Alaska. Geomorphology 138:1–14. doi:10.1016/j.geomorph.2011.08.007

Kioka A, Ashi J (2015) Episodic massive mud eruptions from submarine mud volcanoes examined through topographical signatures. Geophys Res Lett 42: 8406–8414. doi:10.1002/2015GL065713

Kirk RL, Howington-Kraus E, Rosiek MR, Anderson JA, Archinal BA, Becker KJ, Cook DA, Galuszka DM, Geissler PE, Hare TM, Holmberg IM, Keszthelyi LP, Redding BL, Delamere WA, Gallagher D, Chapel JD, Eliason EM, King R, McEwen AS (2008) Ultrahigh resolution topographic mapping of Mars with MRO HiRISE stereo images: meter-scale slopes of candidate Phoenix landing sites. Journal of Geophysical Research: Planets 113:E00A24. doi:10.1029/2007JE003000

Kirkham C, Cartwright J, Hermanrud C, Jebsen C (2017) The spatial, temporal and volumetric analysis of a large mud volcano province within the Eastern Mediterranean. Mar Pet Geol 81:1–16. doi:10.1016/j.marpetgeo.2016.12.026

Kneissl T, van Gasselt S, Neukum G (2011) Map-projection-independent crater size-frequency determination in GIS environments—new software tool for ArcGIS. Planet Space Sci 59:1243–1254. doi:10.1016/j.pss.2010.03.015

Kopf AJ (2002) Significance of mud volcanism. Rev Geophys 40:2-1–2-52. doi:10.1029/2000RG000093

Mackay JR (1979) Pingos of the Tuktoyaktuk peninsula area, Northwest territories. Géog Phys Quatern 33:3–61. doi:10.7202/1000322ar

Malin MC, Bell JF, Cantor BA, Caplinger MA, Calvin WM, Clancy RT, Edgett KS, Edwards L, Haberle RM, James PB, Lee SW, Ravine MA, Thomas PC, Wolff MJ (2007) Context Camera investigation on board the Mars Reconnaissance Orbiter. Journal of Geophysical Research: Planets 112:E05S04. doi:10.1029/2006JE002808

McEwen AS, Eliason EM, Bergstrom JW, Bridges NT, Hansen CJ, Delamere WA, Grant JA, Gulick VC, Herkenhoff KE, Keszthelyi L, Kirk RL, Mellon MT, Squyres SW, Thomas N, Weitz CM (2007) Mars Reconnaissance Orbiter's High Resolution Imaging Science Experiment (HiRISE). Journal of Geophysical Research: Planets 112:E05S02. doi:10.1029/2005JE002605

Meresse S, Costard F, Mangold N, Masson P, Neukum G (2008) Formation and evolution of the chaotic terrains by subsidence and magmatism: Hydraotes Chaos, Mars. Icarus 194:487–500. doi:10.1016/j.icarus.2007.10.023

Michael GG, Neukum G (2010) Planetary surface dating from crater size–frequency distribution measurements: partial resurfacing events and statistical age uncertainty. Earth Planet Sci Lett 294:223–229. doi:10.1016/j.epsl.2009.12.041

Michalski JR, Dobrea EZN, Niles PB, Cuadros J (2017) Ancient hydrothermal seafloor deposits in Eridania basin on Mars. Nat Commun 8:15978. doi:10.1038/ncomms15978

Milkov AV (2000) Worldwide distribution of submarine mud volcanoes and associated gas hydrates. Mar Geol 167:29–42. doi:10.1016/S0025-3227(00)00022-0

Moratto ZM, Broxton MJ, Beyer RA, Lundy M, Husmann K (2010) Ames Stereo Pipeline, NASA's open source automated stereogrammetry software. In: Abstracts of the 41st Lunar and Planetary Science Conference, 2010, The Woodlands, Texas, p 2364

Murton BJ, Biggs J (2003) Numerical modelling of mud volcanoes and their flows using constraints from the Gulf of Cadiz. Mar Geol 195:223–236. doi:10.1016/S0025-3227(02)00690-4

Noguchi R, Kurita K (2015) Unique characteristics of cones in Central Elysium Planitia, Mars. Planet Space Sci 111:44–54. doi:10.1016/j.pss.2015.03.007

Oehler DZ, Allen CC (2010) Evidence for pervasive mud volcanism in Acidalia Planitia, Mars. Icarus 208:636–657. doi:10.1016/j.icarus.2010.03.031

Okubo CH (2016) Morphologic evidence of subsurface sediment mobilization and mud volcanism in Candor and Coprates Chasmata, Valles Marineris, Mars. Icarus 269:23–37. doi:10.1016/j.icarus.2015.12.051

Pike RJ (1978) Volcanoes on the inner planets: some preliminary comparisons of gross topography. In: Proceedings of the 9th Lunar and Planetary Science Conference, 1978, Houston, Texas, pp 3239–3273.

Pike RJ, Clow GD (1981) Revised classification of terrestrial volcanoes and catalog of topographic dimensions, with new results of edifice volume. U. S. Geological Survey Open-File Report 81-1038

Platz T, Michael G, Tanaka KL, Skinner JA, Fortezzo CM (2013) Crater-based dating of geological units on Mars: methods and application for the new global geological map. Icarus 225:806–827. doi:10.1016/j.icarus.2013.04.021

Pondrelli M, Rossi AP, Ori GG, Van Gasselt S, Praeg D, Ceramicola S (2011) Mud volcanoes in the geologic record of Mars: the case of Firsoff crater. Earth Planet Sci Lett 304:511–519. doi:10.1016/j.epsl.2011.02.027

Rodríguez SR, Morales-Barrera W, Layer P, González-Mercado E (2010) A quaternary monogenetic volcanic field in the Xalapa region, eastern trans-Mexican volcanic belt: geology, distribution and morphology of the volcanic vents. J Volcanol Geotherm Res 197:149–166. doi:10.1016/j.jvolgeores.2009.08.003

Shean DE, Alexandrov O, Moratto ZM, Smith BE, Joughin IR, Porter C, Morin P (2016) An automated, open-source pipeline for mass production of digital elevation models (DEMs) from very-high-resolution commercial stereo satellite imagery. ISPRS J Photogramm Remote Sens 116:101–117. doi:1016/j.isprsjprs.2016.03.012

Smith DE, Zuber MT, Frey HV, Garvin JB, Head JW, Muhleman DO, Pettengill GH, Phillips RJ, Solomon SC, Zwally HJ (2001) Mars Orbiter Laser Altimeter: experiment summary after the first year of global mapping of Mars. Journal of Geophysical Research: Planets 106:23689–23722. doi:10.1029/2000JE001364

Tanaka KL, Chapman MG (1990) The relation of catastrophic flooding of Mangala Valles, Mars, to faulting of Memnonia Fossae and Tharsis Volcanism. J Geophys Res Solid Earth 95:14315–14323. doi:10.1029/JB095iB09p14315

Tanaka KL, Skinner JA, Hare TM (2005) Geologic map of the northern plains of Mars. In: U.S. Geological Survey Scientific Investigations Map 2888

Tanaka KL, Skinner JAJ, Dohm JM, Irwin Iii RP, Kolb EJ, Fortezzo CM, Platz T, Michael GG, Hare T (2014) Geologic map of Mars, Scale 1:20,000,000. In: U.S. Geological Survey Scientific Investigations Map 3292.

Wilson L, Head JW (1994) Mars: review and analysis of volcanic eruption theory and relationships to observed landforms. Rev Geophys 32:221–263. doi:10.1029/94RG01113

Wilson L, Head JW (2002) Tharsis-radial graben systems as the surface manifestation of plume-related dike intrusion complexes: models and implications. Journal of Geophysical Research: Planets 107:1-1–1-24. doi:10.1029/2001JE001593

Wood CA (1979) Monogenetic volcanoes of the terrestrial planets. In: Proceedings of the 10th Lunar and Planetary Science Conference, 1979, Houston, Texas, pp 2815–2840.

Wray JJ, Milliken RE, Dundas CM, Swayze GA, Andrews-Hanna JC, Baldridge AM, Chojnacki M, Bishop JL, Ehlmann BL, Murchie SL, Clark RN, Seelos FP, Tornabene LL, Squyres SW (2011) Columbus crater and other possible groundwater-fed paleolakes of Terra Sirenum, Mars. Journal of Geophysical Research: Planets 116:E01001. doi:10.1029/2010JE003694

Yusifov M, Rabinowitz PD (2004) Classification of mud volcanoes in the South Caspian Basin, offshore Azerbaijan. Mar Pet Geol 21:965–975. doi:10.1016/j.marpetgeo.2004.06.002

8

Feasibility of retrieving dust properties and total column water vapor from solar spectra measured using a lander camera on Mars

Naohiro Manago[1*], Katsuyuki Noguchi[2], George L. Hashimoto[3], Hiroki Senshu[4], Naohito Otobe[5], Makoto Suzuki[6] and Hiroaki Kuze[1]

Abstract

Dust and water vapor are important constituents in the Martian atmosphere, exerting significant influence on the heat balance of the atmosphere and surface. We have developed a method to retrieve optical and physical properties of Martian dust from spectral intensities of direct and scattered solar radiation to be measured using a multi-wavelength environmental camera onboard a Mars lander. Martian dust is assumed to be composed of silicate-like substrate and hematite-like inclusion, having spheroidal shape with a monomodal gamma size distribution. Error analysis based on simulated data reveals that appropriate combinations of three bands centered at 450, 550, and 675 nm wavelengths and 4 scattering angles of 3°, 10°, 50°, and 120° lead to good retrieval of four dust parameters, namely, aerosol optical depth, effective radius and variance of size distribution, and volume mixing ratio of hematite. Retrieval error increases when some of the observational parameters such as color ratio or aureole are omitted from the retrieval. Also, the capability of retrieving total column water vapor is examined through observations of direct and scattered solar radiation intensities at 925, 935, and 972 nm. The simulation and error analysis presented here will be useful for designing an environmental camera that can elucidate the dust and water vapor properties in a future Mars lander mission.

Keywords: Mars atmosphere, Dust particles, Total column water vapor, Direct solar radiation, Scattered solar radiation, Radiative transfer simulation, Inverse analysis

Introduction

Dust floating in the Martian atmosphere plays an important role in determining the circulation of atmosphere, since dust is an important heat source in the atmosphere due to the absorption of solar radiation (Gierasch and Goody 1972; Moriyama 1975; Smith 2008). Accurate knowledge of the temporal and spatial distribution of dust particles, as well as their optical parameters, is indispensable for understanding their impact on the heat balance of the planet. Also, it is known that a trace amount of water vapor exists in the Martian atmosphere (Fedorova et al. 2009; Smith 2009; Smith, et al. 2009; Whiteway et al. 2009). The condensation of water vapor leads to the formation of water ice clouds, which can also affect the heat balance of the Martian atmosphere (Flasar and Goody 1976; Haberle et al. 2001; Hinson and Wilson 2004).

Observations of air-borne dust and water vapor in the Martian atmosphere have been performed by Mars orbiters and landers as well as ground-based observations (Smith 2008). As compared with orbiter observations, lander measurements are less likely to be influenced by changes in surface albedo. The dust source is on the surface of Mars, and the concentration of dust is highest near the surface. Thus, compared with orbiters, more accurate measurement can be done from landers. Also, local parameters such as pressure and temperature can be measured precisely from landers. Among the various

* Correspondence: manago.naohiro@chiba-u.jp
[1]Center for Environmental Remote Sensing, Chiba University, 1-33 Yayoi-cho, Inage-ku, Chiba-shi, Chiba 263-8522, Japan
Full list of author information is available at the end of the article

observational methods used so far, the measurements of direct solar radiation (DSR) and scattered solar radiation (SSR) with cameras onboard Mars landers have brought a wealth of information on dust properties. The optical depth of dust can be retrieved as a function of wavelength from DSR measurement, while other optical properties and information on the composition of dust particles can be extracted by coupled analysis of DSR and SSR intensities.

The first observation of dust in the Martian atmosphere was made with the Imaging Camera (Patterson et al. 1977) onboard the Viking Lander (Pollack et al. 1977, 1979, 1995). The optical depth of dust was measured with the Imager for Mars Pathfinder (IMP) experiment (Smith et al. 1997), which revealed seasonal variation similar to the result of the Viking observation, with limited inter-annual variations (Smith and Lemmon 1999; Tomasko et al. 1999; Markiewicz et al. 1999). Subsequently, similar observations were made with Panchromatic Cameras (Pancam, onboard the Spirit and Opportunity rovers) (Bell et al. 2003; Bell et al. 2004a, 2004b; Lemmon et al. 2004), a Surface Stereo Imager (SSI) (onboard the Phoenix) (Lemmon et al. 2008; Moores et al. 2010), and Mastcam (onboard Curiosity) (Malin et al. 2010; Moore et al. 2016). The first observation of water vapor in the Martian atmosphere was carried out with the IMP (Titov et al. 1999), in which 6 spectral bands around 940 nm were employed.

The effective radius (r_{eff}) and its variance (v_{eff}) are the most important parameters that determine dust's optical behavior. As reviewed by Tomasko et al. (1999), the most plausible values for r_{eff} and v_{eff} from early observations are in a range of 1.2–1.8 μm and 0.2–1.0, respectively. In subsequent missions such as Spirit and Opportunity, the value of r_{eff} was estimated to be ~1.5 μm (with the value of v_{eff} fixed at 0.2). The lidar observation of Phoenix reported values between 1.2 and 1.4 μm (Komguem et al. 2013). The variation in the r_{eff} estimated by several missions may indicate that there are temporal and spatial variations of dust properties. Thus, it is desirable to carry out long-term observations at various locations on the Mars surface to monitor changes in dust properties.

The most dominant composition of Martian dust is considered to be basalt, and its weathering products due presumably to past volcanic activity (Dabrowska et al. 2015). The reflectance spectrum in the wavelength range of 0.3–2.5 μm suggests that the rock type is most likely palagonites, the partially recrystallized products of basalt-glass influenced by wind erosion (Dlugach et al. 2003). The value of the real part of the refractive index (RI) is ~1.5 at a wavelength of 0.5 μm, which is close to the value of silicate. The imaginary part, on the other hand, is negligibly small in visible and near-infrared wavelengths, but its value substantially increases toward the shorter wavelengths, suggesting the contribution of iron oxide components (Dlugach et al. 2003). Thus, it is possible that dust particles contain small amounts of crystalline ferric materials such as goethite, hematite, and/or maghemite.

In this paper, we examine the feasibility of retrieving dust properties and total column water vapor (TCWV) in the Martian atmosphere with a camera system onboard a lander. Such camera observation can be realized by modifying a lander camera system with the addition of optical filters. We develop a dust model in which dust properties are parameterized with a small number of "dust model parameters", including optical depth, effective radius/variance of dust size distribution, volume mixing ratio of hematite, complex RI of silicate, and scale height of dust. Since Martian dust's properties are simpler than those of terrestrial aerosols, we can safely assume a simple dust model. A relatively small number (seven) of dust model parameters are used in this model, which helps to alleviate the underconstrained (ill-posed) nature of the inverse problem. These parameter values are optimized by fitting a small number of "observational parameters" based on DSR and SSR measurements. Only three wavelength bands are utilized for dust observations, which helps keep camera hardware size and data transmission requirements small. Another set of two or three wavelengths around 935 nm is postulated for extracting TCWV from DSR/SSR measurements. We also examine the feasibility of such a camera measurement on the basis of sensor specifications currently available.

Methods/Experimental

Dust model based on non-spherical shape and monomodal size distribution

First, we describe the model used to parameterize the optical and physical properties of Martian dust. Basically, the present model is relevant to the physical conditions of dust particles including their composition, shape, and spatial distribution in the Martian atmosphere. The dust composition can be parameterized with a complex RI that depends on substance and wavelength, λ. From past studies (e.g., Hamilton and Christensen 2005 and references therein), it is plausible that Martian dust is mostly composed of silicates, with small amounts of iron oxides such as hematite. Thus, in the present analysis, we assume a two-component dust model with silicate substrate and hematite inclusion. Similar treatment of a multi-component aerosol was employed in the analysis of the terrestrial atmosphere (Manago and Kuze 2010; Manago et al. 2011). The complex RI value of silicate ($n_0 = m_0 - k_0 i$, i: imaginary unit) can be characterized with $m_0 \sim 1.5$ and $k_0 \ll 1$, representing small absorption of visible light (e.g., Table 1 in Dabrowska et al. 2015, Table 5 in Levoni et al. 1997), while that of hematite is parameterized as $n_A(\lambda) = m(\lambda) - k(\lambda)i$, in which we should consider the large absorbance of blue light. Referring to Querry (1985) for the wavelength dependence of m (λ) and k (λ), the RI of their mixture, n, can be calculated using Maxwell-Garnett theory (Ehlers et al. 2014) as

$$n^2 = n_0^2 \frac{n_A^2 + 2n_0^2 + 2\rho_A\left(n_A^2 - n_0^2\right)}{n_A^2 + 2n_0^2 - \rho_A\left(n_A^2 - n_0^2\right)}. \quad (1)$$

Here, ρ_A is the volume mixing ratio of hematite. Figure 1 shows the wavelength dependence of dust RI reported in Wolff et al. (2009). Figure 1 also shows an example of the fitting we did using Eq. (1), resulting in $n_0 = 1.48 - 0.000942i$ ($m_0 = 1.48$ and $k_0 = 0.000942$) for the complex RI of silicate and $\rho_A = 1.77\%$ for the volume mixing ratio of hematite. The agreement between the fitting curves and literature curves is satisfactory when considering the small sensitivity of radiance values to m_0 and k_0, and the derived values of m_0 and k_0 are consistent with reported values for silicate. The advantage of our dust model is that it can reproduce the important features of measured data in spite of the simplicity of the model, being based on just three parameters, m_0, k_0, and ρ_A. We have assumed internal mixing in Fig. 1 as the Maxwell-Garnett theory is applicable to dust particles having internal mixing. Although we tested another model assuming external mixing of silicate and hematite particles, the result exhibited much poorer agreement as compared with internal mixing treatment illustrated in Fig. 1.

Dust particles in the Martian atmosphere can possibly have various shapes, but limited information from direct observations (e.g., Pike et al. 2011) makes it difficult to theoretically formulate their shapes precisely. Moreover, it is difficult to reproduce the effects of non-spherical shapes in a radiative transfer simulation for the Martian atmosphere. To cope with these difficulties, here we propose the use of equivalent volume spheroids with random rotation axes as employed by Dubovik et al.

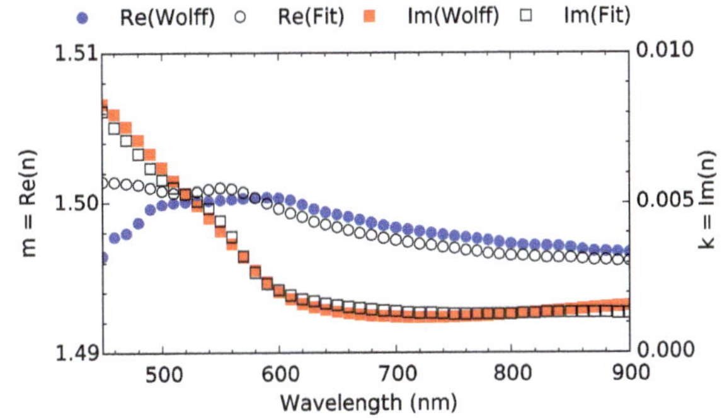

Fig. 1 Complex refractive index of Martian dust plotted against wavelength. Observational results (*filled symbols*) reported by Wolff et al. (2009) are compared with our fitting results (*open symbols*). In this example, the fitting leads to a silicate complex refractive index of $n_0 = 1.48 - 0.000942i$ and a hematite volume mixing ratio of $\rho_A = 1.77\%$

(2006) to estimate the influence of particle shapes for feldspar samples collected from terrestrial soil. Similarly, here we fix the aspect ratio distribution in such a way that the distribution is zero at aspect ratios less than 1.44, and gradually increases in the aspect ratio range between 1.44 and 3. A detailed discussion on the non-spherical shape effect will be given later. In general, when Maxwell-Garnett theory is applied to spheroids, the resulting RIs exhibit dependence on polarization. Such polarization dependence disappears only for the special case of spherical particles. Nevertheless, we assume the polarization-independent form of Eq. (1) for simplicity, though the light scattering calculations are implemented assuming non-spherical dust shapes.

In this research, we assume that the dust size distribution can be expressed with a monomodal gamma distribution including only coarse mode particles (Hansen and Travis 1974; Tomasko et al 1999). This assumption is based on the fact that for the optical properties of desert dust on the Earth, submicron (fine mode) particles can be ignored, and the size distribution can be treated as virtually monomodal, since most of particles are supermicron and in coarse mode (Dubovik et al. 2006). For completeness, the sensitivity of our model to the existence of fine mode particles will be discussed later. A monomodal gamma distribution can be expressed as

$$\frac{\mathrm{d}N}{\mathrm{d}\ln r} = \frac{r^{\xi}}{(r_{\mathrm{eff}}v_{\mathrm{eff}})^{\xi}\Gamma(\xi)} \exp\left[-\frac{r}{r_{\mathrm{eff}}v_{\mathrm{eff}}}\right], \quad (2)$$

with

$$\xi = \frac{1}{v_{\mathrm{eff}}} - 2. \quad (3)$$

Here, r is the particle radius and N is the number of particles. The gamma function (Γ) is used to normalize

Table 1 Dust model parameters (X_1–X_7)

Parameter (X)		X_a	X_{min}	X_{max}	ΔX	$\Delta X/X_a$	Reference
X_1	τ	0.4	0.1	1.0	0.9	2.25	1
X_2	r_{eff}	1.6 μm	1.2 μm	2.0 μm	0.8 μm	0.5	2
X_3	v_{eff}	0.2	0.1	0.4	0.3	1.5	2
X_4	ρ_A	0.03	0	0.05	0.05	1.67	3
X_5	m_0	1.5	1.4	1.6	0.2	0.133	4
X_6	k_0	0.001	0	0.002	0.002	2	4
X_7	z_s	14 km	5 km	20 km	15 km	1.07	5

References (1) Smith and Lemmon 1999; (2) Tomasko et al. 1999; (3) Ehlers et al. 2014; (4) Dabrowska et al. 2015; (5) Hoekzema et al. 2010.

the integral of Eq. (2). Parameters r_{eff} and v_{eff} are the effective radius and effective variance, respectively, and they are defined as

$$r_{\text{eff}} = \int_{-\infty}^{\infty} r^3 \frac{dN}{d\ln r} d\ln r \Big/ \left[\int_{-\infty}^{\infty} r^2 \frac{dN}{d\ln r} d\ln r\right] \tag{4}$$

and

$$v_{\text{eff}} = \int_{-\infty}^{\infty} r^2 (r - r_{\text{eff}})^2 \frac{dN}{d\ln r} d\ln r \Big/ \left[r_{\text{eff}}^2 \int_{-\infty}^{\infty} r^2 \frac{dN}{d\ln r} d\ln r\right]. \tag{5}$$

Since our observation is made optically, the most relevant parameter is the dust extinction coefficient (α_e), which is governed by the size distribution and extinction cross-section of dust particles. We consider the extinction value at wavelength 550 nm, the peak wavelength of solar radiation. It is assumed that α_e is homogeneous horizontally, and it decreases exponentially with increasing altitude, z:

$$\alpha_e(z) = \frac{\tau}{z_s} \exp\left(-\frac{z}{z_s}\right). \tag{6}$$

Here, τ is the dust optical depth at wavelength 550 nm and z_s is the scale height. The typical scale height of aerosols on the Earth is ~2 km. On the other hand, the scale height of Martian dust is typically 10–15 km (Thomas et al. 1999), which is similar to the scale height of pressure (Hoekzema et al. 2010), since the lower atmosphere is well-mixed on Mars. Besides, the variation range of optical depth of Martian dust is larger than that of terrestrial aerosols: namely, on Mars, typical values of τ are 0.2 (non-dimensional) for relatively clear periods, 1–2 for hot southern summers, and 4–5 during major dust storms (Savijärvi 2014).

In summary, our dust model can be described with only seven parameters: real (m_0) and imaginary (k_0) parts of RI of silicate substrate, volume mixing ratio of hematite inclusion (ρ_A), effective radius (r_{eff}) and effective variance (v_{eff}) of size distribution, and optical depth (τ) and scale height (z_s) of extinction coefficient at wavelength 550 nm. Hereafter, these model parameters will be collectively denoted as X. Table 1 lists "a priori" (i.e., the most plausible value before observation, X_a), minimum values (X_{min}), maximum values (X_{max}), and variation ranges ($\Delta X = X_{max} - X_{min}$) of the seven parameters utilized in the present model. When evaluating the impact of changes to dust model parameters, it is desirable that the parameters be normalized beforehand so that their variabilities will be on the same scale. Here, we use dust model parameters divided by the variation range (ΔX) listed in Table 1. These normalized dust model parameters will be collectively denoted as x. The 7th column of Table 1 shows the ratio between ΔX and X_a, which will be used in our later analysis.

Radiative transfer simulation of Martian atmosphere

In this research, a radiative transfer simulation code, MODTRAN5 (Berk et al. 2006) is employed to calculate the theoretical spectra of DSR and SSR that would be observed by a Mars lander. Although MODTRAN is a simulation code originally developed for the terrestrial atmosphere, it is applicable to Martian atmosphere by changing parameters such as distance from the Sun and the concentrations of gas components. The main parameters to be configured for Mars are listed in Table 2. The atmosphere between 0 and 100 km is divided into 36 layers of stratified spherical atmosphere, with the assumption of horizontal uniformity. The use of the DISORT algorithm (Stamnes et al. 1988) is desirable for the detailed analysis of SSR in the observation data. In the present simulation study, however, we use Isaacs' two stream algorithm (Isaacs et al. 1987), which is much faster and the resulting differences in SSR values are estimated to be less than 20% (Isaacs et al. 1987). The single scattering algorithm includes spherical geometry of ray transmission, while multiple scattering contributions are calculated using plane parallel geometry (the solar

Table 2 Main parameters of MODTRAN to be configured for Martian atmosphere

Item	Value
Solar constant	608 W/m^2
Radius of Mars	3396 km
Vertical profile of pressure	Mean value of MCD[a] (7.6 hPa @ground)
Vertical profile of temperature	Mean value of MCD (267 K @ground)
CO_2 volume mixing ratio	95.32%
O_2 volume mixing ratio	0.13%
O_3 volume mixing ratio	30 ppb
H_2O volume mixing ratio	300 ppm (<10 km), 0 (>10 km)
Volume mixing ratio of other gases	0
Vertical profile of dust extinction coefficient	Calculated from the dust model
Wavelength dependence of dust extinction	Calculated from the dust model
Wavelength dependence of dust absorption	Calculated from the dust model
Wavelength/scattering angle dependence of dust phase function	Calculated from the dust model
Surface albedo	Desert model (0.21@450 nm, 0.34@675 nm)

[a]MCD stands for the Mars Climate Database (Millour et al. 2012)

illumination impinging upon each atmospheric level is determined with spherical refractive geometry). The extinction/absorption coefficients and phase function are calculated from the dust model by using a Fortran program developed by Dubovik et al. (2006); this software calculates optical parameters of spheroidal aerosols by using a look-up table prepared by coupling the T-matrix (Mishchenko and Travis 1998) and approximate geometric-optics integral-equation (Yang and Liou 1996) methods.

Observational parameters for dust retrieval

Spectral irradiance/radiance of DSR/SSR in the visible and near infrared spectral regions is to be measured with an environmental monitor camera onboard a Mars lander. Radiance values of SSR are measured at particular scattering angles (χ); the angle between the original solar ray (hitting the scattering particle) and the line of sight (LOS) from the particle to the observer). Here χ = 0 corresponds to scattering in the forward direction. The expressions DSR(λ) and SSR(λ, χ) can be used to explicitly specify wavelength or scattering angle. To reduce the total volume of data that must be transmitted during the Mars observation mission, it is vitally important to minimize the number of quantities to be measured (i.e., the number of wavelengths and scattering angles) when designing an observation strategy.

Figure 2 summarizes the results of MODTRAN simulation, showing how the values of DSR/SSR depend on wavelength and scattering angle for various values of dust parameters (τ, r_{eff}, v_{eff}, ρ_A, m_0, k_0, and z_s). In Fig. 2a, b, DSR/SSR values are plotted for dust optical depths (τ) between 0.1 and 1. At all wavelengths between 400 and 900 nm, decrease in DSR and increase in SSR are observed with increasing τ. Among the seven dust model parameters, DSR shows almost no dependence on parameters other than τ, whereas SSR changes with τ, r_{eff}, v_{eff}, ρ_A, m_0, and k_0 (Fig. 2b–g). Figure 2c shows that the SSR intensity of forward scattering ($\chi = 3°$) increases at all wavelengths with increasing r_{eff}, while Fig. 2d shows that the forward scattering intensity decreases over the absorption band of hematite (blue range around 450 nm) with increasing v_{eff}.

Moreover, Fig. 2e, f shows that the SSR intensity of backward scattering ($\chi = 120°$) decreases over the hematite absorption band with increasing ρ_A and increasing m_0, while Fig. 2g shows that the backward scattering intensity decreases at all wavelengths with increasing k_0. This is due to the fact that absorption is stronger at larger scattering angles due to longer effective path length. As seen from Fig. 2h, SSR is hardly affected by z_s, the dust scale height in the Martian atmosphere.

Thus, in order to derive all the necessary dust information while minimizing redundancy, we selected the wavelengths and scattering angles listed in Table 3. The wavelength band width is chosen to be 10 nm. The measurements of DSR spectra are made at the three wavelengths of (λ_0, λ_B, λ_R) = (550, 450, 675) (in nm), while SSR spectra are measured at the two wavelengths of (λ_B, λ_R) for the four scattering angles (χ_0, χ_A, χ_F, χ_B) = (10°, 3°, 50°, 120°). As a result, 11 values are measured as listed in Table 4. The reason for the choices of $\lambda_B = 450$ nm within the hematite absorption band and $\lambda_R = 675$ nm outside the band is that these wavelengths are close to the peak wavelength of SSR radiance (~550 nm) with reasonable sensitivity of a silicon sensor. As shown in Fig. 3, the measurement of SSR is carried out on the almucantar plane ($\theta_{\mathrm{los}} = \theta_{\mathrm{sun}}$), where LOS zenith angle (θ_{los}) is the same as the solar zenith angle (θ_{sun}). When $\chi > 2\theta_{\mathrm{sun}}$, on the other hand, SSR is measured on the principal plane ($\theta_{\mathrm{los}} = \chi - \theta_{\mathrm{sun}}$, $\phi_{\mathrm{los}} = \phi_{\mathrm{sun}} + 180°$), where LOS azimuth angle (ϕ_{los}) is on the opposite side of the solar azimuth direction (ϕ_{sun}).

So far, we have considered how to choose the minimal set of quantities to be measured that satisfies the necessary and sufficient conditions for estimating the dust model parameters. Additionally, we have to avoid the influence of measurement error as far as possible. For this purpose, the use of relative values of spectral irradiance/radiance, instead of their absolute values, is advantageous for improving the accuracy of the resulting dust parameters, since precise calibration of a lander camera is difficult and the relative values are less sensitive to calibration errors than absolute values. Therefore, in this research, we calculated the appropriate ratios among the 11 measured quantities listed in Table 4, and we propose the final observational parameter set as listed in Table 5. In the second column of this table, the capital letters, D, A, F, and B indicate direct solar radiation, aureole (i.e., scattered solar radiation in close vicinity to the Sun), forward scattering, and backward scattering, respectively, and the subscripts 0, C, B, and R indicate values at the reference wavelength (550 nm), color ratio (i.e., intensity ratio of blue to red), values at the blue wavelength (450 nm), and values at the red wavelength (675 nm), respectively.

Hereafter, raw values of measured quantities and observational parameters will be written in upper-case letters as P and Y, respectively. As in the case of the dust model parameters, we use normalized values of P and Y in order to evaluate their sensitivities. Since relative errors of spectral measurements are expected to be comparable with each other, we use measured quantities divided by the values calculated with a priori model parameters. These normalized measured

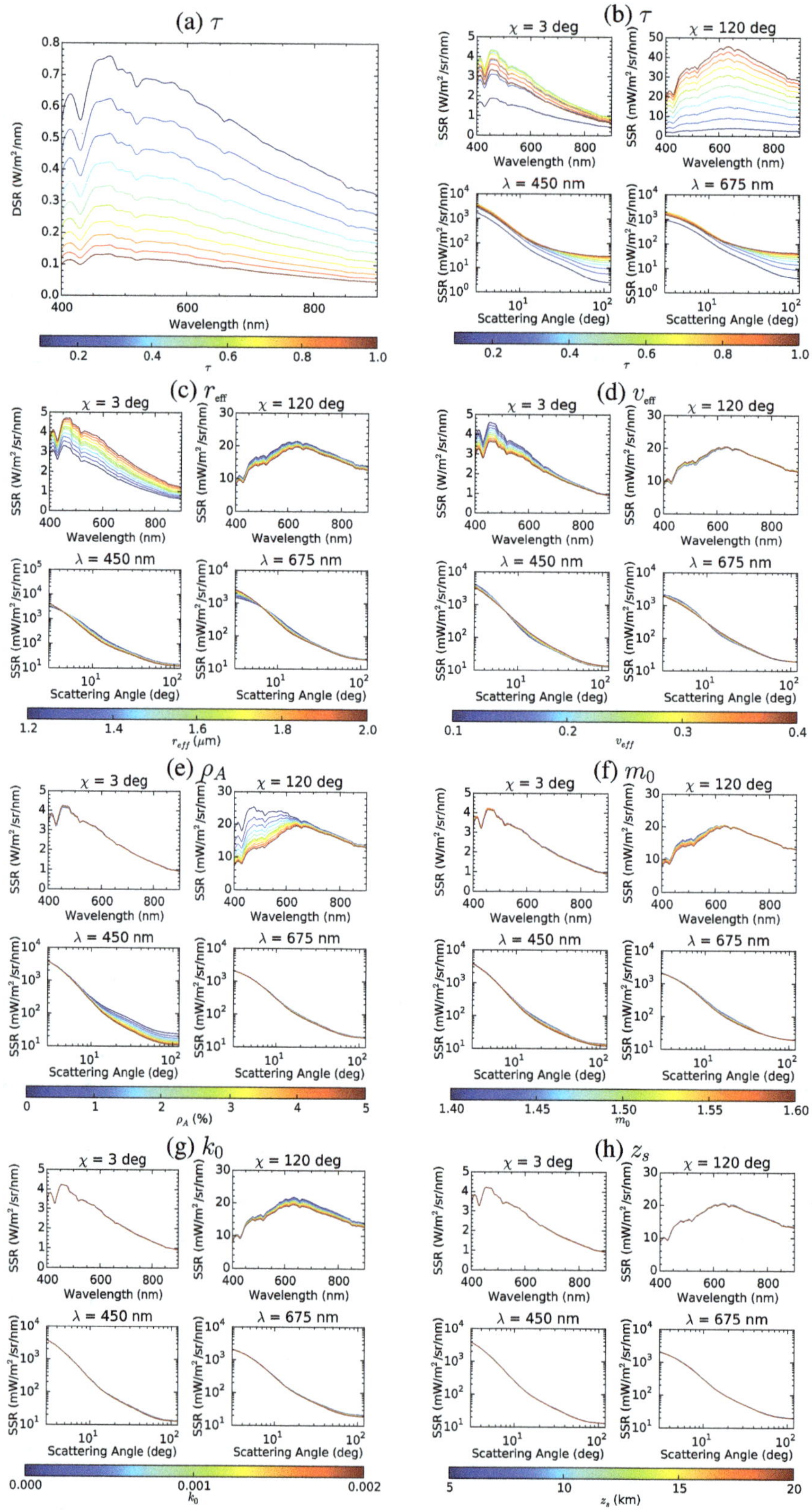

Fig. 2 Sensitivity of (**a**) DSR and (**b**–**h**) SSR to dust model parameters. DSR and SSR intensities for various values of dust model parameters (*a*: τ, *b*: τ, *c*: r_{eff}, *d*: v_{eff}, *e*: ρ_A, *f*: m_0, *g*: k_0, *h*: z_s) calculated with a radiative transfer code (MODTRAN5) are plotted as functions of wavelength or scattering angle

Table 3 Wavelengths and scattering angles chosen for dust observation

Parameter		Description	Value	DSR	SSR
Wavelength	λ_0	Reference wavelength	550 nm	✓	
	λ_B	Inside absorption band of hematite	450 nm	✓	✓
	λ_R	Outside absorption band of hematite	675 nm	✓	✓
Scattering angle	χ_0	Reference scattering angle	10°		✓
	χ_A	Aureole scattering angle	3°		✓
	χ_F	Forward scattering angle	50°		✓
	χ_B	Backward scattering angle	120°		✓

quantities will be denoted in lower-case letters as p, and observational parameters calculated with the normalized measured quantities will be denoted as y.

Observational parameters for water vapor retrieval

Water vapor absorption bands are located inside the sensitive range (350–1000 nm) of conventional complementary metal-oxide semiconductor (CMOS) sensors to be employed for our environmental monitor camera. By measuring the absorbance in DSR or SSR spectra, we can measure TCWV (c_w) through the Martian atmosphere, where the water vapor concentration is very limited (~300 ppm). Among the near-infrared absorption bands, it would be reasonable to employ the strongest absorption band around 930–970 nm. Here, we assume the atmospheric model parameters listed in Table 2, and we calculate the value of c_w by assuming that the volume mixing ratio (VMR) of water vapor is constant below 10 km. The a priori value of c_w is 1.43 mg/cm^2 for a water vapor VMR of 300 ppm. Figure 4 shows transmittance spectra simulated at a solar zenith angle (SZA) of 80° while changing the water vapor VMR as 100, 300, and 500 ppm.

In this study, we calculate values of c_w from the ratio of DSR or SSR at wavelengths inside and outside the water vapor absorption band. Here, we consider the wavelengths and the scattering angles listed in Table 6. The wavelength band width is chosen to be 5 nm. Since the effective optical path length is larger at larger scattering angles, the line-of-sight directions of $\theta_{los} = 85°$ and $\phi_{los} = \phi_{sun} + 180°$ have been chosen for the SSR measurements.

The measured quantities ($P_{12} - P_{17}$) are listed in Table 7, and they are appropriately combined to define the observational parameters as listed in Table 8. When only two wavelengths are available, the outside wavelength should be chosen to be $\lambda_S = 925$ nm,

Table 4 Measured quantities (P_1–P_{11}) of DSR and SSR for dust retrieval

Parameter (P)	Measured quantity	Parameter (P)	Measured quantity
P_1	DSR(λ_0)	P_7	SSR(λ_B, χ_B)
P_2	DSR(λ_B)	P_8	SSR(λ_R, χ_0)
P_3	DSR(λ_R)	P_9	SSR(λ_R, χ_A)
P_4	SSR(λ_B, χ_0)	P_{10}	SSR(λ_R, χ_F)
P_5	SSR(λ_B, χ_A)	P_{11}	SSR(λ_R, χ_B)
P_6	SSR(λ_B, χ_F)		

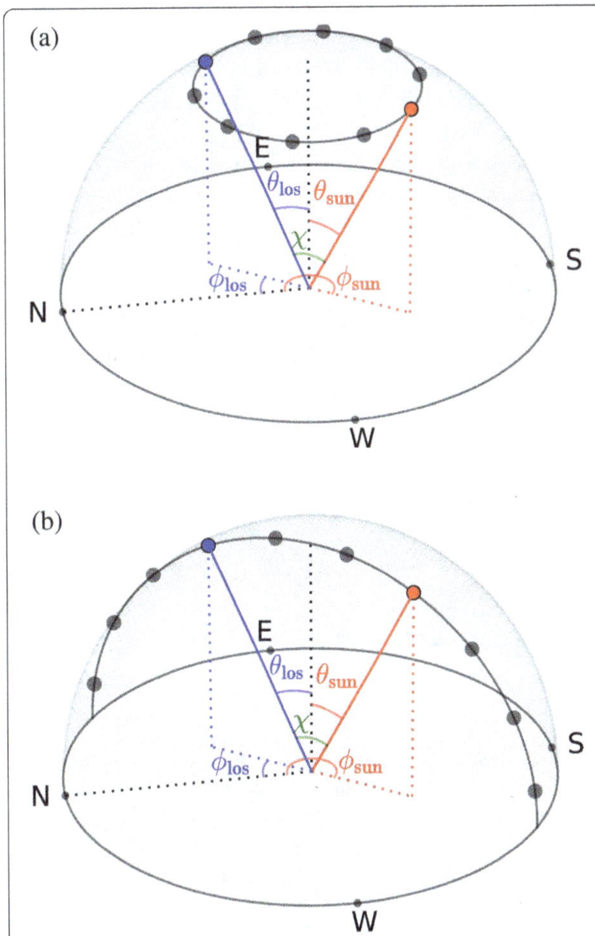

Fig. 3 Schematic figures of observation geometry on (**a**) almucantar plane and (**b**) principle plane

Table 5 Observational parameters (Y_1–Y_{11}) used for dust retrieval

Parameter (Y)		Relationship with measured quantities	
Y_1	D_O	P_1	$DSR(\lambda_O)$
Y_2	D_C	P_2/P_3	$DSR(\lambda_B)/DSR(\lambda_R)$
Y_3	A_C	P_5/P_9	$SSR(\lambda_B, \chi_A)/SSR(\lambda_R, \chi_A)$
Y_4	A_B	P_5/P_4	$SSR(\lambda_B, \chi_A)/SSR(\lambda_B, \chi_O)$
Y_5	A_R	P_9/P_8	$SSR(\lambda_R, \chi_A)/SSR(\lambda_R, \chi_O)$
Y_6	F_C	P_6/P_{10}	$SSR(\lambda_B, \chi_F)/SSR(\lambda_R, \chi_F)$
Y_7	F_B	P_6/P_4	$SSR(\lambda_B, \chi_F)/SSR(\lambda_B, \chi_O)$
Y_8	F_R	P_{10}/P_8	$SSR(\lambda_R, \chi_F)/SSR(\lambda_R, \chi_O)$
Y_9	B_C	P_7/P_{11}	$SSR(\lambda_B, \chi_B)/SSR(\lambda_R, \chi_B)$
Y_{10}	B_B	P_7/P_4	$SSR(\lambda_B, \chi_B)/SSR(\lambda_B, \chi_O)$
Y_{11}	B_R	P_{11}/P_8	$SSR(\lambda_R, \chi_B)/SSR(\lambda_R, \chi_O)$

which is closer to $\lambda_W = 935$ nm. Then, c_w is optimized by matching the calculated ratio of $D_2 = \mathrm{DSR}(\lambda_W)/\mathrm{DSR}(\lambda_S)$ or similar ratio of SSR (S_2) to the measured one. Alternatively, when three wavelengths are available, non-absorptive wavelengths are chosen at wavelengths shorter ($\lambda_S = 925$ nm) and longer ($\lambda_L = 972$ nm) than the absorption band. By linear interpolation, a reference intensity at λ_W ($\mathrm{DSR}_0(\lambda_W) = [(\lambda_L - \lambda_W)\mathrm{DSR}(\lambda_S) + (\lambda_W - \lambda_S)\mathrm{DSR}(\lambda_L)]/(\lambda_L - \lambda_S)$ and similar value for SSR) can be obtained that is free of water vapor absorption. Then, c_w is found by matching the calculated ratio $D_3 = \mathrm{DSR}(\lambda_W)/\mathrm{DSR}_0(\lambda_W)$ or a similar one for SSR (S_3) to the measured ratio. Using three wavelengths, more precise estimation of water vapor absorption is possible with better knowledge of the spectral intensity covering the entire range around the absorption band.

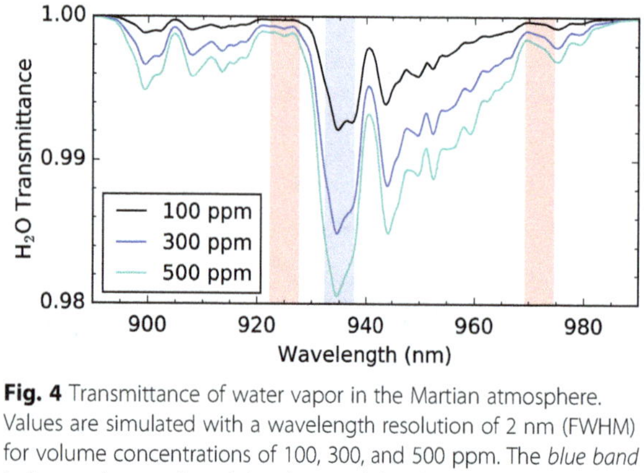

Fig. 4 Transmittance of water vapor in the Martian atmosphere. Values are simulated with a wavelength resolution of 2 nm (FWHM) for volume concentrations of 100, 300, and 500 ppm. The *blue band* indicates the wavelength band around the absorption peak centered at 935 nm with a band width of 5 nm, and the *two red bands* indicate the wavelength band outside the absorption band centered at 935 and 972 nm with a band width of 5 nm

Retrieval of model parameters

We consider the optimization of model parameters (including both dust model parameters and TCWV) through a least squares analysis. We assume a form of

$$\boldsymbol{y} = \boldsymbol{F}(\boldsymbol{x}) + \boldsymbol{\varepsilon}_y, \tag{7}$$

where $\boldsymbol{x}$ and $\boldsymbol{y}$ are vectors composed of the normalized values of model parameters and observational parameters, respectively, $\boldsymbol{F}(\boldsymbol{x})$ is a forward model function, and $\boldsymbol{\varepsilon}_y$ is measurement noise. The Taylor series expansion of Eq. (7) around the a priori value $\boldsymbol{x}_a$ leads to the linear approximation given as

$$\boldsymbol{y} = \boldsymbol{F}(\boldsymbol{x}_a) + \boldsymbol{K}_x(\boldsymbol{x} - \boldsymbol{x}_a) + \boldsymbol{\varepsilon}_y. \tag{8}$$

Here, $\boldsymbol{K}_x = \partial(\boldsymbol{x})/\partial\boldsymbol{x}$ is the Jacobian matrix. If we ignore measurement noise, Eq. (8) can be solved using a pseudo-inverse matrix of $\boldsymbol{K}_x$. In reality, however, measurement noise is not negligibly small. In order to avoid the instability of the solution due to a lack of sufficient information from the observation, we consider the following cost function that includes the residuals of both the observational and model parameters:

$$c(\boldsymbol{x}) = (\boldsymbol{y} - \boldsymbol{F}(\boldsymbol{x}))^{\mathrm{T}} \boldsymbol{S}_y^{-1} (\boldsymbol{y} - \boldsymbol{F}(\boldsymbol{x})) + (\boldsymbol{x} - \boldsymbol{x}_a)^{\mathrm{T}} \boldsymbol{S}_a^{-1} (\boldsymbol{x} - \boldsymbol{x}_a). \tag{9}$$

Here, $\boldsymbol{S}_y$ and $\boldsymbol{S}_a$ are the variance-covariance matrices of the observational parameters and a priori of the model parameters, respectively. The (i, j) element of $\boldsymbol{S}_y$ is the covariance of y_i and y_j, which is $\langle (y_i - \langle y_i \rangle)(y_j - \langle y_j \rangle) \rangle$, where $\langle y_i \rangle$ stands for the expectation value of y_i ($1 \le i \le 11$, see Table 5 for the case of dust retrieval). The i-th diagonal element of $\boldsymbol{S}_y$ represents the variance of y_i, i.e. $\langle (y_i - \langle y_i \rangle)^2 \rangle$. Similarly, the off-diagonal and diagonal elements of $\boldsymbol{S}_a$ represent the covariance and variance of a priori, respectively. We optimize the model parameters by minimizing the cost function with iterative calculation based on the non-linear least square method (Rodgers, 2000).

Linear approximation for error estimation

In the following discussion on the retrieval error, for simplicity, we use a linear approximation (the Jacobian is calculated with the a priori). The expectation value (retrieved solution, $\boldsymbol{x}_r$) of the model parameters can be given as

$$\boldsymbol{x}_r = \boldsymbol{x}_a + \left(\boldsymbol{K}_x^{\mathrm{T}} \boldsymbol{S}_y^{-1} \boldsymbol{K}_x + \boldsymbol{S}_a^{-1}\right)^{-1} \boldsymbol{K}_x^{\mathrm{T}} \boldsymbol{S}_y^{-1} (\boldsymbol{y} - \boldsymbol{F}(\boldsymbol{x}_a)). \tag{10}$$

Here, we have assumed that the errors of observational and model parameters follow Gaussian distributions.

Table 6 Wavelengths and a scattering angle chosen for TCWV measurement

Parameter		Description	Value
Wavelength	λ_W	Peak wavelength of water vapor absorption	935 nm
	λ_S	Shorter wavelength outside absorption band of water vapor	925 nm
	λ_L	Longer wavelength outside absorption band of water vapor	972 nm
Scattering angle	χ_W	Scattering angle for TCWV measurement	$85° + \theta_{sun}$

Defining a gain matrix as $G_y = \partial x_r/\partial y = (K_x^T S_y^{-1} K_x + S_a^{-1})^{-1} K_x^T S_y^{-1}$, we can rewrite Eq. (10) as

$$x_r = x_a + G_y(y - F(x_a)). \tag{11}$$

Furthermore, using $A = \partial x_r/\partial y = G_y K_x = (K_x^T S_y^{-1} K_x + S_a^{-1})^{-1} K_x^T S_y^{-1} K_x$ and $G_a = \partial x_r/\partial x_a = (I_n - A) = (K_x^T S_y^{-1} K_x + S_a^{-1})^{-1} S_a^{-1}$, we can rewrite Eq. (11) as

$$x_r = Ax + G_a x_a. \tag{12}$$

In Eq. (12), the right-hand side can be regarded as a weighted mean of the model parameter obtained from the true state (x) and the a priori (x_a), with weights of $K_x^T S_y^{-1} K_x$ and S_a^{-1}, respectively. Here, A is a matrix called the averaging kernel matrix (AKM), its diagonal elements indicating the extent to which the true model parameters affect the retrieved solution.

Adjustment of regularization strength

The weight of the a priori is determined by the balance between S_y and S_a. From a statistical point of view, the optimum value of S_a would be the actual variance-covariance matrix of a priori (S_e). However, it is difficult to estimate S_e accurately. In some cases, better solutions can be obtained by using S_a that are different from S_e and adjusting the strength of regularization (i.e., constraint from the a priori data). Here, we introduce a regularization parameter γ as

$$S_{a^{-1}} = \gamma S_e^{-1}. \tag{13}$$

The regularization strength is increased by increasing the value of γ. When we increase the regularization parameter, at some point the weight on the a priori becomes non-negligible and the retrieved parameter tends to be drawn toward its a priori value. This results in a gradual increase in the residuals of the observational parameters, since in general, a priori values are different from true values. In accordance with Tikhonov regularization (Doicu et al. 2010), a good regularization strength is achieved by making it stronger within the range in which the residuals of observational parameters, i.e., the first term of the cost function (9), are almost constant.

Table 7 Measured quantities (P_{12}–P_{17}) of DSR and SSR for TCWV retrieval

Parameter (P)	Measured quantity	Parameter (P)	Measured quantity
P_{12}	DSR(λ_W)	P_{15}	SSR(λ_W, χ_W)
P_{13}	DSR(λ_S)	P_{16}	SSR(λ_S, χ_W)
P_{14}	DSR(λ_L)	P_{17}	SSR(λ_L, χ_W)

Error estimation of model parameters

First, we consider the errors in the retrieved model parameters propagated from both measured values and a priori values. A variance-covariance matrix of the model parameters due to propagation error of measurement (S_m) can be calculated as

$$S_m = G_y S_y G_y^{\mathrm{T}}. \tag{14}$$

Similarly, a variance-covariance matrix of model parameters due to a priori error (S_s) can be obtained using $G_a = I_n - A$ as

$$S_s = G_a S_e G_a^{\mathrm{T}}. \tag{15}$$

The combined variance-covariance matrix of model parameters including both measurement error and a priori error is given as

$$S_c = S_m + S_s. \tag{16}$$

The equation $S_c = (K_x^T S_y^{-1} K_x + S_a^{-1})^{-1}$ holds when γ = 1. If we define error ratio $\boldsymbol{\varepsilon}_r$ as

$$\begin{aligned}\boldsymbol{\varepsilon}_r &= \left[\mathrm{diag}\left(S_s S_c^{-1}\right)\right]^{1/2} \\ &= \left[\mathrm{diag}\left(G_a\gamma(A + G_a\gamma)^{-1}\right)\right]^{1/2},\end{aligned} \tag{17}$$

then we have $\boldsymbol{\varepsilon}_r = [\mathrm{diag}(G_a)]^{1/2}$ when γ = 1. Each element of $\boldsymbol{\varepsilon}_r$ changes from 0 to 1 when γ changes from 0 to ∞. In addition to the propagation errors described above, imperfectness of the forward model can cause systematic errors. Systematic errors of retrieved model parameters ($\boldsymbol{\varepsilon}_x$) due to estimation errors of the forward model in the observational parameters ($\boldsymbol{\varepsilon}_f$) can be estimated as

$$\boldsymbol{\varepsilon}_x = G_y \boldsymbol{\varepsilon}_f. \tag{18}$$

Results and discussion

Sensitivity analysis of observational parameters

In order to examine the sensitivity of observational parameters (Y in Table 5) to the dust model parameters (X in Table 1), we changed the values of X from the

Table 8 Observational parameters (Y_{12}–Y_{15}) for TCWV retrieval

Parameter (Y)		Relationship with measured quantities	
Y_{12}	D_2	P_{12}/P_{13}	$DSR(\lambda_W)/DSR(\lambda_S)$
Y_{13}	S_2	P_{15}/P_{16}	$SSR(\lambda_W, \chi_W)/SSR(\lambda_S, \chi_W)$
Y_{14}	D_3	$P_{12}/[\alpha P_{13} + \beta P_{14}]$	$DSR(\lambda_W)/[\alpha DSR(\lambda_S) + \beta DSR(\lambda_L)]$
Y_{15}	S_3	$P_{15}/[\alpha P_{16} + \beta P_{17}]$	$SSR(\lambda_W, \chi_W)/[\alpha SSR(\lambda_S, \chi_W) + \beta SSR(\lambda_L, \chi_W)]$

$\alpha = (\lambda_L - \lambda_W)/(\lambda_L - \lambda_S)$, $\beta = (\lambda_W - \lambda_S)/(\lambda_L - \lambda_S)$

minimum to the maximum (see Table 1) and calculated the variation range of $\boldsymbol{Y}$. Only one dust model parameter was changed at a time, with other parameters fixed to their a priori values. The results of this sensitivity analysis are summarized in Fig. 5a–n.

Figure 5a–g shows the results for normalized observational parameters (y) obtained with the SZA fixed at 60°, which is a typical value as indicated in the results shown below. The scales of the vertical axes are the same for these panels except for Fig. 5a. Figure 5a–g indicates that the present choice of observational parameters has led to reasonably large sensitivity (more than 30% of observational parameters calculated with the a priori) to τ, r_{eff}, v_{eff}, and ρ_A, relatively small sensitivity to m_0 and k_0, and much less sensitivity to z_s. In Fig. 5a–d, most of the resulting curves show linear (or nearly linear) responses, though some non-linear features are seen as in the case of D_0 vs. τ in Fig. 5a.

In Fig. 5h–n, the results for variation range (Δy) are plotted as functions of SZA. Here, Δy stands for the maximum minus the minimum value ($\Delta y > 0$) if the observational parameter increases with increasing model parameter, otherwise the minimum minus the maximum ($\Delta y < 0$). For instance, $\Delta D_0(\tau)$ in Fig. 5h shows the SZA dependence of the difference between the minimum D_0 and maximum D_0 when τ is changed between 0.1 and 1, as indicated in Fig. 5a (for SZA = 60°). Such graphs are useful for estimating how the sensitivity changes with the time of DSR/SSR observations. Although in most cases no dependence on SZA can be seen, in some cases such as D_0 vs. τ, a certain increase in variation (sensitivity) is found with increasing SZA. This is ascribable to the longer path of DSR in the dust layer under large SZA conditions, namely, at dusk and dawn.

For better retrieval of each model parameter, it is required for an observational parameter to have a large sensitivity to that model parameter. It is also desirable that it exhibit sensitivity solely to that specific model parameter, not being affected by other model parameters. For example, $D_0 = DSR(\lambda_0)$ has sensitivity solely to τ; thus, the aerosol optical depth, τ, is well retrieved from the DSR measurement. Similarly, the dust effective radius, r_{eff}, can be retrieved from $A_R = SSR(\lambda_R, \chi_A)/SSR(\lambda_R, \chi_0)$, the hematite volume mixing ratio, ρ_A, from $F_C = SSR(\lambda_B, \chi_F)/SSR(\lambda_R, \chi_F)$, and if r_{eff} and ρ_A are already known, the width parameter of size distribution, v_{eff}, can be obtained from $A_B = SSR(\lambda_B, \chi_A)/SSR(\lambda_B, \chi_0)$. Therefore, the combination of observational parameters listed in Table 4 can provide information sufficient for retrieving τ, r_{eff}, ρ_A, and v_{eff}. Although only limited information on dust scale height (z_s) can be obtained from our retrieval, it is necessary to include z_s in our retrieval procedure so as to examine the influence of z_s on the retrieval of other dust model parameters. A detailed discussion on the information content of each dust model parameter will be given later.

Error estimation for dust measurement

In the following discussion, for simplicity, it is assumed that all measured quantities (P in Table 4) have the same relative error, and the propagation errors of observational parameters (Y in Table 5) are estimated by giving a reasonably small relative error of 1%. Since random noise is considered in the measured quantities, the variance-covariance matrix of measured quantities ($\boldsymbol{S}_p$) is assumed to be diagonal. Errors in the normalized measured quantities (p) propagate to the normalized observational parameters (y) according to the Jacobian matrix $\boldsymbol{K}_p = \partial \boldsymbol{y}/\partial \boldsymbol{p}$ and the variance-covariance matrix can be calculated as $\boldsymbol{S}_y = \boldsymbol{K}_p \boldsymbol{S}_p \boldsymbol{K}_p^{\mathrm{T}}$. Although this matrix can be non-diagonal, we retain only the diagonal elements. If non-diagonal elements are included, the determinant becomes zero (and hence, $\boldsymbol{S}_y^{-1}$ cannot be calculated), since we have assumed the same relative errors (1%) of measured quantities. The uncertainties of the dust model parameters are not yet precisely known. Thus, we assume the standard deviations to be half of the variation range in Table 1, and the variance-covariance matrix $\boldsymbol{S}_e$ is constructed by considering diagonal elements only (i.e., $[\text{diag}(\boldsymbol{S}_e)]^{1/2} = \Delta \boldsymbol{x}/2$).

The relation between the regularization parameter (γ) of dust model parameters and the diagonal elements of AKM is plotted in Fig. 6. The averaging kernel close to 1.0 corresponds to the case that the model parameter is mostly determined by the observed DSR/SSR data while the averaging kernel being close to 0.0 is the case of the model parameter being mainly determined by the a priori data. Since observational parameters have large sensitivity to the four dust model parameters (τ, r_{eff}, v_{eff}, and ρ_A), the residuals of observational parameters

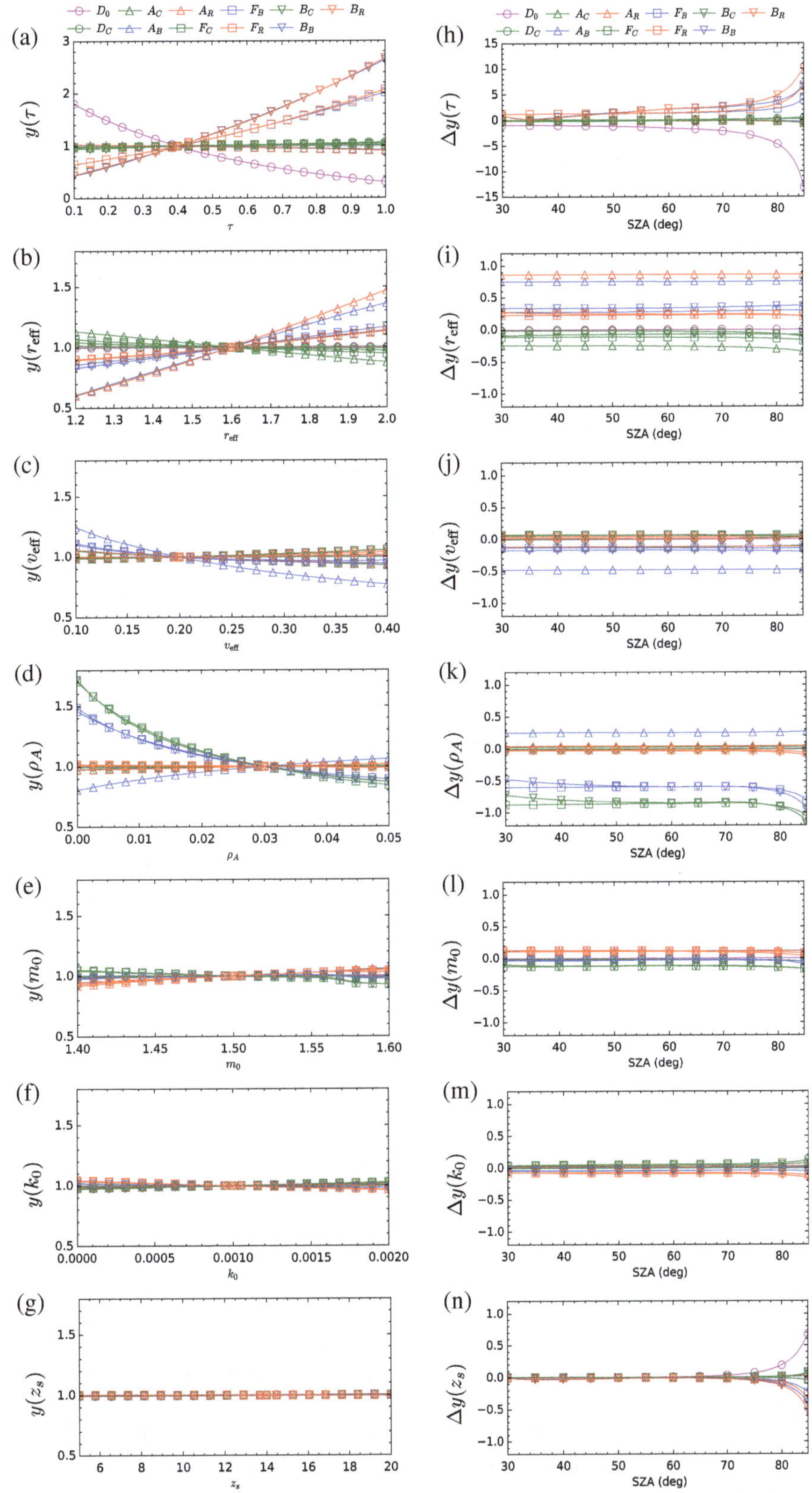

Fig. 5 Sensitivity of observational parameters to dust model parameters. Parameters y and Δy represent the normalized observational parameter, and the difference between the maximum and minimum values of y when the dust model parameter is changed from its minimum to maximum value, respectively. See Table 5 for the definitions of D_0, D_C. Panels (**a**) - (**g**) show the relation between $y(X_i)$ and X_i for X_1 - X_7, respectively, and panels (**h**) - (**n**) show that between $\Delta y(X_i)$ and SZA for X_1 - X_7, respectively

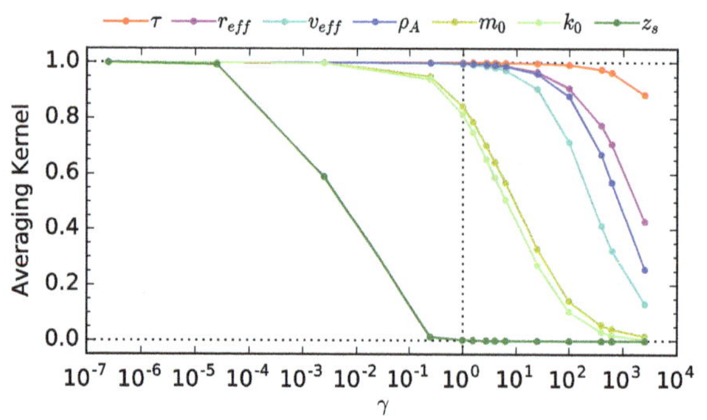

Fig. 6 Relation between regularization parameter (γ) of dust model parameters and diagonal elements of AKM. This graph shows an increased importance of a priori parameters for increasing γ. An SZA of 60° has been assumed for this calculation

increase when γ increases in the range of γ > 1. Although a matrix form of γ can potentially be adopted so as to enable the separate adjustment of regularization strength for each dust model parameter, here, we use the same regularization parameter of γ = 1 for all dust model parameters for the sake of simplicity. When γ = 1, z_s is almost confined to its a priori value (14 km).

Consideration on cases with insufficient data availability

In the actual lander mission, there might be cases in which DSR/SSR cannot be observed at some specific scattering angles or accurate color ratios cannot be obtained due to reasons such as the limited dynamic range of the sensor, bad observation conditions, difficulty in calibration, etc. In view of such possibilities, here, we estimate propagation errors for the following six cases: case 1. Full utilization of all observational parameters; case 2. DSRs (D_0, D_C) are not available; case 3. Aureoles (A_C, A_B, A_R) are not available; case 4. Forward scattering values (F_C, F_B, F_R) are not available; case 5. Backward scattering values (B_C, B_B, B_R) are not available; and case 6. Color ratios (D_C, A_C, F_C, B_C) are not available.

First, the diagonal elements of AKM for these six cases are shown in Fig. 7 (left scale). In case 1, the averaging kernels of τ, r_{eff}, v_{eff}, and ρ_A are almost in unity and those of m_0 and k_0 are about 0.8. Thus, observation information is well reflected in the retrieved values of these dust model parameters. On the other hand, the averaging kernels of z_s are much smaller and nearly zero at SZA > 55°, which means that z_s is almost confined to its a priori value. In other cases, averaging kernels are smaller as compared with case 1 and dependence on a priori is larger especially in cases 3 and 6.

Combined propagation errors of the dust model parameters ($\boldsymbol{E}_c = [\text{diag}(\boldsymbol{S}_c)]^{1/2}$) are shown in Fig. 8. For the evaluation of retrieval errors, errors relative to the a priori values are intuitively more understandable than those relative to the variation range, and hence, here errors are expressed as values relative to both variation ranges (ΔX) and a priori values (X_a). Errors relative to a priori values are obtained from errors relative to variation range multiplied by $\Delta X/X_a$ in Table 1. The left and right scales of Fig. 8 are normalized so that $\Delta x = 1$ and $x_a = 1$, respectively. In case 1, the combined propagation errors normalized with Δx (normalized with x_a) are as follows; τ: 0.5–2% (1–4%), r_{eff}: 3% (2%), v_{eff}: 4% (5%), and ρ_A: 2–3% (4–6%). Compared with case 1, the total propagation errors are larger especially in the following cases; τ: case 2, r_{eff}: cases 3 and 6, v_{eff}: case 3, and ρ_A: cases 2, 3, and 6. As for m_0 and k_0, the combined propagation errors of both parameters in case 1 are about 20% of Δx, which is 3 and 40% of x_a for m_0 and k_0, respectively. The errors of m_0 and k_0 are larger especially in case 3 (without aureole) and case 6 (without color ratio). The combined propagation error of z_s is close to the a priori error ($\Delta\boldsymbol{x}/2$) since z_s is almost confined to its a priori value. The percentage of a priori error in the combined propagation error is larger if the error ratio shown in Fig. 8 (right scale) is larger. The above propagation errors have been estimated with a relative measurement error of 1%, and it is expected that the propagation error is proportional to the measurement error if it is sufficiently small. Note, however, that the percentage of a priori error changes if the weight ratio of $\boldsymbol{K}_x^{\text{T}}\boldsymbol{S}_y^{-1}\boldsymbol{K}_x$ and $\mathbf{S}_a^{-1}$ changes.

Once a retrieval error occurs in a dust model parameter, it may propagate to the other model parameters. This effect becomes significant if the non-diagonal elements of AKM are not negligible. The non-diagonal elements of AKM in the above six cases are shown in Fig. 9. Since we have assumed that all of the diagonal elements of $\boldsymbol{S}_a$ are the same (i.e., $[\text{diag}(\boldsymbol{S}_a)]^{1/2} = \Delta\boldsymbol{x}/2$), AKM ($A_{\text{ij}} = \partial x_{r,i}/\partial x_j$) is symmetric and we show only elements which satisfy $i > j$ and $A_{ij} > 0.02$ at one of the considered SZAs. We can see that m_0 and k_0 (x_r:k_0, x: m_0) are easily influenced by each other (especially in case 6). Also, in case 3, the correlation becomes larger between m_0 and size distribution parameters (r_{eff} and v_{eff}).

Non-spherical dust shape

Previous studies on terrestrial aerosols have shown that the differences in the shapes of coarse mode particles play an important role in the spectra of DSR and SSR, since scattering patterns are dependent on the spherical (hygroscopic) or non-spherical (such as desert dust) shapes of aerosol particles. Dubovik et al. (2006) reported that the light scattering properties of desert dust aerosols in the terrestrial atmosphere can

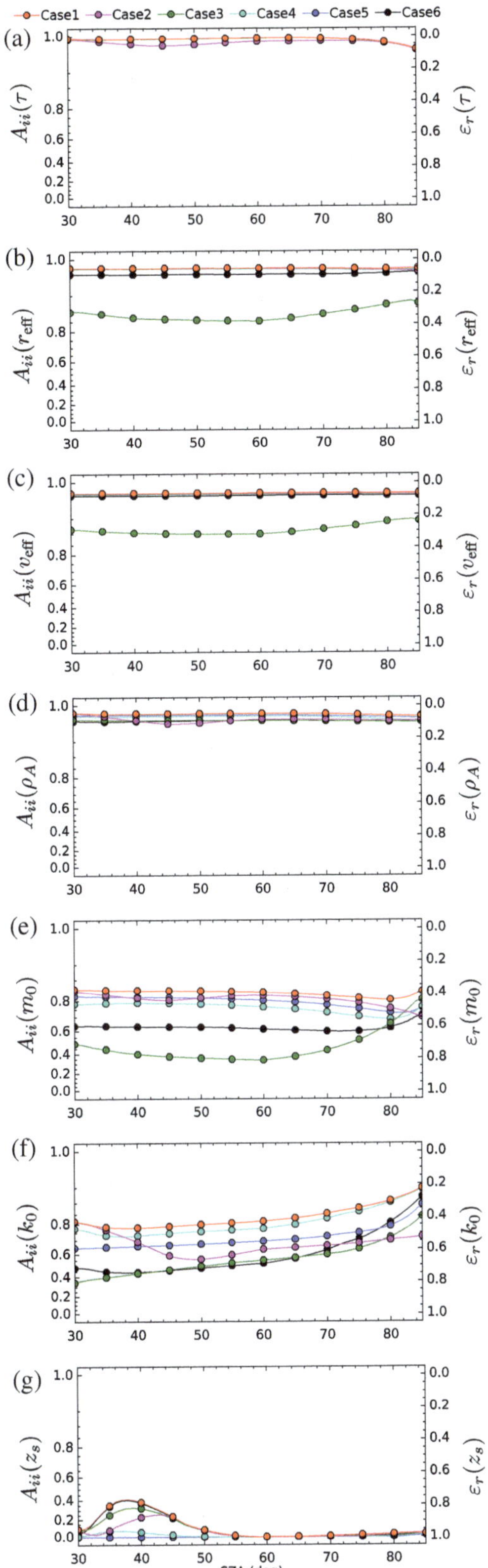

Fig. 7 Diagonal elements of AKM and error ratio of dust model parameters. The left and right scales are for AKM and error ratio, respectively. Case 1, with all observation parameters listed in Table 5; case 2, without DSR; case 3, without aureole; case 4, without forward scattering; case 5 without backward scattering; and case 6, without color ratio. Averaging kernel closer to unity means that information from observations is more significant. Panels (**a**) - (**g**) show the relation between A_{ii} / $\varepsilon_{r,i}$ and SZA for X_1 - X_7, respectively

be modeled by considering a mixture of spherical and non-spherical aerosol types. For non-spherical particles, their aspect ratio distribution was estimated from a feldspar sample (Dubovik et al. 2006). Following their treatment, we assume in the present analysis that the optical properties of Martian dust can be modeled by adjusting the volume mixing ratio (ρ_N) of non-spherical particles to spherical particles. Variation ranges for all the observational parameters listed in Table 5 are examined by changing the value of ρ_N between 0 and 1. The results are plotted in Fig. 10a as functions of SZA. Among the 11 observational parameters, $A_B = \mathrm{SSR}(\lambda_B, \chi_A)/\mathrm{SSR}(\lambda_B, \chi_0)$ exhibits the largest sensitivity to ρ_N ($\Delta y(\rho_N) = 0.15$). Relatively large dependence on SZA is seen for parameters related to backscattering (B_R, B_C, and B_B). Figure 10b shows the systematic errors of retrieved dust model parameters when the wrong assumption of $\rho_N = 0$ (spherical) is adopted to analyze the observational data of dust particles that are totally non-spherical ($\rho_N = 1$). The parameters except ν_{eff} and k_0 have small systematic errors (less than 15%). The parameters ν_{eff} and k_0 show errors as large as 25 and 80% of their variation ranges (~40% and 160% of their a priori values), respectively. Thus, it is desirable to take the non-sphericity of dust particles into account in the retrieval of dust model parameters as with our dust model.

Bimodal size distribution

We have so far assumed a monomodal size distribution based on coarse mode dust particles. From the simultaneous observation in the ultraviolet and infrared ranges by SPICAM on Mars Express, however, Fedorova et al. (2014) reported the existence of fine mode (radius 0.04–0.07 μm) as well as coarse mode (average radius 0.7 μm) particles in the higher altitudes, ranging between 10 and 80 km. Therefore, in this part, we examine the influence of additional fine mode particles in our model calculation. In order to reduce the number of dust model parameters, we assume that the complex RI of the fine mode is the same as that of the coarse mode particles. For the coarse mode, we employ the gamma distribution of Eq. (2) with $r_{\mathrm{eff}} = 1.6$ μm and $\nu_{\mathrm{eff}} = 0.2$ as before. For the fine mode, we follow the model given by Fedorova et al. (2014), namely, a log-normal distribution with effective

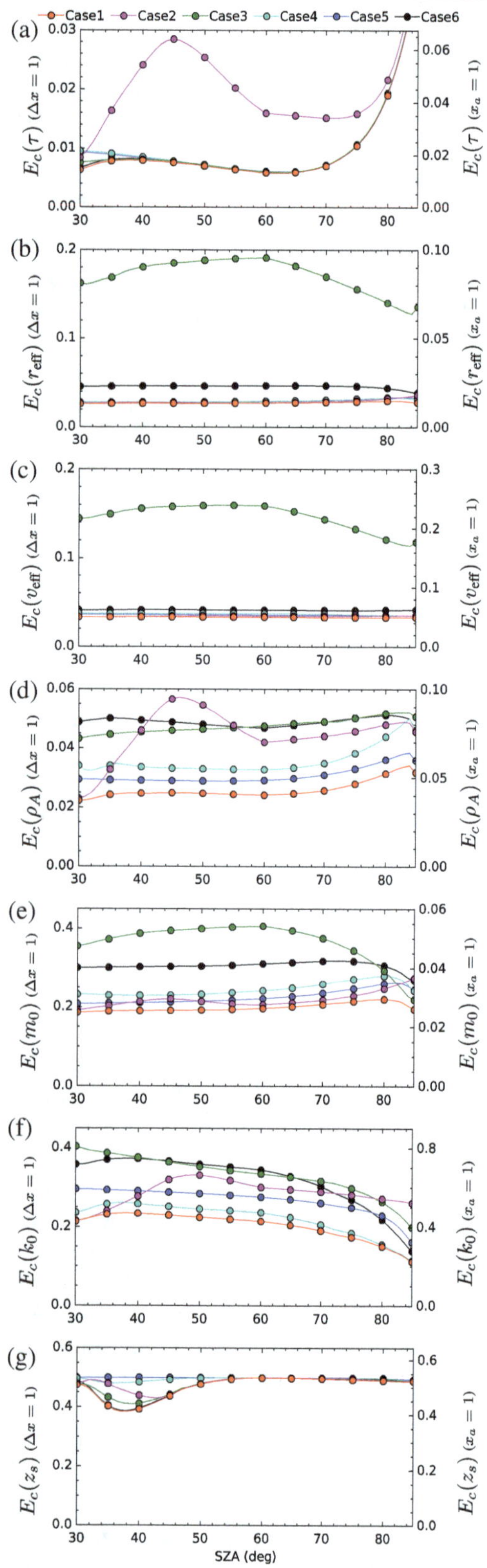

Fig. 8 Combined propagation errors of dust model parameters. The definitions of cases 1–6 are the same as in Fig. 7. The left and right scales are normalized so that $\Delta x = 1$ and $x_a = 1$, respectively, and panels (**a**) - (**g**) show the relation between propagation error and SZA for X_1 - X_7, respectively

radius $r_2 = 0.06$ μm and effective variance $\nu_2 = 0.1$. A log-normal distribution is given as

$$\frac{dN}{d\ln r} = \frac{1}{\sigma\sqrt{2\pi}}\exp\left[-\frac{(\ln r - \ln r_{\mathrm{mod}})^2}{2\sigma^2}\right]. \tag{19}$$

Here, r_{mod} is the mode radius and σ is the standard deviation. The values of mode radius and standard derivation can be calculated from the effective radius and variance as $r_{\mathrm{mod}} = r_2(\nu_2 + 1)^{-5/2}$ and $\sigma = [\ln(\nu_2 + 1)]^{1/2}$, respectively. We parameterize the volume mixing ratio of the fine mode (ρ_F) between 0 (the a priori value) and 0.01.

The sensitivities of dust observation parameters to the volume mixing ratio of fine mode dust (ρ_F) are shown in Fig. 11a. Observational parameters such as F_B, F_C, B_B, and B_C (i.e., parameters related to SSR in the blue band except aureole) are particularly sensitive to ρ_F.

Figure 11b shows the systematic errors of retrieved dust model parameters when the wrong assumption of $\rho_F = 0$ (monomodal) is adopted, while the true value of ρ_F is 0.01 (slightly bimodal). Figure 11b indicates that the magnitudes of error in ρ_A and k_0 are at most 10% of their variation range (~20% of their a priori values) even when the volume mixing ratio of fine mode is as large as 1%. Since the presence of fine mode particles in the lower Mars atmosphere has not been reported conclusively, it would be justified to assume the monomodal size distribution as in the present analysis. Nevertheless, to validate such a monomodal assumption, it is desirable to check the fitting result of real data to be obtained in future Mars lander missions.

Sensitivity analysis and error estimation for TCWV measurement

In a similar way as applied to the dust model parameters, sensitivity analysis for water vapor retrieval is made by changing the value of c_w between 0.476 and 4.76 mg/cm^2 (corresponding to 100 and 1000 ppm, respectively). Figure 12 shows the resulting curves of the variation ranges (sensitivities) of the observational parameters plotted against SZA. The SZA dependencies of the sensitivities of $D_2 = \mathrm{DSR}(\lambda_W)/\mathrm{DSR}(\lambda_S)$ and $D_3 = \mathrm{DSR}(\lambda_W)/\mathrm{DSR}_0(\lambda_W)$ to c_w are very similar. While D_3 is sensitive only to c_w, D_2 exhibits large sensitivity not only to c_w but also to τ. Thus, D_3 is a better parameter than D_2 for the c_w retrieval, as expected from the better knowledge of the spectral intensity around the absorption band. As for SSR, both $S_2 = \mathrm{SSR}(\lambda_W)/\mathrm{SSR}(\lambda_S)$ and $S_3 = \mathrm{SSR}(\lambda_W)/\mathrm{SSR}_0(\lambda_W)$ exhibit similar SZA dependence of the sensitivities to c_w, and the advantage of using S_2 or S_3 for the retrieval of c_w is that the magnitudes of the sensitivities are approximately three times larger than those of D_2 and D_3. A drawback, however, is that they show

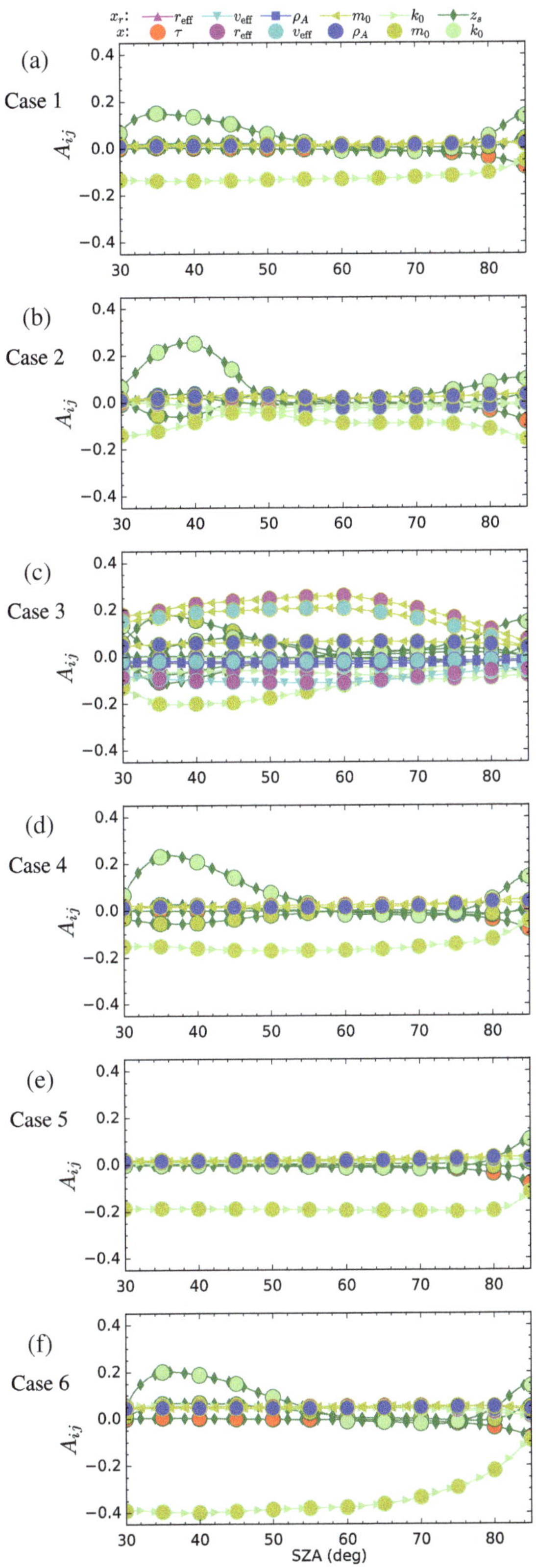

Fig. 9 Non-diagonal elements of AKM ($A_{ij} = \partial x_{r,i}/\partial x_j$) of dust model parameters. The *lines with small symbols* indicate the type of retrieved dust parameters (x_r), while the *large circles* indicate the type of true dust parameters (x). Panels (**a**) - (**f**) show the relation between A_{ij} and SZA for cases 1-6, respectively. The definitions of cases 1–6 are the same as in Fig. 7. These graphs are useful for the examination of error propagation from one dust parameter to another (see text)

sensitivities to τ and z_s in addition to c_w. Especially, the dependence on z_s can be problematic, since this parameter cannot be measured accurately through dust observations. This problem is hardly improved even with three wavelengths, since the difference of τ or z_s results in a change of effective optical path length due to multiple scattering, and hence, the change in the amount of water vapor absorption itself. It should be pointed out, however, that we may be able to retrieve the value of z_s if both DSR and SSR can be measured simultaneously. Moreover, a better prospect for z_s retrieval may possibly be achieved by measuring carbon dioxide (CO_2) absorption with the MAX-DOAS technique (Irie et al. 2008), since it would be relatively easy to estimate the vertical distribution of CO_2, and the molecule has strong absorption bands in the near infrared region (e.g., ~1050 nm).

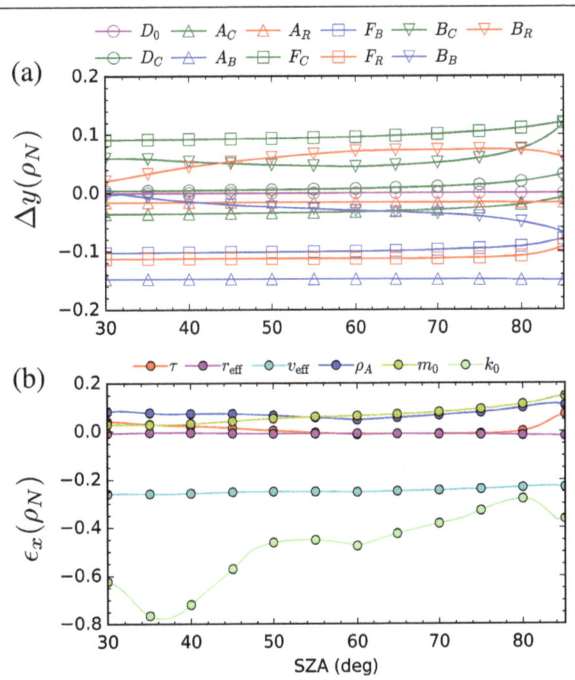

Fig. 10 a Sensitivity of dust observation parameters to the volume mixing ratio of non-spherical dust (ρ_N). Parameter Δy represents the difference between the maximum and minimum values of the normalized observational parameters when ρ_N is changed from 0 to 1. **b** Systematic error due to non-spherical dust shape. Systematic errors of normalized dust model parameters are estimated from differences between observational parameters calculated with true (1) and wrong (0) values of the volume mixing ratio of non-spherical dust (ρ_N)

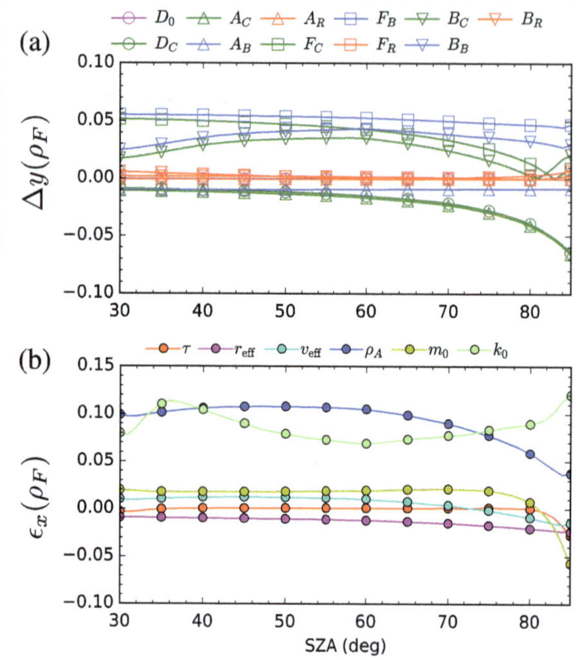

Fig. 11 a Sensitivity of dust observation parameters to the volume mixing ratio of fine mode dust (ρ_F). Parameter Δy represents the difference between the maximum and minimum values of the normalized observational parameters when ρ_N is changed from 0 to 0.01. **b** Systematic error due to ignorance of bimodal size distribution. The systematic errors of the normalized dust model parameters are estimated from the differences between the observational parameters calculated with true (0.01) and wrong (0) values of the fine mode dust volume mixing ratio (ρ_F)

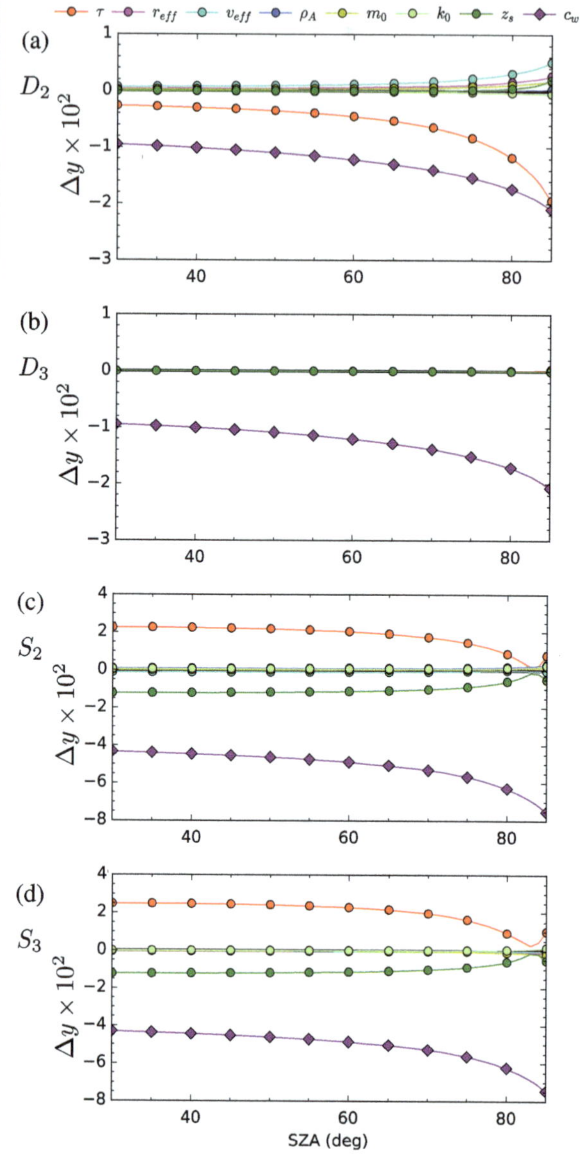

Fig. 12 Sensitivities of water vapor observation parameters to TCWV and dust model parameters. Panels (**a**) - (**d**) show the relation between $\Delta y(c_w)$ and SZA for D_2, D_3, S_2, and S_3, respectively

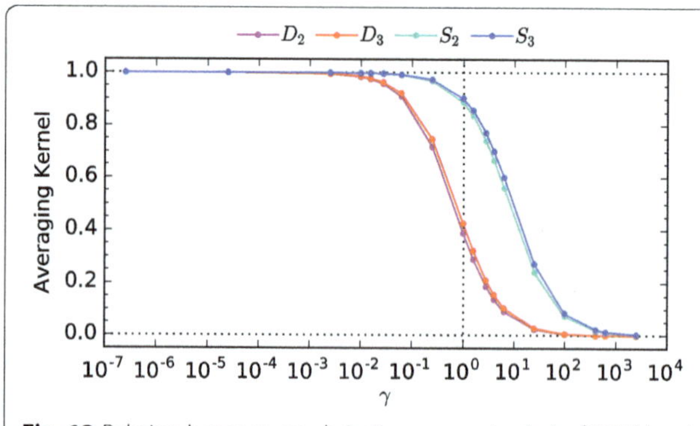

Fig. 13 Relation between regularization parameter (γ_w) of TCWV and diagonal elements of AKM

Here, we estimate the retrieval error of TCWV (c_w) when c_w and dust model parameters are retrieved simultaneously. We assume a relative measurement error of 1%, and a regularization parameter $\gamma_d = 1$ is employed for dust model parameters. Figure 13 shows the relation between the regularization parameter (γ_w) of TCWV and the diagonal elements of AKM. Here, SZA is fixed at 80°. When $\gamma_w = 1$, the diagonal elements of AKM with D_2 or D_3 are about 0.4 and the influence of the a priori information is significant. Hence, we take $\gamma_w = 0.06$, at which the diagonal elements of AKM are larger than 0.9. Weakening the regularization strength results in vulnerability to the random error of measurements, but such influence can be reduced by increasing averaging or exposure time of the camera measurement.

We consider four cases in which one of D_2, D_3, S_2, or S_3 is used for the retrieval of c_w. The diagonal elements of AKM for the four different cases are shown in Fig. 14 (left scale). The difference between two and three wavelengths turns out to be insignificant. In the cases of DSR (D_2, D_3), the averaging kernels are around 0.7–0.9, while in the cases of SSR (S_2, S_3) the averaging kernels are about 0.98.

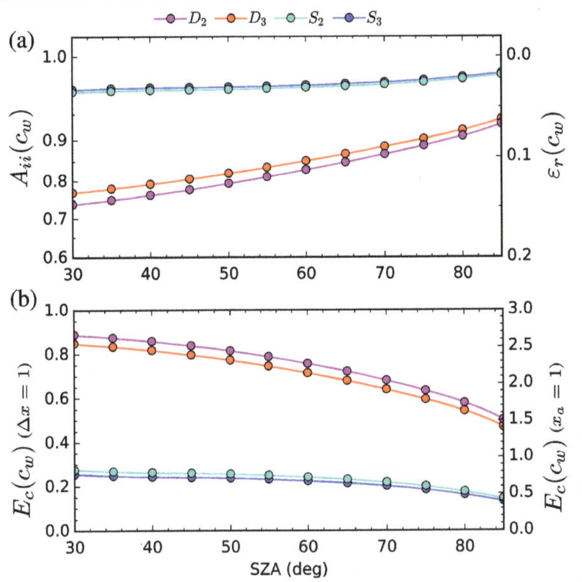

Fig. 14 a Diagonal elements of AKM and error ratio of TCWV. As compared with DSR, SSR results show stronger dependence on observation data of c_w rather than a priori data. **b** Combined propagation errors of TCWV. The left and right scales are normalized so that $\Delta x = 1$ and $x_a = 1$, respectively

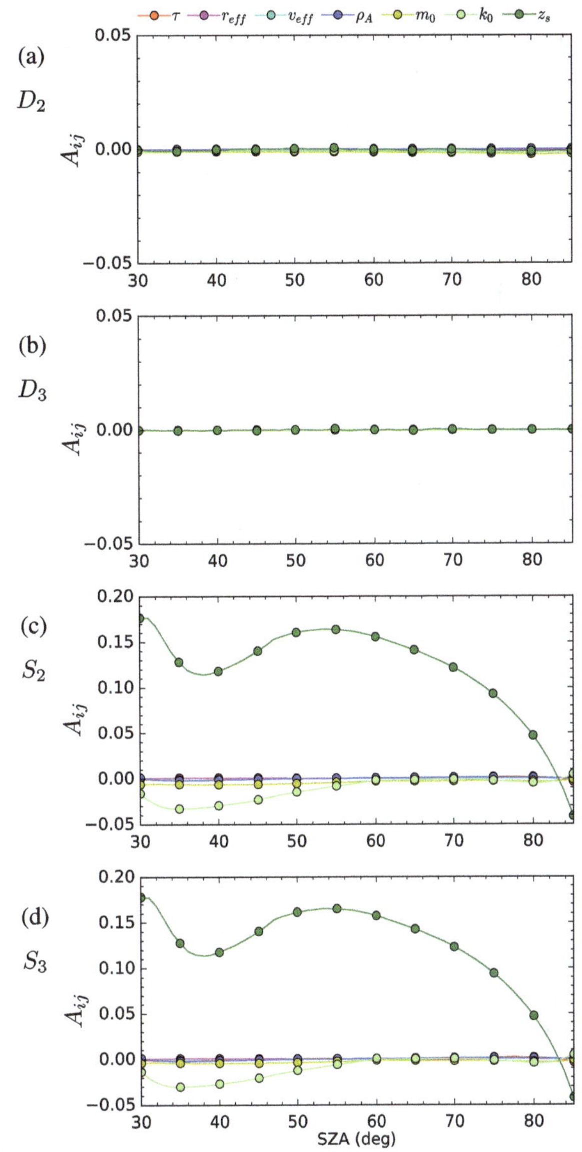

Fig. 15 Non-diagonal elements of AKM of TCWV. Panels (**a**) - (**d**) show the relation between A_{ij} and SZA for D_2, D_3, S_2, and S_3, respectively

The combined propagation errors are shown in Fig. 14b. They include the contributions from both measurement errors and a priori errors according to the error ratio shown in Fig. 14a (right scale). Again, the difference between two and three wavelengths is insignificant. Looking around SZA = 80°, the propagation errors normalized by x_a are 1.7 and 0.5 when we use DSR and SSR, respectively. This means that if we can suppress the measurement error down to 1%, SSR-based retrieval of c_w can achieve a precision of 50% of its a priori value.

As seen above, it is expected that the combined propagation errors are smaller if we use SSR rather than DSR. In the case of SSR, on the other hand, the amplitudes of non-diagonal elements of AKM are larger (Fig. 15), and the retrieved c_w is more affected by the retrieval error of dust model parameters (especially z_s), which is consistent with the results of the present sensitivity analysis (Fig. 12). However, the error of c_w due to the error in z_s is at most ~15% of the variation range (50% of the a priori value). Also, the influence of the water vapor profile on SSR is expected to be small for any particular value of c_w. Therefore, it is advantageous to use SSR, and if available, using both SSR and DSR is desirable.

Applicability of the present algorithm

In order to check the practical applicability of our algorithms, we estimate the measurement noise assuming sensor specifications that can be realized with the current technology. We employ radiative transfer simulation to estimate the minimum signal strength of SSR. The simulation conditions assumed for minimizing SSR intensity are an SZA of 80°, scattering angle of 120°, and dust optical depth of 0.1. Then, signal-to-noise (S/N) ratio is estimated from the SSR intensity. Only the shot noise of the sensor has been considered in this calculation, on the assumption that the dark noise and readout noise are negligibly small.

If we assume a focal length of 10 mm, an array with 500 pixels, and a field-of-view (FOV) angle of 30°, the sizes of a sensor element and array width are calculated to be approximately 10 and 5.2 mm, respectively. A two-dimensional image sensor that fulfills such specifications is easily available. Even when the signal strength is at its minimum and the exposure time is as short as 1 s, it is expected that the random noise is less than 0.5%, and hence, an S/N ratio over 200 can be attained. Actually, the exposure time can be much longer than 1 s, since a temporal resolution of 1 min, for example, is still acceptable. Therefore, the proposed scheme of DSR/SSR observation is practical and achievable by means of hardware existing today.

Conclusions

We have constructed a Martian dust model in order to retrieve optical/physical properties of Martian dust from DSR/SSR spectra to be measured by a multi-wavelength environmental camera onboard a Mars lander. We have assumed that Martian dust is composed of silicate-like substrate and hematite-like inclusion. The dust shape is assumed to be spheroid with a monomodal gamma size distribution and a fixed aspect ratio distribution estimated from a terrestrial feldspar sample. The spatial distribution of dust is homogeneous horizontally, and its concentration decreases exponentially with increasing altitude. Our Martian dust model is parameterized with seven parameters, namely, real (m_0) and imaginary (k_0) parts of silicate RI, volume mixing ratio of hematite (ρ_A), effective radius (r_{eff}) and variance (v_{eff}) of size distribution, optical depth (τ), and scale height (z_s) of extinction coefficient at a wavelength of 550 nm. On the other hand, we consider 11 observational parameters, namely, intensity of DSR at 550 nm (D_0), color ratio of DSR between blue and red bands (D_C), relative intensities of aureole, forward scattering, and backward scattering with respect to the SSR at the reference scattering angle for the blue and red bands (A_B, A_R, F_B, F_R, B_B, B_R), and color ratio of SSR (A_C, F_C, B_C).

From the result of error analysis using simulated data, relative errors (combined propagation error from both measurement and a priori data) of dust model parameters, comparing them to a measurement error of 1%, have been estimated as τ: 1–4%, r_{eff}: 2%, v_{eff}: 5%, ρ_A: 4–6%, m_0: 3%, and k_0: 40% of their a priori values. The value of z_s is almost confined to its a priori value, due to the lack of sensitivity in our observation scheme. Increase of retrieval error is observed when some of the observational parameters are omitted from the retrieval procedure. Such observational parameters are τ: DSR (D_0, D_C), r_{eff}: aureole (A_C, A_B, A_R) and color ratio (D_C, A_C, F_C, B_C), v_{eff}: aureole, ρ_A: DSR, aureole and color ratio, m_0 and k_0: aureole and color ratio. If spherical shape is employed to retrieve the dust model parameters from the observational parameters calculated for non-spherical shape, we expect a systematic error of v_{eff} as large as 40% of its a priori value. If a monomodal gamma distribution of coarse particles is employed in finding model parameters from observational ones but the true size distribution is bimodal containing 1% fine mode particles by volume, we expect a systematic error of ρ_A as large as 20% of its a priori value.

Water vapor has absorption bands inside the sensitive range of a CMOS sensor, and we can measure TCWV (c_w) from absorbed amount in DSR or SSR intensity. We considered four observational parameters of spectral ratio (D_2, D_3, S_2, S_3) calculated with DSR or SSR at two or three wavelengths. Using simulation, it has been confirmed that if we use SSR instead of DSR for the observational parameter, larger sensitivity can be expected, with the relative error of c_w against 1% of measurement error being 50% of its a priori value.

The measurement noise estimated with currently available sensor specifications is sufficiently small, and thus, we conclude that the proposed scheme of dust retrieval with DSR/SSR observation is practical.

Abbreviations

AKM: Averaging Kernel Matrix; CMOS: Complementary metal-oxide semiconductor; DSR: Direct solar radiation; FOV: Field of view; LOS: Line of sight; MAX-DOAS: Multi-axis differential optical absorption spectroscopy; RI: Refractive index; SSR: Scattered solar radiation; SZA: Solar zenith angle; TCWV: Total column water vapor; VMR: Volume mixing ratio

Funding

This work was supported by the joint research program of CEReS, Chiba university (2015, 2016).

Authors' contributions

NM, KN, GLH, and MS proposed the topic, and conceived and designed the study. NM carried out the simulation study. NM and KN wrote the manuscript. HK contributed to editing the manuscript. HS and NO provided ideas for the study. All authors read and approved the final manuscript.

Competing interests

The authors declare that they have no competing interest.

Author details

[1]Center for Environmental Remote Sensing, Chiba University, 1-33 Yayoi-cho, Inage-ku, Chiba-shi, Chiba 263-8522, Japan. [2]Department of Chemistry, Biology, and Environmental Science, Faculty of Science, Nara Women's University, Kita-uoya Nishi-machi, Nara-shi, Nara 630-8506, Japan. [3]Department of Earth Sciences, Okayama University, 3-1-1 Tsushimanaka, Kita-ku, Okayama-shi, Okayama 700-8530, Japan. [4]Planetary Exploration Research Center, Chiba Institute of Technology, 2-17-1 Tsudanuma, Narashino-shi, Chiba 275-0016, Japan. [5]Department of Earth System Science, Faculty of Science, Fukuoka University, 8-19-1 Nanakuma, Jonan-ku, Fukuoka-shi, Fukuoka 814-0180, Japan. [6]Institute of Space and Astronautical Science, Japan Aerospace Exploration Agency, 3-1-1 Yoshinodai, Chuo-ku, Sagamihara-shi, Kanagawa 252-5210, Japan.

References

Bell JF, Squyres SW, Herkenhoff KE, Maki JN, Arneson HM, Brown D, Collins SA, Dingizian A, Elliot ST, Hagerott EC, Hayes AG, Johnson MJ, Johnson JR, Joseph J, Kinch K, Lemmon MT, Morris RV, Scherr L, Schwochert M, Shepard MK, Smith GH, Sohl-Dickstein JN, Sullivan RJ, Sullivan WT, Wadsworth M (2003) Mars Exploration Rover Athena Panoramic Camera (Pancam) investigation. J Geophys Res Planets 108(E12):8063. doi:10.1029/2003JE002070

Bell JF, Squyres SW, Arvidson RE, Arneson HM, Bass D, Calvin W, Farrand WH, Goetz W, Golombek M, Greeley R, Grotzinger J, Guinness E, Hayes AG, Hubbard MYH, Herkenhoff KE, Johnson MJ, Johnson JR, Joseph J, Kinch KM, Lemmon MT, Li R, Madsen MB, Maki JN, Malin M, McCartney E, McLennan S, McSween HY, Ming DW, Morris RV, Dobrea EZN et al (2004a) Pancam multispectral imaging results from the Opportunity Rover at Meridiani Planum. Science 306(5702):1703–1709. doi:10.1126/science.1105245

Bell JF, Squyres SW, Arvidson RE, Arneson HM, Bass D, Blaney D, Cabrol N, Calvin W, Farmer J, Farrand WH, Goetz W, Golombek M, Grant JA, Greeley R, Guinness E, Hayes AG, Hubbard MYH, Herkenhoff KE, Johnson MJ, Johnson JR, Joseph J, Kinch KM, Lemmon MT, Li R, Madsen MB, Maki JN, Malin M, McCartney E, McLennan S, McSween HY et al (2004b) Pancam multispectral imaging results from the Spirit Rover at Gusev Crater. Science 305(5685):800–806. doi:10.1126/science.1100175

Berk A, Anderson GP, Acharya PK, Bernstein LS, Muratov L, Lee J, Fox M, Adler-Golden SM, Chetwynd JH Jr, Hoke ML, Lockwood RB, Gardner JA, Cooley TW, Borel CC, Lewis PE, Shettle EP (2006) MODTRAN5: 2006 update. In: Shen SS, Lewis PE (eds) Proceedings of SPIE 6233, Orlando, 17 April 2006.. doi:10.1117/12.665077

Dabrowska DD, Muñoz O, Moreno F, Ramos JL, Martínez-Frías J, Wurm G (2015) Scattering matrices of Martian dust analogs at 488 nm and 647 nm. Icarus 250:83–94. doi:10.1016/j.icarus.2014.11.024

Dlugach ZM, Korablev OI, Morozhenko AV, Moroz VI, Petrova EV, Rodin AV (2003) Physical properties of dust in the Martian atmosphere: analysis of contradictions and possible ways of their resolution. Sol Syst Res 37(1):1–19. doi:10.1023/A:1022395404115

Doicu A, Trautmann T, Schreler F (2010) Numerical regularization for atmospheric inverse problems. Springer, Heidelberg

Dubovik O, Sinyuk A, Lapyonok T, Holben BN, Mishchenko M, Yang P, Eck TF, Volten H, Muñoz O, Veihelmann B, van der Zande WJ, Leon J-F, Sorokin M, Slutsker I (2006) Application of spheroid models to account for aerosol particle nonsphericity in remote sensing of desert dust. J Geophys Res Atmos 111(D11). doi:10.1029/2005JD006619.

Ehlers K, Chakrabarty R, Moosmüller H (2014) Blue moons and Martian sunsets. Appl Opt 53(9):1808–1819. doi:10.1364/AO.53.001808

Fedorova AA, Korablev OI, Bertaux J-L, Rodin AV, Montmessin F, Belyaev DA, Reberac A (2009) Solar infrared occultation observations by SPICAM experiment on Mars-Express: simultaneous measurements of the vertical distributions of H2O, CO2 and aerosol. Icarus 200(1):96–117. doi:10.1016/j.icarus.2008.11.006

Fedorova AA, Montmessin F, Rodin AV, Korablev OI, Määttänen A, Maltagliati L, Bertaux J-L (2014) Evidence for a bimodal size distribution for the suspended aerosol particles on Mars. Icarus 231:239–260. doi:10.1016/j.icarus.2013.12.015

Flasar FM, Goody RM (1976) Diurnal behaviour of water on mars. Planet Space Sci 24(2):161–181. doi:10.1016/0032-0633(76)90103-3

Gierasch PJ, Goody RM (1972) The effect of dust on the temperature of the Martian atmosphere. J Atmos Sci 29(2):400–402

Haberle RM, McKay CP, Schaeffer J, Cabrol NA, Grin EA, Zent AP, Quinn R (2001) On the possibility of liquid water on present-day Mars. J Geophys Res Planets 106(E10):23317–23326. doi:10.1029/2000JE001360

Hamilton VE, Christensen PR (2005) Evidence for extensive, olivine-rich bedrock on Mars. Geology 33(6):433–436. doi:10.1130/G21258.1

Hansen JE, Travis LD (1974) Light scattering in planetary atmospheres. Space Sci Rev 16(4):527–610. doi:10.1007/BF00168069

Hinson DP, Wilson RJ (2004) Temperature inversions, thermal tides, and water ice clouds in the Martian tropics. J Geophys Res Planets 109(E1). doi:10.1029/2003JE002129. E01002.

Hoekzema NM, Garcia-Comas M, Stenzel OJ, Grieger B, Markiewicz WJ, Gwinner K, Keller HU (2010) Optical depth and its scale-height in Valles Marineris from HRSC stereo images. Earth Planet Sci Lett 294(3–4):534–540. doi:10.1016/j.epsl.2010.02.009

Irie H, Kanaya Y, Akimoto H, Iwabuchi H, Shimizu A, Aoki K (2008) First retrieval of tropospheric aerosol profiles using MAX-DOAS and comparison with lidar and sky radiometer measurements. Atmos Chem Phys 8(2):341–350

Isaacs RG, Wang W-C, Worsham RD, Goldenberg S (1987) Multiple scattering lowtran and fascode models. Appl Opt 26(7):1272–1281. doi:10.1364/AO.26.001272

Komguem L, Whiteway JA, Dickinson C, Daly M, Lemmon MT (2013) Phoenix LIDAR measurements of Mars atmospheric dust. Icarus 223(2):649–653. doi:10.1016/j.icarus.2013.01.020

Lemmon MT, Wolff MJ, Smith MD, Clancy RT, Banfield D, Landis GA, Ghosh A, Smith PH, Spanovich N, Whitney B, Whelley P, Greeley R, Thompson S, Bell JF, Squyres SW (2004) Atmospheric imaging results from the Mars exploration rovers: Spirit and Opportunity. Science 306(5702):1753–1756. doi:10.1126/science.1104474

Lemmon MT, Smith PH, Shinohara C, Tanner R, Woida P, Shaw A, Hughes J, Reynolds R, Woida R, Penegor J, Oquest C, Hviid SF, Madsen MB, Olsen M, Leer K, Drube L, Morris RV, Britt DT (2008) The Phoenix Surface Stereo Imager (SSI) Investigation. Lunar and planetary science conference, League City, pp 10–14, March 2008

Levoni C, Cervino M, Guzzi R, Torricella F (1997) Atmospheric aerosol optical properties: a database of radiative characteristics of different components and classes. Appl Opt 36(30):8031–8041

Malin MC, Caplinger MA, Edgett KS, Ghaemi FT, Ravine MA, Schaffner JA, Baker JM, Bardis JD, Dibiase DR, Maki JN, Willson RG, Bell JF, Dietrich WE, Edwards LJ, Hallet B, Herkenhoff KE, Heydari E, Kah LC, Lemmon MT, Minitti ME, Olson TS, Parker TJ, Rowland SK, Schieber J, Sullivan RJ, Sumner DY, Thomas PC, Yingst RA (2010) The Mars Science Laboratory (MSL) Mast-mounted Cameras (Mastcams) flight instruments. Lunar and planetary science conference, Woodlands, pp 1–5, March 2010

Manago N, Kuze H (2010) Determination of tropospheric aerosol characteristics by spectral measurements of solar radiation using a compact, stand-alone spectroradiometer. Appl Opt 49(8):1446–1458. doi:10.1364/AO.49.001446

Manago N, Miyazawa S, Bannu KH (2011) Seasonal variation of tropospheric aerosol properties by direct and scattered solar radiation spectroscopy. J Quant Spectrosc Radiat Transf 112(2):285–291. doi:10.1016/j.jqsrt.2010.06.015

Markiewicz WJ, Sablotny RM, Keller HU, Thomas N, Titov D, Smith PH (1999) Optical properties of the Martian aerosols as derived from Imager for Mars Pathfinder midday sky brightness data. J Geophys Res Planets 104(E4):9009–9017. doi:10.1029/1998JE900033

Millour E, Forget F, Spiga A, Colaitis A, Navarro T, Madeleine J-B, Chauffray J-Y, Montabone L, Lopez-Valverde MA, Gonzalez-Galindo F, Lefevre F, Montmessin F, Lewis SR, Read PL, Desjean M-C, Huot J-P (2012) Mars Climate Database version 5. European planetary science congress, Madrid, pp 23–28, September 2012

Mishchenko MI, Travis LD (1998) Capabilities and limitations of a current FORTRAN implementation of the T-matrix method for randomly oriented, rotationally symmetric scatterers. J Quant Spectrosc Radiat Transf 60(3):309–324. doi:10.1016/S0022-4073(98)00008-9

Moore CA, Moores JE, Lemmon MT, Rafkin SCR, Francis R, Pla-García J, Haberle RM, Zorzano M-P, Martín-Torres FJ, Burton JR (2016) A full Martian year of line-of-sight extinction within Gale Crater, Mars as acquired by the MSL Navcam through sol 900. Icarus 264:102–108. doi:10.1016/j.icarus.2015.09.001

Moores JE, Lemmon MT, Smith PH, Komguem L, Whiteway JA (2010) Atmospheric dynamics at the Phoenix landing site as seen by the Surface Stereo Imager. J Geophys Res Planets 115(E1). doi:10.1029/2009JE003409. E00E08

Moriyama S (1975) Effects of dust on radiation transfer in the Martian atmosphere, (2). J Meteor Soc Japan 53(3):214–221

Patterson WR, Huck FO, Wall SD, Wolf MR (1977) Calibration and performance of the Viking Lander cameras. J Geophys Res 82(28):4391–4400. doi:10.1029/JS082i028p04391

Pike WT, Staufer U, Hecht MH, Goetz W, Parrat D, Sykulska-Lawrence H, Vijendran S, Madsen MB (2011) Quantification of the dry history of the Martian soil inferred from in situ microscopy. Geophys Res Lett 38(24). doi:10.1029/2011GL049896

Pollack JB, Colburn D, Kahn R, Hunter J, Van Camp W, Carlston CE, Wolf MR (1977) Properties of aerosols in the Martian atmosphere, as inferred from Viking Lander imaging data. J Geophys Res 82(28):4479–4496. doi:10.1029/JS082i028p04479

Pollack JB, Colburn DS, Flasar FM, Kahn R, Carlston CE, Pidek D (1979) Properties and effects of dust particles suspended in the Martian atmosphere. J Geophys Res Solid Earth 84(B6):2929–2945. doi:10.1029/JB084iB06p02929

Pollack JB, Ockert-Bell ME, Shepard MK (1995) Viking Lander image analysis of Martian atmospheric dust. J Geophys Res Planets 100(E3):5235–5250. doi:10.1029/94JE02640

Querry MR (1985) Optical constants., Technical Report AD-A158-623, DTIC Document

Rodgers CD (2000) Inverse methods for atmospheric sounding: theory and practice vol 2. World scientific, Singapore

Savijärvi H (2014) A toy climate model for Mars. Icarus 242:105–111. doi:10.1016/j.icarus.2014.07.029

Smith MD (2008) Spacecraft observations of the Martian atmosphere. Annu Rev Earth Planet Sci 36(1):191–219. doi:10.1146/annurev.earth.36.031207.124334

Smith MD (2009) THEMIS observations of Mars aerosol optical depth from 2002–2008. Icarus 202(2):444–452. doi:10.1016/j.icarus.2009.03.027

Smith PH, Lemmon M (1999) Opacity of the Martian atmosphere measured by the imager for Mars Pathfinder. J Geophys Res Planets 104(E4):8975–8985. doi:10.1029/1998JE900017

Smith PH, Tomasko MG, Britt D, Crowe DG, Reid R, Keller HU, Thomas N, Gliem F, Rueffer P, Sullivan R, Greeley R, Knudsen JM, Madsen MB, Gunnlaugsson HP, Hviid SF, Goetz W, Soderblom LA, Gaddis L, Kirk R (1997) The imager for Mars Pathfinder experiment. J Geophys Res Planets 102(E2):4003–4025. doi:10.1029/96JE03568

Smith PH, Tamppari LK, Arvidson RE, Bass D, Blaney D, Boynton WV, Carswell A, Catling DC, Clark BC, Duck T, DeJong E, Fisher D, Goetz W, Gunnlaugsson HP, Hecht MH, Hipkin V, Hoffman J, Hviid SF, Keller HU, Kounaves SP, Lange CF, Lemmon MT, Madsen MB, Markiewicz WJ, Marshall J, McKay CP, Mellon MT, Ming DW, Morris RV, Pike WT, Renno N, Staufer U, Stoker C, Taylor P, Whiteway JA, Zent AP (2009) H2O at the Phoenix landing site. Science 325(5936):58–61. doi:10.1126/science.1172339

Stamnes K, Tsay S-C, Wiscombe W, Jayaweera K (1988) Numerically stable algorithm for discrete-ordinate-method radiative transfer in multiple scattering and emitting layered media. Appl Opt 27(12):2502–2509. doi:10.1364/AO.27.002502

Thomas N, Markiewicz WJ, Sablotny RM, Wuttke MW, Keller HU, Johnson JR, Reid RJ, Smith PH (1999) The color of the Martian sky and its influence on the illumination of the Martian surface. J Geophys Res Planets 104(E4):8795–8808. doi:10.1029/98JE02556

Titov DV, Markiewicz WJ, Thomas N, Keller HU, Sablotny RM, Tomasko MG, Lemmon MT, Smith PH (1999) Measurements of the atmospheric water vapor on Mars by the imager for Mars Pathfinder. J Geophys Res Planets 104(E4):9019–9026. doi:10.1029/1998JE900046

Tomasko MG, Doose LR, Lemmon M, Smith PH, Wegryn E (1999) Properties of dust in the Martian atmosphere from the imager on Mars Pathfinder. J Geophys Res Planets 104(E4):8987–9007. doi:10.1029/1998JE900016

Whiteway JA, Komguem L, Dickinson C, Cook C, Illnicki M, Seabrook J, Popovici V, Duck TJ, Davy R, Taylor PA, Pathak J, Fisher D, Carswell AI, Daly M, Hipkin V, Zent AP, Hecht MH, Wood SE, Tamppari LK, Renno N, Moores JE, Lemmon MT, Daerden F, Smith PH (2009) Mars water-ice clouds and precipitation. Science 325(5936):68–70. doi:10.1126/science.1172344

Wolff MJ, Smith MD, Clancy RT, Arvidson R, Kahre M, Seelos F, Murchie S, Savijärvi H (2009) Wavelength dependence of dust aerosol single scattering albedo as observed by the Compact Reconnaissance Imaging Spectrometer. J Geophys Res Planets 114(E2). doi:10.1029/2009JE003350. E00D04.

Yang P, Liou KN (1996) Geometric-optics–integral-equation method for light scattering by nonspherical ice crystals. Appl Opt 35(33):6568–6584. doi:10.1364/AO.35.006568

9

Water on Mars: do submarine cyclic steps exist on the red planet?

Svetlana Kostic[1*] and Isaac B. Smith[2]

Abstract

Submarine cyclic steps have a significant impact on the evolution of the ocean floor on Earth. These upper-flow regime bedforms should be even more common in Martian reduced-gravity environments. We postulate that, if early Mars had an ocean, cyclic steps would have formed on Martian deltas spreading along the putative ancient shorelines in the northern plains. The main challenge to identifying submarine cyclic steps on Mars is geologic activity, in particular impacts, volcanism, and aeolian erosion, which have erased or modify much of the geomorphic evidence. Even though many Martian deltaic deposits do not bear much resemblance to the original depositional surfaces, we hypothesize that cementation and consolidation of marine sediments had facilitated the preservation of at least some Martian cyclic steps on deltaic deposits. We combine data from several orbital missions at Mars with a literature review and numerical analysis to inspect bedforms near the foreset-bottomset transition of 17 documented deltaic deposits. The most promising piece of evidence for Martian submarine cyclic steps is a linear series of rhythmic long-wavelength bedforms within a channel incised into the upper delta front of the Aeolis Mensae Delta. These bedforms are remarkably similar to their counterparts observed in channels on deltas and deep-water fans on Earth. Moreover, we speculate that two fields of cyclic steps, likely emanating from a crater upstream of the Aeolis Mensae Delta, are a potential example of contemporary cyclic steps and attest to the sporadic presence of flowing water on present-day Mars.

Keywords: Water on Mars, Martian ocean, Delta, Martian submarine cyclic steps, Foreset deposit, Foreset-bottomset transition, Aeolian bedforms, Reduced gravity, Inverted channel, Consolidation of submarine sediments

Introduction

Ancient ocean on Mars

The paradigm that early Mars could have supported a complex hydrological system and an ancient ocean located in the low-lying northern plains (Fig. 1) is one of the most challenged concepts in Mars research (Baker et al. 1991; Parker et al. 1989, 1993; Malin and Edgett 1999; Head et al. 1999; Ivanov and Head 2001, Fairen et al., 2003; Carr and Head 2003; Ghatan and Zimbelman 2006; McEwen et al. 2007). Yet, the evidence for the northern hemispheric Martian ocean is mounting steadily. Among most compelling is the recent work of Citron et al. (2018) and Villanueva et al. (2015). The putative shorelines in the northern plains (e.g., Parker et al. 1989; Carr and Head 2003) fail to follow an equipotential surface, which has been commonly used to challenge the existence of an early ocean (e.g., Malin and Edgett 1999). Citron et al. (2018) confirmed that variations in shoreline topography could be explained by the emplacement of Tharsis (Perron et al. 2007), a volcanic province that dominates the gravity and topography of Mars. However, their work demonstrates that the shorelines must have formed before and during the emplacement of Tharsis. The results indicate a close relationship between the evolution of the Martian ocean and the initiation and decline of Tharsis volcanism. More importantly, they imply that the Martian ocean formed concurrent with the valley networks (Bouley et al. 2016), which is significantly earlier than previously thought and have broad implications for the geology, hydrological cycle, and climate of early Mars. Villanueva et al. (2015) used observations of the present-day deuterium (hydrogen with an extra neutron) to hydrogen ration (D/H) to model how much water has escaped into space over Martian history. They suggest that early Mars (4.5

* Correspondence: skostic@sdsu.edu
[1]Computational Science Research Center, San Diego State University, San Diego, CA, USA
Full list of author information is available at the end of the article

Fig. 1 Rendition of the ancient ocean occupying nearly half of Martian northern hemisphere (NASA's Goddard Space Flight Center public domain image)

billion years ago) had a global equivalent water layer of at least 137 m, enough to fill the many basins on the planet. Evidence from the surface of Mars today implies that this water at least temporarily pooled to form an ocean in the low-lying northern plains (Fig. 1). The primordial ocean could have held more water than Earth's Arctic Ocean and covered 19% of the planet's surface (in contrast to the Atlantic Ocean, which occupies 17% of the Earth's surface).

Other earlier observations and important evidence for the ocean encompass the interpretation of ancient shorelines from Viking Orbiter images (Parker et al. 1993), the topography of deformed shorelines (Perron et al. 2007), global distribution of Martian deltas and valleys within and along the margins of the northern lowlands (Di Achille and Hynek 2010), a branching network of inverted channel and channel-lobe deposits at Aeolis Dorsa (DiBiase et al. 2013), and tsunami deposits with runout lobes and backwash erosion (Rodriguez et al. 2016; Costard et al. 2017). For years, the controversial paleo-shoreline interpretation (Parker et al. 1989, 1993) has driven research in a very specific direction of documenting terrestrial analogs for the deposits located on the northern plains, where the hypothetical ocean existed. Some of the most convincing evidence includes a resemblance between teardrop-shaped islands in Martian outflow channels and erosional shadow remnants documented in deep-water environments on Earth (Moscardelli and Wood 2011), terrestrial submarine analogs for streamlined forms on Mars (Burr 2011), similarities between large-scale polygonal terrains on Mars and deep-water polygonal fault systems on Earth (Cooke et al. 2011; Allen et al. 2013; Moscardelli et al. 2012), comparisons between high-albedo mounds in Acidalia Planitia and terrestrial submarine mud volcanoes (Allen et al. 2013; Oehler and Allen 2012), and resemblance between boulder-size rocks contained within the Vastitas Borealis Formation on the northern plains of Mars and blocks and boulders that have been documented in deep-water terrestrial environments and reported in outcrop (Moscardelli 2014).

The work presented herein provides an additional line of evidence for the northern hemispheric ocean by recognizing a new class of bedforms on Mars called submarine cyclic steps. Submarine cyclic steps (Fig. 2) are rhythmic, upstream-migrating bedforms bounded by internal hydraulic jumps in overriding turbidity currents (Kostic and Parker 2006). They signify one manifestation of fundamental morphodynamic instability of Froude-supercritical flow over an erodible bed (Kostic 2011). We use submarine analogs on Earth and compare their morphology and geometry to properties of channelized features on Martian deltas in order to establish that submarine cyclic steps exist on Mars. We postulate that these bedforms are created by the same processes (i.e., sediment gravity flows) on our planet and on Mars. Channelized cyclic steps are particularly instrumental in the analysis presented herein as they are net-erosional features, which cannot be mistaken for aeolian net-depositional bedforms.

Research objective, hypotheses, and challenges

The focused objective of this research is to evaluate whether submarine cyclic steps are present on 17 deltaic deposits documented by Di Achille and Hynek (2010), which spread along the putative shorelines of a primordial northern ocean (Fig. 3). The presence of a slope break is a necessary, but not sufficient, condition for the spontaneous evolution of an erodible plane bed into cyclic steps on deep-sea fan-deltas on Earth (Kostic 2011). The foreset-bottomset transition (Fig. 4) associated with a transition from a steep delta front or foreset deposit to a mildly sloping bottomset deposit constitutes such a slope break. Thus, we hypothesize that the foreset-bottomset

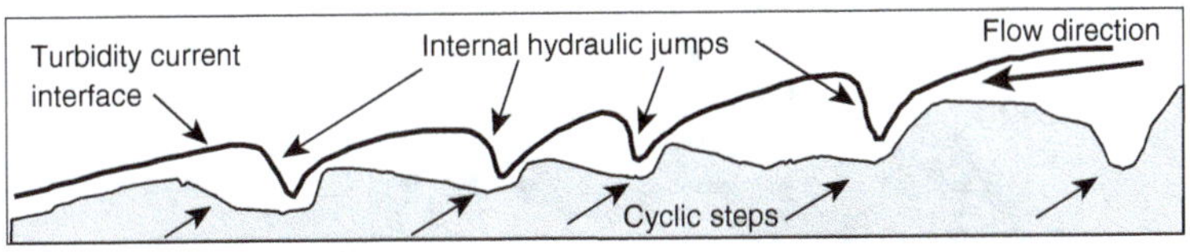

Fig. 2 Schematics of submarine cyclic steps bounded by internal hydraulic jumps (modified from Fildani et al. 2006)

transition on Martian deltas could have triggered the formation of submarine cyclic steps analogous to their terrestrial counterparts (e.g., Normandeau et al. 2016) in both process and form.

The main challenge of this research is that the current Mars topography provides an imperfect record of its depositional/erosional history because of resurfacing caused by impacts, volcanism, and mass wasting, as well as ongoing aeolian and likely occasion fluvial modification. Widespread aeolian surface modification is a particularly important factor in ongoing surface processes (Irwin III and Zimbelman 2012). Assuming e.g. a late Hesperian ocean was present (Fairen et al. 2003), it is questionable whether the morphologic signature of cyclic steps (O(1–10^2 m) height, O(10–10^3 m) wavelength) would be preserved. We postulate that Martian northern deltas are likely to host remnants of ancient cyclic steps, because marine bed sediments tend to develop strength over time through consolidation, and thus resist erosion. Furthermore, we assume that remnants of cyclic steps are more likely to be preserved on deltas displaying inverted channels due to cementation (DiBiase et al. 2013) as these deltas are more resistant to weathering and erosion.

Submarine cyclic steps as new analog for Martian streamlined forms emplaced by turbidity currents

Cyclic steps emplaced by turbidity currents are common features on the ocean floor of Earth (e.g., Kostic 2011). They were first discovered in numerical experiments on internal hydraulic jumps (Kostic and Parker 2006). Since then, they have swiftly gained recognition as important building blocks of deep-water depositional systems in both unconfined and confined submarine environments (e.g., Fildani et al. 2006, 2013; Kostic 2011, 2014; Cartigny et al. 2011; Covault et al. 2014, 2017; Hamilton et al. 2015; Postma et al. 2016).

Submarine cyclic steps should be more prevalent on Mars than on our planet. This becomes apparent from the densimetric Froude number Fr_d for turbidity currents that can be written as:

$$Fr_d = U/\sqrt{RCgH} \tag{1}$$

where H is an appropriate measure of turbidity current thickness, U is the layer-averaged flow velocity, g is the acceleration of gravity, C is the layer-averaged volume concentration of suspended sediment carried by the turbidity current ($C << 1$), and R is the submerged specific gravity of the sediment (which is close to 1.65 for most natural sediments). For the same values of flow velocity U, concentration of suspended sediment C, and turbidity current thickness H, turbidity currents on Mars should be intrinsically more biased toward supercritical flows than their analogs on Earth due to the significantly lower gravitational acceleration ($g = 3.7$ m/s^2, which is a little over 1/3 of the acceleration of gravity on Earth). By extension, it should be more likely for slope breaks on Mars (such as e.g. the foreset-bottomset transition) to

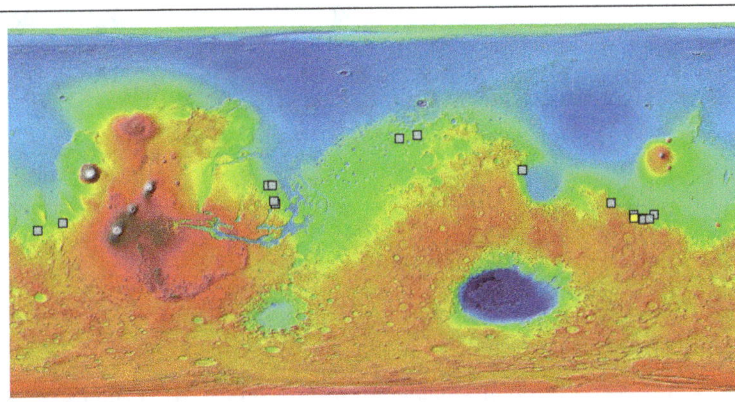

Fig. 3 MOLA topography of the low-laying northern plains of Mars with white squares showing remnants of 17 deltaic deposits identified by Di Achille and Hynek (2010). The yellow square indicates the location of Aeolis Mensae Delta

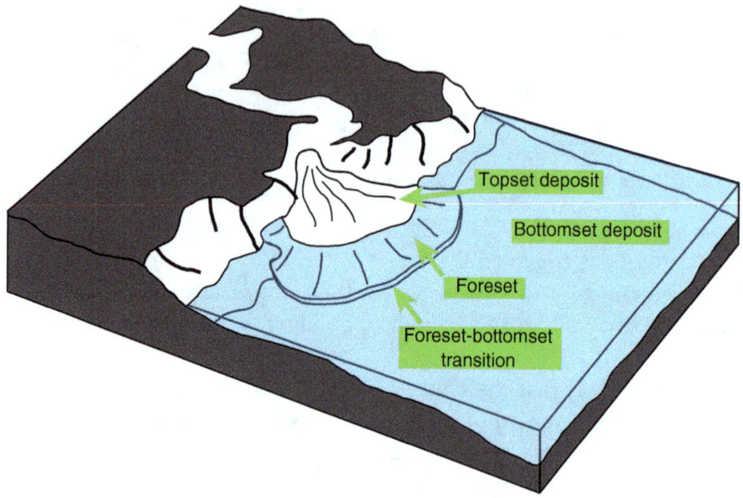

Fig. 4 Structure of a typical deltaic deposit on Mars (modified from Di Achille and Hynek 2010). Schematic diagram illustrates **a** topset (also called delta plain), foreset (also called delta front), and bottomset deposit (also called delta base), as well as the foreset-bottomset transition. Under the right conditions, the foreset-bottomset transition generates submarine cyclic steps

trigger internal hydraulic jumps and associated cyclic steps under the right conditions.

Formative conditions for Martian submarine cyclic steps

The dimensionless parameters that control the formation of submarine cyclic steps are summarized in Kostic (2011). Two of them, i.e., the dimensionless fall velocity v_s/U_o and the drop in bed Shields number $\Delta\tau^*$ across the jump, are particularly important for the analysis at hand. The dimensionless fall velocity v_s/U_o defines the ratio of the sediment fall velocity v_s to layer-averaged inflow velocity U_o. This parameter can be useful in estimating the cutoff size of sediment that causes turbidity currents to undergo internal hydraulic jumps (Kostic 2011). Supercritical turbidity currents driven by sediment fine enough to satisfy the approximate condition

$$\frac{v_s}{U_o} < 3 \times 10^{-3} \tag{2}$$

regularly display an internal hydraulic jump induced by the slope break, whereas the coarse-grained turbidity currents with the ratio

$$\frac{v_s}{U_o} > 5 \times 10^{-3} \tag{3}$$

do not undergo a transition to subcritical flow due to a rapid rate of sediment deposition on the bed (Kostic and Parker 2006; Kostic 2011). When an internal hydraulic jump occurs in response to a slope break, it is not always strong enough to leave a clear depositional signal in associated depositional record. A drop in the bed Shields number $\Delta\tau^*$ quantifies the corresponding drop in mobility of bed sediment across the jump, such that:

$$\Delta\tau^* = \left(\frac{\Delta u_*}{v_s} R_f\right)^2 \tag{4}$$

where Δu_* denotes the drop in shear velocity across the jump and R_f is a dimensionless particle fall velocity, given by:

$$R_f = v_s / \sqrt{RgD} \tag{5}$$

with D being the mean grain size of sediment. The drop in bed Shields number across the jump needs to satisfy the following crude criterion to yield a detectable depositional signal (Kostic and Parker 2006; Kostic 2011):

$$\Delta\tau^* > 3 \tag{6}$$

The analysis in Table 1 was carried out for median grain size taking values of 20, 50, 80, 100, 150, and 200 μm. The results imply that the formation of cyclic steps on Mars can be triggered by flows that are about 60% slower than their counterparts on Earth (i.e., the ratio k in Table 1). For example, a turbidity current driven by 50-μm sediment on Mars should have the inflow velocity of the order of 30 cm/s or higher (in contrast to over 70 cm/s on Earth) to trigger submarine cyclic steps. For a 100-μm turbidity current, the inflow velocity higher than 1.0 m/s (in contrast to 2.5 m/s on Earth) is necessary for steps to emerge.

Other results from Table 1 that should interest the planetary science community can be summarized as following: If the incoming flow velocity is less than 17 cm/s for a 50-μm current and 62 cm/s for a 100-μm current, the turbidity current after a slope break remains supercritical and cyclic steps never form. For higher underflow velocity of the order of 3–4 m/s the cutoff size for

Table 1 Calculated formative conditions for submarine cyclic steps on Earth and Mars

	Ds (μm)	20	50	80	100	150	200
On Earth							
	v_s (cm/s)	0.04	0.22	0.51	0.74	1.43	2.20
Always steps	U_o (cm/s) ^{a} >	11.8	72.6	169.5	247.9	476.5	734.6
No steps	U_o (cm/s) ^{b} <	7.1	43.5	101. 7	148.7	285.9	440.8
	R_f	0.020	0.077	0.141	0.185	0.290	0.388
	Δu^*(cm/s)^c >	3.1	4.9	6.2	7.0	8.5	9.8
On Mars							
	v_s (cm/s)	0.01	0.08	0.21	0.31	0.62	0.98
Always steps	U_o (cm/s) ^{a} >	4.3	28.2	68.6	102.5	204.9	325.1
No steps	U_o (cm/s) ^{b} <	2.6	16.9	41.2	61.5	122.9	195.1
	R_f	0.012	0.048	0.093	0.124	0.203	0.279
	Δu^*(cm/s) ^{c} >	1.9	3.0	3.8	4.3	5.2	6.1
Comparison							
Always steps	k ^{d}	0.64	0.61	0.60	0.59	0.57	0.56
	m ^{e}	0.39	0.39	0.39	0.39	0.39	0.39

Note: ^{a}Eq. (2), ^{b}Eq. (3), ^{c}Eq. (6), $^d(U_o^{Earth} - U_o^{Mars})/U_o^{Earth}$, $^e(\Delta u^{*Earth} - \Delta u^{*Mars})/\Delta u^{*Earth}$

submarine cyclic steps is roughly between 100 and 150 μm on Earth, and between 200 and 300 μm on Mars. The required drop in shear velocity across the jump that yields detectable cyclic steps on Mars is 40% lower than on Earth (i.e., the ratio m in Table 1).

Examples of potential terrestrial analogs in channels

The focus here is on cyclic steps in channels (Fildani et al. 2006; Smith et al. 2007; Lamb et al. 2008; Maier et al. 2012; Fildani et al. 2013; Covault et al. 2014; Tubau et al. 2015) and their role in channel initiation and maintenance. In particular, we review cyclic steps in the Monterey East Channel (Fildani et al. 2006) and San Mateo Channel (Covault et al. 2014) in California as potential terrestrial analogs for channelized cyclic steps on Martian deltas. The Monterey East system in central California is a unique deep-water environment to study cyclic steps as a means of channel initiation on modern submarine fans. Fildani et al. (2006) combined high-resolution acoustic data with numerical modeling to investigate the Monterey East Channel at the Monterey Fan. This fan segment is characterized by a linear train of large crescent-shaped scours incised into large-scale sediment waves at the outer levee of the Monterey Channel at the Shepard Meander (Fig. 5a). The scours in the channel range from 3 to 5 km in width, 3 to 6 km in length, and 80 to 200 m in depth. Fildani et al. (2006) offer evidence that these features constitute net-erosional cyclic steps carved by supercritical turbidity currents stripped from the main flow in the Monterey Channel moving through the meander. The formative supercritical flows were postulated to have focused through a breach in the levee, likely triggered by slumping of the steep meander wall. More importantly, net-erosional cyclic steps were recognized for the first time as a means of the early stage of an eventual avulsion from the present-day Monterey Channel and initiation of a new channel named the Monterey East Channel.

The San Mateo Channel located on a tectonically active slope offshore of southern California is another valuable deep-water environment to study channel initiation, filling, and maintenance by way of generation of cyclic steps. Covault et al. (2014) combined numerical experiments with interpretations of sea-floor and shallow subsurface stratigraphic imagery to examine a series of flute-shaped bedforms along the preset-day channel thalweg (Fig. 5b). These features, ranging in wavelength from 0.3 to 1.1 km and in height from 20 to 50 m, were identified as net-depositional cyclic steps emplaced by turbidity currents and credited with the channel maintenance. Moreover, numerical simulations suggested that an incipient, proto-San Mateo Channel was originally initiated by way of a series of net-erosional cyclic steps, which nucleated out of sea-floor perturbations across the tectonically active lower slope.

The Gravel River emptying into fjord-Lake Walker in Canada is selected as an additional delta analog for channelized cyclic steps on Martian deltas. The mean foreset slope of the Gravel River Delta is 4°, although the upper slope is significantly steeper, reaching 12°. Normandeau et al. (2016) reported noticeable bedforms covering an approximate area of 633,500 m^2 of the delta

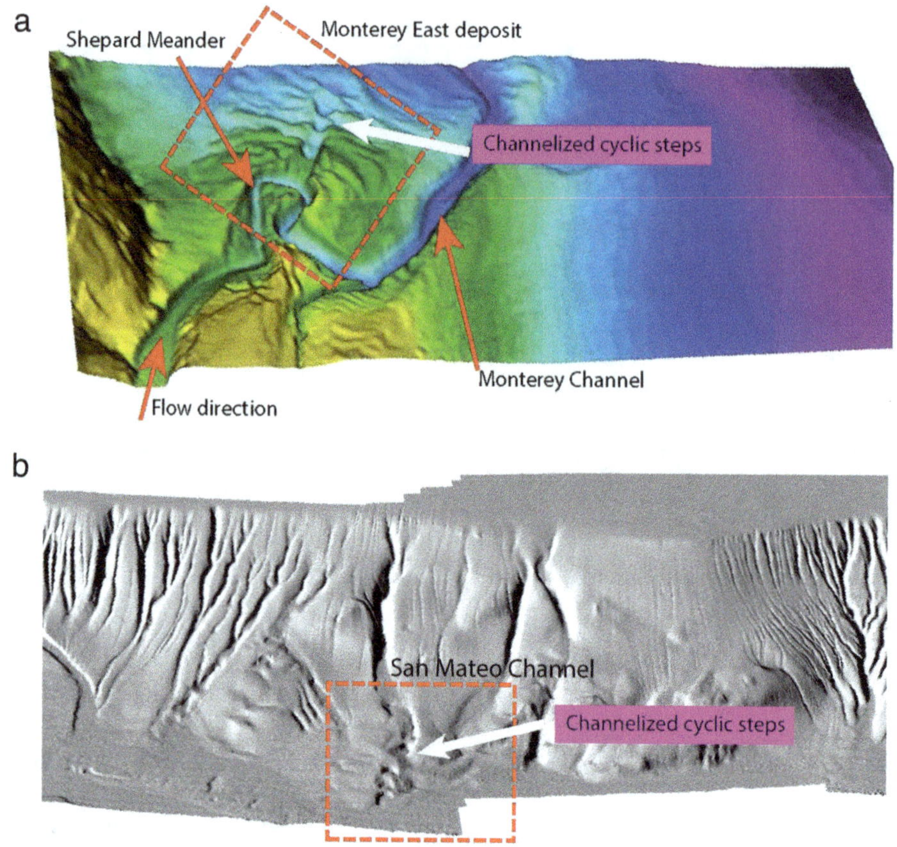

Fig. 5 Cyclic steps in submarine channels on Earth as analogs for bedforms on Martian deltas. **a** Large-scale net-depositional cyclic steps on the outer levee of the Shepard Meander dissected by a linear train of net-erosional cyclic steps in the Monterey East Channel, California, USA (image courtesy A. Fildani; high vertical exaggeration is to enhance bedforms); **b** A series of cyclic steps along the thalweg of the San Mateo Channel, California, USA (modified from Covault et al. 2014)

front (Fig. 6). These features have a mean wavelength of 35 m and a mean height of 2.2 m. They are mostly channelized between depths of 25 and 100 m, but also occur on the open delta plain. Normandeau et al. (2016) used the shape of the bedforms and the relationships between their morphological properties to demonstrate that these features were likely cyclic steps emplaced by supercritical turbidity currents. The underflows on the Gravel River Delta were attributed to hyperpycnal flows, even though there was also some evidence of delta-front slope failures.

All three selected terrestrial analogs are useful in their own way for identification and classification of channelized cyclic steps on Martian deltas. The Monterey East Channel analog establishes cyclic steps as a means of initiation of a new channel. The San Mateo Channel analog distinguishes between net-erosional cyclic steps as a means of channel initiation and net-depositional cyclic steps as a means of channel maintenance. The Gravel River Delta, even though not in a marine setting, illustrates well the effect of the foreset-bottomset transition on the formation of channelized cyclic steps on the foreset deposit.

Can cyclic steps form on Mars?

Aeolian cyclic steps

Cyclic steps were already discovered on Mars in an unexpected environment: the polar ice caps (Smith et al. 2013). They are attributed to atmospheric processes, i.e., katabatic winds and asymmetric ice accumulation. Katabatic winds have been studied extensively on Earth (Ball 1956; Pettré and André 1991) and have analogous properties to turbidity currents in submarine environments. They experience disruptive katabatic jumps that behave similarly to internal hydraulic jumps in turbidity currents. At the Martian polar caps, the cold air on top of the polar plateau densifies and begins to flow downhill, beneath the ambient air, as seasonal katabatic winds reaching speeds upwards of 20 m/s (Smith et al. 2013). The winds are disturbed by katabatic jumps similar to terrestrial katabatic jumps, but with one notable difference. Air is more compressible on Mars, so the fluid at the katabatic jump decompresses, lowering the atmospheric temperature enough to create water-ice clouds instead of preserving the water vapor (Smith et al. 2013). Additionally, the winds are unimpeded by

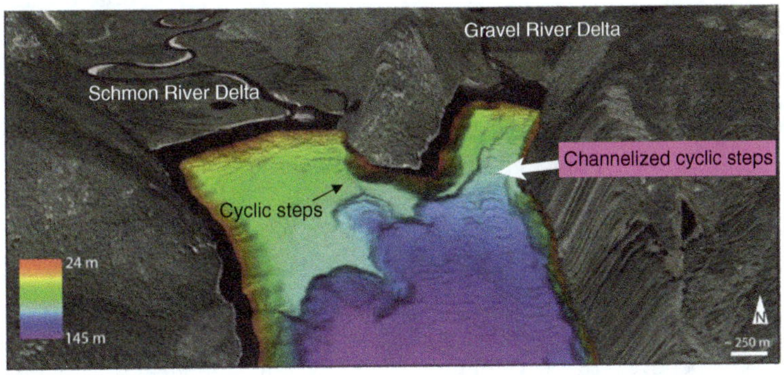

Fig. 6 Cyclic steps observed on deltas on Earth. Channelized and unconfined cyclic steps on foreset deposits of fjord-lake deltas at the Gravel River and Schmon River, Québec, Canada, (modified from Normandeau et al. 2016)

regional topography, so the southward slope drives the primary direction of the winds, and the Coriolis force steers them clockwise, setting up a recognizable spiraling pattern of polar troughs emplaced by katabatic winds.

Figure 7 illustrates curvilinear depressions called spiral troughs of the Norther Polar Layered Deposits (NPLD). These aeolian features are stratigraphy consistent with net-depositional cyclic steps and have a characteristic wavelength from 20 to 70 km and wave heights of the order of 400 to 1000 m (Smith and Holt 2010; Smith et al. 2013). They have migrated approximately 50 km upslope in two million years (Fig. 7b; Smith et al. 2013). The scale of aeolian cyclic steps, which is well beyond the scale of submarine cyclic steps, and their signature spiral pattern make them easily distinguishable from submarine cyclic steps.

Aeolian transverse ridges

Mars exhibits many aeolian bedforms besides polar spiral troughs. First, large dunes of black olivine- and pyroxene-rich sand are found in many dune fields across the planet. These have identical forms to dunes on Earth, namely barchanoid and parabolic shapes and longitudinal and latitudinal dunes with fingering modes. Some confined locations, i.e., within large craters, may contain star dunes hundred kilometers tall. In many locations, the dunes have been observed to migrate with the same behavior and sediment flux of terrestrial dunes

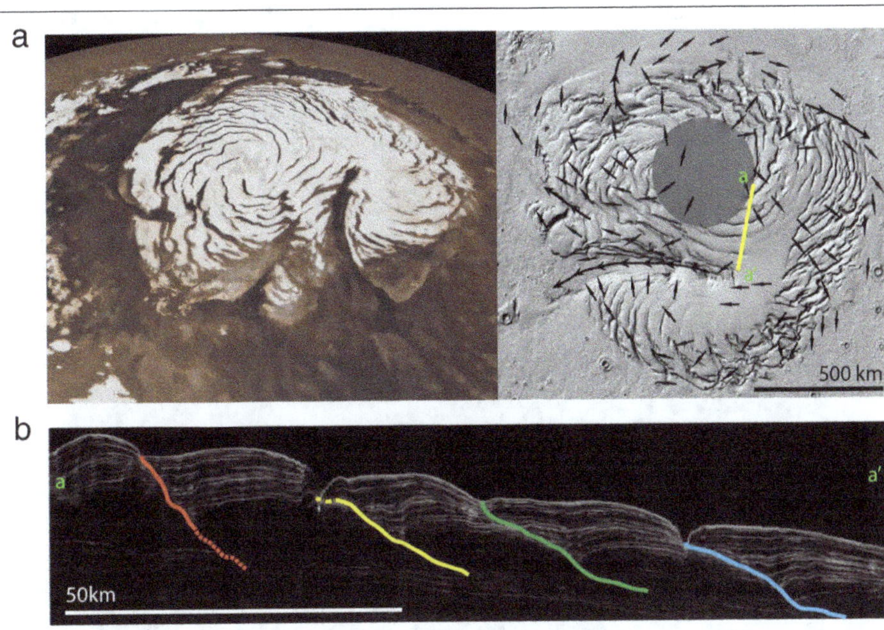

Fig. 7 Aeolian cyclic steps on the North Pole of Mars (modified from Smith et al. 2013). Shaded relief surface from MOLA elevation data demonstrates that **a** spiral troughs cover majority of the polar cap (left) and display a recognizable spiraling pattern due to the direction of katabatic winds working together with the Coriolis force (right). SHARAD profile a-a' shows **b** stratigraphy associated with spiral trough migration

(Silvestro et al. 2010; Bridges et al. 2012`, 2012b), except where hindered by seasonal CO_2 frost at the polar caps.

Besides the sand dunes, another aeolian bedform is expressed across much of the Mars surface. These are called transverse aeolian ridges, or TARs (Balme et al. 2008; Zimbelman 2010; Fig. 8). TARs are not well understood. They are mid-way in height between dunes and ripples. Unlike the dark sand that comprises the dunes, TARs frequently consist of light-toned sand due to induration and are immobile. None have been observed to move, even over time spans greater than a decade (Zimbelman 2010). They are believed to be cemented material from a previous period in Martian history.

Both TARs and dunes are always positive relief features, frequently observed in fields among similar features (Figs. 8 and 9). They are common in low places like valley floors. These features can be sometimes confused for net-depositional cyclic steps, especially in unconfined environments. This is in part why this work focuses on channelized net-erosional cyclic steps, which are quite distinctive upper-flow regime bedforms.

Methods/experimental

In seeking cyclic steps in submarine deposits, we analyzed high resolution context imager (CTX, Malin et al. 2007) data over 17 open basin deltas recognized by Di Achille and Hynek (2010). CTX has approximately 6 m/pixel resolution and full coverage over these features. Additionally, data from the High Resolution Imaging Science Experiment (HiRISE, McEwen et al. 2007) was consulted when greater detail was necessary. HiRISE has a resolution of roughly 0.5 m/pixel, but lower coverage. The JMARS geospatial information system platform loads the georeferenced images automatically, and we used this tool to make the mosaics (Christensen et al. 2009). Once the data were loaded, we sought periodic structure in the visible imagery and analyzed CTX mosaics at ever site.

To measure slopes and elevation, we first used gridded data from the Mars Orbital Laser Altimeter (MOLA, Smith et al. 2001) and then digital terrain models (DTMs) generated from stereo pairs of the High Resolution Stereo Camera (HRSC, Neukum and Jaumann 2004). From those products, we were able to extract elevations and traverse distance, and then derive slopes and bedform characteristic wavelengths.

Results

Setting

Aeolis Mensae is located close to the volcanic region of Elysium and near the boundary (i.e., hemispherical topographic dichotomy) between the heavily cratered highlands in the southern hemisphere and the flat, sparsely cratered plains that cover most of the northern hemisphere of Mars (Irwin III et al. 2004). This region of Mars is of particular geological interest as it has been shaped by a powerful combination of tectonic activity and wind activity in the transitional zone between the highlands and lowlands. Thus, the scenery of the region includes many unusual features such as tectonic grabens, mesas, and sinuous and linear ridges. Numerous tectonic grabens (or fracture zones) crossing this region are the result of the Martian crust stretching apart under tectonic activity early in Mars' history. The occasional stranded blocks called freestanding mesas rise above the surrounding terrain as remnants of the crust's expansion. They have been eroded by wind, and possibly water and ice for over millions of years, and will disappear in the distant future. Several sinuous ridges in the region exhibit strong characteristics of ancient channels that have become inverted, as illustrated in Fig. 9. This positive relief in channels is attributed to early sediment cementation (DiBiase et al. 2013), which makes channels more resistant to weathering and erosion than surrounding floodplain material. Linear ridges, also called yardangs, frequently observed on the sides of inverted channels indicate the course of the prevailing wind over a long period of time.

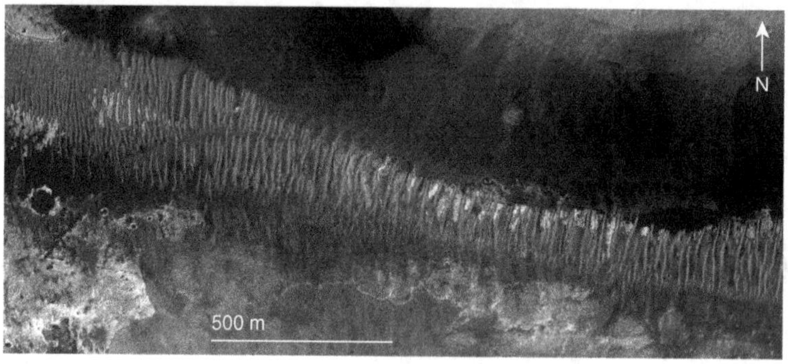

Fig. 8 Aeolian dunes in a channel on Mars

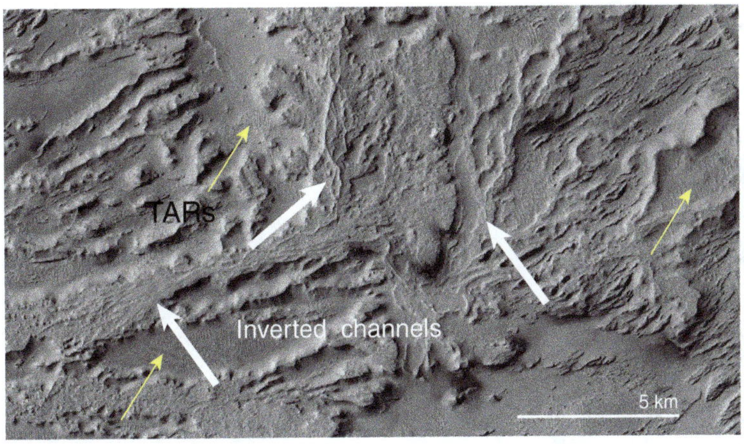

Fig. 9 Aeolis Mensae perspective, illustrating TARs and inverted channels

Aeolis Mensae Delta

We examine a deltaic deposit centered at 132.861° E, − 5.149° N near the rim of Robert Sharp Crater, west of Gale Crater. The Aeolis Mensae Delta resides downstream of a canyon that enters the crater from the southwest. The mouth of the canyon leading to the delta is at − 2357 m elevation (below mean elevation of the planet), and the topset deposit varies in elevation around − 2580 m.

The catchment of water is from the south, where an extensive valley network drained water into the canyon that fed the delta (Fig. 10a, b). Because the setting is within a crater, the rim acted as the boundary of the basin, and the shoreline would have been on steep slopes. The delta measures nearly 20 km across and 11.5 km from mouth to foreset-bottomset transition. A massive mesa, taller than the mean elevation of the delta, blocked the delta from prograding toward the east (Fig. 10b).

The upper foreset deposit of the Aeolis Mensae Delta is steep, with a mean slope of 2°. It should be noted that the local slopes are greater than the mean slope due to a poor resolution of interpolated data. For example, local foreset slopes in the vicinity of field 1 and field 2 (Fig. 10b) are of the order of 3° and 3–5°, respectively. There is a distinct break in slope between upper and lower foreset beds, with the lower delta front being quite short, i.e., spreading over a distance of about 100–110 m.

Aeolis Mensae Delta bedforms

We found evidence of three distinctive fields of bedforms on the upper foreset deposit of the Aeolis Mensae Delta, as illustrated in Fig. 10b. The first field (field 1 in Fig. 10b, c) includes a linear train of crescent-shape scours within a channel leading from the feeder canyon, as well as broader un-channelized constructive bedforms immediately outside of this channel. Further east, there are two more fields of wave-like bedforms (field 2 and field 3 in Figs. 10b and 11), possibly triggered by a small break in the crater rim that acted as a catchment for water.

Our focus here is on the channelized scour-like bedforms within field 1 located in the upper reaches of the delta slope (Fig. 10c). The channel rises 223 m over 4.3 km distance, giving an average slope of 3°. The channel and accompanying channelized bedforms start at 300 m wide and narrow to 115 m wide before disappearing. Bedforms have an average wavelength of 540 m. This geomorphology is coined the Aeolis Mensae Channel and compared to its thoroughly investigated terrestrial analogs, i.e., Monterey East and San Mateo Channel. A relatively discontinuous series of scours in the Monterey East Channel were classified as net-erosional cyclic steps associated with the channel initiation (Fildani et al. 2006), whereas more continuous bedforms in the San Mateo Channel were found to be net-erosional to net-depositional cyclic steps related to the channel maintenance (Covault et al. 2014). In contrast to the Monterey East Channel, the Aeolis Mensae Channel appears to be more continuous, better established, and in that respect more reminiscent of the San Mateo Channel (Fig. 10c). Based on this analogy, Aeolis Mensae Channel is found to be in the late initiation or early maintenance phase by way of submarine cyclic steps emplaced by turbidity currents.

This is the first time that submarine cyclic steps have been recognized on Mars. The bedforms that are confined within the Aeolis Mensae Channel suggest powerful flow events capable of eroding the floor of the primordial northern hemispheric ocean. We postulate that the bedforms of field 1 have been likely formed by hyperpycnal flows and/or slope failures of the prograding delta front. Similar events have been observed in deltas on Earth (Hughes Clarke et al. 2014; Normandeau et al. 2016; Winsemann et al. 2018).

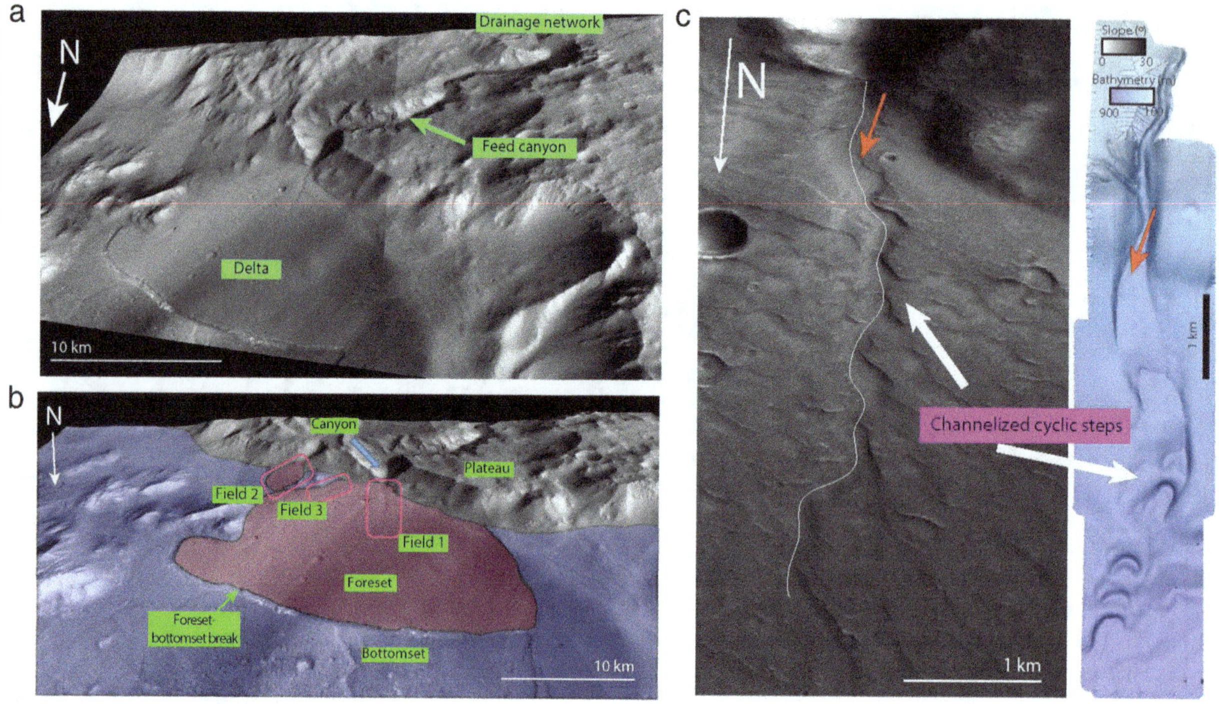

Fig. 10 Newly documented submarine cyclic steps on Mars. A compilation of HiRISE and CTX images shows **a** extensive valley network draining water into the feeder canyon and Aeolis Mensae Delta, **b** fields of cyclic steps on the upper foreset deposit of this delta, and **c** a close-up of field 1 of channelized cyclic steps along the Aeolis Mensae Channel (left) in comparison to San Mateo channelized cyclic steps (right). The white line (left) loosely follows the right side of the Aeolis Mensae Channel

Two additional fields of bedforms on the foreset of the Aeolis Mensae Delta (field 2 and field 3 in Fig. 11) deserve to be noted as well. The slope of field 2 is about 5°, and of field 3 close to 2°. The average wavelength of bedforms is 156 m, and 195 m, respectively. It is highly unlikely for the morphologic signature of net-depositional cyclic steps, such as features in field 2 and field 3, to be widely preserved on Mars due to geologic activity that has erased or modify much of the geomorphic evidence. We speculate that these fields of bedforms could be a first piece of evidence for contemporary cyclic steps generated by the sporadic presence of flowing water on Mars.

Discussion

Cyclic steps on deltas on Earth appear to be far less documented and understood than cyclic steps in overbank regions of deep-sea fans (e.g., Fildani et al. 2006; Lamb et al. 2008). Wave-like undulations on steep forest slopes (Prior and Bornhold, 1989; Gilbert and Crookshanks, 2009; Urgeles et al. 2011) are by far the most commonly reported delta bedforms. The term that is used to

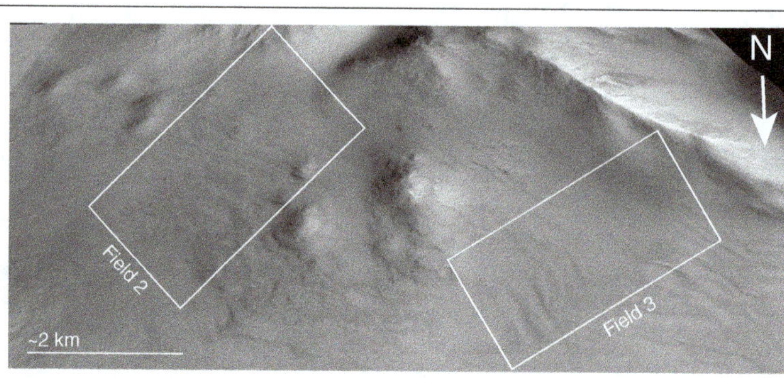

Fig. 11 A close-up of potential contemporary cyclic steps (field 2 and field 3) on the Aeolis Mensae Delta

describe them is sediment waves or sediment undulations. The scientific community is still debating on their origin, and it will take time before many of these features get identified as net-depositional cyclic steps (Kostic 2011, 2014). Urgeles et al. (2011) noted the similarity between cyclic steps tied to channel-levee systems (Fildani et al. 2006) and sediment undulations on some Mediterranean prodeltas and introduced the idea that these undulations could be cyclic steps generated by hyperpycnal flows. A common misconception is that cyclic steps on deltas form only on the steep foreset deposits. However, cyclic steps on deltas get triggered, under the right conditions, by the foreset-bottomset transition, and can emerge upstream, downstream, or at this slope break (Kostic 2011). In addition, the upper foreset slope is commonly steeper than the downstream foreset slope due to failure and/or reworking of delta front (Kostic et al. 2002; Girardclos et al. 2012; Hughes Clarke et al. 2014). This additional slope break can work together with the foreset-bottomset transition to facilitate the flow conditions conducive to cyclic steps (e.g., Kostic 2011). Cyclic sedimentation of delta foreset can also be triggered by the periodic alternation between supercritical and subcritical flow on the delta topset just upstream of the shoreline, as illustrated in the flume experiments of Muto et al. (2012). Undulations on the delta front can be sustained as long as the inflow conditions of water and sediment support the formation of cyclic steps on the alluvial topset deposit. Other examples of cyclicity in delta foreset bedding observed in experiments should be noted as well (Kleinhans, 2005; Kim et al. 2006).

The Squamish Prodelta in British Columbia is possibly the best known natural delta laboratory, and research conducted there has significantly improved our understanding of channelized upper-flow regime bedforms on deltas emplaced by turbidity currents. The Squamish Prodelta team (Hughes Clarke et al. 2012, 2014; Hughes Clarke 2016) conducted repeat monitoring surveys (1–3-day period during 4 months) of channels on the Squamish Prodelta. They established processes of occasional, aperiodic failure of the delta front at the shelf edge and intense channel reworking by what they taught to be cyclic steps in all channels. Covault et al. (2017) applied the morphodynamic model of Kostic (2011) to small sediment waves (30–70 m wavelength; 2–3 m wave height) that dominate the floor of the Northern Channel of the Squamish Prodelta and demonstrated that these channelized features were likely transitional upper-flow regime bedforms consisting of cyclic steps and antidunes rather than cyclic steps. Recent work of Casalbore et al. (2017) reported cyclic steps in gullies on the deltaic deposit of the Mazzarrà River and outer shelf sediments along the NE Sicilian margin (southern Tyrrhenian Sea). Delta waveforms had wavelengths of 34–110 m and wave heights of 0.5–3 m. They were attributed to hyperpycnal flows, but the effect of internal waves or slow deformation processes (i.e., creeps) was also noted. Casalbore et al. (2017) postulated the connection between these bedforms and the onset and growth of the Mazzarrà Delta since the Last Glacial Maximum. Some of the most illustrative examples of terrestrial cyclic steps on the foreset deposit of deltas were recently documented in lakes rather than in marine settings (i.e., Fricke et al. 2015; Normandeau et al. 2016; Winsemann et al. 2018).

Our work showcases marine cyclic steps on deltas as an additional terrestrial analog in support of the hypothetical existence of the relatively shallow northern hemispheric ocean on Mars. This is not the first time that streamlined forms shaped by turbidity currents and submarine debris flows on Earth were invoked as analogs for some Martian features (Komar 1979; Nummedal and Prior 1981). Burr (2011) argued that, due to the lower gravity on Mars, the physical sedimentology of some or all streamlined forms on Mars could to be similar to that of submarine (Komar 1979) rather than terrestrial environments on Earth. Work of Moscardelli and Wood (2011) and Burr (2014) suggests that submarine-style processes on Earth could likely provide a new analog for Martian streamlined forms in general, regardless of their proposed subaerial or submarine context.

Ojha et al. (2015) provided the strongest evidence yet that liquid water flows intermittently on present-day Mars. They used the Compact Reconnaissance Imaging Spectrometer for Mars instrument onboard the Mars Reconnaissance Orbiter (MRO) to analyze four different locations where downhill flows, known as recurring slope lineae (RSL), were present. They uncovered evidence for hydrated salts at all four locations in the seasons when recurring slope lineae are most extensive, which suggests modern-day water activity on Mars.

We speculate that the sporadic presence of flowing water in reduced-gravity environments of present-day Mars could promote the formation of contemporary subaerial cyclic steps in the vicinity of slope breaks. One such example could be two fields of cyclic steps possibly emanating from a catchment/crater upstream of the Aeolis Mensae Delta. Contemporary cyclic steps on Mars may have their terrestrial analogs in the field (e.g., Winterwerp et al. 1992) and in the laboratory (e.g., Taki & Parker, 2005). Also, Sun and Parker (2005) explained theoretically cyclic steps in alluvium under conditions of bed aggradation or degradation, as well as in the absence of net bed level change. These newly identified subaerial cyclic steps on Mars are beyond the scope of work presented herein and deserve further investigation.

Conclusions

Cyclic steps may have a crucial role in the onset, growth, and evolution of deltas on Earth and Mars via processes of channel avulsion, initiation, filling, and maintenance. Our research is the first modest step toward better understanding of marine cyclic steps on Mars. We were able to identify a relatively well-preserved train of cyclic steps along the Aeolis Mensae Channel thalweg on Mars, which is strikingly similar to the linear series of cyclic steps observed in channels on deltas and deep-water fans on Earth. These channelized erosional bedforms have been most likely formed by turbidity currents triggered by hyperpycnal flows and delta-front failure events. Based on the Monterey East and San Mateo terrestrial analogs, the Aeolis Mensae Channel on Mars is estimated to be in the late initiation or early maintenance phase. The analysis also unveiled potential examples of contemporary Martian cyclic steps, which are attributed to the sporadic presence of flowing water on Aeolis Mensae.

Submarine cyclic steps on Mars are a newly recognized phenomenon. The analysis presented here suggests that they can be triggered by flows that are 60% slower than corresponding formative flows on Earth. These important upper-flow regime bedforms warrant future investigation in order to better understand the effect of reduced gravity on both ancient and contemporary streamline forms and sedimentary landforms on Mars.

Abbreviations

CTX: Context imager; DTMs: Digital terrain models; HDO: Deuterated water form; HiRISE: High Resolution Imaging Science Experiment; HRSC: High Resolution Stereo Camera; JMARS: Java Mission-planning and Analysis for Remote Sensing; MOLA: Mars Orbiter Laser Altimeter; MRO: Mars Reconnaissance Orbiter; RSL: Recurring slope lineae; TARs: Transverse aeolian ridges

Acknowledgements

SK gratefully acknowledges the conveners' travel support from the Japan Geoscience Union to attend the JpGU-AGU Joint Meeting 2017 held in Chiba, Japan. The authors thank Jacob Covault for editing the manuscript and providing helpful inputs. The paper benefited from constructive reviews by Miwa Yokokawa and Andrea Fildani.

Funding

This research is an extension of authors' collaboration dating back to 2015 and is currently not supported by any grants.

Authors' contributions

SK proposed the topic, conceived the study, performed the numerical analysis, and wrote the paper. IS compiled the database, prepared the figures, and contributed to the manuscript. Both authors read and approved the final manuscript.

Authors' information

SK developed the first numerical model of submarine upper-flow regime bedforms and identified submarine cyclic steps, which paved the way for a growing recognition of the global presence and importance of these features in terrestrial deep-water environments. IS contributed to the discovery of aeolian cyclic steps on Mars Polar Layer Deposits.

Competing interests

The authors declare that they have no competing interests.

Author details

[1]Computational Science Research Center, San Diego State University, San Diego, CA, USA. [2]Planetary Science Institute, Lakewood, CO, USA.

References

Allen CC, Oehler DZ, Etiope G, Van Rensbergen P, Baciu C, Feyzullayev A, Martinelli G, Tanaka K, Van Rooij D (2013) Fluid expulsion in terrestrial sedimentary basins: a process providing potential analogs for giant polygons and mounds in the martian lowlands. Icarus 224:424–432. https://doi.org/10.1016/j.icarus.2012.09.018

Baker VR, Strom RG, Gulick VC, Kargel JS, Komatsu G, Kale VS (1991) Ancient oceans, ice sheets and the hydrological cycle on Mars. Nature 352(6336):589–594. https://doi.org/10.1038/352589a0

Ball FK (1956) The theory of strong katabatic winds. Aust J Phys 9:373–386

Balme M, Berman DC, Bourke MC, Zimbelman JR (2008) Transverse Aeolian ridges (TARs) on Mars. Geomorphology 101:703–720. https://doi.org/10.1016/j.geomorph.2008.03.011

Bouley S, Baratoux D, Matsuyama I, Forget F, Séjourné A, Turbet M, Costard F (2016) Late Tharsis formation and implications for early Mars. Nature 531:344–347

Bridges NT, Ayoub F, Avouac J-P, Leprince S, Lucas A, Mattson S (2012a) Earth-like sand fluxes on Mars. Nature 485:339–342

Bridges NT, Bourke MC, Geissler PE, Banks ME, Colon C, Diniega S, Golombek MP, Hansen CJ, Mattson S, McEwen AS, Mellon MT, Stantzos N, Thomson BJ (2012b) Planet-wide sand motion on Mars. Geology 40:31–34. https://doi.org/10.1130/G32373.1

Burr DM (2011) Sedimentology in a reduced-gravity environment: submarine analogs for streamlined forms on Mars. Geology 39:703–704. https://doi.org/10.1130/focus072011.1

Carr MH, Head JW (2003) Oceans on Mars: an assessment of the observational evidence and possible fate. J Geophys Res 108. https://doi.org/10.1029/2002JE001963

Casalbore D, Ridente D, Bosman A, Chiocci FL (2017) Depositional and erosional bedforms in Late Pleistocene-Holocene pro-delta deposits of the Gulf of Patti (southern Tyrrhenian margin, Italy). Mar Geol 385:216–227

Christensen P R, Engle E, Anwar S, Dickenshied S, Noss D, Gorelick N, Weiss-Malik M (2009) JMARS – a planetary GIS, American Geophysical Union, Fall Meeting 2009, abstract #IN22A-06

Citron R, Manga M, Hemingway D (2018) Timing of oceans on Mars from shoreline deformation. Nature 555:643–646. https://doi.org/10.1038/nature26144

Cooke M, Islam F, McGill G (2011) Basement controls on the scale of giant polygons in Utopia Planitia, Mars. J Geophys Res Planets 116. https://doi.org/10.1029/2011JE003812

Costard F, Séjourné A, Kelfoun K, Clifford S, Lavigne F, Di Pietro I, Bouley S (2017) Modeling tsunami propagation and the emplacement of thumbprint terrain in an early Mars ocean. J Geophys Res Planets. https://doi.org/10.1002/2016JE005230

Covault JA, Kostic S, Paull CK, Ryan HF, Fildani A (2014) Submarine channel initiation, filling and maintenance from sea-floor geomorphology and morphodynamic modelling of cyclic steps. Sedimentology 61:1031–1054

Covault JA, Kostic S, Paull CK, Sylvester Z, Fildani A (2017) Cyclic Steps and Related Supercritical bedforms: Building Blocks of Submarine Fans and Canyon-Channel Systems, Western North America. Mar Geol 393: 4–20.

Di Achille G, Hynek BM (2010) Ancient ocean on Mars supported by global distribution of deltas and valleys. Nat Geosci 3:459–463. https://doi.org/10.1038/ngeo891

DiBiase RA, Limaye AB, Scheingross JS, Fischer WW, Lamb MP (2013) Deltaic deposits at Aeolis Dorsa: sedimentary evidence for a standing body of water on the northern plains of Mars. J Geophys Res Planets 118:1285–1302. https://doi.org/10.1002/jgre.20100

Fairen AG, Dohm JM, Baker VR, de Pablo MA, Ruiz J, Ferris JC, Anderson RC (2003) Episodic flood inundations of the northern plains of Mars. Icarus 165(1):53–67. https://doi.org/10.1016/s0019-1035(03)00144-1.

Fildani A, Hubbard SM, Covault JA, Maier KL, Romans BW, Traer M, Rowland JC (2013) Erosion at inception of deep-sea channels. Mar Pet Geol 41:48–61

Fildani A, Normark WR, Kostic S, Parker G (2006) Channel formation by flow stripping: large-scale scour features along the Monterey East Channel and their relation to sediment waves. Sedimentology 53:1265–1287

Fricke AT, Sheets BA, Nittrouer CA, Allison MA, Ogston AS (2015) An examination of Froude-supercritical flows and cyclic steps on a subaqueous lacustrine delta, Lake Chelan, Washington, USA. J Sediment Res 85:754–767

Ghatan GJ, Zimbelman JR (2006) Paucity of candidate coastal constructional landforms along proposed shorelines on Mars: implications for a northern lowlands-filling ocean. Icarus 185:171–196. https://doi.org/10.1016/j.icarus.2006.06.007

Gilbert R, Crookshanks S (2009) Sediment waves in a modern high-energy glacilacustrine environment. Sedimentol 56: 645–659

Girardclos S, Hilbe M, Corella JP, Loizeau JL, Kremer K, delSontro T, Arantegui A, Moscariell A, Alraud F, Akhtman Y, Anselmetti FS, Lemmin U (2012) Searching for the Rhone delta channel in Lake Geneva since François-Alphonse Forel. Arch Sci 65:1–16

Hamilton PB, Strom KB, Hoyal DC (2015) Hydraulic and sediment transport properties of autogenic avulsion cycles on submarine fans with supercritical distributaries. Journal of Geophysical Research: Earth Surf 120(7):1369–1389

Head JW, Hiesinger H, Ivanov MA, Kreslavsky MA, Pratt S, Thomson BJ (1999) Possible ancient oceans on Mars: evidence from Mars orbiter laser altimeter data. Science 286:2134–2137. https://doi.org/10.1126/science.286.5447.2134

Hughes Clarke JE (2016) First wide-angle view of channelized turbidity currents links migrating cyclic steps to flow characteristics. Nat Commun 7:11896. https://doi.org/10.1038/ncomms11896

Hughes Clarke JE, Brucker S, Muggah J, Church I, Cartwright D, Kuus P, Hamilton T, Pratomo D, Eisan B (2012) The Squamish ProDelta: monitoring active landslides and turbidity currents. In: Proceedings of Canadian Hydrographic Conference 2012, p 15

Hughes Clarke JE, Vidiera Marques CR, Pratomo D (2014) Imaging active mass-wasting and sediment flows on a Fjord Delta, Squamish, British Columbia. In: Krastel S, Behrmann JH, Völker D, Stipp M, Berndt C, Urgeles R, Chaytor J, Huhn K, Strasser M, Harbitz CB (eds) Submarine mass movements and their consequences, advances in natural and technological hazards research. Springer, Heidelberg, pp 249–260

Irwin III R P, Watters T R, Howard A D, Zimbelman J R (2004) Sedimentary resurfacing and fretted terrain development along the crustal dichotomy boundary, Aeolis Mensae, Mars. J Geophys Res Planets 109, dio: https://doi.org/10.1029/2004JE002248

Irwin RP III, Zimbelman JR (2012) Morphometry of Great Basin pluvial shore landforms: implications for paleolake basins on Mars. J Geophys Res 117. https://doi.org/10.1029/2012je004046

Ivanov MA, Head JW (2001) Chryse Planitia Mars: topographic configuration, outflow channel continuity and sequence, and tests for hypothesized ancient bodies of water using Mars Orbiter Laser Altimeter (MOLA) data. J Geophys Res 106:3275–3295. https://doi.org/10.1029/2000je001257

Kim W, Paola C, Swenson JB, Voller VR (2006) Shoreline response to autogenic processes of sediment storage and release in the fluvial system. J Geophys Res 111 (F4), https://doi.org/10.1029/2006JF000470

Kleinhans MG (2005) Autogenic cyclicity of foreset sorting in experimental Gilbert-type deltas. Sediment Geol 181:215–224

Komar PD (1979) Comparisons of the hydraulics of water flows on Martian outflow channels with flows of similar scale on Earth. Icarus 37:156–181. https://doi.org/10.1016/0019-1035(79)90123-4

Kostic S (2011) Modeling of submarine cyclic steps: controls on their formation, migration, and architecture. Geosphere 7:294–304. https://doi.org/10.1130/GES00601.1

Kostic S (2014) Upper flow regime bedforms on levees and continental slopes: turbidity current flow dynamics in response to fine-grained sediment waves. Geosphere 10:1094–1103. https://doi.org/10.1130/GES01015.1

Kostic S, Parker G (2006) The response of turbidity currents to a canyon–fan transition: internal hydraulic jumps and depositional signatures. J Hydraul Res 44:631–653

Kostic S, Parker G, Marr J (2002) Role of turbidity currents in setting the foreset slope of clinoforms prograding into standing fresh water. J Sediment Res 72: 353–362. https://doi.org/10.1306/081501720353

Lamb MP, Parsons JD, Mullenbach BL, Finlayson DP, Orange DL, Nittrouer CA (2008) Evidence for superelevation, channel incision, and formation of cyclic steps by turbidity currents in Eel Canyon, California. GSA Bull 120:463–475

Maier KL, Fildani A, McHargue T, Paull CK, Graham SA, Caress DW (2012) Deep-water punctuated channel migration: high-resolution subsurface data from the Lucia Channel System, offshore California. J Sediment Res 82:1–8

Malin MC, Edgett KS (1999) Oceans or seas in the Martian northern lowlands: high resolution imaging tests of proposed coastlines. Geophys Res Lett 26:3049–3052

Malin MC et al (2007) Context camera investigation on board the Mars reconnaissance orbiter. J Geophys Res Planets 112(E5). https://doi.org/10.1029/2006JE002808

McEwen AS, Eliason EM, Bergstrom JW, Bridges NT, Hansen CJ, Delamere WA, Grant JA, Gulick VC, Herkenhoff KE, Keszthelyi L, Kirk RL, Mellon MT, Squyres SW, Thomas N, Weitz CM (2007) Mars reconnaissance orbiter's high resolution imaging science experiment (HiRISE). J Geophys Res Planets 112 (E5), https://doi.org/10.1029/2005JE002605

Moscardelli L (2014) Boulders of the Vastitas Borealis Formation: potential origin and implications for an ancient martian ocean. GSA Today 24. https://doi.org/10.1130/GSATG197A.1

Moscardelli L, Dooley T, Dunlap D, Jackson M, Wood L (2012) Deep-water polygonal fault systems as terrestrial analogs for large-scale Martian polygonal terrains. GSA Today 22:4–9. https://doi.org/10.1130/GSATG147A.1

Moscardelli L, Wood L (2011) Deep-water erosional remnants in eastern offshore Trinidad as terrestrial analogs for teardrop-shaped islands on Mars: implications for outflow channel formation. Geology 39:699–702. https://doi.org/10.1130/G31949.1

Muto T, Yamagishi C, Sekiguchi T, Yokokawa M, Parker G (2012) The hydraulic autogenesis of distinct cyclicity in delta foreset bedding: flume experiments. J Sediment Res 82(7):545–558

Neukum G, Jaumann R (2004) HRSC: the high resolution stereo camera of Mars express. In: Mars Express: the Scientific Payload, vol 1240, pp 17–35

Normandeau A, Lajeunesse P, Poiré AG, Francus P (2016) Morphological expression of bedforms formed by supercritical sediment density flows in four fjord-lake deltas of the southeastern Canadian Shield (Eastern Canada). Sedimentology 63: 2106–2129. https://doi.org/10.1111/sed.12298

Nummedal D, Prior DB (1981) Generation of Martian chaos and channels by debris flows. Icarus 45:77–86. https://doi.org/10.1016/0019-1035(81)90007-5

Oehler DZ, Allen CC (2012) Focusing the search for biosignatures on Mars: facies prediction with an example from Acidalia Planitia. In: Grotzinger J, Milliken R (eds) Sedimentary Geology of Mars: SEPM Special Publication, vol 102, pp 183–194

Ojha L, Wilhelm MB, Murchie SL, McEwen AS, Wray JJ, Hanley J, Massé M, Chojnacki M (2015) Spectral evidence for hydrated salts in recurring slope lineae on Mars AOP. Nat Geosci. https://doi.org/10.1038/ngeo2546

Parker TJ, Gorsline DS, Saunders RS, Pieri DC, Schneeberger DM (1993) Coastal geomorphology of the Martian northern plains. J Geophys Res 98:11061–11078. https://doi.org/10.1029/93JE00618

Parker TJ, Saunders RS, Schneeberger DM (1989) Transitional morphology in west Deuteronilus Mesae, Mars-implications for modification of the lowland upland boundary. Icarus 82:111–145. https://doi.org/10.1016/0019-1035(89)90027-4

Perron JT, Matsuyama I, Richards MA, Mitrovica JX, Manga M (2007) Evidence for an ancient martian ocean in the topography of deformed shorelines. Nature 447:840–843. https://doi.org/10.1038/nature05873

Pettré P, André JC (1991) Surface-pressure change through Loewe's phenomena and katabatic flow jumps - study of two cases in Adelie Land, Antarctica. J Atmos Sci 48:557–571

Postma G, Hoyal D, Abreu V, Cartigny M, Demko T, Fedele J, Kleverlaan K, Pederson KH (2016) Morphodynamics of supercritical turbidity currents in the channel-lobe transition zone. In: Lamarche G, Mountjoy J (eds) Submarine mass movements and their consequences. Springer, Dordrecht: 469–478.

Prior DB, Bornhold BD (1989) Submarine sedimentation on a developing Holocene fan delta. Sedimentol 36(6):1053–1076.

Rodriguez JAP, Fairén AG, Tanaka KL, Zarroca M, Linares R, Platz T, Komatsu G, Miyamoto H, Kargel JS, Yan J, Gulick V, Higuchi K, Baker VR, Glines N (2016) Tsunami waves extensively resurfaced the shorelines of an early Martian ocean. Sci Rep 6. https://doi.org/10.1038/srep25106

Silvestro S, Fenton LK, Vaz DA, Bridges NT, Ori GG (2010) Ripple migration and dune activity on Mars: evidence for dynamic wind processes. Geophys Res Lett 37. https://doi.org/10.1029/2010GL044743

Smith DE, Zuber MT, Frey HV, Garvin JB, Head JW, Muhleman DO, Pettengill GH, Phillips RJ, Solomon SC, Zwally HJ (2001) Mars Orbiter Laser Altimeter: experiment summary after the first year of global mapping of Mars. J Geophys Res Planets 106:23–689

Smith DP, Kvitek R, Iampietro PJ, Wong K (2007) Twenty-nine months of geomorphic change in upper Monterey Canyon (2002–2005). Mar Geol 236:79–94

Smith IB, Holt JW (2010) Onset and migration of spiral troughs on Mars revealed by orbital radar. Nature 465:450–453. https://doi.org/10.1038/nature09049

Smith IB, Holt JW, Spiga A, Howard AD, Parker G (2013) The spiral troughs of Mars as cyclic steps. J Geophys Res Planets. https://doi.org/10.1002/jgre.20142

Sun T, Parker G (2005) Transportational cyclic steps created by flow over an erodible bed. Part 2. Theory and numerical simulation. J Hydraul Res 43:502–514

Taki K, Parker G (2005) Transportational cyclic steps created by flow over an erodible bed. Part 1. Experiments. J Hydraul Res 43:488–501

Tubau X, Paull CK, Lastras G, Caress DW, Canals M, Lundsten E, Anderson K, Gwiazda R, Amblàs D (2015) Submarine canyons of Santa Monica Bay, Southern California: variability in morphology and sedimentary processes. Mar Geol 365:61–79

Urgeles R, Cattaneo A, Puig P, Liquete C, De Mol B, Amblàs D, Sultan N, Trincardi F (2011) A review of undulated sediment features on Mediterranean prodeltas: distinguishing sediment transport structures from sediment deformation. Mar Geophys Res 32:49–69

Villanueva GL, Mumma MJ, Novak RE, Käufl HU, Hartogh P, Encrenaz T, Tokunaga A, Khayat A, Smith MD (2015) Strong water isotopic anomalies in the martian atmosphere: probing current and ancient reservoirs. Science 348:218–221. https://doi.org/10.1126/science.aaa3630

Winsemann J, Lang J, Polom U, Loewer M, Igel J, Pollolk L, Brandes C (2018) Ice-marginal forced-regressive deltas in glacial lake basins: geomorphology, facies variability and large-scale depositional architecture. Boreas. https://doi.org/10.1111/bor.12317

Winterwerp JC, Bakker WT, Mastbergen DR, van Rossum H (1992) Hyperconcentrated sand-water mixture flows over erodible bed. J Hydraul Eng 118(11):1508–1525

Zimbelman JR (2010) Transverse aeolian ridges on Mars: first results from HiRISE images. Geomorphology 121:22–29. https://doi.org/10.1016/j.geomorph.2009.05.012

10

Iron snow, crystal floats, and inner-core growth: modes of core solidification and implications for dynamos in terrestrial planets and moons

Doris Breuer*, Tina Rueckriemen and Tilman Spohn

Abstract

Recent planetary space missions, new experimental data, and advanced numerical techniques have helped to improve our understanding of the deep interiors of the terrestrial planets and moons. In the present review, we summarize recent insights into the state and composition of their iron (Fe)-rich cores, as well as recent findings about the magnetic field evolution of Mercury, the Moon, Mars, and Ganymede. Crystallizing processes in iron-rich cores that differ from the classical Earth case (i.e., Fe snow and iron sulfide (FeS) crystallization) have been identified and found to be important in the cores of terrestrial bodies. The Fe snow regime occurs at pressures lower than that in the Earth's core on the iron-rich side of the eutectic, where iron freezes first close to the core–mantle boundary rather than in the center. FeS crystallization, instead, occurs on the sulfur-rich side of the eutectic. Depending on the core temperature profile and the pressure range considered, FeS crystallizes either in the core center or close to the core–mantle boundary. The consequences of the various crystallizing mechanisms for core dynamics and magnetic field generation are discussed. For the Moon, revised paleomagnetic data obtained with advanced techniques suggest a peculiar history of its internal dynamo, with an early strong field persisting between 4.25 and 3.5 Ga, and subsequently a much weaker field. In addition, the long-lasting dynamo and the possible presence of an inner core, as inferred from a revised interpretation of Apollo seismic data, suggest core crystallization as a viable process of magnetic field generation for a substantial period during lunar evolution. The present-day magnetic fields of Mercury and Ganymede (if they occur on the iron-rich side of the Fe–FeS eutectic) and the related dynamo action are likely generated in the Fe snow regime and seem to be recent features. An earlier dynamo in Mercury would have been powered differently. For Mercury, MESSENGER data further suggest core formation under reducing conditions that may have resulted in an Fe–S–Si composition, further complicating the core crystallization process. Mars, with its early and strong paleo-field, likely has not yet started to freeze out an inner iron core.

Keywords: Core crystallization, Dynamo generation, Iron snow, Terrestrial planets, Thermal evolution

Review

Introduction

Our view of the deep interiors of the terrestrial planets and moons, and their magnetic fields, has changed substantially over the past decade as a consequence of insights from recent space missions, new experimental data that have become available, and the results of more detailed and sophisticated numerical models. Because the evolution of a planet's magnetic field is closely linked to its thermal evolution, constraints on the dynamo in iron (Fe)-rich cores will help provide a better understanding of a planet's evolution in general. For instance, the MESSENGER mission has collected new magnetic field data for Mercury (e.g., Anderson et al. 2011; Johnson et al. 2015), and lunar samples have been reanalyzed with improved paleomagnetic methods (for a review, see Weiss and Tikoo 2014 among others). Space missions have further helped to better constrain the surface compositions of planets and to provide better constraints on geochemical models of their interiors. MESSENGER spectroscopic

* Correspondence: doris.breuer@dlr.de
DLR, Institute of Planetary Research, Rutherfordstrasse 2, 12489 Berlin, Germany

data show that the planet is more volatile rich than previously thought and additionally suggest that it was formed and differentiated under reducing conditions (Nittler et al. 2011; Weider et al. 2012). Therefore, the core is now thought to be composed of Fe–Si–S (e.g., Malavergne et al. 2010; Chabot et al. 2014) rather than Fe–S, which is generally taken as a good model for Earth-like planetary cores. Experimental data relevant to Fe–FeS eutectic melting relations (Fei et al. 1997, 2000; Chudinovskikh and Boehler 2007; Stewart et al. 2007; Chen et al. 2008; Buono and Walker 2011) suggest that crystallization in the core can proceed substantially differently in small planets compared with the Earth's core. In small planets, iron may start crystallizing at the core–mantle boundary (CMB) rather than in the center, and iron snow may form (e.g., Hauck et al. 2006). As a consequence, a dynamo that is tied to crystallization may work differently in these planets than it would in Earth. Moreover, for a core richer in sulfur than the eutectic composition, iron sulfide (FeS) will crystallize first and also result in a different crystallization process as compared with crystallization on the iron-rich side of the eutectic.

In the present paper, we review our present understanding of the composition, as well as the thermal and magnetic evolution, of the iron-rich cores of terrestrial planets and moons for which evidence of present or past magnetic fields is available. These are (in order of distance from the Sun) Mercury, the Moon, Mars, and Ganymede. Other planets, such as Venus, may have had magnetic fields. It has been speculated that Venus had a magnetic field driven by a thermal dynamo that vanished as the planet cooled (Stevenson et al. 1983; Gaidos et al. 2010), or it may have transited from plate tectonics to stagnant lid, losing its dynamo (Stevenson 2002). Another potential candidate for the presence of a magnetic field is the Jovian moon Io, although strong tidal heating may have prevented the formation of a dynamo (Wienbruch and Spohn 1995). This may also have been true for early Europa. It is likely that other large moons such as Callisto and Titan did not form an iron-rich core at all (e.g., Anderson et al. 2001; Iess et al. 2010). We begin by presenting the general concepts of thermal and compositional dynamos, along with a discussion of differences in crystallization at low and high pressures, and then proceed to discuss new findings and their implications for Mercury, the Moon, Mars, and Ganymede.

Thermal and chemical dynamos

Present-day magnetic fields of terrestrial planets and moons are either generated in their iron-rich cores (e.g., for the Earth, Mercury, and Ganymede), induced in electrically conducting layers (e.g., in the subsurface oceans of icy satellites), or are due to remanently magnetized crustal rock (e.g., for the Moon and Mars). The latter suggests a self-generated magnetic field during the early evolution of the planet, at the time of crustal formation. In any case, at the time of dynamo action, the presence of a convective (partially) fluid core is required, either an entirely fluid core or partially fluid with a solid inner core. We distinguish between thermal and chemical (compositional) dynamos.

For a fully fluid core, thermal buoyancy may be available to drive the dynamo if a sufficiently large superadiabatic temperature difference exists between core and mantle. The heat flux from the core needs to exceed the heat flux conducted along the core adiabat. The latter heat flux provides an important and necessary criterion for thermal convection in the core. The heat flux (q_{crit}) along the adiabat is given by

$$q_{\text{crit}} = k_c \frac{\mathrm{d}T}{\mathrm{d}r}\Big|_{\text{ad}} = k_c \frac{\mathrm{d}T}{\mathrm{d}P}\Big|_{\text{ad}} \frac{\mathrm{d}P}{\mathrm{d}r} = k_c \frac{\alpha_c g_c T}{C_p}, \tag{1}$$

where k_c is the thermal conductivity in the core, $\mathrm{d}T/\mathrm{d}r|_{\text{ad}}$ the adiabatic temperature gradient, P the pressure, T the temperature, r the radius, α_c the thermal expansivity, C_p the heat capacity at constant pressure p, g_c the gravity, and $\mathrm{d}T/\mathrm{d}P|_{\text{ad}} = \alpha_c T/\rho C_p$, with ρ representing the density. Values for the critical heat flux that have been used in the literature vary between 5 and 20 mW m^{-2} for Mars (Nimmo and Stevenson 2000) and Mercury (Stevenson et al. 1983; Schubert et al. 1988) and between 1 and 10 mW m^{-2} for the Moon (Zhang et al. 2013; Laneuville et al. 2014; Evans et al. 2014) and Ganymede (Hauck et al. 2006; Bland et al. 2008; Kimura et al. 2009; Rückriemen et al. 2015).

The large range of values for the critical heat flux originates mostly from uncertainties in the thermal conductivity and the thermal expansivity of the iron-rich alloy at core pressures (Stacey and Anderson 2001). Recent ab initio calculations of transport properties suggest that the thermal conductivity in the Earth's core is a factor of two higher than previously thought, and thus upward revisions to the thermal conductivity in other terrestrial planets are also being discussed. Revised values of the thermal conductivity in the Earth's core range from 90 to 150 W m^{-1} K^{-1} (de Koker et al. 2012; Pozzo et al. 2012). Similar values also result from measurements of the electrical conductivity and from using the Wiedemann–Franz law to calculate the thermal conductivity (Gomi et al. 2013). Recently, Deng et al. (2013) suggested a thermal conductivity for Mercury's core of 113–125 W m^{-1} K^{-1} and a critical heat flux of 45–100 mW m^{-2} from measurements of the electrical conductivity of iron. Note, however, that ab initio calculations have recently been criticized by Zhang et al. (2015), who argued that such studies overestimate the thermal conductivity, because the effects of electron–electron scattering are not included in the numerical calculations.

The thermal conductivity likely decreases with increasing light-element abundances (e.g., Stacey and Anderson 2001; de Koker et al. 2012), although Ohta et al. (2015), using recent measurements of the electrical resistivity of iron-rich alloys, reported that the effect is not sufficiently large to reduce the core thermal conductivity to more conventional values of about 50 W m^{-1} K^{-1} (Stacey and Anderson 2001). For the discussion in this paper, we emphasize the implications of larger values for the thermal conductivity, although we note, of course, that the matter is not yet entirely settled.

The thermal expansivity of liquid iron at ambient pressure ranges from 8.2×10^{-5} K^{-1} to 13.2×10^{-5} K^{-1} (Hixson et al. 1990; Nasch and Steinemann 1995; Assael et al. 2006). The expansivity also depends slightly on the volatile content (Kaiura and Toguri 1979). With increasing pressure the thermal expansivity decreases, and a value of about 10^{-5} K^{-1} to 3×10^{-5} K^{-1} is proposed for the Earth's core (Boehler et al. 1990; Kandpal and Gupta 2007).

It is important to note that the heat flux from the core is controlled mainly by the thermal evolution of the overlying mantle. The mantle needs to remove heat from the core at a rate that exceeds the critical heat flow for the core to become convective. If the mantle removes heat at a rate below the critical heat flow, the core will be stratified thermally stably, cooling conductively, and dynamo action driven by thermal convection is not possible.

Compositional convection can occur when a buoyant chemical component is released during the freezing of a fluid core with a non-eutectic composition (Braginsky 1964). The existence of light alloying elements in the core is cosmochemically reasonable and has been postulated for all terrestrial planets and moons, but the nature and the concentrations of the suggested light elements vary, as we discuss below. In any case, these elements will, other than reducing the average core density, significantly reduce the melting temperature of the core alloy (e.g., Usselman 1975; Fei et al. 1997). Although the effects of a variety of elements are conceivable, the physics of crystallization in the core has been studied mostly for Fe–FeS systems, both experimentally and theoretically.

Core crystallization processes

Recent experimental work has shown that the rate of change of the eutectic temperature T_e with pressure for Fe–FeS will change sign at a pressure of 14 GPa: it is negative for lower pressures and positive otherwise (e.g., Fei et al. 1997, 2000; Chudinovskikh and Boehler 2007; Stewart et al. 2007; Chen et al. 2008; Buono and Walker 2011). As a consequence, crystallization in a cooling core will proceed very differently in a large planet, where the pressure is significantly higher than ~40 GPa, compared with a small planet where the pressure is lower. While the former case is applicable to the Earth's core, the latter is applicable to Mars as well as to smaller planets and moons (see Table 1). The case of the Earth's core has been studied extensively in past decades, while the smaller planets are the subjects of ongoing research.

Inner-core growth

The conventional model of inner-core growth on the Fe-rich side of the eutectic is applicable to cases where the (positive) slope of the core liquidus is steeper than that of the adiabat, $dT/dP_{melt} > dT/dP_{ad}$ (e.g., Stevenson et al. 1983; see Fig. 1a). As a consequence, when the core has cooled sufficiently to reach the liquidus at the center, iron will crystallize and an inner-core nucleus will form; upon further cooling, the core will freeze from the inside out.

Table 1 Parameters of the interior structure

	Earth	Moon	Mercury	Mars	Ganymede
Planetary radius (km)	6371[a]	1737[b]	2439[c]	3389[d,e]	2632[f]
Surface gravity (m s^{-2})	9.8[a]	1.6[a]	3.7[a]	3.7[a]	1.4[a]
Core radius (km)	2890[a]	250–430[b]	1965–2050[g,h]	1370–1870[d,e]	650–880[f]
R_c/R_p	0.45[a]	0.14–0.25[b]	0.8–0.84[g,h]	0.4–0.55[d,e]	0.25–0.33[f]
P_{cm} (GPa)	139[a]	4.5–5[i]	2.8–3.3[h]	20–25[d,e]	5.5–7[f]
P_c (GPa)	357[a]	5–6[i]	33–37[h]	38–42[d,e]	9.5–11[f]
Gravity at CMB (m s^{-2})	10.3[a]	0.7[b]	4[h]	3[d,e]	1.3[f]

Parameters of the interior structure for the Moon, Mercury, Mars, and Ganymede. Earth is included for comparison
R_c/R_p is the relative core radius, P_{cm} is the density at the core–mantle boundary, P_c is the central pressure
[a]Lodders and Fegley (1998)
[b]Sohl and Schubert (2007)
[c]Anderson et al. (1987)
[d]Sohl and Spohn (1997)
[e]Sohl et al. (2005)
[f]Sohl et al. (2002)
[g]Hauck et al. (2013)
[h]Rivoldini and Van Hoolst (2013)
[i]Sohl et al. (2009)

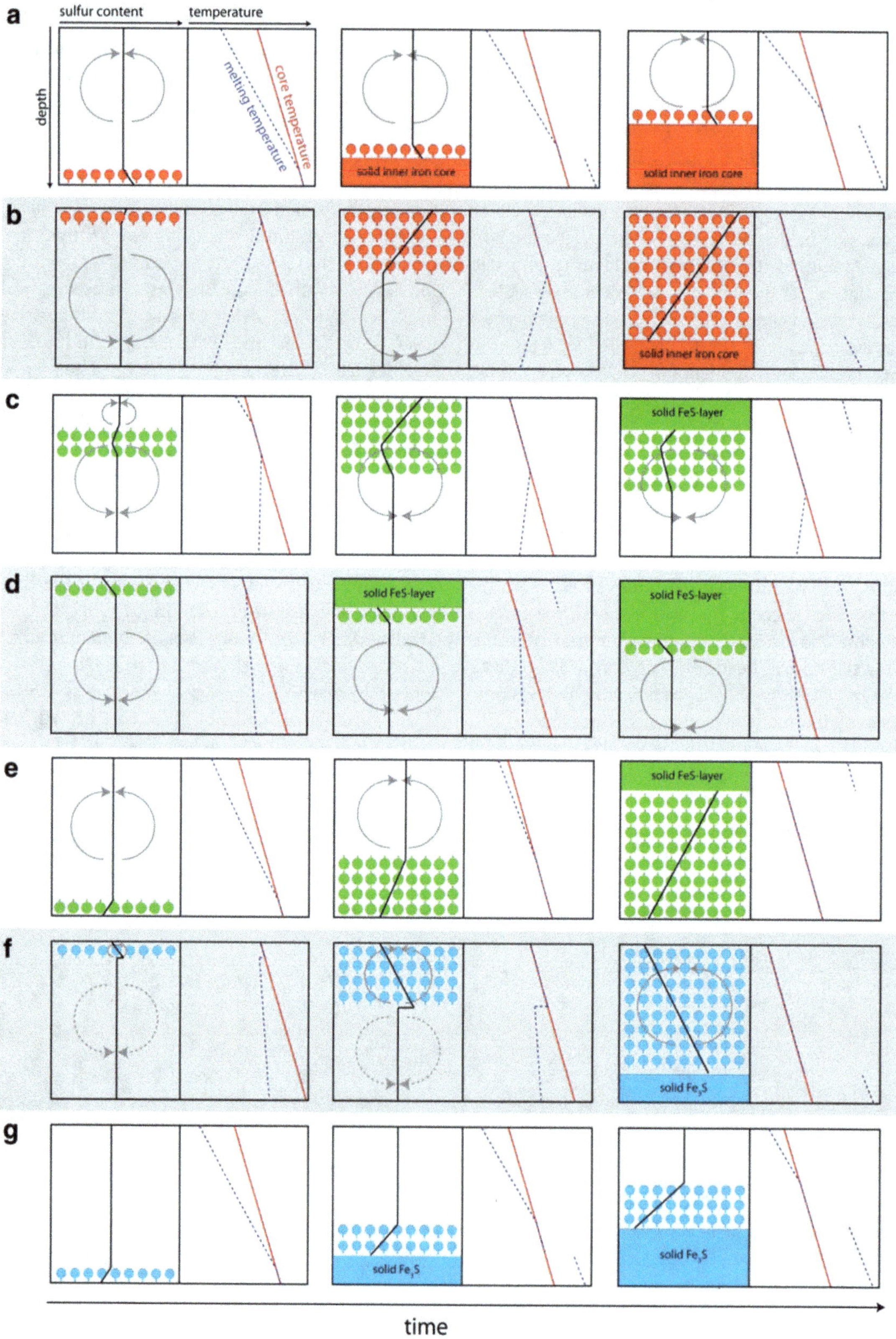

Fig. 1 (See legend on next page.)

(See figure on previous page.)
Fig. 1 Crystallization scenarios in the Fe–FeS system. Crystallization scenarios in the Fe–FeS system. **a** Earth-like, Fe-rich inner core grows from the center. Sulfur is enriched in the outer core and drives chemical convection. **b** Iron snow forms at the CMB, sinks, and remelts at depth and drives chemical convection. The stable snow zone grows in time. When it reaches the center, an inner solid core will form. **c** Floating FeS crystals form a stable zone growing toward the CMB where eventually a solid FeS layer will form. The fluid below is enriched in iron and unstable to convection. **d** A solid FeS layer grows from the CMB. Expulsion of Fe results in chemical convection in the fluid below as in (**c**). **e** FeS crystals rise from the center and remelt at lower depths. The liquid above the FeS zone is convectively unstable. When the FeS crystal zone reaches the CMB, a solid FeS layer will form. **f** Fe_3S snow forms at the CMB and the chemically unstable snow zone grows until it comprises the entire core where a solid Fe_3S inner core will form. The layer below is chemically homogeneous but may convect thermally. **g** A solid Fe_3S inner core grows with time without the release of chemical buoyancy to the outer core and convection above. *Red, green*, and *blue dots* indicate solid iron, solid FeS, and Fe_3S, respectively. *Short dashes* show the direction of sinking or rising. *Red solid lines* are the core temperature, *blue dashed lines* the core melting temperature, and *black solid lines* the concentration of sulfur, respectively. *Solid arrows* indicate chemical and *dashed arrows* thermal convection, respectively. For further explanation, see text

The outer core will be increasingly depleted in iron and enriched in sulfur until the eutectic composition is reached. At that point, the core will continue to freeze in a eutectic composition. Figure 2 shows, for various initial sulfur concentrations, the enrichment of sulfur in the outer fluid core and the associated inner-core size, assuming that the inner solid core is free of sulfur. The maximum relative inner-core size, which is attained when the outer core has a eutectic composition, decreases with increasing initial sulfur concentration.

During crystallization, the concentration of sulfur will not be homogeneous; instead, a boundary layer will form just above the inner core, where the sulfur concentration is above average. The fluid in this layer will be lighter than in the remainder of the fluid core, and compositional convection will be induced to homogenize the outer-core composition. Chemical convection and the associated generation of a magnetic field in the core will occur if the temperature in the fluid (outer) core lies between that of the solidus and the liquidus of the core alloy. Inner-core growth will induce outer-core convection even if the heat flow through the CMB is lower than the critical heat flow for a thermal dynamo, as discussed above (e.g., Stevenson et al. 1983). Thus, the temperature gradient may be even shallower than adiabatic, and compositional convection may carry heat downward, against the temperature gradient. The buoyancy release associated with this compositional change exceeds the work done against the stabilizing thermal stratification.

Iron snow regime

Compositional convection, however, can be very different from the scenario described above in smaller planetary bodies or for higher sulfur contents than characteristic of the eutectic composition. Recent experimental studies (e.g.,

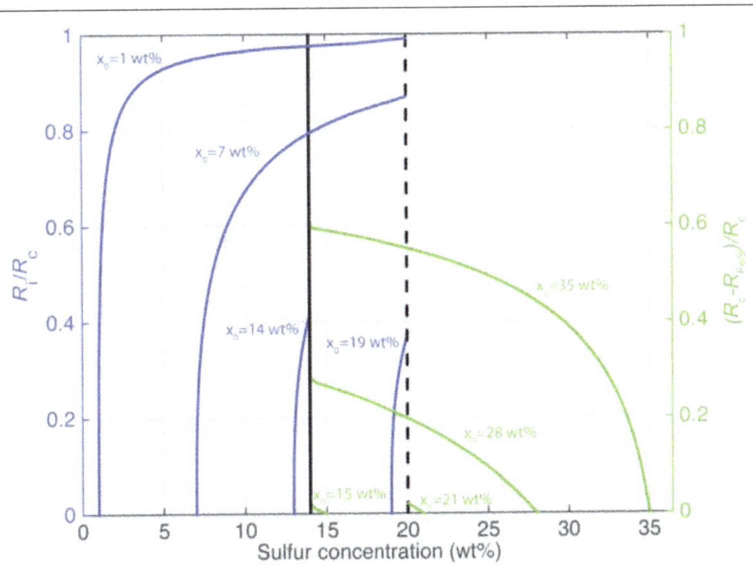

Fig. 2 Radius of solid inner Fe core and thickness of solid FeS layer. Relative radius of the solid inner Fe core (*blue lines*) and thickness of the solid FeS layer (*green lines*) as a function of the sulfur content in the fluid part of the core and depending on the initial sulfur content. R_i is the inner-core radius, R_c the core radius, and R_{FeS} the radius of the lower boundary to the solid FeS layer. Shown are initial sulfur concentrations of 1, 7, 14, 15, 19, 21, 28, and 35 wt.%. The *solid* and *dashed vertical lines* indicate the eutectic sulfur concentration at different pressures (14 wt.% corresponds to ~30 GPa and 20 wt.% to ~10 GPa). Left of the eutectic a solid inner Fe core and right of the eutectic a solid FeS layer are formed. The relative inner-core radius and FeS layer thickness are smaller the closer the sulfur content is to the eutectic

Fei et al. 1997, 2000; Li et al. 2001; Chudinovskikh and Boehler 2007; Stewart et al. 2007; Chen et al. 2008; Buono and Walker 2011) have revealed two important aspects of the Fe–FeS phase diagram: (1) at pressures below 14 GPa, the eutectic melting temperature decreases with increasing pressure; and (2) the S content at the eutectic decreases with increasing pressure up to a pressure of 60 GPa (Stewart et al. 2007; Chudinovskikh and Boehler 2007; Morard et al. 2008; Figs. 3 and 4). Note that the exact decline of the eutectic concentration with pressure is somewhat uncertain. While Fei et al. (1997, 2000) found that the eutectic concentration remains constant between ~7 and 10 GPa, the data of Morard et al. (2007) suggest a constant eutectic concentration between slightly higher pressures of ~11 and 15 GPa (see Fig. 5 in Morard et al. 2007). These uncertainties also impact the slope of the melting temperature, which is typically obtained through linear interpolation of the melting temperature of the end-members (Fe–FeS eutectic and pure Fe or pure FeS).

A melting temperature profile with a negative slope (or a shallower melting temperature profile than that of the core temperature, $dT/dP_{melt} < dT/dP_{ad}$) will have profound implications. Upon cooling, Fe will precipitate at the CMB rather than in the center and may fall as iron snow, since the solid iron particles are heavier than the surrounding Fe–FeS fluid (Fig. 1b). A snow zone forms, limited from below, at a depth where the core temperature is higher than the melting temperature, and the iron snow particles remelt. Owing to the solidification of iron, the concentration of sulfur increases in the snow zone, and the liquidus of the fluid decreases there (e.g., Buono and Walker 2011). Under the assumption of thermodynamic equilibrium, iron continues to solidify and the local sulfur concentration in the liquid increases until the liquidus reaches the core temperature. Thus, in the snow zone, the liquidus becomes collinear with the core temperature and reverses its slope, so that the liquidus temperature increases with increasing depth. The increase of the liquidus with depth in the snow zone is inevitably accompanied by a decrease in the sulfur concentration with depth, which implies the presence of a stable chemical gradient across the snow zone (Hauck et al. 2006; Williams 2009; Rückriemen et al. 2015).

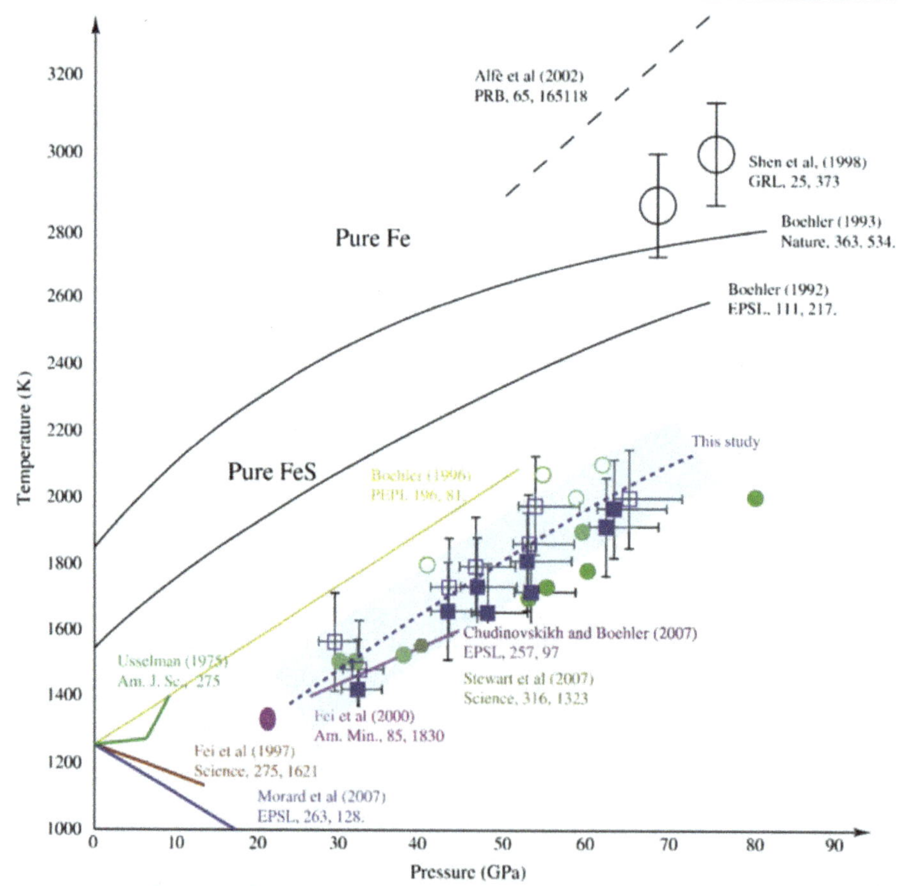

Fig. 3 Pressure dependence of the eutectic temperature in the Fe–FeS system. Pressure dependence of the eutectic temperature in the Fe–FeS system. *Empty squares* indicate the presence of melt, while *filled squares* indicate the presence of Fe and Fe_3S solid phases only. The *solid line* represents linear interpolation for this data set. Other melting data are from Usselman (1975), Boehler (1993), Boehler (1996), Fei et al. (1997, 2000), Shen et al. (1998), Alfè et al. (2002b), Campbell et al. (2007), Chudinovskikh and Boehler (2007), Morard et al. (2007), and Stewart et al. (2007). Figure after Morard et al. (2008)

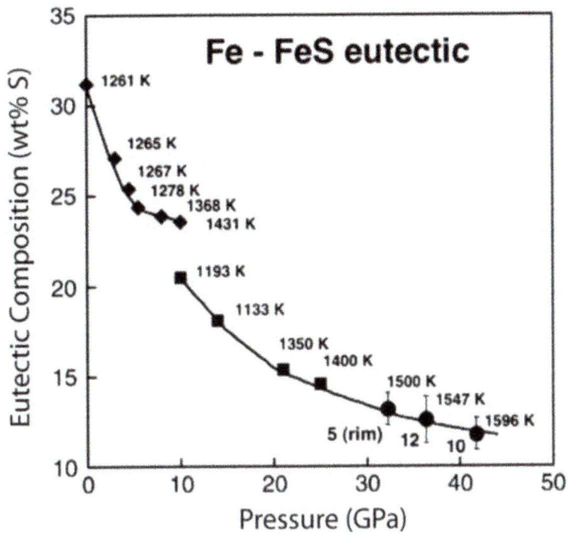

Fig. 4 Pressure dependence of Fe–FeS eutectic composition. Pressure dependence of the Fe–FeS eutectic composition (wt.% S); *circles*: Chudinovskikh and Boehler (2007); *diamonds*: Usselman (1975); *squares*: Fei et al. (1997, 2000) and Li et al. (2001). Figure after Chudinovskikh and Boehler (2007). The eutectic melting temperature is also given

The remelting of iron particles below the snow zone results in a local enrichment of iron in the Fe–FeS melt. This creates a gravitationally unstable situation, where a heavier layer is formed on top of a lighter one. The Fe-rich layer initiates compositional overturn, which leads to a well-mixed state of the deeper, entirely liquid core below the snow zone. With ongoing cooling, the snow zone grows at the expense of the deeper, well-mixed fluid core. Finally, an inner core begins to grow when the snow zone encompasses the entire core, because the sinking iron particles can no longer remelt.

This so-called iron snow regime has already been suggested to apply to the present cores of Ganymede (Hauck et al. 2006; Christensen 2015; Rückriemen et al. 2015) and Mercury (Chen et al. 2008; Vilim et al. 2010; Wicht and Heyner 2014). It has also been suggested that Mars' core may enter the iron snow regime (Stewart et al. 2007) and that the lunar core may have gone through an iron snow episode (Zhang et al. 2013; Laneuville et al. 2014). Although the thermodynamics of the iron snow regime seems to be reasonably well developed, the questions of how and where a magnetic field can be generated in an iron snow core remain debated. Hauck et al. (2006) suggested that the power dissipated by the sinking iron snow zone is sufficient to drive a dynamo. Their work is supported by that of Zhan and Schubert (2012), who obtained a multipolar magnetic field using a magneto-hydrodynamic model where the fluid motions within the snow zone resembled internally heated convection. In contrast to these studies, Bland et al. (2008) argued that compositional-driven convection in the Fe snow zone would not provide enough power to generate a magnetic field, since the latent heat is released close to the CMB and cannot be used to efficiently drive the dynamo. In

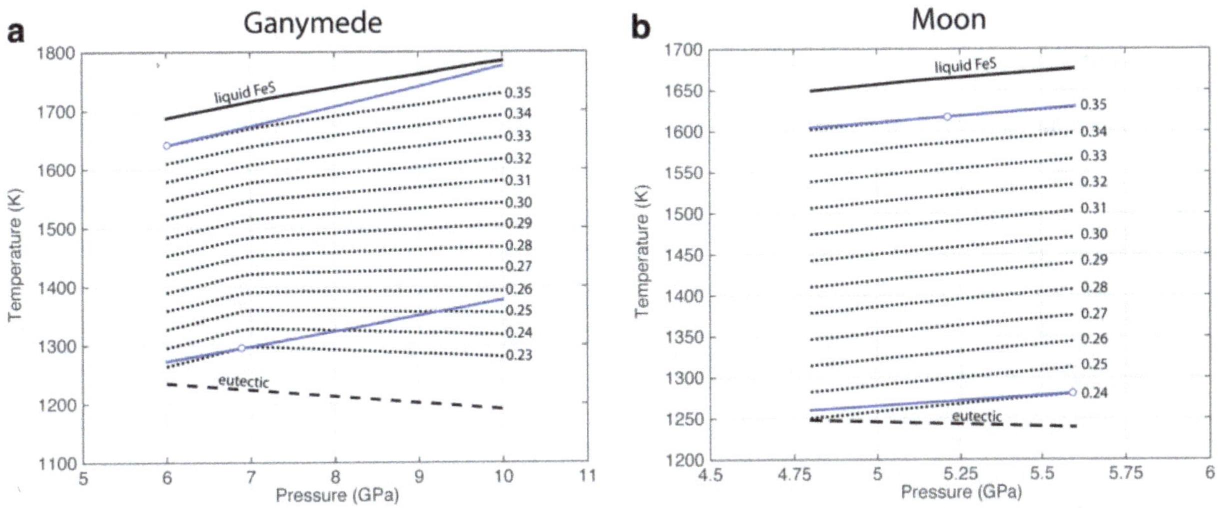

Fig. 5 Melting temperature on the sulfur-rich side of eutectic. Melting temperature of FeS (*solid black line*) and Fe–FeS eutectic (*dashed black line*) as a function of pressure. FeS data are from Boehler (1992) and Fe–FeS eutectic from Fei et al. (1997, 2000). *Dotted lines* are melting temperatures for different sulfur contents, obtained by linear interpolation between the melting temperatures of FeS and Fe–FeS eutectic. **a** corresponds to the pressure range of Ganymede's core and (**b**) corresponds to the pressure range of the Moon's core. *Blue solid lines* correspond to adiabatic temperature profiles and *white circles* indicate the point where melting and core temperatures coincide first during cooling, i.e., where crystallization of FeS starts. The adiabatic temperature profiles have been calculated for $\alpha = 9 \times 10^{-5}$ K^{-1}, $\rho = 5500$ kg m^{-3}, cp = 830 J kg^{-1}, $g = 1$ m s^{-2} (Ganymede), and $\alpha = 9 \times 10^{-5}$ K^{-1}, $\rho = 5500$ kg m^{-3}, cp = 830 J kg^{-1}, $g = 0.6$ m s^{-2} (Moon). Depending on the sulfur content, for Ganymede, FeS starts to crystallize in the core or at the CMB and for the Moon, FeS starts to crystallize in the core or in the core center

addition, Rückriemen et al. (2015) argued that the nature of the convection induced by sedimentation in an otherwise chemically stably stratified layer is likely of small scale and thus lacks an important criterion (i.e., large-scale convection) that is necessary for dynamo action (Christensen and Wicht 2007). Instead, it is argued that the dynamo may be located in the iron-rich layer below the snow zone. Here, vigorous convection caused by the release of buoyancy upon remelting of iron may drive the dynamo (Vilim et al. 2010; Christensen 2015; Rückriemen et al. 2015). However, this dynamo would only be active during the period between the formation of the snow zone and the time when it reached the center of the core.

Most dynamo models for iron snow cores have thus far been calculated for Ganymede, and these will be discussed in more detail in the section about Ganymede below. Note, however, that the iron snow scenario described above assumes thermodynamic equilibrium. It might also be possible that the iron particles sink so rapidly that they fail to remelt (completely) before reaching the center. Whether or not a dynamo may then still be possible remains an open question.

Precipitation of FeS

If the core is sulfur rich (i.e., if its sulfur content is greater than the eutectic concentration), solid FeS (or Fe_3S_2 for pressures between 14 and 20 GPa, and Fe_3S for even higher pressures) will precipitate within the core (Fei et al. 2000; Hauck et al. 2006; Campbell et al. 2007; Stewart et al. 2007; Morard et al. 2008). Figure 5 shows the melting temperature as a function of pressure for various sulfur concentrations, assuming a linear interpolation of the melting temperature between pure FeS and the eutectic melting temperature. The pressure ranges shown represent conditions in Ganymede and the Moon. For Ganymede (Fig. 5a), a comparison with the adiabatic temperature profile shows that for concentrations close to the eutectic, freezing will start at about 7 GPa; i.e., within the core but close to the CMB. Because of the high sulfur content of FeS (~36.5 wt.% S), even solid FeS will be less dense than the residual liquid, provided that the composition of the liquid is not close to FeS. The increasing concentration of solid FeS will remelt at shallower depths, thus increasing the S concentration and the liquidus temperature there (on the sulfur-rich side of the eutectic, the melting temperature increases with the sulfur content; see Fig. 1c). Over time, a layer forms where solid FeS rises (FeS floating zone) and where the temperature is equal to the melting temperature. The floating zone can be divided into two parts. The upper part, where $dT/dP_{melt} > dT/dP_{ad}$, exhibits a chemically stable gradient, and the lower part, where $dT/dP_{melt} < dT/dP_{ad}$, is chemically unstable, including the fluid layer below. The fluid of the lower floating zone is left enriched in iron and tends to remix with the deeper fluid core, leading to a decrease in the melting temperature there. The floating zone grows toward the CMB until an FeS layer forms at the CMB. The subsequent crystallization of the core proceeds similarly to the growth of the Earth's inner core but in reverse as the solid FeS layer grows in thickness toward the center. The continuous enrichment of Fe in the fluid inner core will maintain chemical convection and may drive a dynamo.

For sulfur concentrations greater than about 31 wt.%, crystallization of FeS will start immediately at the CMB and an FeS layer will grow in time from there (see Fig. 1d). This is similar to the scenario shown in Fig. 1c when the floating zone has reached the CMB. Figure 2 (right-hand side) shows the depletion of sulfur in the fluid inner core as a consequence of the growth of the FeS layer for various initial sulfur concentrations. The FeS outer core will grow in thickness until the eutectic composition is reached in the inner core. From that point in time, the core will freeze in eutectic composition. The maximum thickness of the FeS outer core, expressed in core radii, decreases with decreasing initial sulfur concentration. Hauck et al. (2006) argue that FeS crystallization will start in the center of Ganymede (rather than in the upper core, as argued above). This is possible, but it requires a conductive core with a suitably small temperature gradient at the onset of freezing (see Fig. 5a).

For the Moon, we do indeed find that FeS would start crystallizing at the center of the core. In this case, an FeS floating zone develops with a stable chemical gradient where FeS crystallizes and rises upward toward the outer core (see Fig. 1e), just like in an inverted iron snow scenario. When FeS remelts, it releases buoyancy, which may drive chemical convection in the outer core. With core cooling, the FeS floating zone grows toward the CMB. When this zone eventually comprises the entire core, solid FeS remelting will cease and a solid FeS layer begins to grow at the CMB. This scenario results in dynamics similar to that of the iron snow regime on the Fe-rich side of the eutectic, although the direction of growth of the floating zone is reversed, with solid particles rising instead of sinking. There may thus be a dynamo in the fluid outer core above the floating zone rather than one in a fluid inner core below the iron snow zone. The dynamo would be active as long as the floating zone grows.

For core pressures higher than about 21 GPa, solid Fe_3S forms for S contents higher than the eutectic (Fei et al. 2000; Campbell et al. 2007; Stewart et al. 2007; Morard et al. 2008). The slope of the eutectic temperature is estimated at 15 K GPa^{-1} (Campbell et al. 2007; Morard et al. 2008). For Mars, precipitation of Fe_3S would first occur at the CMB, because $dT/dP_{melt} < dT/dP_{ad}$ is given for most plausible parameter variations to determine the adiabatic temperature profile. Fe_3S is denser than the residual liquid (Stewart et al. 2007), so it sinks into the core and remelts where $T_c > T_{melt}$ (Fig. 1f). The crystallization process of Fe_3S is similar to that in the iron snow regime (Fig. 1b), because an Fe_3S snow region forms at the CMB. However, there are important differences. The density profile of the liquid in the Fe_3S snow region is unstable. The sinking and remelting of Fe_3S results in an increase in S with increasing depth, which results in turn in a decreasing density with depth. In the Fe_3S snow region, therefore, the material continuously crystallizes and remelts because of this unstable situation. Below the snow zone, compositional convection is absent, since the remelting of Fe_3S results in a local increase of S and leads to the formation of a lower density fluid layer on top of the denser liquid core below. This situation is stable. However, thermally driven convection in the bottom fluid layer may still be possible, since the heat flow from the lower core is adiabatic owing to the convective snow zone. Over time, the Fe_3S snow region comprises the entire core and a solid inner Fe_3S core forms. For this crystallization scenario, we can find two convection layers and, therefore, two dynamo regions if thermally driven convection is efficient in the liquid inner core. Alternatively, the dynamo region is located in the upper core, attains its maximum extent when the snow zone reaches the center, and then shrinks again owing to the growth of the solid inner core.

If Fe_3S crystallizes first in the core center (see Fig. 1g), which is possible for a conductive temperature profile, the dense crystals remain in the center to form an inner core, as in the conventional Earth scenario. Compositional convection in the outer liquid core is unlikely to occur, since the liquid from which Fe_3S crystallizes will be depleted in S and its density will increase. A fluid layer forming in the solid Fe_3S core, which is denser than the remaining fluid core, does not induce an upward flow through compositional buoyancy forces, and dynamo action is less efficient or even unlikely.

Scaling laws and power for dynamo action

Scaling laws have been derived in recent years for hydromagnetic dynamos which relate magnetic induction to the buoyancy flux, a measure of the convective power available for the dynamo, and the dimensions of the dynamo region (e.g., Christensen and Tilgner 2004; Christensen and Aubert 2006; Olson and Christensen 2006; Aubert et al. 2009). These scaling laws suggest that the magnetic field strength is related to the convective power available to balance ohmic dissipation and that it is not determined by a force balance between the Coriolis and Lorentz forces, as in the older scaling laws (e.g., Busse 1976; Stevenson 1979; Curtis and Ness 1986). A scaling based on ohmic dissipation has already been used by Stevenson et al. (1983). A detailed discussion of the various scaling laws can be found in Christensen (2010).

The scaling laws derived from dynamo simulations are characterized by diffusivities and viscosities that are many orders of magnitude too large. Such large values are chosen to dampen any small-scale flow that cannot be resolved with the present numerical resolution and available computational power. The scaling of Christensen and Aubert (2006) attempts to circumvent this problem by avoiding diffusivity and viscosity. While a weak dependence on the magnetic Prandtl number leads to a slightly better fit to the simulated data (e.g., Christensen 2010), the main advantage of the present scaling is that it satisfactorily covers a wide range of parameters, from planetary dynamos to dynamos of fully convective rapidly rotating stars. Future simulations for more extreme parameters than those explored by Christensen and Aubert (2006) will further clarify the applicability of the proposed scaling law.

Accordingly, the surface value of the magnetic induction can be calculated using

$$B_{\text{surf}} = f\mu_0^{1/2}\rho_c^{\,1/6}(\phi D)^{1/3}\left(\frac{R_c}{R_p}\right)^3, \tag{2}$$

where f is a constant prefactor that includes the partitioning of energy between the poloidal and toroidal components, μ_0 is the permeability of free space, ρ_c the core density, ϕ the volumetric ohmic dissipation rate, D the thickness of the fluid core, R_c the core radius, and R_p the planetary radius. The final term shows that the decrease of the field strength is proportional to at least the third power of the distance to the surface.

To calculate the ohmic dissipation rate, an entropy balance is required for the core. Ohmic and viscous dissipation do not enter the core's global energy budget, but, as irreversible processes, both are sources of entropy. In the fluid regions of the core, viscous dissipation can be neglected compared with ohmic dissipation. The entropy budget for the core (for a review, see Nimmo 2007) reads

$$E_\Phi = E_S + E_g + E_L + E_{\text{rad}} - E_{\text{ad}}, \tag{3}$$

where E_Φ is the rate of entropy production through ohmic dissipation, E_S is the entropy production rate owing to secular cooling, and E_g owing to the release of chemical buoyancy. E_L, E_{rad}, and E_{ad} are the entropy production rates owing to latent heat release, radioactive heating, and conduction along the adiabat, respectively. Note that ohmic dissipation and heat conduction are actually entropy sinks. The individual contributions to the entropy balance are given by

$$E_S = -\int \rho_c C_c \left(\frac{1}{T_c} - \frac{1}{T}\right) \frac{dT_c}{dt} dV; \tag{4}$$

$$E_g = \frac{Q_g}{T_c} = \frac{1}{T_c} \int \psi \frac{\partial \rho_c}{\partial t} dV; \tag{5}$$

$$E_L = L_H \left(\frac{1}{T_c} - \frac{1}{T}\right) \frac{dm_i}{dt}; \tag{6}$$

$$E_{\text{rad}} = \int \rho_c h \left(\frac{1}{T_c} - \frac{1}{T}\right) dV; \tag{7}$$

$$E_{\text{ad}} = \int k_c \left(\frac{\nabla T}{T}\right)^2 dV. \tag{8}$$

Here, T_c is the temperature at the CMB, Q_g is the heat carried by chemical convection, Ψ the gravitational potential, C the concentration of the light alloying element, L_H the latent heat, dm_i/dt the rate of change of solidified mass, and h the internal heating rate. For the classical case of inner-core growth, as applicable to the Earth, the terms E_S, E_g, and E_L can be expressed as functions of the radial growth rate of the inner core (e.g., Labrosse 2003; Nimmo 2007).

Since dissipation always produces entropy, E_Φ has to be greater than zero for a dynamo to occur. The entropy balance also illustrates the importance of core freezing for a dynamo because compositional buoyancy is not limited by the thermodynamic (Carnot) efficiency factor $(^1/_{T_c} - ^1/_T)$. For a purely thermal dynamo ($E_g = E_L = 0$) and assuming no internal heat sources in the core, the heat flux from the core needs to be greater than the heat flux along the core adiabat.

The volumetric ohmic dissipation rate is then calculated as

$$\phi = \frac{E_\Phi T}{V_{\text{fl}}}, \tag{9}$$

where V_{fl} is the volume of the fluid core and T the characteristic core temperature where most entropy is produced.

In addition to constraints to the power available to generate a dynamo, the magnetic Reynolds number must be greater than a numerically defined critical value of about 50 (Christensen and Aubert 2006). The magnetic Reynolds number characterizes the importance of magnetic advection versus magnetic diffusion and is defined as $\text{Re}_m = vL/\eta$, where v is the characteristic velocity, L the characteristic length (the thickness of the convective shell, $D = L$), and η the magnetic diffusivity. The characteristic velocity can be expressed through a scaling law,

$$v \approx 0.7 \left(\frac{D}{\Omega}\right)^{1/5} \left(\frac{\Phi}{\rho_c}\right)^{2/5}, \tag{10}$$

where Ω is the rotation rate and ρ_c is the average density of the fluid core.

In conclusion, whether a terrestrial planet features core convection and whether it can have a thermal or a chemical dynamo in these models depends mainly on the cooling rate of the core, the core temperatures, and the core melting temperatures. Thermal evolution models suggest that the existence of a present-day magnetic field in a terrestrial planet requires a crystallizing core. A purely thermally driven dynamo is unlikely for any present-day terrestrial planet, since the models typically predict a slowly cooling core with a core heat flux that is smaller than the heat flux along the core adiabat. In fact, a thermal dynamo is, in general, difficult to obtain for stagnant-lid planets. For these, superheating of the core with respect to the mantle must be postulated (e.g., Stevenson et al. 1983; Breuer and Spohn 2003; Laneuville et al. 2014). If superheating can be obtained (e.g., upon core formation), the thermal dynamo typically shuts off very early during the evolution, since the heat flow from the core decreases rapidly during the first few hundred million years. Considering the high values for the thermal conductivity recently suggested, early thermal dynamos would either be even more short-lived or entirely absent. Whether an inner core can grow (and by which crystallizing mechanism) and whether a compositional dynamo can be initiated depends strongly on the temperature evolution, the core pressure, and the core composition (melting temperature).

Core composition

Cosmochemical abundances suggest that the bulk of the core is made of iron, with some Ni and light elements. In addition to cosmochemical arguments, light elements are required to balance the density deficit of the Earth's outer fluid core compared with pure fluid iron and of the solid inner core with respect to pure solid iron (e.g., McDonough 2014). It is commonly assumed that sulfur is the most likely candidate for the light elements in the cores of terrestrial planets and moons (e.g., Ringwood 1977; McCammon et al. 1983). First, it is abundant in the solar system and is a siderophile (an element that is preferentially enriched in the iron melt), and it is thus likely enriched in the core upon core formation. Second, iron meteorites consist of pure Fe–Ni, with varying amounts of sulfides (FeS). These meteorites are believed to be remnants of iron-rich cores from planetesimals that have differentiated during their early evolution and are considered building blocks of the terrestrial planets. The Fe–FeS system is a eutectic system. It is usually assumed that the compositions of most terrestrial planets are on the Fe-rich side of the eutectic composition (e.g., Lodders and Fegley 1998), while for the large Jovian moons (e.g., Scott et al. 2002) and Mars (e.g., Stewart et al. 2007), this is under debate.

Many models of core processes simply assume an Fe–FeS composition. This assumption is almost certainly an oversimplification. We know from the Earth's core that light elements such as Si, O, C, and H are likely incorporated in the core (for recent reviews, see, e.g., Hirose et al. 2013; McDonough 2014; Nimmo 2015). The actual composition of the core depends strongly on the primordial composition of the planet's forming material and the partition coefficient of light elements between the solid and liquid phases, which again depends on temperature and pressure conditions, as well as on the oxidation state during core formation: the abundance of light elements in the iron-rich cores will, therefore, vary among planets and moons. Even for the Earth's core, the core composition is not well-known and compositional models for the core (e.g., a silicon- versus an oxygen-bearing core) are offered as competing hypotheses (e.g., McDonough 2014). In contrast to other terrestrial bodies, compositional models for the Earth's core can be constrained by seismological data (e.g., Badro et al. 2014).

Knowing the light elements and their concentrations is important for modeling the thermal and the magnetic field evolution, for two main reasons. First, the melting relations pertaining to the core depend on the nature and concentrations of the light elements. For instance, the eutectic temperatures in the Fe–FeS and Fe–Fe_3C systems are about 1000 and 500 K lower, respectively, than the melting temperature of pure iron at 25 GPa (Chabot et al. 2008; Buono and Walker 2011). Second, the buoyancy forces powering the dynamo depend strongly on the density difference between the solid and the fluid phases. In the Fe–FeS system, for instance, pure Fe is crystallizing (including a small amount of sulfur at higher pressure; Li et al. 2001), thereby maximizing the effects of the enrichment of light elements in the outer core and the density difference between the outer and the inner cores. In the Fe–Si system, alloys of Fe and Si form a solid solution with maximum compositional differences of ~2 wt.% Si between the liquid and solid at 21 GPa and temperatures <2000 K (Kuwayama and Hirose 2004), and only a small density difference is expected, which is mainly due to the volume change upon crystallization. For more complex, multicomponent systems, melting temperatures and density differences are less well studied. First steps have been taken to explore the Fe–Si–S system (e.g., Sanloup and Fei 2004; Siebert et al. 2004; Malavergne et al. 2007) and the Fe–O–S system (e.g., Naldrett 1969; Urakawa et al. 1987; Tsuno and Ohtani 2009).

Another important issue for magnetic field generation is the presence of radioactive elements in the core, since these can provide additional power for the dynamo (see Eq. 6). In general, uranium, thorium, and potassium are lithophile elements that will not easily partition into the core (for reviews, see Roberts et al. 2003; McDonough 2014). Thermal models usually neglect them, although at least potassium has been discussed (e.g., Lewis 1971; Murthy 1991; Breuer and Spohn 1993; Buffett 2002). Recent experimental studies found that K increasingly partitions into Fe liquids at high pressures and temperatures with increasing concentrations of S or Ni (Gessmann and Wood 2002; Murthy et al. 2003; Lee et al. 2004). For Mars, a value of about 150 ppm K in the core has been suggested (Murthy et al. 2003), which has only minor effects on the planet's thermochemical evolution, however (Williams and Nimmo 2004). In addition, at very low oxygen fugacities, as suggested for Mercury's interior, elements that are typically considered lithophile can become more siderophile or chalcophile. This has been shown for U (Malavergne et al. 2007) and is also possible for Th (Malavergne et al. 2010).

Mercury

The MESSENGER mission, which orbited Mercury between March 2011 and the end of April 2015, confirmed the presence of an internally generated magnetic field (Anderson et al. 2011), which had already been detected during the first and third close encounters of Mariner 10 in 1974 and 1975 (Ness et al. 1974, 1975, 1976; Connerney and Ness 1988; Russell et al. 1988). Mercury's magnetic field features a peculiarly small dipole moment, which is a factor of about 10^4 smaller than that of the Earth. MESSENGER found the equivalent dipole to be offset toward the north by about 0.2 planetary radii, suggesting a strong quadrupole contribution, and a small deviation from axial symmetry (Anderson et al. 2012; Johnson et al. 2012). Moreover, evidence for remanent crustal magnetization was found as the spacecraft was spiraling toward Mercury to hit the surface on 30 April 2015. The cratering record of the magnetized surface suggests that a magnetizing field has existed since at least 3.7–3.9 Ga (Johnson et al. 2015).

Since the discovery of Mercury's magnetic field, it has mostly been argued that a hydromagnetic dynamo driven by chemical convection in an outer liquid core provides the best explanation, although alternative dynamo mechanisms have been suggested as well (see below). It is generally accepted that Mercury has an iron-rich core. The high density of the planet suggests that this core must be comparatively large, about 0.8 planetary radii. Recent MESSENGER gravity data (Smith et al. 2012) and the amplitude of libration measurements with Earth-based radar (Margot et al. 2007, 2012) have been used to calculate improved values of the planet's moment of inertia factor, C, and the ratio of the moment of inertia of the solid part of the planet overlying the liquid core to that of the whole planet, C_m/C, following Peale (1976). From these, improved estimates of the core radius range between 1840 and 2040 km (Smith et al. 2012; Hauck et al. 2013; Rivoldini and Van Hoolst 2013). The value of $C_m/C = 0.431 \pm 0.025$ also supports the presence of a fluid outer core (Peale 1976; Margot et al. 2012). In a recent study, these data have been used to constrain the size of the putative inner core (Dumberry and Rivoldini 2015).

Assuming an Fe–FeS core composition, the largest inner-core radius consistent with the geodetic observations is 1325 ± 250 km.

Prior to the MESSENGER observations, the planet's bulk composition, including that of the core, was poorly constrained. Generally, a low volatile content was assumed because of Mercury's close orbital position to the Sun and its unusually large metal-to-silicate ratio suggested by the average density. Most models that aim to explain the large core radius predict a nearly complete loss of volatiles; i.e., a loss of silicates through evaporation (Fegley and Cameron 1987), high temperature equilibrium condensation of nebular materials (Lewis 1973), a giant impact removing part of the silicate planet (Cameron et al. 1988), or meteoritic mixing of refractory and volatile condensates (Morgan and Anders 1980). Measurements of the elemental ratios K/Th and K/U with the MESSENGER gamma-ray spectrometer (Peplowski et al. 2011, 2012), however, argue for a volatile content similar to that of the other terrestrial planets and thus may also allow for a higher volatile content in Mercury's core than previously assumed. Note that this assumption is still uncertain, because up to 10 % of Mercury's U inventory, and possibly also a significant amount of Th, may have partitioned into the core (McCubbin et al. 2012) and thereby changed the elemental ratios K/Th and K/U if Mercury's core formed under extreme reducing conditions.

As for the other terrestrial planets, sulfur has been the most likely candidate for the light element in the iron core (e.g., Ringwood 1977; McCammon et al. 1983). However, the model of an Fe–FeS core has recently been questioned, since the MESSENGER x-ray spectrometer (XRS) data (Nittler et al. 2011; Weider et al. 2012) indicate strongly reducing conditions, which has been used to suggest a core containing Si as well as S (e.g., Malavergne et al. 2010; Chabot et al. 2014). The presence of silicon may have major implications for the core's structure and evolution (Smith et al. 2012; Hauck et al. 2013) and for the magnetic field evolution (see Fig. 6).

Thermal evolution models (Schubert et al. 1988; Hauck et al. 2004; Grott et al. 2011; Tosi et al. 2013) suggested that a typical evolution of the magnetic field starts with a thermally driven dynamo, followed by a chemically driven dynamo after the start of inner-core growth. Thermally driven dynamo action may cease before the inner core begins to freeze, in which case there would be a gap in the magnetic field history. Whether such a gap occurs and its length in time would depend on the value of the heat flow along the core adiabat and the core's volatile content. The larger the heat flux along the core adiabat becomes, the shorter the lifetime of a thermally driven dynamo is. Recent laboratory measurements for iron suggested a thermal conductivity of 113–125 W m^{-1} K^{-1} for Mercury core pressure and temperature conditions and a heat flow along the adiabat of 45–100 mW m^{-2} (Deng et al. 2013). This high value suggests a rapid decline of a thermal dynamo, within less than a few tens of Ma (e.g., Grott et al. 2011; Tosi et al. 2013): see Fig. 7c. Note, however, that volatiles in the core reduce the thermal conductivity and thus the heat flow along the adiabat (de Koker et al. 2012). This, in turn, increases the lifetime of an early thermal dynamo. On the other hand, the higher the volatile content is, the later the onset of inner-core growth occurs (e.g.,

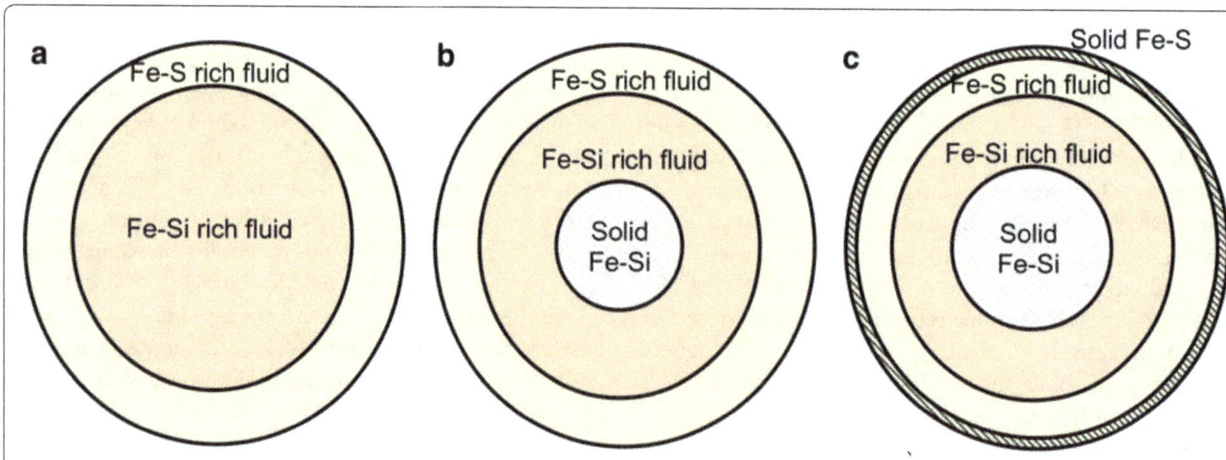

Fig. 6 Structure of an Fe–S–Si core in Mercury. Cross-section through Mercury's core, illustrating possible layered Fe–S–Si cores, depending on the temperature distribution (temperature decreases from **a** to **c**). **a** The immiscibility of S- and Si-rich liquids (Sanloup et al. 2000; Morard and Katsura 2010) up to 14 GPa and the lower density Fe–S-rich phase compared with the Fe–Si-rich phase (Badro et al. 2007) leads to a segregation of S-rich liquids (with almost no Si) near the top of the core, with more Si-rich liquid materials (with almost no Si) at greater depths. For pressures higher than 14 GPa, a liquid layer with Fe–Si–S can, in principle, be present, but such a configuration is unstable, since Si liquids that are located higher up are denser and sink into the core, resulting in the final structure shown. **b** The smaller melting point depression of Fe–Si alloys (e.g., Kuwayama and Hirose 2004) compared with Fe–S alloys (e.g., Usselman 1975; Fei et al. 1997; Li et al. 2001; Stewart et al. 2007; Chen et al. 2008) suggests that a solid Fe–Si core grows first. Alloys of Fe and Si form a solid solution with small compositional differences between the liquid and solid (Kuwayama and Hirose 2004). **c** A solid FeS layer forms at the CMB even when the temperature decreases below 1600–1700 K

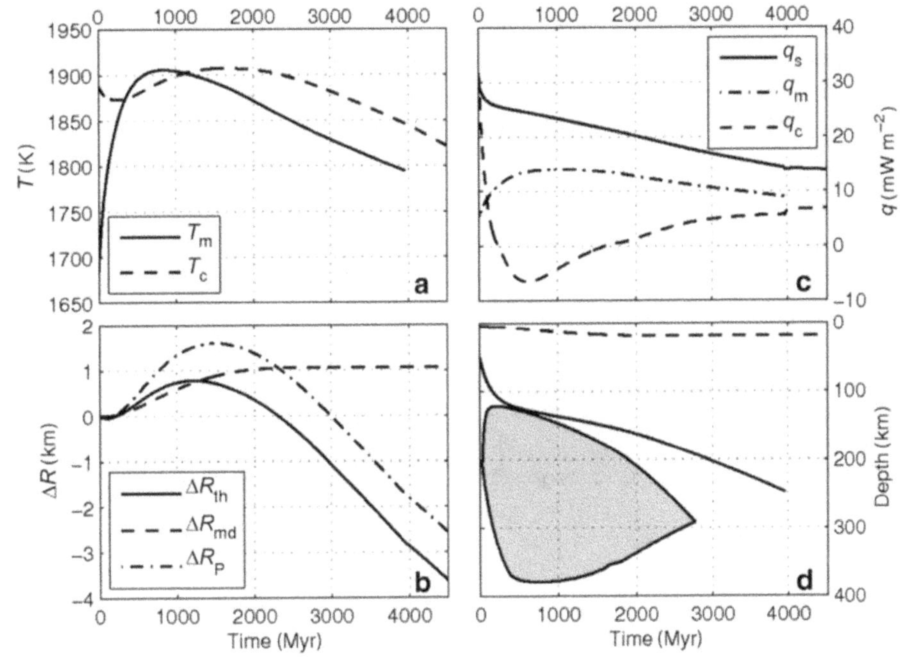

Fig. 7 Thermochemical evolution of Mercury. Representative **a** time evolution of mantle temperature and CMB temperature; **b** time evolution of the planetary radius change caused by thermal expansion/contraction of the mantle and core (*solid line*), mantle differentiation (*dashed line*), and sum of the two contributions (*dash-dotted line*); **c** time evolution of the surface heat flux (*solid line*), mantle heat flux (*dash-dotted line*), and core heat flux (*dashed line*); **d** time evolution of the thickness of the secondary crust (*dashed line*), the stagnant lid (*solid line*), and the region where partial melting occurs (*shaded gray*). Figure adapted from Tosi et al. (2013)

Schubert et al. 1988; Hauck et al. 2004; Grott et al. 2011; Tosi et al. 2013). For example, the core starts to freeze after 0.75, 0.8, and 2.5 Gyr for an initial concentration of 0.2, 1, and 5 wt.% S, respectively, assuming the same mantle parameters (Schubert et al. 1988).

The existence of a solid inner core places constraints on the volatile content of the core and an upper limit to the concentration can be derived depending on the radioactive heat source density in the mantle, the thermal history, the nature of the volatiles, and the details of the melting relations. The amount of contraction of the planet derived from the distribution, lengths, and heights of lobate scarps (e.g., Strom et al. 1975; Byrne et al. 2014), on the other hand, places constraints on the minimum amount of volatiles in the core, as well as on the cooling rate of the planet since the late heavy bombardment (LHB; Hauck et al. 2004; Breuer et al. 2007; Grott et al. 2011; Tosi et al. 2013). Before MESSENGER, the distribution and morphology of the scarps were thought to imply an average contraction of the planet's radius by only 1–2 km (Strom et al. 1975) since the LHB. From MESSENGER camera observations, a larger value of 4.2–6.3 km has recently been inferred (Byrne et al. 2014).

The most likely source of global contraction is a combination of thermal contraction (caused by cooling) and a volume change through solidification of parts of the core, with core freezing contributing more effectively. Complete solidification of a purely iron core would shrink the planet by about 17 km in radius (Solomon 1976). This effect can be considerably amplified if a light alloying element such as sulfur is present in the core, which is expelled from the inner core upon solidification, increasing the density difference between the liquid and solid core phases. The radius of the inner core can thus be taken as roughly proportional to the amount of planetary contraction (Hauck et al. 2004; Breuer et al. 2007; Grott et al. 2011; Tosi et al. 2013). Considering a value of ~7 km (Byrne et al. 2014) and assuming that it is entirely caused by the volume change owing to inner-core growth, an inner-core radius of ~800–1000 km (0.4–0.5 R_c; Grott et al. 2011) is obtained. The inner solid core is likely smaller, however, because of secular cooling contributing to the radial contraction. A larger solid inner core can only be reconciled with the observed contraction data if most of the inner core formed during the first 500 Ma before the LHB (e.g., Schubert et al. 1988), thereby erasing the geological evidence of early contraction.

Models that consider the combined thermal, magmatic, and tectonic evolution of Mercury (Hauck et al. 2004) suggest a bulk core sulfur content between 6.5 and 7.5 wt.% and a late formation of the inner core since 1 Ga. These authors further argued that acceptable models will require a dry-olivine mantle rheology and that heat production is provided by thorium only, as predicted by the formation model with late evaporation of silicates (Fegley and Cameron 1987). The more recent models of Grott et al. (2011) and Tosi et al. (2013) predicted similar values of the sulfur content (6–10 wt.%) to explain the global contraction, even when considering a non-refractory composition consistent with the MESSENGER gamma-ray spectrometer data (Peplowski et al. 2011, 2012) if the insulating effect of a regolith layer is included (see Fig. 7b). Combined, the results of thermochemical evolution models suggest that dynamo action in Mercury to produce the present field is recent and that the observed early magnetic field episode was likely brief.

The classical hydromagnetic dynamo, as assumed in the thermal evolution models above, has problems to explain the strength of Mercury's magnetic field, however. Estimates from scaling laws such as Eq. 9 above suggest a much stronger field than observed (Stevenson et al. 1983; Schubert et al. 1988; Wicht et al. 2007). Other models that predict a strong quadrupole component (e.g., Christensen 2006; Christensen and Wicht 2008) imply a large dipole tilt, which has not been observed. A recent review of the various dynamo models for Mercury can be found in Wicht and Heyner (2014). We only provide a summary here.

Thin-shell dynamo models with a large inner core (relative inner-core size larger than about 0.8) can produce magnetic fields with Mercury-like field strengths (Heimpel et al. 2005; Stanley et al. 2005), but these models are not consistent with geodetic and tectonic observations (Grott et al. 2011; Dumberry and Rivoldini 2015). Alternatively, a deep dynamo with a small inner core (of less than 1000 km in radius) has been proposed by Heimpel et al. (2005), Christensen (2006), and Christensen and Wicht (2008). In the latter two models, the dynamo operates below a stable conducting layer in the outermost core. The field is then strongly attenuated by the electromagnetic skin effect of the stagnant (i.e., no vertical movement) but electrically conducting layer. However, the stagnant layer may be destabilized by double diffusive effects, which can only be avoided if the sulfur concentration in Mercury's core is very small, only a fraction of a percent (Manglik et al. 2010). A core with such a low sulfur concentration is again not likely to be consistent with the geodetic and tectonic evidence (Hauck et al. 2004; Grott et al. 2011; Tosi et al. 2013).

These dynamo models assume an Earth-like dynamo with inner-core freezing starting at the center. However, as discussed before, recent data about the melting relations of Fe–FeS at moderate pressures suggest a negative liquidus temperature gradient in the outer layers of the Hermean core. The present core would then most likely be precipitating solid iron in the form of snow at a single depth or in two distinct zones (Chen et al. 2008). Such a scenario predicts a stably stratified layer in the outermost core, as required to explain the geometry and magnetic field strength through the electromagnetic skin effect (Christensen 2006; Christensen and Wicht 2008; Vilim et al. 2010). However, dynamo models that include an iron snow layer are strongly time dependent and generate Mercury-like magnetic fields only for a fraction of the total simulation time (Wicht and Heyner 2014). Large-scale variations in the CMB heat flux may help to stabilize strong Mercury-like quadrupole terms (Cao et al. 2014), but these are not consistent with the current understanding of Mercury's lower mantle (e.g., Tosi et al. 2013).

Motivated by the findings produced by the MESSENGER XRS instrument (Nittler et al. 2011), Si has been suggested as a core phase in addition to S (Malavergne et al. 2010), which may have strong implications for the core structure and evolution (Smith et al. 2012; Hauck et al. 2013). High-pressure experiments with the Fe–S–Si ternary system (e.g., Malavergne et al. 2010) have revealed an immiscibility of S- and Si-rich liquids up to a pressure of ~14 GPa (Sanloup and Fei 2004; Morard and Katsura 2010) and smaller density differences between liquid and solid phases compared with those in the Fe–S binary system (Kuwayama and Hirose 2004). These findings suggest a compositional stratification in the core as S-rich liquids would segregate toward the CMB, leaving more Si-rich materials at greater depths (Malavergne et al. 2010; Smith et al. 2012; Hauck et al. 2013; see Fig. 6). Hauck et al. (2013) even suggest that the outer FeS layer is solid and part of the solid outer shell (crust, mantle, and upper core), whose thickness has been estimated from the moment-of-inertia data. This would imply a mantle even thinner than 400 km and require core–mantle temperatures lower than that for the liquidus of pure FeS (1600–1700 K, depending on the pressure at the CMB; Boehler 1992). These temperatures are not consistent with the most recent thermal evolution models (Tosi et al. 2013; see Fig. 7a). The implications of an Fe–S–Si system for the dynamo are, however, unknown.

The present magnetic field data do not allow us to conclusively prove or reject the various dynamo models and thermal history calculations. More details about the magnetic field are expected from the upcoming BepiColombo mission, and new constraints may be derived for the dynamo. BepiColombo consists of two orbiters in different orbits (characterized by small and large eccentricities), both equipped with magnetic field sensors that may help to better disentangle the magnetic field of the planet.

The moon

Lunar paleomagnetic data suggest a magnetizing field of up to ~100 μT between at least 4.25 and 3.56 Ga b.p., a field similar in strength to the present terrestrial magnetic field (Cisowski and Fuller 1986; Garrick-Bethell et al. 2009; Shea et al. 2012; Tikoo et al. 2012; Suavet et al. 2013). The field strength decreased by an order of magnitude between 3.56 and ~3.3 Ga b.p. (Tikoo et al. 2012, 2014) and possibly persisted until 3.19–1.3 Ga b.p. (Tikoo et al. 2014) or even until 200 Ma b.p. (Wieczorek et al. 2006) at a level of a few nano-Teslas (see Fig. 8). The paleomagnetic data younger than 3.19 Ga are, however, also consistent with no field, and further laboratory analysis is needed. Alternative explanations to a dynamo-generated global field for the lunar magnetic record suggest that at least part of the remanent magnetization was acquired from plasma clouds that had been generated by impacts antipodal to the magnetic anomaly (Hide 1972; Hood and Vickery 1984; Hood and Huang 1991; Hood et al. 2001). A comprehensive review of the Apollo-era measurements and recent paleomagnetic studies can be found in Weiss and Tikoo (2014).

A lunar dynamo requires an iron-rich core, whose existence is largely uncontroversial. A core of 250–450 km radius is consistent with both the lunar polar moment of inertia (Konopliv et al. 1998) and electromagnetic sounding data from the Lunar Prospector (Hood et al. 1999; Shimizu et al. 2013). Reanalyses of Apollo seismic data (Garcia et al. 2011; Weber et al. 2011) have confirmed the existence of a core with a radius between 340 and 420 km but disagree on the existence of a solid inner core. Recent analysis of lunar laser ranging data by Williams and Boggs (2015) required a region of high dissipation with a radius ≥535 km, consistent with a (partially molten?) high-dissipation layer on top of a core (see also Weber et al. 2011).

The paleomagnetic data show characteristic features that any magnetic field generation model needs to explain. First, there is no remanently magnetized rock older than 4.25 Ga.

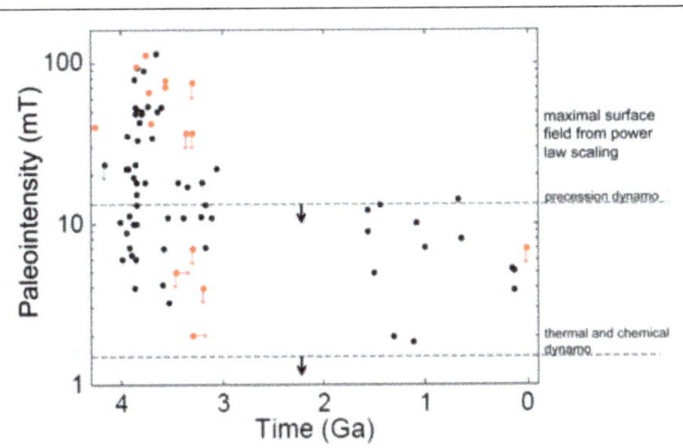

Fig. 8 Paleomagnetic data of the Moon. Summary of lunar paleomagnetic data. *Black points* represent Apollo-era measurements (Fuller and Cisowski 1987; Wieczorek et al. 2006) and *red points* indicate modern, more reliable data produced using the isothermal remnant magnetization (IRM) and Thellier–Thellier methods (Weiss and Tikoo 2014). *Arrows* indicate data points that only provide an upper limit to the field strength and/or age. The two *horizontal dashed lines* represent the calculated maximum surface field obtained from the power scaling law (Eq. 9) for both precession dynamo (Dwyer et al. 2011) and thermal and/or chemical dynamo (Evans et al. 2014; Laneuville et al. 2014). Figure adapted from Scheinberg et al. (2015)

Two alternative explanations have been suggested. (1) The remanent magnetization of older crustal rock could have been destroyed by impact gardening and/or volcanic activity in the first 250 Ma failed to produce enough rock that could have acquired significant remanent magnetization. Alternatively, (2) the internal field may have been triggered only about 250 Ma after core formation. Second, the paleointensity between 3.56 and 3.85 Ga b.p., inferred for the magnetizing field from the measured magnetization, was about two orders of magnitude higher than suggested by estimates from magnetic field scaling laws for typical thermal or chemical dynamos (e.g., Evans et al. 2014; Laneuville et al. 2014; Weiss and Tikoo 2014). Finally, no lunar dynamo model exists to date that is capable of explaining the rapid decline of the field intensity with time until 3.3 Ga b.p. while continuing to produce a weak field thereafter (but still one order of magnitude larger than that from conventional dynamo scaling models). The latter predict fields of about 1 μT or smaller.

Most previous thermal and magnetic field evolution models have considered a thermally driven dynamo during the early evolution of the Moon (Konrad and Spohn 1997; Spohn et al. 2001; Laneuville et al. 2014). These models show that a thermally driven dynamo would have been active since core formation and may have lasted a maximum of 500 Myr, until 4 Ga b.p., if the core was superheated by a few hundred degrees with respect to the mantle. These thermal evolution models, however, neglect any chemical stratification of at least the lower mantle. A recent study by Evans et al. (2014) showed a long-standing thermal dynamo of up to 2 Gyr if the lower mantle rheology is weakened by water, but they also assume a low thermal conductivity of the iron-rich core of 25 W m^{-1} K^{-1} and an adiabatic heat flux of 2.4 mW m^{-2}.

Models of a late onset of dynamo action invoked the thermal and chemical consequences of a magma ocean (Stegman et al. 2003; Zhang et al. 2013). After crystallization of the magma ocean, a dense ilmenite and pyroxene cumulate layer, located just below the enriched KREEP layer, sinks to the deep lunar mantle and thermally insulates the core. (The KREEP layer is enriched in potassium (K), rare earth elements (REE), and phosphorous (P).) The core is thus prevented from cooling convectively and also from developing a dynamo. Eventually, because of the heating by radiogenic elements, the dense lower mantle becomes buoyant and convection currents rise toward the surface. With the removal of the thermal blanket, the core begins to convect vigorously, cooling and producing a short-lived thermal dynamo, although the heat flux from the core is less than 8 mW m^{-2} in that scenario and, therefore, is just at the limit of or even below the critical heat flux along the adiabat. The period it takes the thermal blanket to heat up and rise back toward the surface is essential in these models and is broadly consistent with the time of the eruption of the mare basalts and the magnetization of lunar rocks between 4.25 Ga and 3.55 Ga b.p.

Other models explain the magnetic signatures by mechanically stirring the core through either precession (Williams et al. 2001; Dwyer et al. 2011) or changes in the lunar rotation rate from large impacts (Le Bars et al. 2011). In particular, the precession dynamo (see Fig. 9) can be long lasting (until 3.4–2.5 Ga b.p.) and may be capable of producing the strong surface magnetization, since the power caused by differential motion can be much larger compared with the power driving thermal or chemical convection (Tian et al. 2014). Although the precession dynamo can be viable for the strong early lunar field, the possible long-lasting and weak field until recent times may still require an alternative explanation.

A recent interpretation of Apollo seismic data suggested that the core may be stratified with a liquid outer and a solid inner core (Weber et al. 2011) of 240 km radius. The core radius is 330 km in this model. If correct, a chemical dynamo may become a viable alternative, a possibility that has been discarded by previous evolution models (e.g., Konrad and Spohn 1997). These latter authors argued that the absence of a present-day magnetic field would be difficult to reconcile with a growing inner core and a chemical dynamo supporting inner-core growth would be difficult to stop in a continuously cooling planet. This is consistent with the conclusion of Laneuville et al. (2014) but inconsistent with that of Scheinberg et al. (2015). In the former study, however, the authors neglected the partial derivative of the liquidus temperature owing to light-element enrichment in the outer core, similar to the approach of Nimmo (2007). Considering this contribution, sufficient power is indeed not available to drive a present-day dynamo, even with a growing inner core (Scheinberg

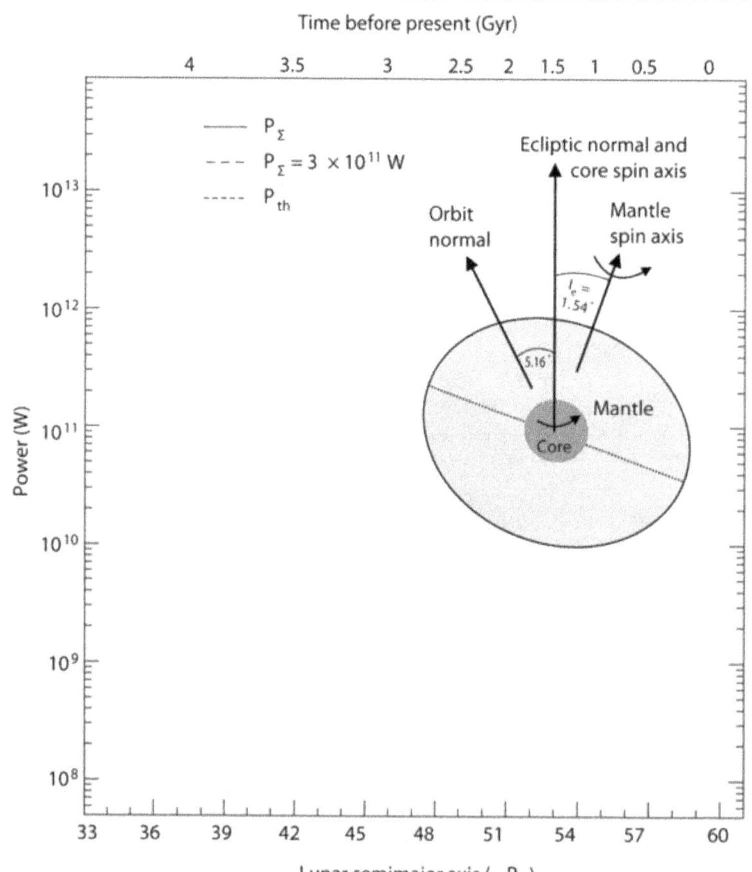

Fig. 9 Precession dynamo: the total power deposited into the lunar core. The *solid line* shows the total power deposited into the lunar core, P_Σ, as a function of semimajor axis (bottom *x* axis) for a precession dynamo: continuous mechanical stirring caused by the differential motion, due to Earth-driven precession of the lunar spin axis, between the solid silicate mantle and the liquid core below. The adiabatic threshold value is marked as P_{th} (*dashed line*). The dash-dotted line is for $P_\Sigma = 3 \times 10^{11}$ W, which represents an estimate of the power available to drive the dynamo in the Earth. The time before the present is plotted at the top. There is 7×10^{28} J of energy available for dynamo generation between the time of the Cassini state transition (at a lunar semimajor axis of 34 R_E) and the cessation of the dynamo (when the power decreases to below P_{th}, i.e., ~2.8 Ga b.p.). The results suggest that a mechanically driven dynamo could persist for at least ~1.6 Gyr. The inset shows the geometry of the situation. The mean motion, *n*, of an orbit is related to the semimajor axis, *a*, through $n = (GM_e/a^3)^{0.5}$, where M_e is the mass of Earth. Figure after Dwyer et al. (2011)

et al. 2015). The growth rate of the core and, therefore, the compositional buoyancy becomes too small to overcome the stable temperature gradient for subadiabatic core cooling. These models, which assume Earth-like core crystallization, further suggest a field strength that is two orders of magnitude too low to explain the observed early strong magnetization and one order too low to explain the weaker, younger magnetization (Laneuville et al. 2014). Alternatively, for a sulfur content above about 8 wt.%, iron snow fall is a likely mechanism for core solidification (Laneuville et al. 2014), as has been suggested for Ganymede (e.g., Hauck et al. 2006), Mars (Stewart et al. 2007), and Mercury (Chen et al. 2008). In the iron snow regime, the formation and growth of the inner core would mark the demise of dynamo action (Christensen 2015; Rückriemen et al. 2015).

Mars

Since the Mars Global Surveyor (MGS) mission (Acuña et al. 1998, 1999, Mitchell et al. (2001); Connerney et al. 1999), it has been known that the crust of Mars is remanently magnetized (see Fig. 10) and that the planet must have had an active magnetic field of significant strength some time during its history, most likely in the Early Noachian. The magnetization of up to ~10–30 A m^{-1}, comparable in magnitude to the remanent magnetization of fresh extrusive basalt near the oceanic ridge axes of Earth (Bleil and Petersen 1983), suggests a magnetizing field of 0.1 to 10 times that of the present Earth (Mitchell et al. 1999, 2001; Ness et al. 1999), depending on the mineralogy and the magnetic microstructure of the Martian crust. The magnetic history of the planet has been reviewed in a number of publications (e.g., Connerney et al. 2004; Breuer et al. 2010; Breuer and Moore 2015; Connerney 2015), some of which discuss the subject in comparison with the histories of other terrestrial planets, and we will therefore only briefly summarize the present knowledge.

The most popular model to explain an early magnetic field for Mars and the absence of a present dynamo field goes back to the work of Stevenson et al. (1983) and Schubert and Spohn (1990) who proposed, on the basis of their thermal history models, that the Martian core may be lacking an inner core and that an early thermally driven dynamo had ceased to operate after a few hundred million years of evolution. The model was subsequently refined (e.g., Spohn et al. 1998; Hauck and Phillips 2002; Breuer and Spohn 2003, 2006; Williams and Nimmo 2004; Michel and Forni 2011) and is roughly consistent with a dynamo lifetime of 500 Ma, as inferred from the magnetization pattern (Fig. 11). The timing has been questioned, however, by Schubert et al. (2000) who argue for younger magnetized rock on the surface, as suggested by the timing of magnetized volcanic units by Hood et al. (2010). The hypothesis of a magnetic field at epochs later than the Early Noachian is supported by the magnetization, albeit weak, of Martian meteorites with formation ages of 1.3 Ga to 180 Ma. Their magnetization is consistent with a weak surface field of a few hundred to a few thousand nano-Teslas (Collinson 1997; Rochette et al. 2001). Although it cannot be excluded that the dynamo became somehow reactivated, it is more likely that the young Martian meteorites were magnetized by the older, strongly magnetized crust rock (for a review, see Connerney et al. 2004).

The hypothesis of the absence of an inner core can be tested based on geochemical evidence for the composition of the Martian core. A molten core for the entire evolution is possible if the core contains at least ~5 wt.% of sulfur (Williams and Nimmo 2004), although the more recent Fe–FeS melting data of Stewart et al. (2007) suggest a sulfur threshold of at least 10 wt.% or even higher, depending on the efficiency of core cooling (see Fig. 12). In the latter study, it was also suggested that iron snow would become active in the future of Mars because of further cooling. A core sulfur concentration of 14.2 wt.% has been suggested (Dreibus and Wänke 1985; McSween 1985) based on chemical analysis of SNC meteorites. The estimate assumed, however, that the all available sulfur went into the core during core formation. Recent experiments showed that it is impossible to simultaneously account for the Martian mantle depletion of moderately siderophile elements if the Martian core sulfur content exceeds 10.5 wt.% under reducing conditions; i.e., one logarithmic unit below the iron-wüstite buffer (Rai and van Westrenen 2013). Note that these high sulfur concentrations also indicate the partitioning of oxygen into the core (Tsuno et al. 2011).

Alternative scenarios that explain the magnetic field history invoke changes in tectonic style and in the heat transport mechanism, from plate tectonics to stagnant lid (Nimmo and Stevenson 2000; Stevenson 2001; Breuer and Spohn 2003; see Fig. 11), or consider the effects of impacts. Some of the latter models suggest that a late impact may have heated the mantle to the extent that convection and dynamo action in the core stopped (e.g., Roberts et al. 2009; Arkani-Hamed and Olson 2010), while Reese and Solomatov (2010) argued that a late large impact may have triggered the onset of the dynamo. Which of these conflicting scenarios is more likely depends mainly on the dynamics in the core during the merging of an iron diapir with a preexisting core, a process that is poorly understood (Monteux et al. 2013).

The dichotomy of the observed magnetization, with the southern hemisphere being more strongly magnetized than the northern hemisphere (e.g., Langlais et al. 2004; Morschhauser et al. 2014), is an important feature that must be explained. Two general types of models have been suggested. A once more or less homogeneous magnetization may have been destroyed in the

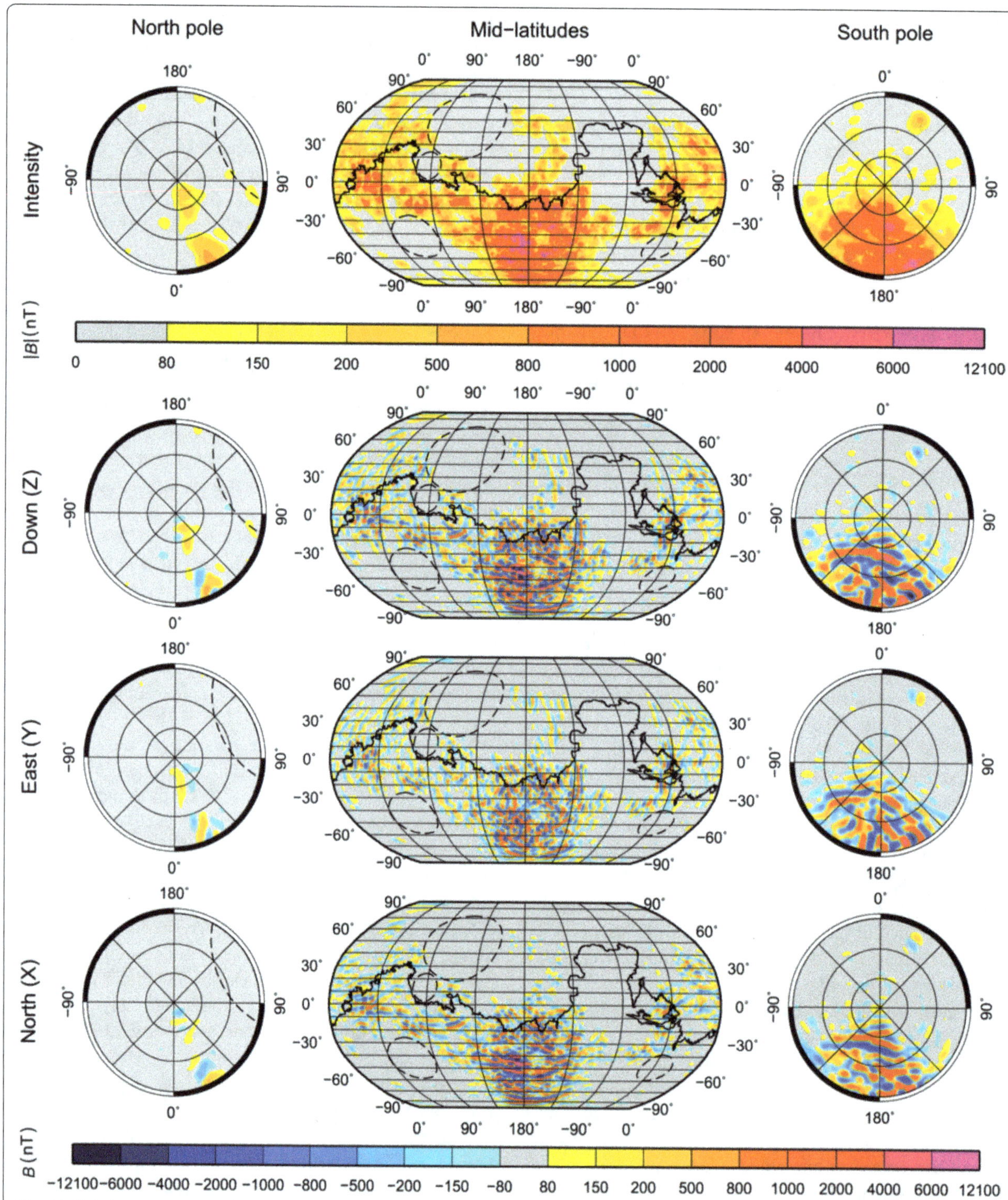

Fig. 10 Global magnetic map of Mars. Map of the lithospheric part of the magnetic field model evaluated at the mean planetary radius of 3393.5 km. (*top* to *bottom*) The magnetic field intensity and the vertically down (*Z*), horizontally east (*Y*), and north (*X*) components are shown. (*left* to *right*) The north pole, mid-latitudes, and south pole are presented. Major impact basins are indicated by *dashed lines*, while the *thick black line* represents the zero level of the Mars Observer Laser Altimeter (MOLA) topography, indicating the topographic dichotomy boundary (after Morschhauser et al. 2014)

northern hemisphere by, for example, hydrothermal alteration (Solomon et al. 2005) or impacts (e.g., Rochette et al. 2003; Mohit and Arkani-Hamed 2004). The alternative explanation suggests that the dynamo itself produced a hemispherical field owing to lateral heat flow variations at the CMB (Stanley et al. 2008; Amit et al. 2011). This

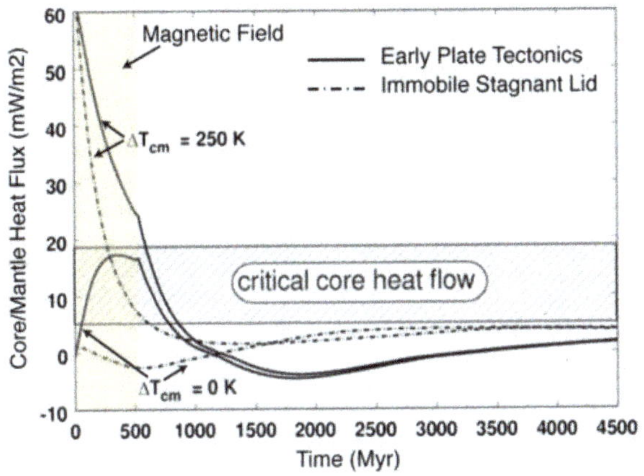

Fig. 11 Evolution of Martian core heat flow. Core–mantle heat flow as a function of time for models with early plate tectonics (*solid line*) and stagnant-lid convection throughout their entire evolution (*dashed line*), with initial temperature differences across the CMB of $\Delta T_{cm} = 0$ K and $\Delta T_{cm} = 250$ K. The *gray region* marks the range of the critical core–mantle heat flux along the adiabat and the *orange region* the suggested time period of the global magnetic field (after Breuer and Spohn 2003)

hemispherical dynamo would produce strong magnetic fields in the southern hemisphere if the heat flow from the core below the southern hemisphere was stronger than that below the northern hemisphere. A recent numerical model by Dietrich and Wicht (2013) showed, however, that these hemispherical dynamos reverse on timescales of about 10 kyr: too quickly to allow for the unidirectional magnetization of a thick crustal layer required to explain the strong magnetization in the southern hemisphere.

Ganymede

In 1996, the Galileo mission measured a magnetic field around the Jovian moon Ganymede (Kivelson et al. 1996), which was interpreted as a field of internal origin embedded in the Jovian magnetic field. The field would have an equatorial surface field strength of approximately 719 nT (Kivelson et al. 2002). The discovery of a self-sustained magnetic field at Ganymede came as a surprise, since it was commonly believed that the icy moon—with an average density of 1936 kg m^{-3}—was composed of a rocky core surrounded by a large mantle of water or water ice with an ice surface. The magnetic field has an anomalously low quadrupole moment relative to its dipole moment (Kivelson et al. 2002), placing a strong constraint on the dynamo mechanism.

The origin of Ganymede's magnetic field is most likely found in an iron-rich core, because alternative mechanisms such as remanent magnetization of a silicate layer containing magnetic minerals below the ice mantle or a dynamo in an electrically conducting, salty subsurface ocean are implausible (Schubert et al. 1996). For remanent magnetization, both the concentration of magnetic minerals and the external field (i.e., the present Jovian magnetic field at Ganymede's orbit, ~100 nT) are too low to reasonably explain the observed field. On the other hand, the flow velocity in its ocean must be implausibly large, on the order of 1 m s^{-1}, to explain the observed magnetic field through a dynamo in a subsurface ocean.

The existence of an iron-rich core is further supported by the small mean moment-of-inertia factor (MoI = 0.3105) derived from Galileo gravity and rotation data (e.g., Anderson et al. 1996; Sohl et al. 2002; Schubert et al. 2004). Fixing the

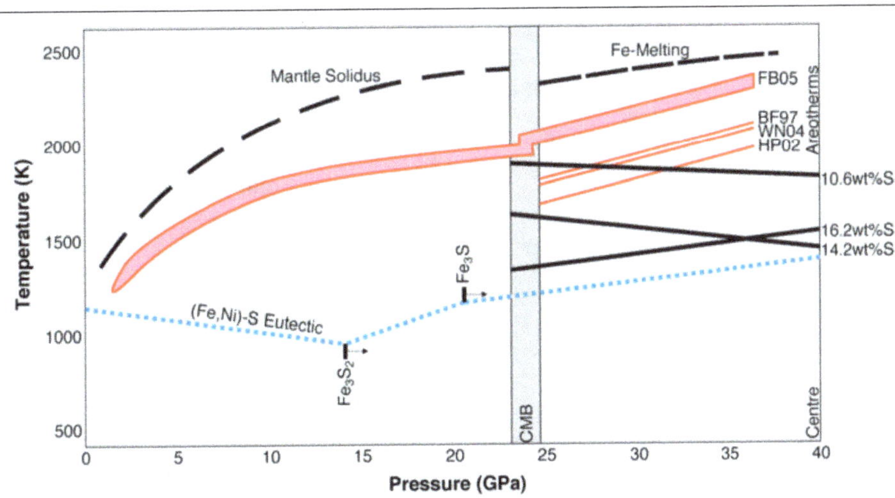

Fig. 12 Melting temperature of Mars core. Effect of pressure on the (Fe, Ni)–S system. Eutectic melting temperatures for (Fe, Ni)–S with Fe:Ni = 10.6:1 are shown as *dotted blue lines*. Eutectic temperature inflections around 14 and 20 GPa are owing to the stabilization of Fe_3S_2 and Fe_3S above these pressures. Liquidus temperatures for core S contents of 10.6, 14.2, and 16.2 wt.% are presented as *solid black lines*. The 16.2 wt.% curve has a positive slope, because its S content is higher than eutectic. Shown is the present-day mantle temperature profile and core adiabat of Fei and Bertka (2005), as well as the core adiabat based on CMB temperature determinations by Bertka and Fei (1997), Williams and Nimmo (2004), and Hauck and Phillips (2002) using parameters from Williams and Nimmo (2004). The pressure of the CMB, the mantle solidus, and the melting curve of pure Fe (Fei and Bertka 2005) are given for reference (after Stewart et al. 2007)

silicate and ice-shell densities to plausible values and allowing for core compositions ranging from pure iron to pure FeS, Sohl et al. (2002) found core radii between $^1/_4$ and $^1/_3$ Ganymede radii for a three-layer model (see Fig. 13). As a light and siderophile element, sulfur is the main candidate (e.g., Schubert et al. 2004). Assuming a composition identical to a type I carbonaceous chondrite (CI), experiments by Scott et al. (2002) indicate that the alloying of iron and sulfur may form FeS-dominated cores. However, because of the unknown oxidation state of Ganymede's interior during core differentiation, it is difficult to say whether the core composition is likely to be of Fe- or FeS-rich eutectic composition.

Compositional-driven convection is the preferred mechanism for driving Ganymede's dynamo (Hauck et al. 2006; Bland et al. 2008) although Kimura et al. (2009) have suggested a model including a present-day thermal dynamo. This model could work if Ganymede accreted cold and was subsequently heated by the decay of radioactive elements. They assumed a thermal expansivity of $1.7\,10^{-5}$ K^{-1}, which is, unfortunately, not consistent with recent data for liquid iron at Ganymede's core pressure (e.g., Williams 2009), and they used a heat flux along the adiabat as small as 1 mW m^{-2} to meet the necessary condition for a thermal dynamo.

The melting behavior of Fe–FeS alloys at Ganymede's core pressure, between 6 and 10 GPa (Sohl et al. 2002),

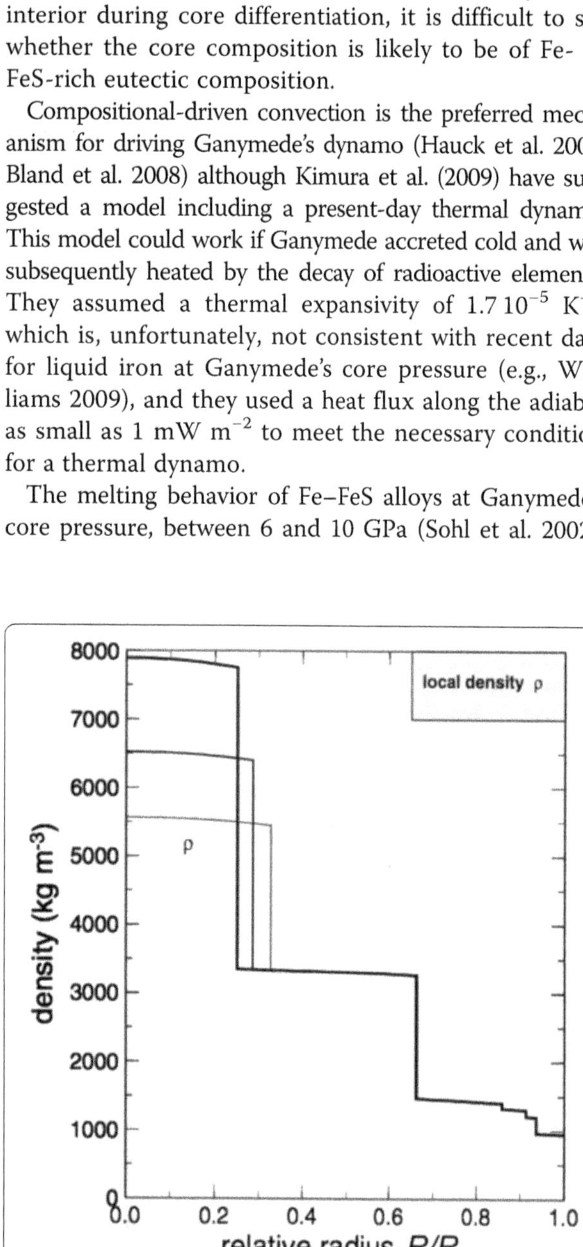

Fig. 13 Interior structure of Ganymede. Radial distribution of density as a function of the relative radius (R/R_p) for variable core compositions of Ganymede (*bold curve* 100 wt.% Fe; *semi-bold curve* 50 wt.% Fe and 50 wt.% FeS; *light curve* 100 wt.% FeS). Figure after Sohl et al. (2002)

motivated Hauck et al. (2006) to consider the iron snow model for Ganymede. In addition, they introduced an FeS flotation model (see above). Assuming a linear interpolation between the iron liquidus and the Fe–Fes eutectic temperature, they found that Fe snow can occur for bulk core sulfur concentrations greater than 6 wt.% and FeS flotation for bulk core sulfur contents above 22.5 wt.%. Recent data by Buono and Walker (2011), however, suggest that the iron snow regime should work even at sulfur concentrations below 6 wt.%. Because of the pressure gradient in the core, it is possible that the iron snow regime is established in the outer core, while an inner core begins to freeze independently at the center for certain ranges of sulfur concentration (Rückriemen et al. 2015). With cooling, the crystallization mechanisms can evolve further, depending on the bulk sulfur content (see Fig. 14).

The implications of the Fe snow regime for the core dynamics and magnetic field evolution are controversial. Hauck et al. (2006) proposed that the sinking iron particles may cause motions within the fluid of the stable snow zone and, thereby, provide the driving force for the magnetic field of Ganymede. By applying the magnetic field strength scaling law of Christensen and Aubert (2006), they found surface magnetic field strengths between 600 and 1600 nT. However, Hauck et al. (2006) failed to explain why particle settling in a stably stratified fluid would cause the large-scale fluid motions that are necessary for a dynamo (Christensen and Wicht 2007). Zhan and Schubert (2012) used a magneto-hydrodynamic dynamo model and assumed that the sinking of iron crystals in the snow zone would be equivalent to internally heated convection. They found multipolar solutions for the magnetic field, at variance with the observed strong dipolarity of the field (Kivelson et al. 2002). Their main assumption that Fe snow resembles internally heated convection is questionable, because they did not model the particular flow. These authors suggested that dipole-dominated magnetic fields are only possible for FeS floating dynamos and inner-core growth, the latter implying a low concentration of sulfur.

Zhan and Schubert's (2012) conclusions are consistent with the thermal and magnetic field evolution models of Bland et al. (2008). The former authors suggested that the power requirement for dynamo action can only be satisfied if the sulfur mass fraction in Ganymede's core is either very low (<3 wt.%, with freezing starting at the core center) or very high (>22.5 wt.%; i.e., solidification of FeS deep within the core). They further argued that the conditions for magnetic field generation cannot be met in the iron snow regime, because the latent heat released by the solidification of iron would readily be removed by the mantle above and thus be unavailable to power the dynamo. However, the assumption of latent heat being released close to the CMB is only correct initially, when the

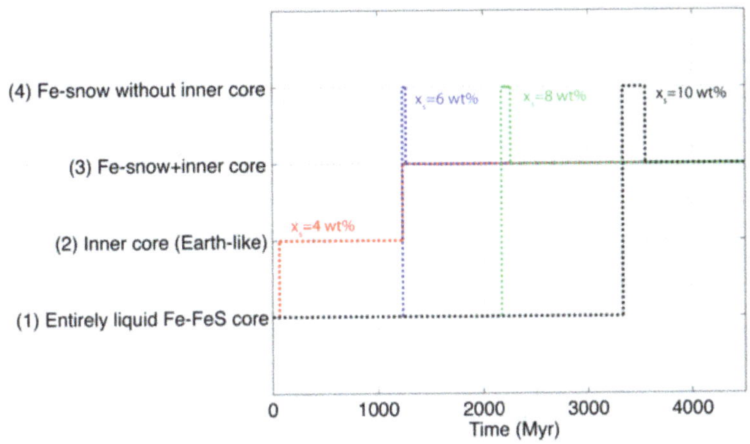

Fig. 14 Evolution of crystallization regimes in Ganymede's core. Evolution of the core state for four bulk core sulfur contents and CI chondritic heat production. The model with 4 wt.% sulfur passes through a period with an Earth-like core evolution prior to beginning precipitation of Fe snow. With increasing sulfur content, crystallization starts later because of the lower melting temperature and commences directly with Fe snow until a solid inner core starts to grow (after Hauck et al. 2006)

snow zone starts to grow. For continuous cooling, the snow zone extends into deeper core regions, thereby triggering the release of latent heat in the deeper core as well.

In a recent study, Christensen (2015) proposed that the deeper, entirely fluid core below the zone snow could be the locale of the dynamo. Compositional convection would be driven by the sinking and remelting of iron crystals in the lower core region underlying an electrically conducting snow zone with a strongly stabilizing density gradient. For plausible values of the buoyancy flux from Rückriemen et al. (2015) and an upper stable layer that is thinner than the fluid layer below, the model can reproduce Ganymede's observed dipole. Such a dynamo would be restricted in its lifetime by the time the snow zone takes to grow through the core. Rückriemen et al. (2015) found values of 320–800 Myr, depending mainly on the initial core sulfur content and the heat flow from the core. To best explain the present magnetic field with such a model, high sulfur concentrations (but lower than the eutectic concentration) were favored. The dynamo would then start late in the evolution and could easily last to the present day. Moreover, the model would predict the absence of a solid inner core.

Magnetic field generation for a core composition on the sulfur-rich side of the eutectic has been studied less often. FeS floating, where FeS crystals form first in the core center and rise, is considered a viable mechanism for dynamo action in Ganymede (Hauck et al. 2006; Bland et al. 2008; Zhan and Schubert 2012). Thermal evolution models showed that the power requirements for a dynamo can be met (Bland et al. 2008) and magneto-hydrodynamic models showed that dipole-dominated magnetic fields are possible with this specific freezing mechanism, since buoyancy sources are located at the bottom of the convective zone, similar to the Earth-like case (Zhan and Schubert 2012). However, at Ganymede's core pressures, FeS crystallization at the CMB or in the core is more likely (see Fig. 2) than in the center, and a solid outer-core layer would grow inward, similar to the Earth's inner-core growth but in reverse (see Fig. 1c, d). Such a dynamo could be long lasting (at least longer than the dynamo in the Fe snow regime), because compositional convection in a cooling Ganymede would only stop when the eutectic composition is reached in the liquid inner core. However, whether enough power for dynamo action can be generated with this scenario still needs to be tested.

Conclusions

Our perception of the cores of terrestrial planets and moons and their crystallization have changed substantially during the past few years. Improved Fe–FeS phase diagrams and melting relations have motivated novel models of core dynamics and dynamos, in particular for planetary bodies with masses significantly smaller than the Earth's. The most important findings are that the melting temperature at the eutectic decreases with increasing pressure up to 14 GPa and that the eutectic composition decreases with increasing pressure up to about 60 GPa (see Fig. 3). Both effects imply Fe crystallization at the CMB and iron snow fall for wide ranges of sulfur concentrations if $dT/dP_{ad} > dT/dP_{melt}$. Iron snow can provide a viable dynamo mechanism for the Moon, Mercury, and Ganymede and would even be likely for Martian core pressures up to 40 GPa, suggesting that dynamo generation may occur in the future through this mechanism. In iron-rich cores that comprise a large pressure range, as for Mercury, even several zones of iron snow can emerge. The lunar case is less clear, however, since Fe may start to crystallize in the center for a sulfur concentration below about 8 wt.%, while for higher sulfur contents, iron snow would be the dominant mechanism.

Crystallization of FeS for sulfur-rich compositions is a less well-developed scenario, but it would also differ from the classical Earth case of solid inner-core growth from the center. Here, FeS would precipitate either at the CMB, in the core, or at the core center, depending on the sulfur concentration and the temperature profile. The situation on the sulfur-rich side of the eutectic can be even more complex, because the solid phase depends on pressure and may change to Fe_3S_2 or Fe_3S, crystal phases that differ in density from FeS. In summary, a rich variety of crystallization scenarios and dynamos have become apparent and it has become clear that the Earth-like dynamo is probably not typical. While we observe this for the comparatively simple Fe–FeS system, we note that more realistic models incorporating other light elements should be even more varied and complex.

The models we have reviewed in this paper further show that better knowledge of the adiabatic temperature gradient and the adiabatic heat flux is required. First, the relation between the slopes of the adiabatic temperature and the melting temperature as a function of pressure determines the crystallization process; i.e., whether the first crystals form at the CMB, in the center, or somewhere in the core. Note, however, that it is not obvious that the core temperature must vary along an adiabat at the time of first crystal formation. For thermal evolution models, we typically attain a subadiabatic core heat flux after a few hundred million years of evolution and prior to crystallization. The core is then likely thermally stable, exhibiting a conductive core temperature profile. Second, there is an inverse correlation between the heat flux along the adiabat and the lifetime of the dynamo. In the contemporary literature, we find differences in the dynamo lifetime for either a thermal or compositional dynamo from 100 Ma to several billion years, depending on the parameters adopted for the thermal conductivity and thermal expansivity and, thus, for the adiabatic heat flux.

In conclusion, the crystallization processes in the cores of the terrestrial planets and moons are diverse and imply diversity in the mechanisms and efficiency of dynamo action. To improve our understanding, better constraints on the thermal properties, the melting relations, as well as the compositions of the cores are required through experimental work on phase diagrams and transport properties, numerical modeling, and space mission data to constrain the core compositions. Open questions to be answered are manifold. A few that result immediately from the present discussion are as follows: Can a dynamo be generated in the stable iron snow zone or only in the fluid core below? Is the concept of thermochemical equilibrium for iron snow appropriate, and what are the consequences if iron does not remelt during descent? What is the composition of the core of the terrestrial planets and moons and how does the composition alter the thermal properties and the melting temperatures?

Abbreviations
CMB: core–mantle boundary; LHB: late heavy bombardment.

Competing interests
The authors declare that they have no competing interests.

Authors' contributions
DB proposed the topic. TS and TR collaborated with the corresponding author on the composition of the manuscript. All authors read and approved the final manuscript.

Authors' information
All authors are from the DLR (German Aerospace Agency), Institute of Planetary Research, Rutherfordstrasse 2, 12489 Berlin, Germany. DB is the head of the "Planetary Physics" department, TR is pursuing her Ph.D. degree in the "Planetary Physics" department, and TS is the director of the Institute of Planetary Research.

Acknowledgements
We thank Sabine Stanley and an anonymous reviewer for excellent reviews and constructive comments. This work has been supported by the DFG (German Research Foundation) as part of the priority program "Planetary magnetism," SPP 1488.

References
Acuña MH, Connerney JEP, Ness NF, Lin RP, Mitchell D, Carlson CW, et al. Magnetic field and plasma observations at Mars: initial results of the Mars Global Surveyor Mission. Science. 1998;279:1676–80.

Acuña MH, Connerney JEP, Ness NF, Lin RP, Mitchell D, Carlson CW, et al. Global distribution of crustal magnetism discovered by the Mars Global Surveyor MAG/ER Experiment. Science. 1999;284:790_3.

Acuña MH, Connerney JEP, Wasilewski P, Lin RP, Mitchell D, Anderson KA, et al. Magnetic field of Mars: summary of results from the aerobraking and mapping orbits. J Geophys Res. 2001;106:23403–18.

Alfè D, Gillan MJ, Price GD. Complementary approaches to the ab initio calculation of melting properties. J Chem Phys. 2002;116:6170–7.

Amit H, Christensen UR, Langlais B. The influence of degree-1 mantle heterogeneity on the past dynamo of Mars. Phys Earth Planet Inter. 2011;189:63–79.

Anderson JD, Colombo G, Esposito PB, Lau EL, Trager GB. The mass, gravity field, and ephemeris of Mercury. Icarus. 1987; 71:337-349.Anderson JD, Jurgens RF, Lau EL, Slade III MA, Schubert G. Shape and orientation of Mercury from radar ranging data. Icarus. 1996;124:690–7.

Anderson JD, Jacobson RA, Elrath TP, Moore WB, Schubert G, Thomas PC. Shape, mean radius, gravity field, and interior structure of Callisto. Nature. 2001;384:157–61.

Anderson BJ, Johnson CL, Korth H, Purucker ME, Winslow RM, Slavin JA, et al. The global magnetic field of Mercury from MESSENGER orbital observations. Science. 2011;333:1859–62.

Anderson BJ, Johnson CL, Korth H, Winslow RM, Borovsky JE, Purucker ME, et al. Low-degree structure in Mercury's planetary magnetic field. J Geophys Res. 2012. doi:10.1029/2012JE004159.

Arkani-Hamed J, Olson P. Giant impact stratification of the martian core. Geophys Res Lett. 2010. doi:10.1029/2009GL041417.

Assael MJ, Kakosimos K, Banish RM, Brillo J, Egry I, Brooks R, et al. Reference data for the density and viscosity of liquid aluminum and iron. J Phys Chem Ref Data. 2006;35:285_300.

Aubert J, Labrosse S, Poitou C. Modelling the palaeo-evolution of the geodynamo. Geophys J Int. 2009;179:1414–28.

Badro J, Fiquet G, Guyot F, Gregoryanz E, Occelli F, Antonangeli D, et al. Effect of light elements on the sound velocities in solid iron: implications for the composition of Earth's core. Earth Planet Sci Lett. 2007;254:233–8.

Badro J, Côté AS, Brodholt JP. A seismologically consistent compositional model of Earth's core. Proc Natl Acad Sci. 2014;111:7542–5.
Bertka CM, Fei Y. Mineralogy of the Martian interior up to core-mantle boundary pressures. J Geophys Res. 1997;102:5251–64.
Bland MT, Showman AP, Tobie G. The production of Ganymede's magnetic field. Icarus. 2008;198:384–399.
Bleil U, Petersen N. Variations in magnetization intensity and low-temperature titanomagnetite oxidation of ocean floor basalts. Nature. 1983;301:384–8.
Boehler R. Melting of the Fe FeO and the Fe FeS systems at high pressure: constraints on core temperatures. Earth Planet Sci Lett. 1992;111:217–27.
Boehler R. Temperatures in the Earth's core from melting-point measurements of iron at high static pressures. Nature. 1993;363:534–6.
Boehler R. Fe-FeS eutectic temperatures to 620 kbar. Phys Earth Planet Inter. 1996;96:181–6.
Boehler R, Bargen NV, Chopelas A. Melting, thermal expansion, and phase transitions of iron at high pressures. J Geophys Res. 1990;95:21731–6.
Braginsky SI. Magnetohydrodynamics of the Earth core. Geomagn Aeron. 1964;4:698–712.
Breuer D, Moore WB. Dynamics and thermal history of the terrestrial planets, the Moon, and Io. In: Spohn T, Schubert G, editors. Treatise on Geophysics, vol. 10. 2nd ed. Oxford: Elsevier; 2015. p. 255–305.
Breuer D, Spohn T. Cooling of the Earth, Urey ratios, and the problem of potassium in the core. Geophys Res Lett. 1993;20:1655–8.
Breuer D, Spohn T. Early plate tectonics versus single-plate tectonics: evidence from the magnetic field history and crust evolution. J Geophys Res. 2003;108:5072. doi:10.1029/20002JE001999.
Breuer D, Spohn T. Viscosity of the Martian mantle and its initial temperature: constraints from crust formation history and the evolution of the magnetic field. Planet Sci Lett. 2006;54:153–69.
Breuer D, Hauck SA, Buske M, Pauer M, Spohn T. Interior evolution of Mercury. Space Sci Rev. 2007;132:229–60.
Breuer D, Labrosse S, Spohn T. Thermal evolution and magnetic field generation in terrestrial planets and satellites. Space Sci Rev. 2010;152:449–500.
Buffett BA. Estimates of heat flow in the deep mantle based on the power requirements for the geodynamo. Geophys Res Lett. 2002. doi:10.1029/2001GL014649.
Buono AS, Walker D. The Fe-rich liquidus in the Fe–FeS system from 1 bar to 10 GPa. Geochim Cosmochim Acta. 2011;75:2072–87.
Busse FH. Generation of planetary magnetism by convection. Phys Earth Planet Inter. 1976;12:350–8.
Byrne PK, Klimczak C, Şengör AC, Solomon SC, Watters TR, Hauck SA. Mercury/'s global contraction much greater than earlier estimates. Nat Geosci. 2014;7:301–7.
Cameron AGW, Fegley Jr B, Benz W, Slattery WL. The strange density of Mercury: theoretical considerations. In: Vilas F, Chapman CR, Matthews MS, editors. Mercury. Tucson: Univ. Arizona Press; 1988. p. 692–708.
Campbell AJ, Seagle CT, Heinz DL, Shen G, Prakapenka VB. Partial melting in the iron–sulfur system at high pressure: a synchrotron X-ray diffraction study. Phys Earth Planet Inter. 2007;162:119–28.
Cao H, Aurnou JM, Wicht J, Dietrich W, Soderlund KM, Russell CT. A dynamo explanation for Mercury's anomalous magnetic field. Geophys Res Lett. 2014;41:4127–34.
Chabot NL, Campbell AJ, McDonough WF, Draper DS, Agee CB, Humayun M, et al. The Fe–C system at 5GPa and implications for Earth's core. Geochim Cosmochim Acta. 2008;72:4146–58.
Chabot NL, Wollack EA, Klima RL, Minitti ME. Experimental constraints on Mercury's core composition. Earth Planet Sci Lett. 2014;390:199–208.
Chen B, Li J, Hauck II SA. Non-ideal liquidus curve in the Fe-S system and Mercury's snowing core. Geophys Res Lett. 2008;35:L07201. doi:10.1029/2008GL033311.
Christensen UR. A deep dynamo generating Mercury's magnetic field. Nature. 2006;444:1056–8.
Christensen UR. Dynamo scaling laws and applications to the planets. Space Sci Rev. 2010;152:565–90.
Christensen UR. Iron snow dynamo models for Ganymede. Icarus. 2015;247:248–59.
Christensen UR, Aubert J. Scaling properties of convection-driven dynamos in rotating spherical shells and application to planetary magnetic fields. Geophys J Int. 2006;166:97–114.
Christensen UR, Tilgner A. Power requirement of the geodynamo from ohmic losses in numerical and laboratory dynamos. Nature. 2004;429:169–71.
Christensen UR, Wicht J. Numerical dynamo simulations. In: Schubert G, editor. Core dynamics, Treatise on Geophysics, vol. 10. Oxford: Elsevier; 2007. p. 245–82.
Christensen UR, Wicht J. Models of magnetic field generation in partly stable planetary cores: applications to Mercury and Saturn. Icarus. 2008;196:16–34.
Chudinovskikh L, Boehler R. Eutectic melting in the system Fe–S to 44 GPa. Earth Planet Sci Lett. 2007;257:97–103.
Cisowski SM, Fuller M. Lunar paleointensities via the IRMs normalization method and the early magnetic history of the moon. In: Hartmann WK, Phillips RJ, Taylor GJ, editors. Origin of the Moon. Tucson: Univ. Arizona Press; 1986. p. 411–24.
Collinson DW. Magnetic properties of Martian meteorites: implications for an ancient Martian magnetic field. Meteoritics. 1997;803:32–6.
Connerney JEP. Planetary magnetism. In: Spohn T, Schubert G, editors. Treatise on Geophysics, 2nd ed, Vol. 10, Planets and Moons. Amsterdam: Elsevier; 2015.
Connerney JEP, Ness NF. Mercury's magnetic field and interior. In: Vilas F, Chapman CR, Matthews MS, editors. Mercury. Tucson: Univ. of Arizona Press; 1988. p. 494–513.
Connerney JEP, Acuña MH, Wasilewski P, Ness NF, Rème H, Mazelle C, et al. Magnetic lineations in the ancient crust of Mars. Science. 1999;284:794–8.
Connerney JEP, Acuna MH, Ness NF, Spohn T, Schubert G. Mars crustal magnetism. Space Sci Rev. 2004;111:1–32.
Curtis SA, Ness NF. Magnetostrophic balance in planetary dynamos: predictions for Neptune's magnetosphere. J Geophys Res. 1986;91:11003–8.
de Koker N, Steinle-Neumann G, Vlček V. Electrical resistivity and thermal conductivity of liquid Fe alloys at high P and T, and heat flux in Earth's core. Proc Natl Acad Sci U S A. 2012;109:4070–3.
Deng L, Seagle C, Fei Y, Shahar A. High pressure and temperature electrical resistivity of iron and implications for planetary cores. Geophys Res Lett. 2013;40:33–7.
Dietrich W, Wicht J. A hemispherical dynamo model: implications for the Martian crustal magnetization. Phys Earth Planet Inter. 2013;217:10–21.
Dreibus G, Wänke H. Mars: a volatile rich planet. meteoritics. 1985;20:367–82.
Dumberry M, Rivoldini A. Mercury's inner core size and core-crystallization regime. Icarus. 2015;248:254–68.
Dwyer CA, Stevenson DJ, Nimmo F. A long-lived lunar dynamo driven by continuous mechanical stirring. Nature. 2011;479:212–4.
Evans AJ, Zuber MT, Weiss BP, Tikoo SM. A wet, heterogeneous lunar interior: lower mantle and core dynamo evolution. J Geophys Res. 2014;119:1061–77.
Fegley Jr B, Cameron AGW. A vaporization model for iron/silicate fractionation in the Mercury protoplanet. Earth Planet Sci Lett. 1987;82:207–22.
Fei Y, Bertka C. The interior of Mars. Science. 2005;308:1120–1.
Fei Y, Bertka CM, Finger LW. High-pressure iron-sulfur compound, Fe_3S_2, and melting relations in the Fe-FeS system. Science. 1997;275:1621–3.
Fei Y, Li J, Bertka CM, Prewitt CT. Structure type and bulk modulus of Fe_3S, a new iron-sulfur compound. Am Minerol. 2000;85:1830–3.
Fuller M, Cisowski SM. Lunar paleomagnetism. Geomagn. 1987;2:307–455.
Gaidos E, Conrad CP, Manga M, Hernlund J. Thermodynamic limits on magnetodynamos in rocky exoplanets. Astrophys J. 2010;718:596–609.
Garcia R, Gagnepain-Beyneix J, Chevrot S, Lognonne P. Very preliminary reference Moon model. Phys Earth Planet Inter. 2011;188:96–113.
Garrick-Bethell I, Weiss BP, Shuster DL, Buz J. Early lunar magnetism. Science. 2009;323:356–9.
Gessmann CK, Wood BJ. Potassium in the Earth's core? Earth planet. Sci Lett. 2002;200:63–78.
Gomi H, Ohta K, Hirose K, Labrosse S, Caracas R, Verstraete MJ, et al. The high conductivity of iron and thermal evolution of the Earth's core. Phys Earth Planet Inter. 2013;224:88–103.
Grott M, Breuer D, Laneuville M. Thermo-chemical evolution and global contraction of Mercury. Earth Planet Sci Lett. 2011;307:135–46.
Hauck SA, Phillips RJ. Thermal and crustal evolution of Mars. J Geophys Res. 2002. doi:10.1029/2001JE001801.
Hauck SA, Dombard AJ, Phillips RJ, Solomon SC. Internal and tectonic evolution of Mercury. Earth Planet Sci Lett. 2004;222:713–28.
Hauck SA, Aurnou JM, Dombard AJ. Sulfur's impact on core evolution and magnetic field generation on Ganymede. J Geophys Res. 2006. doi:10.1029/2005JE002557.
Hauck II SA, Margot J-L, Solomon SC, Lemoine FG, Mazarico E, Peale SJ, et al. The curious case of Mercury's internal structure. J Geophys Res. 2013;118:1303–22.
Heimpel MH, Aurnou JM, Al-Shamali FM, Gomez Perez N. A numerical study of dynamo action as function of spherical shell geometry. Earth Planet Sci Lett. 2005;236:542–57.
Hide R. Comments on the Moon's magnetism. Moon. 1972;4:39.

Hirose K, Labrosse S, Hernlund J. Composition and state of the core. Annu Rev Earth Planet Sci. 2013;41:657–91.

Hixson RS, Winkler MA, Hodgdon ML. Sound speed and thermophysical properties of liquid iron and nickel. Phys Rev B. 1990;42:6485–91.

Hood LL, Huang Z. Formation of magnetic anomalies antipodal to lunar impact basins: two dimensional model calculations. J Geophys Res. 1991;96:9837–46.

Hood LL, Vickery A. Magnetic field amplification and generation in hypervelocity meteorid impacts with application to lunar paleomagnetism. J Geophys Res. 1984;89:C211–23.

Hood LL, Mitchell DL, Lin RP, Acuna MH, Binder AB. Initial measurements of the lunar induced magnetic dipole moment using Lunar Prospector magnetometer data. Geophys Res Lett. 1999;106:27825–39.

Hood LL, Zakharian A, Halekas J, Mitchell DL, Lin RP, Acuna MH. Initial mapping and interpretation of lunar crustal magnetic anomalies using Lunar Prospector magnetometer data. J Geophys Res. 2001;106:27825–39.

Hood LL, Harrison KP, Langlais B, Lillis RJ, Poulet F, Williams DA. Magnetic anomalies near Apollinaris Patera and the Medusae Fossae Formation in Lucus Planum, Mars. Icarus. 2010. doi:10.1016/j.icarus.2010.01.009.

Iess L, Rappaport NJ, Jacobson RA, Racioppa P, Stevenson DJ, Tortora P, et al. Gravity field, shape, and moment of inertia of Titan. Science. 2010;327:1367–9.

Johnson CL, Purucker ME, Korth H, Anderson BJ, Winslow RM, Al Asad MM, et al. MESSENGER observations of Mercury's magnetic field structure. J Geophys Res. 2012. doi:10.1029/2012JE004217.

Johnson CL, Phillips RJ, Purucker ME, Anderson BJ, Byrne PK, Denevi BW, et al. Low-altitude magnetic field measurements by MESSENGER reveal Mercury's ancient crustal field. Science. 2015;348:892–5.

Kaiura GH, Toguri JM. Densities of the molten FeS, FeS–Cu2S and Fe–S–O systems-utilizing a bottom-balance Archimedean technique. Can Metall Q. 1979;18:155–64.

Kandpal D, Gupta BRK. Analysis of thermal expansivity of iron (Fe) metal at ultra high temperature and pressure. Pramana. 2007;68:129–34.

Kimura J, Nakagawa T, Kurita K. Size and compositional constraints of Ganymede's metallic core for driving an active dynamo. Icarus. 2009;202:216–24.

Kivelson MG, Khurana KK, Russell CT, Walker RJ, Warnecke J, Coroniti FV, et al. Discovery of Ganymede's magnetic field by the Galileo spacecraft. Nature. 1996;384:537–41.

Kivelson MG, Khurana KK, Volwerk M. The permanent and inductive magnetic moments of Ganymede. Icarus. 2002;157:507–22.

Konopliv AS, Binder AB, Hood LL, Kucinskas AB, Sjogren WL, Williams JG. Improved gravity field of the Moon from Lunar Prospector. Science. 1998;281:1476–80.

Konrad W, Spohn T. Thermal history of the Moon: implications for an early core dynamo and post-accretional magmatism. Adv Space Rev. 1997;19:1511–21.

Kuwayama Y, Hirose K. Phase relations in the system Fe-FeSi at 21 GPa. Am Minerol. 2004;89:273–6.

Labrosse S. Thermal and magnetic evolution of the Earth's core. Phys Earth Planet Inter. 2003;140:127–43.

Laneuville M, Wieczorek MA, Breuer D, Aubert J, Morard G, Rückriemen T. A long-lived lunar dynamo powered by core crystallization. Earth Planet Sci Lett. 2014;401:251–60.

Langlais B, Purucker ME, Mandea M. Crustal magnetic field of Mars. J Geophys Res. 2004;109:E02008. doi:10.1029/2003JE002048.

Le Bars M, Wieczorek MA, Karatekin O, Cebron D, Laneuville M. An impact-driven dynamo for the early Moon. Nature. 2011;479:215–8.

Lee KK, Steinle-Neumann G, Jeanloz R. Ab-initio high-pressure alloying of iron and potassium: implications for the Earth's core. Geophys Res Lett. 2004;31:L11603. doi:10.1029/2004GL019839.

Lewis JS. Satellites of the outer planets: their physical and chemical nature. Icarus. 1971;15:174–85.

Lewis JS. Chemistry of the planets. Annu Rev Phys Chem. 1973;24:339–51.

Li J, Fei Y, Mao HK, Hirose K, Shieh SR. Sulfur in the Earth's inner core. Earth Planet Sci Lett. 2001;193:509–14.

Lodders K, Fegley B. The planetary scientist's companion. New York: Oxford University Press; 1998.

Malavergne V, Tarrida M, Combes R, Bureau H, Jones J, Schwandt C. New high-pressure and high-temperature metal/silicate partitioning of U and Pb: Implications for the cores of the Earth and Mars. Geochim Cosmochim Acta. 2007;71:2635–55.

Malavergne V, Toplis MJ, Berthet S, Jones J. Highly reducing conditions during core formation on Mercury: implications for internal structure and the origin of a magnetic field. Icarus. 2010;206:199–209.

Manglik A, Wicht J, Christensen UR. A dynamo model with double diffusive convection for Mercury's core. Earth Planet Sci Lett. 2010;289:619–28.

Margot JL, Peale SJ, Jurgens RF, Slade MA, Holin IV. Large longitude libration of Mercury reveals a molten core. Science. 2007;316:710–4.

Margot JL, Peale SJ, Solomon SC, Hauck SA, Ghigo FD, Jurgens RF, et al. Mercury's moment of inertia from spin and gravity data. J Geophys Res. 2012. doi:10.1029/2012JE004161.

McCammon CA, Ringwood AE, Jackson I. Thermodynamic of the system Fe-FeO-MgO at high pressure and temperature and a model for the formation of the Earth's core. Geophys J Roy Astron Soc. 1983;72:577–95.

McCubbin FM, Riner MA, Vander Kaaden KE, Burkemper LK. Is Mercury a volatile-rich planet? Geophys Res Lett. 2012;39:L09202. doi:10.1029/2012GL051711.

McDonough WF. Compositional model for the Earth's core. In: Carlson RW, Holland H, Turekian K, editors. Treatise on Geochemistry, vol. 3. 2nd ed. Amsterdam: Elsevier; 2014. p. 559–77.

McSween Jr HY. SNC meteorites: clues to Martian petrologic evolution? Rev Geophys. 1985;23:391–416.

Michel N, Forni O. Mars mantle convection: influence of phase transitions with core cooling. Planet Space Sci. 2011;59:741–8.

Mitchell DL, Lin RP, Rème H, Acuña MH, Cloutier PA, Ness NF. Crystal magnetospheres observed in the Martian night hemisphere. Am. Astronomical Society, DPS Meeting #31, abstracts #59.04, Bulletin of the American Astronomical Society. 1999:31;1584.

Mitchell DL, Lin RP, Mazelle C, Rème H, Cloutier PA, Connerney JEP, et al. Probing Mars' crustal magnetic field and ionosphere with the MGS Electron Reflectometer. J Geophys Res. 2001;106:23419–28.

Mohit PS, Arkani-Hamed J. Impact demagnetization of the Martian crust. Icarus. 2004;168:305–17.

Monteux J, Jellinek AM, Johnson CL. Dynamics of core merging after a mega-impact with applications to Mars' early dynamo. Icarus. 2013;226:20–32.

Morard G, Katsura T. Pressure–temperature cartography of Fe–S–Si immiscible system. Geochim Cosmochim Acta. 2010;74:3659–67.

Morard G, Sanloup C, Fiquet G, Mezouar M, Rey N, Poloni R, et al. Structure of eutectic Fe–FeS melts to pressures up to 17 GPa: implications for planetary cores. Earth Planet Sci Lett. 2007;263:128–39.

Morard G, Andrault D, Guignot N, Sanloup C, Mezouar M, Petitgirard S, et al. In situ determination of Fe-Fe3S phase diagram and liquid structural properties up to 65 GPa. Earth Planet Sci Lett. 2008;272:620–62.

Morgan JW, Anders E. Chemical composition of earth, Venus, and Mercury. Proc Natl Acad Sci U S A. 1980;77:6973–7.

Morschhauser A, Lesur V, Grott M. A spherical harmonic model of the lithospheric magnetic field of Mars. J Geophys Res. 2014;119:1162–88.

Murthy VR. Early differentiation of the Earth and the problem of mantle siderophile elements: a new approach. Science. 1991;253:303–6.

Murthy VR, Van Westrenen W, Fei Y. Experimental evidence that potassium is a substantial radioactive heat source in planetary cores. Nature. 2003;423:163–5.

Naldrett AJ. A portion of the system Fe–S–O between 900 and 1080 C and its application to sulfide ore magmas. J Petrol. 1969;10:171–201.

Nasch PM, Steinemann SG. Density and thermal expansion of molten manganese, iron, nickel, copper, aluminum and tin by means of the gamma-ray attenuation technique. Phys Chem Liq. 1995;29:43–58.

Ness NF, Behannon KW, Lepping RP, Schatten KH, Whang YC. Magnetic field observations near Mercury: preliminary results from Mariner 10. Science. 1974;185:151–60.

Ness NF, Behannon KW, Lepping RP, Whang YC. Magnetic field of Mercury. J Geophys Res. 1975;80:2708–16.

Ness NF, Behannon KW, Lepping RP, Whang YC. Observations of Mercury's magnetic field. Icarus. 1976;28:479–88.

Ness NF, Acuña MH, Connerney J, Wasilewski P, Mazelle C, Sauvaud J, et al. MGS magnetic fields and electron reflectometer investigation: discovery of paleomagnetic fields due to crustal remanence. Adv Space Res. 1999;23:1879–86.

Nimmo F. Energetics of the core. In: Schubert G, editor. Treatise on Geophysics, vol. 8. 1st ed. Amsterdam: Elsevier; 2007. p. 31–65.

Nimmo F. Thermal and compositional evolution of the core. In: Schubert G, editor. Treatise on Geophysics, vol. 8. 2nd ed. Amsterdam: Elsevier; 2015. p. 201–19.

Nimmo F, Stevenson D. Influence of early plate tectonics on the thermal evolution and magnetic field of Mars. J Geophys Res. 2000;105:11969–79.
Nittler LR, Starr RD, Weider SZ, McCoy TJ, Boynton WV, Ebel DS, et al. The major-element composition of Mercury's surface from MESSENGER X-ray spectrometry. Science. 2011;333:1847–950.
Ohta K, Suehiro S, Hirose K, Ohishi, Y. The electrical resistivity of iron alloys at Earth's core conditions. Goldschmidt 2015 Conf. Abstr. Book, 2326. 2015.
Olson P, Christensen UR. Dipole moment scaling for convection-driven planetary dynamos. Earth Planet Sci Lett. 2006;250:561–71.
Peale SJ. Inferences from the dynamical history of Mercury's rotation. Icarus. 1976;28:459–67.
Peplowski PN, Evans LG, Hauck SA, McCoy TJ, Boynton WV, Gillis-Davis JJ, et al. Radioactive elements on Mercury's surface from MESSENGER: implications for the planet's formation and evolution. Science. 2011;333:1850–2.
Peplowski PN, Lawrence DJ, Rhodes EA, Sprague AL, McCoy TJ, Denevi BW, et al. Variations in the abundances of potassium and thorium on the surface of Mercury: results from the MESSENGER Gamma-Ray Spectrometer. J Geophys Res. 2012;117:E00L04. doi:10.1029/2012JE004141.
Pozzo M, Davies C, Gubbins D, Alfe D. Thermal and electrical conductivity of iron at Earth's core conditions. Nature. 2012;485:355–9.
Rai N, van Westrenen W. Core-mantle differentiation in Mars. J Geophys Res. 2013;118:1195–203.
Reese CC, Solomatov VS. Early martian dynamo generation due to giant impacts. Icarus. 2010;207:82–97.
Ringwood AE. Composition of the core and implications for the origin of the Earth. Geochem J. 1977;11:111–35.
Rivoldini A, Van Hoolst T. The interior structure of Mercury constrained by the low-degree gravity field and the rotation of Mercury. Earth Planet Sci Lett. 2013;377:62–72.
Roberts PH, Jones CA, Calderwood AR. Energy fluxes and ohmic dissipation in the Earth's core. Earth's core and lower mantle. In: Jones C, Soward AM, Zhang K, editors. Book series the fluid mechanics of astrophysics and geophysics. Vol. 11. London: Taylor & Francis; 2003. p. 100–129.
Roberts JH, Lillis RJ, Manga M. Giant impacts on early Mars and the cessation of the Martian dynamo. J Geophys Res. 2009;114:E04009. doi:10.1029/2008JE003287.
Rochette P, Lorand J-P, Fillion G, Sautter V. Pyrrhotite and the remanent magnetization of SNC meteorites: a changing perspective on Martian magnetism. Earth Planet Sci Lett. 2001;190:1–12.
Rochette P, Fillion G, Ballou R, Brunet F, Ouladdiaf B, Hood L. High pressure magnetic transition in pyrrhotite and impact demagnetization on Mars. Geophys Res Lett 2003:30. doi:10.1029/2003GL017359.
Rückriemen T, Breuer D, Spohn T. The Fe-snow regime in Ganymede's core: a deep seated dynamo below a stable snow zone. J Geophys Res. 2015;120:1095–118.
Russell CT, Baker DN, Slavin JA. The magnetosphere of Mercury. In: Vilas F, Chapman CR, Matthews MS, editors. Mercury. Tucson: Univ. of Arizona Press; 1988. p. 514–61.
Sanloup C, Fei Y. Closure of the Fe-S-Si liquid miscibility gap at high pressure. Phys Earth Planet Inter. 2004;147:57–65.
Sanloup C, Guyot F, Gillet P, Fiquet G, Mezouar M, Martinez I. Density measurements of liquid Fe-S alloys at high-pressure. Geophys Res Lett. 2000;27:811–4.
Scheinberg A, Soderlund KM, Schubert G. Magnetic field generation in the lunar core: the role of inner core growth. Icarus. 2015;254:62–71.
Schubert G, Spohn T. Thermal history of Mars and the sulfur content of its core. J Geophys Res. 1990;95:14095–104.
Schubert G, Ross MN, Stevenson DJ, Spohn T. Mercury's thermal history and the generation of its magnetic field. In: Vilas F, Chapman CR, Matthews MS, editors. Mercury'. Tucson: Univ. of Arizona Press; 1988. p. 514–61.
Schubert G, Zhang K, Kivelson MG, Anderson JD. The magnetic field and internal structure of Ganymede. Nature. 1996;384:544–5.
Schubert G, Russell CT, Moore WB. Timing of the Martian dynamo. Nature. 2000;408:666–7.
Schubert G, Anderson JD, Spohn T, McKinnon WB. Interior composition, structure and dynamics of the Galilean satellites. In: Bagenal F, Dowling T, McKinnon WB, editors. Jupiter: the planet, satellites and magnetosphere. Cambridge: Cambridge Uni. Press; 2004. p. 281–306.
Scott HP, Williams Q, Ryerson FJ. Experimental constraints on the chemical evolution of large icy satellites. Earth Planet Sci Lett. 2002;203:399–412.
Shea EK, Weiss BP, Cassata WS, Shuster DL, Tikoo SM, Gattacceca J, et al. A long-lived lunar core dynamo. Science. 2012;335:453–6.
Shen G, Mao HK, Hemley RJ, Duffy TS, Rivers ML. Melting and crystal structure of iron at high pressures and temperatures. Geophys Res Lett. 1998;25:373–6.
Shimizu H, Matsushima M, Takahashi F, Shibuya H, Tsunakawa H. Constraint on the lunar core size from electromagnetic sounding based on magnetic field observations by an orbiting satellite. Icarus. 2013;222:32–43.
Siebert J, Malavergne V, Guyot F, Combes R, Martinez I. The behaviour of sulphur in metal–silicate core segregation experiments under reducing conditions. Phys Earth Planet Inter. 2004;143:433–43.
Smith D, Zuber M, Phillips R, Hauck SA, Lemoine F, Mazarico E, et al. Gravity field and internal structure of Mercury from MESSENGER. Science. 2012;336:214–7.
Sohl F, Spohn T. The interior structure of Mars: Implications from SNC meteorites. J Geophys Res. 1997;102:1613-1635.Sohl F, Spohn T, Breuer D, Nagel K. Implications from Galileo observations of the interior structure and chemistry of the Galilean satellites. Icarus. 2002;157:104–19.Sohl F, Schubert G. Interior structure, composition, and mineralogy of the terrestrial planets. In: Spohn T, Schubert G, editors. Treatise on Geophysics, vol. 10, Oxford: Elsevier; 2007. p. 27-68.
Sohl F, Schubert G, Spohn T. Geophysical constraints on the composition and structure of the Martian interior. J Geophys Res. 2005. doi:10.1029/2005JE002520.Solomon SC. Some aspects of core formation in Mercury. Icarus. 1976;28:509–21.
Solomon CS, Aharonson O, Aurnou JM, Banerdt WB, Carr MH, Dombard AJ, et al. New perspectives on ancient Mars. Science. 2005;307:1214–20.
Spohn T, Sohl F, Breuer D. Mars. Astron Astrophys Rev. 1998;8:181–235.
Spohn T, Acuña MA, Breuer D, Golombek M, Greeley R, Halliday A, et al. Geophysical constraints on the evolution of Mars. Space Sci Rev. 2001;96:231–62.
Stacey FD, Anderson OL. Electrical and thermal conductivities of Fe–Ni–Si alloy under core conditions. Phys Earth Planet Inter. 2001;124:153–62.
Stanley S, Bloxham J, Hutchinson WE, Zuber MT. Thin shell dynamo models consistent with Mercury's weak observed magnetic field. Earth Planet Sci Lett. 2005;234:27–38.
Stanley S, Elkins-Tanton L, Zuber MT, Parmentier EM. Mars' paleomagnetic field as the result of a single-hemisphere dynamo. Science. 2008;321:1822–5.
Stegman DR, Jellinek AM, Zatman SA, Baumgardner JR, Richards MA. An early lunar core dynamo driven by thermochemical mantle convection. Nature. 2003;421:143–6.
Stevenson DJ. Turbulent thermal convection in the presence of rotation and a magnetic field: a heuristic theory. Geophys Astrophys Fluid Dyn. 1979;12:139–69.
Stevenson DJ. Mars core and magnetism. Nature. 2001;412:214–9.
Stevenson DJ. Possible connections between the history of the Venus magnetic field and observable features. In: AGU Spring Meeting Abstracts (Vol. 1, p. 08). Am. Geophys. U., Spring Meeting 2002 abstract #P21A-08. 2002.
Stevenson DJ, Spohn T, Schubert G. Magnetism and thermal evolution of the terrestrial planets. Icarus. 1983;54:466–89.
Stewart AJ, Schmidt MW, van Westrenen W, Liebske C. Mars: a new core-crystallization regime. Science. 2007;316:1323–5.
Strom RG, Trask NJ, Guest JE. Tectonism and volcanism on Mercury. J Geophys Res. 1975;80:2478–507.
Suavet C, Weiss BP, Cassata WS, Shuster DL, Gattacceca J, Chan L, et al. Persistence and origin of the lunar core dynamo. Proc Natl Acad Sci U S A. 2013;110:8453–8.
Tian BY, Stanley S, Tikoo SM, Weiss BP. AGU Fall meeting, abstract GP54A-04. 2014.
Tikoo SM, Weiss BP, Buz J, Lima EA, Shea EK, Melo G, et al. Magnetic fidelity of lunar samples and implications for an ancient core dynamo. Earth Planet Sci Lett. 2012;337:93–103.
Tikoo SM, Weiss BP, Cassata WS, Shuster DL, Gattacceca J, Lima EA, et al. Decline of the lunar core dynamo. Earth Planet Sci Lett. 2014;404:89–97.
Tosi N, Grott M, Plesa A-C, Breuer D. Thermo-chemical evolution of Mercury's interior. J Geophys Res. 2013;118:2474–87.
Tsuno K, Ohtani E. Eutectic temperatures and melting relations in the Fe–O–S system at high pressures and temperatures. Phys Chem Min. 2009;36:9–17.
Tsuno K, Frost DJ, Rubie DC. The effects of nickel and sulphur on the core-mantle partitioning of oxygen in Earth and Mars. Phys Earth Planet Inter. 2011;185:1–12.
Urakawa S, Kato M, Kumazawa M. Experimental study of the phase relation in the system Fe-Ni-O-S up to 15 GPa. In: Manghnani MH, Syono Y, editors. High-pressure research in mineral physics. Tokyo: Terrapub; 1987. p. 95–111.

Usselman TM. Experimental approach to the state of the core; Part I, The liquidus relations of the Fe-rich portion of the Fe-Ni-S system from 30 to 100 kb. Am J Sci. 1975;275:278–90.
Vilim R, Stanley S, Hauck SA. Iron snow zones as a mechanism for generating Mercury's weak observed magnetic field. J Geophys Res. 2010;115:E11003. doi:10.1029/2009JE003528.
Weber RC, Lin PY, Garnero EJ, Williams Q, Lognonne P. Seismic detection of the lunar core. Science. 2011;331:309–12.
Weider SZ, Nittler LR, Starr RD, McCoy TJ, Stockstill-Cahill KR, Byrne PK, et al. Chemical heterogeneity on Mercury's surface revealed by the MESSENGER X-Ray Spectrometer. J Geophys Res. 2012. doi:10.1029/2012JE004153.
Weiss BP, Tikoo SM. The lunar dynamo. Science. 2014;346:1246753. doi:10.1126/science.1246753.
Wicht J, Heyner D. Mercury's magnetic field in the messenger era. In: Shuanggen J, editor. Planetray geodesy and remote sensing. London: CRC Press; 2014.
Wicht J, Mandea M, Takahashi F, Christensen UR, Matushima M, Langalis B. The origin of Mercury's internal magnetic field. Space Sci Rev. 2007;12:261–90.
Wieczorek MA, Jolliff BL, Khan A, Pritchard ME, Weiss BP, Williams JG, et al. The constitution and structure of the lunar interior. In: Joliff BL, Wieczorek MA, Shearer CK, Neal CR (eds) New Views of the Moon. Rev Min Geochem. 2006;60:221–364.
Wienbruch U, Spohn T. A self-sustained magnetic field on Io? Planet Space Sci. 1995;43:1045–57.
Williams Q. Bottom-up versus top-down solidification of the cores of small solar system bodies: constraints on paradoxical cores. Earth Planet Sci Lett. 2009;284:564–9.
Williams JG, Boggs DH. Tides on the Moon: theory and determination of dissipation. J Geophys Res. 2015;120:689–724.
Williams JG, Nimmo F. Thermal evolution of the Martian core: implications for an early dynamo. Geology. 2004;32:97–100.
Williams JG, Boggs DH, Yoder CF, Ratcliff JT, Dickey JO. Lunar rotational dissipation in solid body and molten core. J Geophys Res. 2001;106:27933–68.
Zhan X, Schubert G. Powering Ganymede's dynamo. J Geophys Res. 2012;117:E08011. doi:10.1029/2012JE004052.
Zhang N, Parmentier EM, Liang Y. Effects of lunar cumulate mantle overturn and megaregolith on the expansion and contraction history of the Moon. Geophys Res Lett. 2013;40:5019–23.
Zhang P, Cohen RE, Haule K. Effects of electron correlations on transport properties of iron at Earth's core conditions. Nature. 2015;517:605–7.

11

The competition between Lorentz and Coriolis forces in planetary dynamos

Krista M. Soderlund[1*], Andrey Sheyko[2], Eric M. King[3] and Jonathan M. Aurnou[4]

Abstract

Fluid motions within planetary cores generate magnetic fields through dynamo action. These core processes are driven by thermo-compositional convection subject to the competing influences of rotation, which tends to organize the flow into axial columns, and the Lorentz force, which tends to inhibit the relative movement of the magnetic field and the fluid. It is often argued that these forces are predominant and approximately equal in planetary cores; we test this hypothesis using a suite of numerical geodynamo models to calculate the Lorentz to Coriolis force ratio directly. Our results show that this ratio can be estimated by $\Lambda_d^* \simeq \Lambda_i Rm^{-1/2}$ (Λ_i is the traditionally defined Elsasser number for imposed magnetic fields and Rm is the system-scale ratio of magnetic induction to magnetic diffusion). Best estimates of core flow speeds and magnetic field strengths predict the geodynamo to be in magnetostrophic balance where the Lorentz and Coriolis forces are comparable. The Lorentz force may also be significant, i.e., within an order of magnitude of the Coriolis force, in the Jovian interior. In contrast, the Lorentz force is likely to be relatively weak in the cores of Saturn, Uranus, Neptune, Ganymede, and Mercury.

Keywords: Planetary dynamos; Core convection; Magnetic fields; Rotation; Magnetostrophic force balance

Background

Magnetic fields are common throughout the solar system and provide a unique perspective on the internal dynamics of planetary interiors. Planetary magnetic fields are driven by the conversion of kinetic energy into magnetic energy; this process is called dynamo action. Kinetic energy is derived from thermo-compositional convection of an electrically conducting fluid although mechanical mechanisms, such as libration and precession, may also drive core flow in smaller bodies (e.g., Le Bars et al. 2015). The geodynamo is the most studied planetary magnetic field, yet the mechanisms that control its strength, morphology, and secular variation are still not well understood. Towards explaining these observations, dimensionless parameters are often used to characterize the force balances present in the core and their link to processes that govern convective dynamics and dynamo action. Two forces that are particularly influential on core processes are the Coriolis force, which tends to organize the core fluid motions, and the Lorentz force which back-reacts on the fluid motions to equilibrate magnetic field growth.

In order to better understand the manifestation of Coriolis and Lorentz forces on core flow, let us consider the Boussinesq momentum equation. In this commonly used approximation, the material properties are assumed to be constant and the density is treated as constant except in the buoyancy force of the momentum equation (e.g., Tritton 1998):

$$\frac{D\mathbf{u}}{Dt} + 2\boldsymbol{\Omega} \times \mathbf{u} = -\nabla\Pi - \alpha T'\mathbf{g} + \nu\nabla^2\mathbf{u} + \frac{1}{\rho_o}\mathbf{J} \times \mathbf{B}. \quad (1)$$

Here, $\mathbf{u}$ is the velocity vector, $\boldsymbol{\Omega}$ is the rotation rate of the reference frame, Π is the non-hydrostatic pressure normalized by the background density ρ_o, $\mathbf{J}$ is the electric current density, $\mathbf{B}$ is the magnetic induction, and $\mathbf{g}$ is the acceleration of gravity. Terms from left to right are inertial acceleration, Coriolis acceleration, pressure gradient, buoyancy, viscous diffusion, and Lorentz force. The inertial term consists of the temporal evolution of the velocity field as well as non-linear advection. The pressure gradient term absorbs the hydrostatic mean gravitational component as well as the centrifugal force. The buoyancy force arises from density differences with respect to the density

*Correspondence: krista@ig.utexas.edu
[1]Institute for Geophysics, John A. & Katherine G. Jackson School of Geosciences, The University of Texas at Austin, J. J. Pickle Research Campus, Building 196 (ROC), 10100 Burnet Road (R2200), Austin, Texas, 78758-4445, USA
Full list of author information is available at the end of the article

of the motionless background state (assumed to be constant density ρ_o). These perturbations, $\rho'/\rho_o = -\alpha T'$, are produced by temperature variations, T', from the background state temperature profile; α is thermal expansivity. Viscous diffusion depends on the kinematic viscosity of the fluid, ν. The Lorentz force is due to interactions between the magnetic field and current density.

Planetary cores are rapidly rotating and the Coriolis force is expected to be large compared to the viscous, inertial, and buoyancy forces at global scales (the former two forces are scale dependent quantities). If we neglect these forces and magnetic fields in the Boussinesq momentum equation (Eq. 1), the system is in geostrophic balance:

$$2\boldsymbol{\Omega} \times \mathbf{u} = -\nabla\Pi. \tag{2}$$

This balance requires horizontal fluid motions to follow lines of constant pressure. Further insight can be obtained by taking the curl of Eq. 2 and using the continuity equation ($\nabla \cdot \mathbf{u} = 0$) to obtain $\boldsymbol{\Omega} \cdot \nabla\mathbf{u} = 0$. If we assume $\boldsymbol{\Omega} = \Omega\hat{\mathbf{z}}$ where Ω is constant, this equation simplifies to

$$\frac{\partial \mathbf{u}}{\partial z} = 0. \tag{3}$$

This is the Taylor-Proudman theorem and states that the fluid motion will be invariant along the direction of the rotation axis. However, in order for convection to occur, the Taylor-Proudman constraint cannot strictly hold as there must be slight deviations from two-dimensionality, at least in the boundary layers, to allow overturning motions in the fluid layer (e.g., Zhang 1992; Olson et al. 1999; Grooms et al. 2010). Further beyond the onset of convection, convective flows break down into an anisotropic rotating turbulence (e.g., Sprague et al. 2006; Julien et al. 2012; Stellmach et al. 2014; Cheng et al. 2015; Ribeiro et al. 2015).

Linear asymptotic analyses of quasigeostrophic convection predict that the azimuthal wavenumber of these columns relative to the thickness of the fluid shell varies as $\ell_U/D \propto E^{1/3}$, where the Ekman number, $E = \nu/\left(2\Omega D^2\right)$, represents the ratio of viscous to Coriolis forces (e.g., Roberts 1968; Julien et al. 1998; Jones et al. 2000; Dormy et al. 2004). Thus, in rapidly rotating systems such as planetary cores (where $E \lesssim 10^{-12}$), it is often predicted that quasigeostrophic convection occurs as tall, thin columns (e.g., Kageyama et al. 2008; cf. Cheng et al. 2015).

In planetary dynamos, however, magnetic fields are also thought to play an important dynamical role on the convection and zonal flows. When strong imposed magnetic fields and rapid rotation are both present, the dominant force balance is magnetostrophic—a balance between the Coriolis, pressure gradient, and Lorentz terms:

$$2\boldsymbol{\Omega} \times \mathbf{u} = -\nabla\Pi + \frac{1}{\rho_o}\mathbf{J} \times \mathbf{B}. \tag{4}$$

In contrast to Eq. 3, the curl of the magnetostrophic balance Eq. 4 yields

$$\frac{\partial \mathbf{u}}{\partial z} = -\frac{1}{2\,\rho_o\,\Omega}\nabla \times (\mathbf{J} \times \mathbf{B}). \tag{5}$$

This equation shows that magnetic fields can relax the Taylor-Proudman constraint, allowing global-scale motions that differ fundamentally from the small-scale axial columns typical of non-magnetic, rapidly rotating convection (e.g., Cardin and Olson 1995; Olson and Glatzmaier 1996; Zhang and Schubert 2000; Roberts and King 2013).

Magnetic fields are typically considered to be important dynamically when the ratio of Lorentz to Coriolis forces is of order unity. This ratio is represented by the Elsasser number, Λ, which can be approximated by assuming a characteristic current density, magnetic induction, density, rotation rate, and velocity scale:

$$\Lambda = \frac{\text{Lorentz}}{\text{Coriolis}} = \frac{\left|\frac{1}{\rho_o}\mathbf{J} \times \mathbf{B}\right|}{|2\boldsymbol{\Omega} \times \mathbf{u}|} \approx \frac{JB}{2\rho_o\Omega U}. \tag{6}$$

In systems where the magnetic field is not strongly time variant (i.e., $\partial\mathbf{B}/\partial t = -\nabla \times \mathbf{E} \sim 0$), the current density can be approximated via Ohm's Law, $J \sim \sigma UB$ where $\sigma = (\mu_o\eta)^{-1}$ is electrical conductivity and η is magnetic diffusivity, such that

$$\Lambda_i = \frac{B^2}{2\rho_o\mu_o\eta\Omega}. \tag{7}$$

This formulation is thus appropriate for steady *imposed* magnetic fields (e.g., Olson and Glatzmaier 1996; King and Aurnou 2015). In dynamos, which tend to exhibit significant temporal variability, it is more appropriate to approximate the current density using Ampere's law under the MHD approximation, $J \sim B/\mu_o\ell_B$ where ℓ_B is the characteristic length scale of magnetic field variations (e.g., Cardin et al. 2002; Soderlund et al. 2012). In this *dynamic* case, Eq. 6 becomes

$$\Lambda_d = \frac{B^2}{2\rho_o\mu_o\Omega U\ell_B} = \frac{\Lambda_i}{Rm}\frac{D}{\ell_B}, \tag{8}$$

indicating that the Lorentz to Coriolis force ratio is length scale dependent. In the latter equality, $Rm = UD/\eta$ is the global ratio of magnetic induction to magnetic diffusion. This parameter must exceed approximately 10 in order for dynamo action to occur (e.g., Roberts 2007). Thus, on global scales, where $\ell_B \simeq D$, the dynamic Elsasser number will always be less than Λ_i by at least an order of magnitude for active dynamos. However, the influence of magnetic fields will become increasingly important at smaller scales since $\Lambda_d \propto \ell_B^{-1}$ in (Eq. 8).

The quantity ℓ_B/D cannot be measured for planetary cores, however. It is, therefore, important to provide a scaling estimate for Λ_d in terms of quantities that can be estimated from observations, such as Λ_i and Rm. The length scale ℓ_B can be approximated using the magnetic induction equation:

$$\frac{\partial \mathbf{B}}{\partial t} = \nabla \times (\mathbf{u} \times \mathbf{B}) + \eta \nabla^2 \mathbf{B} \tag{9}$$

Assuming a balance between induction and diffusion yields $UB/D \sim \eta B/\ell_B^2$. Rearranging, the dimensionless length scale is proportional to the inverse square root of the magnetic Reynolds number:

$$\ell_B/D \sim Rm^{-1/2} \tag{10}$$

(e.g., Galloway 1978; King and Roberts 2013). Using this relationship in Eq. 8, we estimate the dynamic Lorentz to Coriolis force ratio to be

$$\Lambda_d^* \sim \Lambda_i Rm^{-1/2}. \tag{11}$$

This scaling estimate will be referred to as the *modified dynamic* Elsasser number.

In this paper, we will investigate the relative magnitudes of Lorentz and Coriolis forces to better understand the dynamics of planetary cores. Towards this end, we calculate directly the Lorentz to Coriolis force ratio in a suite of geodynamo models described in the "Methods" section, test Eq. 11 to predict this ratio in the "Results" section, and extrapolate the results to Earth and other planetary cores in the "Discussion" section.

Methods

We analyze 34 dynamo models presented in Soderlund et al. (2012, 2014) and five dynamo models presented in Sheyko (2014); these datasets will be referred to as SKA and S14, respectively. These simulations of Boussinesq convection and dynamo action are carried out in thick, rotating spherical shells with thickness D, rotation rate Ω, and no-slip boundaries. Gravity is assumed to vary linearly with radius; g_o is acceleration at the top of the fluid shell. The outer boundary at radius r_o is assumed electrically insulating and the solid inner core at radius r_i has the same electrical conductivity as the convecting fluid region.

The SKA dataset considers three rotation rates (Ekman numbers in the range $10^{-3} \geq E \geq 10^{-5}$), two magnetic diffusivities (magnetic Prandtl numbers of $Pm = 2, 5$), and thermal forcing near onset (Ra_c) to more than 1000 times critical ($1.9 < Ra/Ra_c < 1125$); the shell geometry is fixed to $\chi = r_i/r_o = 0.4$ with isothermal boundaries. This survey is complemented by the S14 dataset focusing on low Ekman, low magnetic Prandtl number simulations: $3 \times 10^{-7} \lesssim E \lesssim 10^{-6}$, $0.05 \leq Pm \leq 1$, and $10^8 \lesssim Ra \lesssim 10^{10}$ with $\chi = 0.35$, an isothermal inner shell boundary, and fixed heat flux on the outer shell boundary. For both datasets, the thermal (κ) and momentum (ν) diffusivities are assumed equal, such that the Prandtl number, $Pr = \nu/\kappa$, is unity. Here, $E = \nu/\left(2\Omega D^2\right)$ is the ratio of viscous to Coriolis forces; $Pm = \nu/\eta$ is the ratio of viscous to magnetic diffusivities, and $Ra = \alpha g_o \Delta T D^3/(\nu\kappa)$ is the ratio of buoyancy due to superadiabatic temperature contrast ΔT to diffusion. The combination of these datasets, given in Table 1, results in one of the broadest surveys of supercriticality, Ekman, and magnetic Prandtl numbers in the literature.

The SKA simulations are carried out on numerical grids that range from 37 fluid shell radial levels and 42 spherical harmonic degrees (near onset at $E = 10^{-3}$) to 73 radial levels and 213 spherical harmonic degrees (highest Ra for $E = 10^{-5}$). Hyperdiffusion is used for the three most extreme cases ($Ra \geq 2 \times 10^8$ for $E \leq 10^{-4}$); additional details are given in Soderlund et al. (2012). The S14 simulations are carried out on numerical grids with 512–528 radial levels and 256 spherical harmonic degrees. All simulations are carried out using pseudo-spectral methods with no azimuthal symmetries assumed (Wicht 2002; Christensen and Wicht, 2007; Sheyko 2014). Once the initial transient has subsided, global diagnostic quantities (Λ_i, Rm, ℓ_B, Λ_d, and Λ_d^*) are time-averaged to minimize statistical errors.

In our models, the imposed Elsasser number is calculated from the dimensionless magnetic energy density, $\mathcal{E}_M$, within the fluid shell of volume V:

$$\Lambda_i = 2PmE\mathcal{E}_M \text{ where } \mathcal{E}_M = \frac{1}{2\,Pm\,E\,V}\int_V \mathbf{B}\cdot\mathbf{B}dV, \tag{12}$$

while the magnetic Reynolds number is calculated from the dimensionless kinetic energy density, $\mathcal{E}_K$:

$$Rm = \sqrt{2\mathcal{E}_K}Pm \text{ where } \mathcal{E}_K = \frac{1}{2V}\int_V \mathbf{u}\cdot\mathbf{u}dV. \tag{13}$$

These quantities are non-dimensionalized using distance, magnetic induction, velocity, and energy density scales of $D = r_o - r_i$, $\sqrt{2\rho\mu_o\eta\Omega}$, ν/D, and $\rho\nu^2/D^2$, respectively. We also characterize ℓ_B as a quarter wavelength of the typical magnetic field wavenumber (Soderlund et al. 2012):

$$\ell_B/D \sim \frac{(\pi/2)}{\sqrt{\overline{l_B}^2 + \overline{m_B}^2}}, \text{ where } \overline{l_B} = \frac{\sum l(\mathbf{B}_l\cdot\mathbf{B}_l)}{2\,Pm\,E\,\mathcal{E}_M} \text{ and}$$
$$\overline{m_B} = \frac{\sum m(\mathbf{B}_m\cdot\mathbf{B}_m)}{2\,Pm\,E\,\mathcal{E}_M}. \tag{14}$$

Here, $\mathbf{B}_l$ and $\mathbf{B}_m$ are magnetic induction at spherical harmonic degree l and order m.

Two approaches are used to calculate the Lorentz to Coriolis force ratio in situ. The first integrates the Lorentz

Table 1 Input (E, Pm, Ra, $Ro_c = \sqrt{RaE^2/Pr}$) and output (Λ_i, Λ_d, Λ_d^*, Λ^I, Λ^P) parameters for our dataset. $Pr = 1$ is fixed for all simulations. Additional simulation information is given in Soderlund et al. (2012, 2014) for cases with $E \geq 10^{-5}$ and Sheyko (2014) for cases with $E \lesssim 10^{-6}$

E	Pm	Ra	Ro_c	Λ_i	Λ_d	Λ_d^*	Λ^I	Λ^P
10^{-3}	5	6.50×10^4	0.26	4.34	0.44	0.47	0.39	0.26
10^{-3}	5	9.70×10^4	0.31	3.91	0.30	0.35	0.43	0.32
10^{-3}	5	1.12×10^5	0.33	3.25	0.20	0.25	0.40	0.29
10^{-3}	5	1.32×10^5	0.36	0.61	0.03	0.04	0.05	0.01
10^{-3}	5	5.00×10^5	0.71	4.14	0.15	0.17	0.29	0.16
10^{-3}	5	1.96×10^6	1.40	23.0	0.51	0.64	0.79	0.46
10^{-3}	5	1.96×10^7	4.43	248	2.62	4.00	4.20	3.12
10^{-4}	2	1.42×10^6	0.12	1.30	0.14	0.12	0.18	0.12
10^{-4}	2	2.12×10^6	0.15	2.10	0.17	0.16	0.21	0.14
10^{-4}	2	2.83×10^6	0.17	2.42	0.17	0.16	0.24	0.16
10^{-4}	2	3.54×10^6	0.19	2.37	0.15	0.14	0.27	0.19
10^{-4}	2	3.68×10^6	0.19	2.36	0.15	0.13	0.21	0.17
10^{-4}	2	3.75×10^6	0.19	2.27	0.14	0.12	0.21	0.15
10^{-4}	2	3.82×10^6	0.20	0.14	0.01	0.007	0.01	0.006
10^{-4}	2	3.96×10^6	0.20	0.18	0.01	0.009	0.02	0.009
10^{-4}	2	4.11×10^6	0.20	0.22	0.01	0.01	0.02	0.009
10^{-4}	2	4.24×10^6	0.21	0.24	0.02	0.01	0.02	0.006
10^{-4}	2	6.00×10^6	0.24	0.68	0.04	0.03	0.06	0.03
10^{-4}	2	8.50×10^6	0.29	1.53	0.07	0.06	0.11	0.06
10^{-4}	2	1.42×10^7	0.38	3.44	0.13	0.11	0.21	0.11
10^{-4}	2	2.10×10^7	0.46	5.48	0.18	0.16	0.28	0.17
10^{-4}	2	4.24×10^7	0.65	11.4	0.29	0.26	0.44	0.27
10^{-4}	2	8.00×10^7	0.89	20.7	0.40	0.39	0.59	0.37
10^{-4}	2	1.42×10^8	1.19	36.3	0.58	0.58	0.81	0.53
10^{-4}	2	4.24×10^8	2.06	106	0.84	1.28	1.18	0.77
10^{-4}	2	8.48×10^8	2.91	197	1.11	2.10	1.63	1.07
10^{-5}	2	3.10×10^7	0.06	4.84	0.30	0.30	0.39	0.33
10^{-5}	2	5.89×10^7	0.08	7.07	0.31	0.31	0.45	0.37
10^{-5}	2	8.20×10^7	0.09	8.28	0.32	0.30	0.45	0.34
10^{-5}	2	8.50×10^7	0.09	8.54	0.32	0.31	0.47	0.39
10^{-5}	2	9.50×10^7	0.10	9.02	0.32	0.30	0.45	0.35
10^{-5}	2	1.05×10^8	0.10	9.63	0.33	0.31	0.47	0.34
10^{-5}	2	1.50×10^8	0.12	12.4	0.36	0.35	0.52	0.37
10^{-5}	2	2.00×10^8	0.14	14.8	0.38	0.37	0.56	0.36
1.2×10^{-6}	1	1.86×10^8	0.02	5.73	0.21	0.28	0.39	0.30
1.2×10^{-6}	0.2	1.86×10^8	0.04	0.07	0.005	0.007	0.01	0.009
1.2×10^{-6}	0.2	9.28×10^8	0.09	0.61	0.03	0.03	0.08	0.04
1.2×10^{-6}	0.2	5.57×10^9	0.02	9.92	0.24	0.25	0.47	0.30
3.0×10^{-7}	0.05	2.23×10^{10}	0.04	3.61	0.13	0.17	0.32	0.24

($\mathbf{F_L}$) and Coriolis ($\mathbf{F_C}$) forces over the spherical shell volume following Soderlund et al. (2012):

$$F_L^I = \int_V \left((\mathbf{F_L} \cdot \hat{\mathbf{r}})^2 + \left(\mathbf{F_L} \cdot \hat{\theta}\right)^2 + \left(\mathbf{F_L} \cdot \hat{\phi}\right)^2 \right)^{1/2} dV \quad (15)$$

$$F_C^I = \int_V \left((\mathbf{F_C} \cdot \hat{\mathbf{r}})^2 + \left(\mathbf{F_C} \cdot \hat{\theta}\right)^2 + \left(\mathbf{F_C} \cdot \hat{\phi}\right)^2 \right)^{1/2} dV \quad (16)$$

where the volume-*integrated* force ratio is $\Lambda^I = F_L^I / F_C^I$. Alternatively, the root mean square (RMS) Lorentz and Coriolis forces are calculated at each grid point to determine the force ratio locally (cf. Dharmaraj and Stanley 2012; Dharmaraj et al. 2014). A representative value is determined by taking the most probable ratio in the histogram; this *probability* approach yields Λ^P (Fig. 1). Histogram bins are uniformly spaced to contain equal ranges of $\log_{10}(F_L^{RMS}/F_C^{RMS})$ values. The probability of each bin is defined as the volume of the fluid shell that is occupied by the respective range of force ratio values relative to the total volume of the fluid shell. For both approaches, velocity and magnetic fields from three random snapshots in time are used to calculate the force ratios for the SKA dataset, with Λ^I and Λ^P being the average values. Single snapshots are used for the S14 dataset.

Results

A Lorentz to Coriolis force ratio histogram is evaluated for a representative dipole-dominated dynamo case at a snapshot in time in Fig. 1. The histogram indicates that the

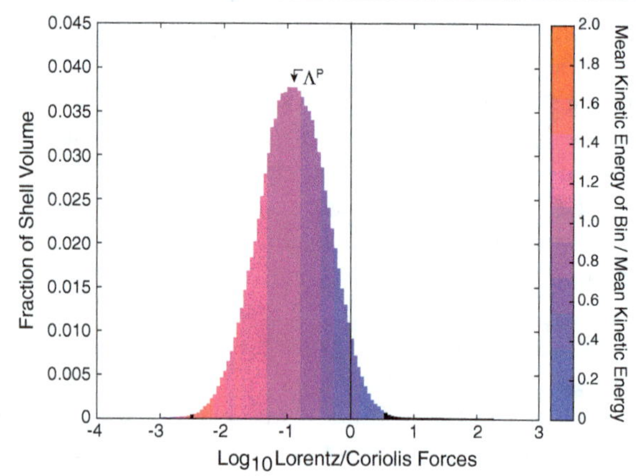

Fig. 1 Histogram of Lorentz to Coriolis force ratios on a point-by-point basis. Histogram showing the distribution of Lorentz to Coriolis forces at each point in our simulation with $E = 10^{-4}$, $Ra = 1.9Ra_c$, $Pr = 1$, and $Pm = 2$. *Color* denotes the mean kinetic energy per force ratio bin relative to the mean kinetic energy of the system. The probability Elsasser number corresponds to the most probable force ratio bin, $\Lambda^P = 0.12$. In contrast, the imposed Elsasser number is $\Lambda_i = 1.31$, the integrated Elsasser number is $\Lambda^I = 0.18$, the dynamic Elsasser number is $\Lambda_d = 0.14$, and the modified dynamic Elsasser number is $\Lambda_d^* = 0.12$

Coriolis force is dominant for the majority of grid points, most frequently by an order of magnitude. Moreover, the points with Lorentz to Coriolis force ratios greater than unity tend to have relatively low kinetic energies, denoted by coloration of the bins. Thus, the Lorentz force is expected to have a secondary influence on the convective (non-zonal) dynamics (Soderlund et al. 2012). This does not mean, however, that the magnetic field cannot have an important, local-scale dynamical impact (cf. Sreenivasan and Jones 2011).

Figure 2 shows the ratio of Lorentz to Coriolis forces in our simulations as characterized by the *integrated* Λ^I, *probability* Λ^P, *imposed* Λ_i, *dynamic* Λ_d, and *modified dynamic* Λ_d^* Elsasser numbers (see Table 2 for a summary of definitions). For comparison, ratios of these parameters are plotted in Fig. 3. The Λ_i Elsasser number, which is traditionally used to test for magnetostrophic balance, substantially overestimates the Lorentz to Coriolis force ratio in all of our dynamo models (i.e., typically by an order of magnitude; see Fig. 3a). In contrast, we find that all of the other Lorentz to Coriolis force ratio definitions (Λ^I, Λ^P, Λ_d, Λ_d^*) agree to within a factor of five; a factor of 1.4 is the mean deviation between definitions. Comparing the in situ calculations, Λ^I exceeds Λ^P in all cases investigated. This result implies that, independent of the Ekman and magnetic Prandtl numbers and in agreement with the literature (e.g., Aubert et al. 2008, Sreenivasan et al. 2014), the Lorentz force is concentrated in discrete regions that

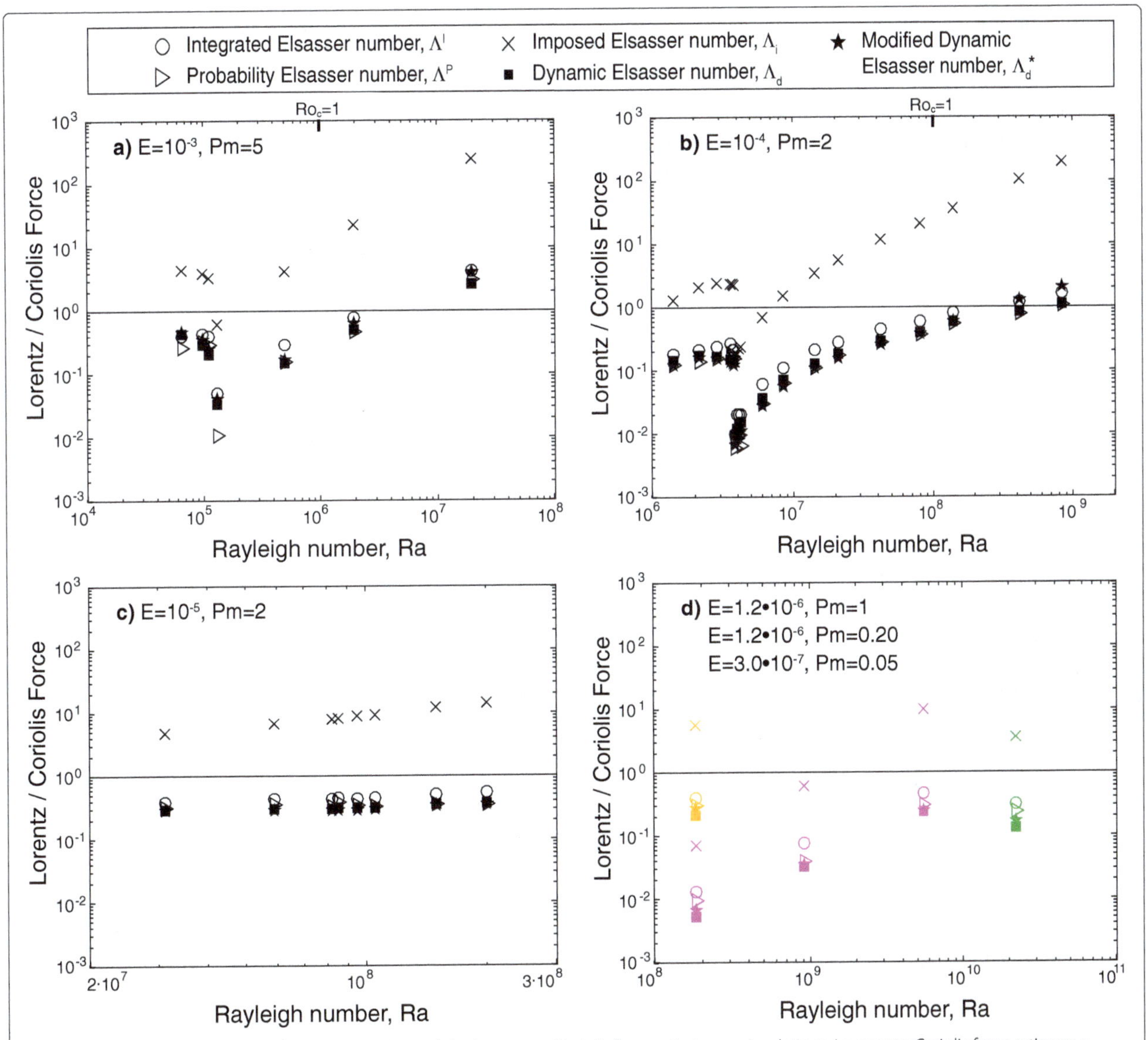

Fig. 2 a–d Direct calculations and parameterizations of the Lorentz to Coriolis force ratio in our simulations. Lorentz to Coriolis force ratios as a function of Rayleigh number for five Ekman numbers and five magnetic Prandtl numbers. The modified dynamic Elsasser number includes the length scale pre-factor: $\Lambda_d^* = 0.77\Lambda_i Rm^{-1/2}$. The Rayleigh numbers corresponding to $Ro_c = \sqrt{RaE^2/Pr} = 1$ are denoted on the top axes

Table 2 Elsasser number definitions: (1) Imposed Λ_i, (2) Dynamic Λ_d, (3) Modified dynamic Λ_d^*, (4) Integrated Λ^I, and (5) Probability Λ^P

Λ_i	Λ_d	Λ_d^*	Λ^I	Λ^P
$\frac{B^2}{2\rho_o\mu_o\eta\Omega}$	$\frac{\Lambda_i}{Rm}\frac{D}{\ell_B}$	$0.77\frac{\Lambda_i}{Rm^{1/2}}$	$\frac{\int_V \lvert\mathbf{F_L}\rvert dV}{\int_V \lvert\mathbf{F_C}\rvert dV}$	$\frac{\lvert\mathbf{F_L}\rvert}{\lvert\mathbf{F_C}\rvert}$, max probability

are more influenced by magnetic fields than the rest of the fluid volume. Similarly, the dynamic Elsasser number is also found to be in better agreement with Λ^P than Λ^I (Fig. 3b, c); the mean (maximum) percent differences between Λ_d and Λ^P are 28 % (68 %) for $E = 10^{-3}$, 18 % (58 %) for $E = 10^{-4}$, 9 % (21 %) for $E = 10^{-5}$, and 49 % (80 %) for $E \lesssim 10^{-5}$.

The Lorentz to Coriolis force ratios in Fig. 2 have distinct minima for cases with $E \geq 10^{-4}$. This sharp drop as the Rayleigh number is increased from onset occurs when the dipole-dominance of the dynamo breaks down. The Lorentz force increases more rapidly than the Coriolis force as Ra is further increased (see Fig. 4a of Soderlund et al. 2012), which explains the subsequent increase in Elsasser numbers.

Before discussing the modified Elsasser number, the assumption that the dimensionless length scale representative of magnetic field gradients in our simulations is approximately equal to the theoretical prediction (Eq. 10) must be assessed. This comparison is made in Fig. 4. Agreement within a factor of two—and typically much less, 16 % difference on average—between the ℓ_B/D values calculated directly from our simulations and the scaling estimate occurs when a pre-factor of 1.3 is included in Eq. 10:

$$\ell_B/D = 1.3Rm^{-1/2}. \tag{17}$$

This pre-factor is determined by minimizing the least squares residual between the calculated and predicted values across our dataset. Roberts and King (2013) also investigated the length scale of magnetic field variations in geodynamo models and found a similar dependence on $Rm^{-1/2}$ with a pre-factor of 3 due to their slightly modified definition of ℓ_B/D. More importantly, they also show that the magnetic length scale is weakly dependent on the magnetic field strength Λ_i, which may explain some of the spread in our model-prediction deviations.

The modified dynamic Elsasser number is given by solid, starred markers in Fig. 2; here, Λ_d^* has been modified compared to Eq. 11 to include the pre-factor derived from the length scale calculations (Eq. 17):

$$\Lambda_d^* = 0.77\Lambda_i Rm^{-1/2}. \tag{18}$$

For all Ekman numbers, the dynamic Λ_d and modified dynamic Λ_d^* Elsasser numbers differ by less than a factor

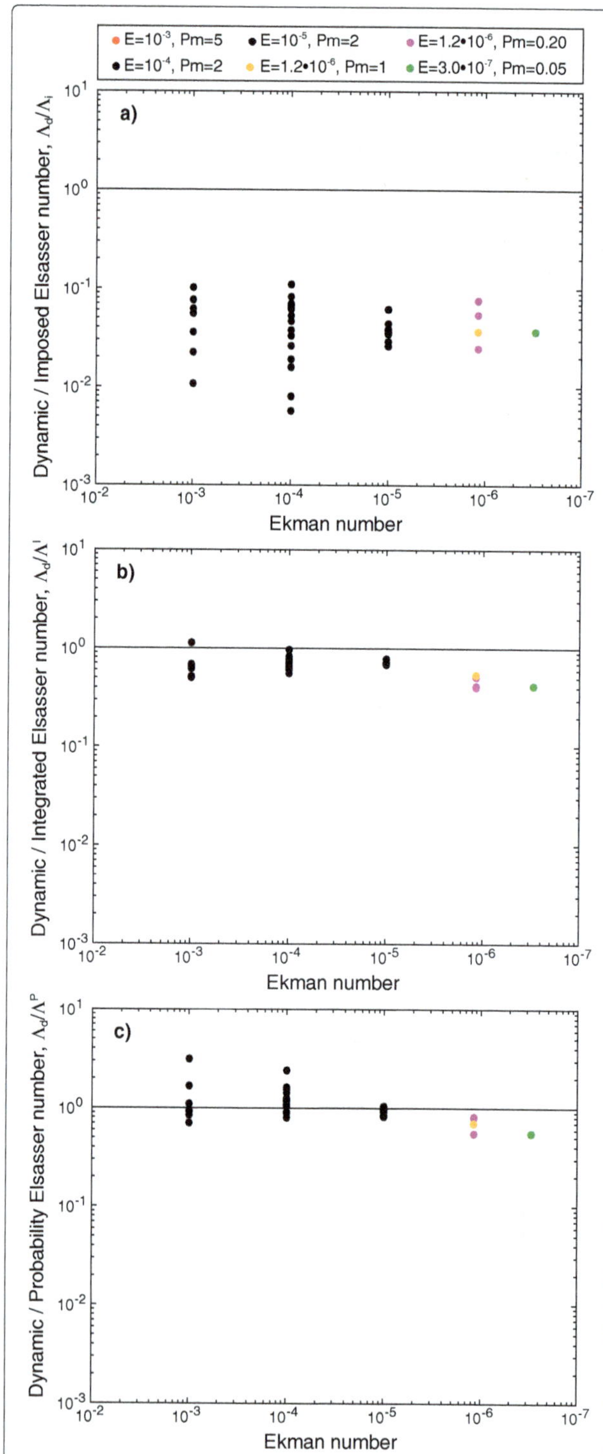

Fig. 3 Comparison of Elsasser number definitions in our simulations. Ratio of the dynamic Elsasser number Λ_d to the **a** imposed Elsasser number Λ_i, **b** integrated Elsasser number Λ^I, and **c** probability Elsasser number Λ^P as a function of Ekman number

of two (Fig. 5): the mean (maximum) percent difference is 23 % (53 %) for $E = 10^{-3}$, 21 % (89 %) for $E = 10^{-4}$, 4 % (6 %) for $E = 10^{-5}$, and 19 % (30 %) for

$E \lesssim 10^{-6}$. The differences between Λ_d^* and the probability Elsasser number Λ^P tend to be comparable: 28 % (73 %) for $E = 10^{-3}$, 16 % (49 %) for $E = 10^{-4}$, 13 % (28 %) for $E = 10^{-5}$, and 25 % (43 %) for $E \lesssim 10^{-6}$. Thus, the modified dynamic Elsasser number Λ_d^* accurately captures the ratio of Lorentz to Coriolis forces in our suite of simulations.

In order to apply the modified dynamic Elsasser number to planetary cores, its applicability to core conditions must first be evaluated. Our datasets cover a broad range of diagnostic physical parameters: magnetic field strengths of $10^{-1} \lesssim \Lambda_i \lesssim 10^2$, magnetic induction to diffusion ratios of $10^2 \lesssim Rm \lesssim 10^4$, and Lorentz to Coriolis force ratios of $10^{-2} \lesssim \Lambda_d \lesssim 1$. However, these ranges span only a fraction of the planetary parameter estimates, which vary over $10^{-5} \lesssim \Lambda_i \lesssim 10^{10}$, $10^2 \lesssim Rm \lesssim 10^5$, and $10^{-6} \lesssim \Lambda_d^* \lesssim 10^7$ as detailed in the "Discussion" section. The most fundamental deviation is for $\Lambda_d > 1$, which implies a different dominant force balance. While our dataset includes cases with $\Lambda_d > 1$, these simulations are strongly driven with inertia exceeding both Lorentz and Coriolis forces (see Fig. 4 of Soderlund et al. 2012), which is not expected for most planets (cf. Soderlund et al. 2013).

Figure 5 shows the ratio of the dynamic to modified dynamic Elsasser numbers as a function of the convective Rossby number, $Ro_c = \left(RaE^2/Pr\right)^{1/2}$, which characterizes the ratio of buoyancy-driven inertial forces to Coriolis forces (e.g., Gilman 1977, Aurnou et al. 2007). The Λ_d^* approximation works well when the Coriolis force dominates; in contrast, the largest discrepancy occurs when $Ro_c > 1$. Geodynamo simulations from Dormy (2014) provide a Λ_d^* test for cases with $\Lambda_d > 1$ and $Ro_c < 1$ at an Ekman number of $E = 1.5 \times 10^{-4}$ and a Prandtl number of unity. Here, large magnetic Prandtl numbers ($Pm \geq 12$) are required to obtain dynamo action since the inertial forces are relatively weak (i.e., quasi-laminar flows). Application of Eq. 18 to his dataset yields $0.85 \leq \Lambda_d/\Lambda_d^* \leq 1.3$, implying that our results are applicable to strong field systems with dominant Lorentz forces.

A discrepancy between input parameters also exists between the simulations and planetary cores. In particular, all numerical dynamo models are necessarily limited to massively overestimated kinematic viscosities, which prohibits simulations with realistic Ekman and magnetic Prandtl numbers: $10^{-19} \lesssim E \lesssim 10^{-12}$ and $10^{-8} \lesssim Pm \lesssim 10^{-6}$ are estimated for planets with active dynamos (e.g., Schubert and Soderlund 2011). In contrast, our dataset considers $10^{-7} \lesssim E \lesssim 10^{-3}$ and $10^{-1} \lesssim Pm \lesssim 10^1$. Importantly, no clear Ekman or magnetic Prandtl number dependence is identified in Fig. 5. We, therefore, hypothesize that the modified dynamic Elsasser number (Eq. 18) can be extrapolated to extreme planetary parameters.

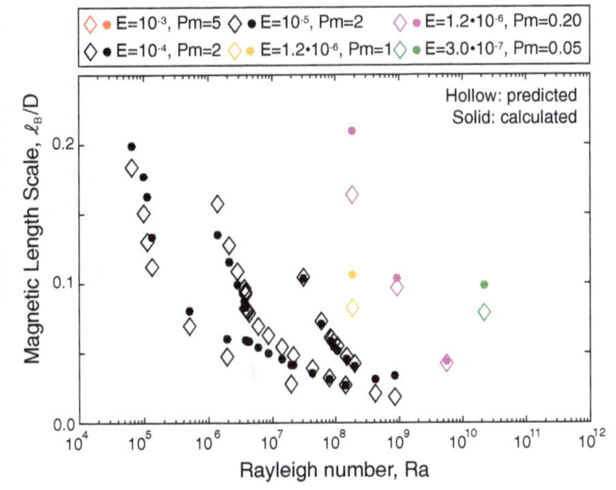

Fig. 4 Calculated and predicted magnetic length scales. Magnetic length scale as a function of the Rayleigh number with Ekman and magnetic Prandtl numbers denoted by *color*. A best-fit between the calculated and predicted values is obtained when a pre-factor is included in (Eq. 10): $\ell_B/D = 1.3Rm^{-1/2}$ (included here)

Discussion

The modified dynamic Elsasser number can be used to estimate the Lorentz to Coriolis force ratio in planetary cores when Λ_i and Rm are known (see Fig. 6 and Table 3). The imposed Elsasser number is traditionally used to quantify magnetic field strengths of planets since the poloidal component (B_p) can be measured by spacecraft and estimates for the other components (density ρ, magnetic diffusivity η, and rotation rate Ω) are relatively

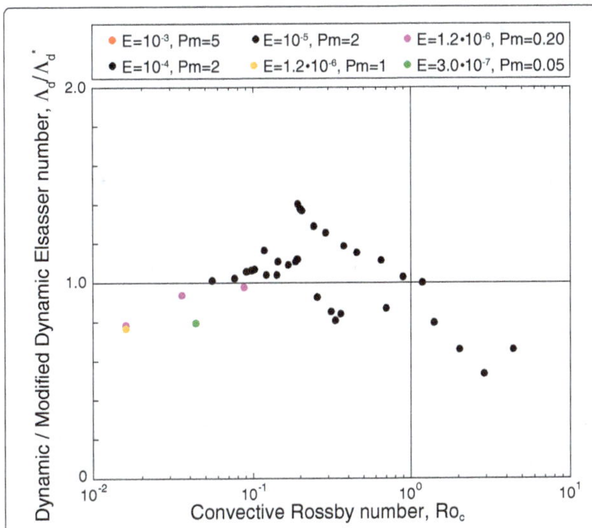

Fig. 5 Comparison of predicted and calculated Elsasser numbers as a function of the relative strengths of buoyancy and Coriolis forces. Ratio of Λ_d/Λ_d^* plotted versus the convective Rossby number, $Ro_c = \sqrt{RaE^2/Pr}$

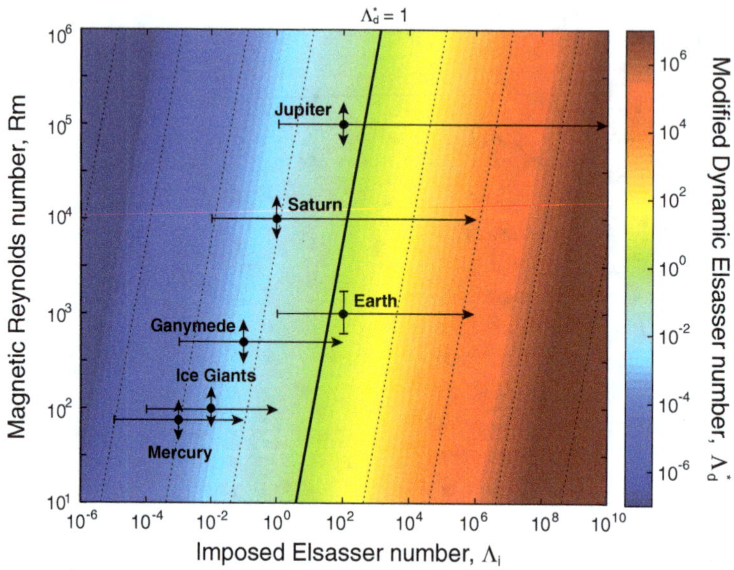

Fig. 6 Lorentz to Coriolis force ratio predictions for planetary cores. Modified dynamic Elsasser number as a function of imposed Elsasser number and magnetic Reynolds number across a range of values that may be relevant to planets in our solar system as well as exoplanets. The approximate ranges of anticipated Λ_i and *Rm* values are given for planets with active dynamos. *Barred line ends* for Λ_i denote the observed magnetic field strengths; *arrowed line ends* denote upper bounds. *Barred line ends* for *Rm* denote constrained bounds; *arrowed line ends* denote a larger range of uncertainties. Black circles denote the most probable Λ_i and *Rm* values

straightforward (see Schubert and Soderlund 2011 for a review); these measurements constitute a lower bound since the toroidal component is not included and small-scale poloidal contributions are below the resolution of detection. Toroidal magnetic fields (B_T) can be amplified by stretching of the poloidal field due to azimuthal differential rotation (zonal flows). This so-called ω-effect can generate toroidal magnetic fields up to $B_T \sim RmB_p$ (Roberts 2007), which sets an Λ_i upper bound. For a best estimate, we assume the total core field strength B to be an order of magnitude larger than the direct measurements (i.e., Λ_i increases by a factor of 100). This assumption is motivated both by geomagnetic data (e.g., Shimizu et al. 1998) and geodynamo modeling results (e.g., Aubert et al. 2009).

Earth: Geodynamo core flow inversions based on secular variation of the magnetic field imply velocities of $U \sim 5 \times 10^{-4}$ m/s (e.g., Holme 2007; Aubert 2013), which corresponds to $Rm \sim 10^3$ assuming $\eta \sim 1$ m^2/s for liquid metals. Ensemble inversions further suggest RMS magnetic field strengths in the cylindrical direction of $B_S \sim 2$ mT (Gillet et al. 2010), while toroidal magnetic field constraints of $1 < B_T/B_S < 100$ have been inferred by Shimizu et al (1998). Thus, $B \sim 10B_S \sim 20$ mT appears to be a reasonable estimate of the core's total field strength such that $\Lambda_i \sim 10^2$. In contrast, direct magnetic field measurements suggest a lower bound of $\Lambda_i \sim 1$ and application of the ω-effect suggests an upper bound of $\Lambda_i \sim 10^6$. We, therefore, predict a Lorentz to Coriolis force ratio of $\Lambda_d^* \sim 1$ for the geodynamo, with a possible range of $10^{-2} \lesssim \Lambda_d^* \lesssim 10^4$. This result suggests that the Lorentz force cannot be neglected dynamically in the core and, more likely, that the core is in magnetostrophic balance.

Table 3 Order of magnitude estimates of the magnetic Reynolds number *Rm*, the imposed Elsasser number Λ_i, and the Lorentz to Coriolis force ratio as estimated by Λ_d^* for planetary cores with active dynamos

	Mercury	Earth	Ganymede	Jupiter	Saturn	Ice Giants
Rm	75	10^3	500	10^5	10^4	10^2
Λ_i	10^{-3}	10^2	10^{-1}	10^2	10^0	10^{-2}
Λ_d^*	10^{-4}	10^0	10^{-3}	10^{-1}	10^{-2}	10^{-3}

Mercury and Ganymede: No observation-based velocity constraints are available currently for Mercury or Ganymede, so dynamo models that capture many of their magnetic field characteristics may serve as a basis for determining the magnetic Reynolds number. For example, Cao et al. (2014) obtain a Mercury-like dynamo when convection is driven by volumetrically distributed buoyancy sources and the core-mantle boundary heat flux peaks at low latitudes; their best fitting case has $Rm = 75$. Similarly, Christensen (2015) find Ganymede-like dynamos driven by iron snow with a stably stratified layer below the outer shell boundary; $Rm \approx 500$ in his optimal simulation. Using the upper and lower bound constraints derived from the ω-effect and observations, respectively, the imposed Elsasser number is estimated to vary between

$\Lambda_i \sim 10^{-5}$ and $\sim 10^{-1}$ for Mercury, which corresponds to $\Lambda_d^* \lesssim 10^{-2}$ and implies that magnetostrophic balance is unlikely in the Hermian core. Conversely, the possible magnetic field strengths for Ganymede range from $\Lambda_i \sim 10^{-3}$ to $\sim 10^2$, yielding $10^{-5} \lesssim \Lambda_d^* \lesssim 1$. However, the best estimate for the imposed Elsasser number is $\Lambda_i \sim 10^{-1}$ such that $\Lambda_d^* \sim 10^{-3}$. As a result, magnetostrophic balance is possible, but unlikely, in Ganymede's core.

Gas Giants: Jovian velocities can be estimated through scaling arguments and potentially secular variation. Jones (2014) suggests flow velocities of $U \gtrsim 10^{-3}$ m/s, leading to values of $Rm \gtrsim 10^5$ given the planet's large size. In this case, magnetic field strengths range from $\Lambda_i \sim 1$ via core field measurements to $\Lambda_i \sim 10^{10}$ when the ω-effect is included, with a best estimate of $\Lambda_i \sim 10^2$. The corresponding Lorentz to Coriolis force ratio then spans the range $10^{-3} \lesssim \Lambda_d^* \lesssim 10^7$, with $\Lambda_d^* \sim 0.1$ being the best estimate. Thus, barring the large uncertainties, the Lorentz force is predicted to be sub-dominant, but not negligible, in Jupiter's core.

Starchenko and Jones (2002) predict similar flow speeds for the Saturnian interior, which corresponds to a lower $Rm \sim 10^4$ value due to the smaller size of the dynamo region. Saturn's magnetic field is also weaker than that of Jupiter; the best estimate imposed Elsasser number is $\Lambda_i \sim 1$ with a feasible range of $10^{-2} \lesssim \Lambda_i \lesssim 10^6$. In the most likely scenario, the Lorentz force is predicted to be two orders of magnitude smaller than the Coriolis force ($\Lambda_d^* \sim 10^{-2}$) with uncertainties extending the possible range to $10^{-4} \lesssim \Lambda_d^* \lesssim 10^4$.

Ice Giants: Uranus and Neptune have relatively weak core magnetic fields with $\Lambda_i \sim 10^{-4}$ based on spacecraft observations, but there are few constraints on the flow speeds. We, therefore, assume $U \sim 10^{-3}$ m/s based on the other planetary estimates such that $Rm \sim 10^2$ to give an upper bound of $\Lambda_i \sim 1$ when the toroidal estimate is included. For these Λ_i estimates, the predicted Lorentz to Coriolis force ratio is always less than unity: $10^{-5} \lesssim \Lambda_d^* \lesssim 10^{-1}$. These results thus imply that the ice giants are not in magnetostrophic balance.

Conclusions

The Lorentz to Coriolis force ratio is an important parameter for the dynamics of planetary cores since dynamos with dominant Coriolis forces at global scales are expected to be driven by fundamentally different archetypes of fluid motions than those with dominant (or co-dominant) Lorentz forces (Roberts and King 2013, Calkins et al. 2015). We hypothesize that a representative global estimate of the Lorentz to Coriolis force ratio can be predicted by $\Lambda_d^* = 0.77\Lambda_i Rm^{-1/2}$. An advantage of this formulation is that it depends on quantities that can be estimated for planetary cores (Table 3). Our results suggest that the Earth's core is likely to be in magnetostrophic balance where the Lorentz and Coriolis forces are comparable. The Lorentz force may also be substantial in Jupiter's core, where it is predicted to be a factor of ten less than the Coriolis force. Magnetic fields become increasingly sub-dominant for the other planets: the Coriolis force is predicted to exceed the Lorentz force by at least two orders of magnitude within the cores of Saturn, Uranus/Neptune, Ganymede, and Mercury.

These conclusions are subject to large uncertainties, however. Core flow speeds are difficult to estimate, while total magnetic field strengths cannot be measured directly. The applicability of Eq. 18 for the Lorentz to Coriolis force ratio may also break down at extreme planetary core conditions that cannot be explored numerically or in the laboratory due to technological limitations. In order to mitigate these uncertainties, we have considered a range of possible core magnetic field strengths (Λ_i) and included state-of-the-art simulations (Sheyko 2014; cf. Nataf and Schraeffer 2015).

Abbreviations

MHD: Magnetohydrodynamics; RMS: Root mean square; SKA: Soderlund et al. (2012, 2014); S14: Sheyko 2014.

Competing interests

The authors declare that they have no competing interests.

Authors' contributions

KMS and JMA designed the study. KMS and AS carried out and analyzed the numerical simulations. All authors interpreted the data. All authors read and approved the final manuscript.

Acknowledgements

The authors thank two anonymous referees for their thoughtful reviews, Hao Cao for helpful suggestions, and Wolfgang Bangerth for enlightening discussions at the 2014 Study of Earth's Deep Interior (SEDI) meeting in Kanagawa, Japan. KMS gratefully acknowledges Japan Geoscience Union the National Science Foundation to attend this symposium as well as research support from the National Science Foundation (grant AST-0909206). JMA acknowledges the support of the National Science Foundation Geophysics Program (grant EAR-1246861). Computational resources supporting this work were provided by the NASA High-End Computing (HEC) Program through the NASA Advanced Supercomputing (NAS) Division at Ames Research Center and by the Swiss National Supercomputing Centre (CSCS) under project ID s225. This is UTIG contribution 2858.

Author details

[1]Institute for Geophysics, John A. & Katherine G. Jackson School of Geosciences, The University of Texas at Austin, J. J. Pickle Research Campus, Building 196 (ROC), 10100 Burnet Road (R2200), Austin, Texas, 78758-4445, USA. [2]Institut für Geophysik, ETH Zürich, NO H 11.2, Sonneggstrasse 5 8092 Zürich, LA59 Switzerland. [3]U.S. Global Development Lab, U.S. Agency for International Development, Washington, DC, 20004, USA. [4]Department of Earth, Planetary, and Space Sciences, University of California, Los Angeles, 595 Charles Young Drive East, Los Angeles, California, 90095-1567, USA.

References

Aubert J, Aurnou JM, Wicht J (2008) The magnetic structure of convection-driven numerical dynamos. Geophys J Int 172:945–956

Aubert J, Labrosse S, Poitou C (2009) Modelling the palaeo-evolution of the geodynamo. Geophys J Int 179:1414–1428

Aubert J (2013) Flow throughout the Earth's core inverted from geomagnetic observations and numerical dynamo models. Geophys J Int 192:537–556

Aurnou JM, Heimpel MH, Wicht J (2007) The effects of vigorous mixing in a convective model of zonal flow on the Ice Giants. Icarus 190:110–126

Calkins MA, Julien K, Tobias SM, Aurnou JM (2015) A multiscale dynamo model driven by quasi-geostrophic convection. J Fluid Mech. doi:10.1017/jfm.2015.464

Cao H, Aurnou JM, Wicht J, Dietrich W, Soderlund KM, Russell CT (2014) A dynamo explanation for Mercury's anomalous magnetic field. Geophys Res Lett 41(12):4127–4134

Cardin P, Olson PL (1995) The influence of toroidal magnetic field on thermal convection in the core. Earth Planet Sci Lett 133:167–181

Cardin P, Brito D, Jault D, Nataf HC, Masson JP (2002) Towards a rapidly rotating liquid sodium dynamo experiment. Magnetohydrodynamics 38:177–189

Cheng JS, Stellmach S, Ribeiro A, Grannon A, King EM, Aurnou JM (2015) Laboratory-numerical models of rapidly rotating convection in planetary cores. Geophys J Int 201:1–17

Christensen U. R, Wicht J (2007) Numerical Dynamo Simulations. In: Schubert G (ed). Treatise on Geophysics, Core Dynamics. Elsevier, Amsterdam Vol. 8. pp 245–282. Chap. 8

Christensen UR (2015) Iron snow dynamo models for Ganymede. Icarus 247:248–259

Dharmaraj G, Stanley S (2012) Effect of inner core conductivity on planetary dynamo models. Phys Earth Planet Int 212–213:1–9

Dharmaraj G, Stanley S, Qu AC (2014) Scaling laws, force balances and dynamo generation mechanisms in numerical dynamo models: influence of boundary conditions. Geophys J Int 199:514–532

Dormy E, Soward AM, Jones CA, Jault D, Cardin P (2004) The onset of thermal convection in rotating spherical shells. J Fluid Mech 501:43–70

Dormy E (2014) Strong field spherical dynamos. arXiv:1412.4090v1

Galloway DJ, Proctor MRE, Weiss NO (1978) Magnetic flux ropes and convection. J Fluid Mech 87:243–261

Gilman PA (1977) Nonlinear dynamics of Boussinesq convection in a deep rotating spherical shell – I. Geophys Astrophys Fluid Dyn 8:93–135

Gillet N, Jault D, Canet E, Fournier A (2010) Fast torsional waves and strong magnetic field within the Earth's core. Nature 465:74–77

Grooms I, Julien K, Weiss JB, Knobloch E (2010) Model of convective Taylor columns in rotating Rayleigh-Bénard convection. Phys Rev Lett 104:224501

Holme R (2007) Large-scale flow in the core. In: Schubert G (ed). Treatise on Geophysics, Core Dynamics. Elsevier, Amsterdam Vol. 8. pp 107–130. Chap. 4

Jones CA, Soward AM, Mussa AI (2000) The onset of thermal convection in a rapidly rotating sphere. J Fluid Mech 405:157–179

Jones CA (2014) A dynamo model of Jupiter's magnetic field. Icarus 241:148–159

Julien K, Knobloch E, Werne J (1998) A new class of equations for rotationally constrained flows. Theoret Comput Fluid Dynamics 11:251–261

Julien K, Rubio AM, Grooms I, Knobloch E (2012) Statistical and physical balances in low Rossby number Rayleigh-Bénard convection. Geophys Astrophys Fluid Dyn 106:392–428

Kageyama A, Miyagoshi T, Sato T (2008) Formation of current coils in geodynamo simulations. Nature 454:1106–1108

King EM, Aurnou JM (2015) Magnetostrophic balance as the optimal state for turbulent magnetoconvection. Proc Natl Acad Sci 112:990–994

Le Bars M, Cebron D, Le Gal P (2015) Flows driven by libration, precession, and tides. Annu Rev Fluid Mech 47:163–193

Nataf H-C, Schaeffer N (2015) Turbulence in the core. In: Olson P, Schubert G (eds). Treatise on Geophysics, 2nd ed., Core Dynamics. Elsevier BV, Amsterdam. pp 161–181. Chap. 6

Olson PL, Glatzmaier GA (1996) Magnetoconvection and thermal coupling of the Earths core and mantle. Philos Trans R Soc London Ser A 354:1413–1424

Olson PL, Christensen UR, Glatzmaier GA (1999) Numerical modeling of the geodynamo: Mechanisms of field generation and equilibration. J Geophys Res 104:10383–10404

Ribeiro A, Fabre G, Guermond JL, Aurnou JM (2015) Canonical models of geophysical and astrophysical flows: Turbulent convection experiments in liquid metals. Metals 5:289–335

Roberts PH (1968) On the thermal instability of a rotating-fluid sphere containing heat sources. Philos Trans R Soc London Ser A 264:93–117

Roberts, PH (2007) Theory of the geodynamo. In: Schubert G (ed). Treatise on Geophysics, Core Dynamics. Elsevier, Amsterdam Vol. 8. pp 67–105. Chap. 3

Roberts PH, King EM (2013) On the genesis of the Earth's magnetism. Rep Prog Phys 76(9):096801

Schubert G, Soderlund KM (2011) Planetary magnetic fields: Observations and models. Phys Earth Planet Int 187:92–108

Sheyko A (2014) Numerical investigations of rotating MHD in a spherical shell. Dissertation, ETH ZURICH

Shimizu H, Koyama T, Utada H (1998) An observational constraint on the strength of the toroidal magnetic field at the CMB by time variation of submarine cable voltages. Geophys Res Lett 25:4023–4026

Soderlund KM, King EM, Aurnou JM (2012) The influence of magnetic fields in planetary dynamo models. Earth Planet Sci Lett 333-334:9–20

Soderlund KM, Heimpel MH, King EM, Aurnou JM (2013) Turbulent models of ice giant internal dynamics: Dynamos, heat transfer, and zonal flows. Icarus 224:97–113

Soderlund KM, King EM, Aurnou JM (2014) Corrigendum to "The influence of magnetic fields in planetary dynamo models". Earth Planet Sci Lett 392:121–123

Sprague M, Julien K, Knobloch E, Werne J (2006) Numerical simulation of an asymptotically reduced system for rotationally constrained convection. J Fluid Mech 551:141–174

Sreenivasan B, Jones CA (2011) Helicity generation and subcritical behavior in rapidly rotating dynamos. J Fluid Mech 688:5–30

Sreenivasan B, Sahoo S, Gaurav D (2014) The role of buoyancy in polarity reversals of the geodynamo. Geophys J Int 199:1698–1708

Starchenko SV, Jones CA (2002) Typical velocities and magnetic field strengths in planetary interiors. Icarus 157:426–435

Stellmach S, Lischper M, Julien K, Vasil GM, Cheng JS, Ribeiro A, King EM, Aurnou JM (2014) Approaching the asymptotic regime of rapidly rotating convection: Boundary layers versus interior dynamics. Phys Rev Lett 113:254501

Tritton DJ (1998) Physical Fluid Dynamics. Oxford University Press, Oxford

Wicht J (2002) Inner-core conductivity in numerical dynamo simulations. Phys Earth Planet Int 132:281–302

Zhang K (1992) Spiraling columnar convection in rapidly rotating spherical fluid shells. J Fluid Mech 236:535–554

Zhang K, Schubert G (2000) Magnetohydrodynamics in rapidly rotating spherical systems. Annu Rev Fluid Mech 32:409–443

12

Asteroid Ryugu before the Hayabusa2 encounter

Koji Wada[1*], Matthias Grott[2], Patrick Michel[3], Kevin J. Walsh[4], Antonella M. Barucci[5], Jens Biele[6], Jürgen Blum[7], Carolyn M. Ernst[8], Jan Thimo Grundmann[9], Bastian Gundlach[7], Axel Hagermann[10], Maximilian Hamm[2], Martin Jutzi[11], Myung-Jin Kim[12], Ekkehard Kührt[2], Lucille Le Corre[13], Guy Libourel[3], Roy Lichtenheldt[14], Alessandro Maturilli[2], Scott R. Messenger[15], Tatsuhiro Michikami[16], Hideaki Miyamoto[17], Stefano Mottola[2], Thomas Müller[18], Akiko M. Nakamura[19], Larry R. Nittler[20], Kazunori Ogawa[19], Tatsuaki Okada[21], Ernesto Palomba[22], Naoya Sakatani[21], Stefan E. Schröder[2], Hiroki Senshu[1], Driss Takir[23], Michael E. Zolensky[15] and International Regolith Science Group (IRSG) in Hayabusa2 project

Abstract

Asteroid (162173) Ryugu is the target object of Hayabusa2, an asteroid exploration and sample return mission led by Japan Aerospace Exploration Agency (JAXA). Ground-based observations indicate that Ryugu is a C-type near-Earth asteroid with a diameter of less than 1 km, but the knowledge of its detailed properties is very limited prior to Hayabusa2 observation. This paper summarizes our best understanding of the physical and dynamical properties of Ryugu based on ground-based remote sensing and theoretical modeling before the Hayabusa2's arrival at the asteroid. This information is used to construct a design reference model of the asteroid that is used for the formulation of mission operation plans in advance of asteroid arrival. Particular attention is given to the surface properties of Ryugu that are relevant to sample acquisition. This reference model helps readers to appropriately interpret the data that will be directly obtained by Hayabusa2 and promotes scientific studies not only for Ryugu itself and other small bodies but also for the solar system evolution that small bodies shed light on.

Keywords: Hayabusa2, Ryugu, Asteroids, Regolith, Physical properties

Introduction

Asteroids and comets are relicts of the early stages of solar system evolution. They contain materials and structures which are relatively primitive compared to those composing the planets and thus provide a window into the earliest stages of planet formation. Investigation of various properties of small bodies thus provides the crucial knowledge needed to understand the origin and evolution of the solar system. The asteroid (162173) Ryugu is a C-type near-Earth asteroid (NEA) with a diameter of less than 1 km. It is the target of Hayabusa2, the second asteroid sample return mission led by the Japan Aerospace Exploration Agency (JAXA) (Tsuda et al. 2013). Ryugu will offer a great opportunity to understand the present status and the evolutionary history of volatile-rich primitive materials in the solar system (Watanabe et al. 2017). This expectation is supported by the fact that Hayabusa, the previous and first asteroid exploration and sample return mission by JAXA, acquired substantial data from remote sensing observations and the returned particles of the S-type NEA (25143) Itokawa (e.g., Fujiwara et al. 2006; Miyamoto et al. 2007; Nagao et al. 2011; Nakamura et al. 2011; Yoshikawa et al. 2015). Although small bodies are full of treasures, only a limited number have been visited by other rendezvous spacecraft so far, such as asteroid (433) Eros (e.g., Cheng 2002) and comet 67P/Churyumov-Gerasimenko (e.g., Sierks et al. 2015). C-type (and related B-type) asteroids are considered to be the most primitive bodies in the inner solar system as they are inferred to be relatively rich in volatile materials like water and organics. The only

* Correspondence: wada@perc.it-chiba.ac.jp
[1]Planetary Exploration Research Center, Chiba Institute of Technology, Tsudanuma 2-17-1, Narashino, Chiba 275-0016, Japan
Full list of author information is available at the end of the article

C-type asteroid visited by a spacecraft so far is the main belt asteroid (253) Mathilde, but it was a flyby observation and full global mapping images were not obtained (Thomas et al. 1999). Thus, we do not have yet any detailed information on the most primitive types of asteroids, other than laboratory analysis of primitive meteorites, but the specific sources or origins of these are unknown.

Hayabusa2 arrives at Ryugu at the end of June 2018 and will stay around Ryugu for 1 year and a half (Tsuda et al. 2013; Watanabe et al. 2017). In addition, the NASA asteroid sample return mission OSIRIS-REx will arrive at a B-type asteroid (101955) Bennu (~ 500 m in diameter) in August of the same year (Lauretta et al. 2017). These will be the first volatile-rich asteroids for which we will obtain high-resolution images of nearly the entire surface and will provide critical insight as to whether observed characteristics of boulders, craters, and surface roughness (in addition to other properties discussed elsewhere in this paper) are representative of similarly sized, primitive asteroids.

Since Ryugu has never been visited before, the limited information before Hayabusa2's arrival should be reviewed and appropriately set out in order to guide the Hayabusa2 mission toward success. For this purpose, the International Regolith Science Group (IRSG) was organized as a branch of the Interdisciplinary Science Team in the Hayabusa2 project. In this paper, a product of an IRSG activity, we review the information about Ryugu before the encounter of Hayabusa2. We focus especially on the properties of the surface regolith layer, i.e., its global properties including surface geological features ("Global properties" section), its thermophysical properties ("Regolith thermophysical properties" section), and its mechanical properties ("Regolith mechanical properties" section). The information described in each section falls into three categories in terms of our present knowledge: the measured data obtained so far, the properties directly derived/deduced from the present data with basic physics, and the predicted properties. As a whole, this paper provides a reference model of Ryugu's regolith with the categorized data/properties. This model will be useful for Hayabusa2 operations such as landing site selection and touchdown operations as well as for interpreting and understanding data that will be obtained by the remote sensing observations of Hayabusa2 and by the analysis of samples returned from the asteroid. One might have a question about an expiration date of this paper after the end of the Hayabusa2 mission. The reference model proposed in this paper, however, will keep its validity even after we have a lot of observational data on Ryugu because the model offers a clear example, in a well-organized form, of what information we can draw about unexplored asteroids and other small bodies and will be a useful guide for future missions planning to visit small, unexplored bodies.

Hayabusa2 carries a suite of instruments for scientific observation. These include three optical navigation cameras (ONC), consisting of a telescopic camera with seven band filters (ONC-T) and two wide field of view cameras (ONC-W1 and W2), a near-infrared spectrometer (NIRS3), a thermal infrared imager (TIR), and a light detection and ranging (LIDAR) instrument. The surface of Ryugu will be investigated with these instruments. For example, global and local mapping of topography, such as craters and boulders, will be carried out by ONC imaging (Kameda et al. 2017). The spectroscopic properties of the surface will be revealed by ONC-T and NIRS3 observations, e.g., for detection of hydrated minerals (Kameda et al. 2015; Iwata et al. 2017). The thermophysical properties such as the thermal inertia derived from the surface temperature distribution will be explored by TIR observations, and these will help to estimate the particle size and the porosity of the regolith layer (Okada et al. 2017; Arai et al. 2017; Takita et al. 2017). LIDAR will be used for the albedo observation and dust detection around Ryugu, in addition to measuring the distance between the Hayabusa2 spacecraft and Ryugu's surface (Namiki et al. 2014; Mizuno et al. 2017; Yamada et al. 2017; Senshu et al. 2017). Two landers (rovers) are also carried by Hayabusa2, called MINERVA-II and MASCOT. In particular, MASCOT, developed by the German Aerospace Centre (DLR) in collaboration with the Centre National d'Etudes Spatiales (CNES), has several instruments for scientific investigation on the surface (Ho et al. 2017): a camera with illumination unit (MasCam, Jaumann et al. 2017), a near- to mid-infrared spectromicroscope (MicrOmega, Bibring et al. 2017), a multi-channel radiometer (MARA, Grott et al. 2017), and a magnetometer (MasMag, Herčík et al. 2017). The close observation of the regolith surface by each of MASCOT's instruments will provide "ground truth" of the remote sensing data obtained by the Hayabusa2 spacecraft, such as the particle size distribution, mineral composition, and thermal inertia. In addition to the above remote sensing data, the mechanical properties of the regolith layer will be investigated by an impact experiment carried out with the so-called Small Carry-on Impactor (SCI) and in situ observation with a deployable camera (DCAM3) (Saiki et al. 2017; Arakawa et al. 2017; Ogawa et al. 2017; Ishibashi et al. 2017; Sawada et al. 2017). If the trajectory of MASCOT during its descent and after rebound on the surface is traceable, its analysis can also be used to estimate the mechanical properties of the regolith layer (e.g., Thuillet et al. 2018). After the end of the mission at Ryugu, and with the benefit of analyses of the returned samples, the reference model built in this paper will be fully checked and refined, and our knowledge of small bodies in the solar system will be considerably improved.

Review

Global properties

Here, the global properties of Ryugu are presented along with the best knowledge of the uncertainties in the measurements. The described properties provide essential inputs for mission planning in advance of Ryugu's arrival. In particular, the design reference mission requires precise knowledge of Ryugu's orbit and dynamical state, size and shape, and near-surface spectral reflectance properties.

Measured quantities

Orbital properties The observations used for orbit determination of Ryugu date from 1986 and include 725 astrometric reports included in the JPL-Horizons database (from 16 May 2017). The derived orbit, computed by JPL, is listed in Table 1. Ryugu's perihelion (q) of 0.963 AU and aphelion (Q) of 1.416 AU classify it as an Apollo-type near-Earth object. Its Earth minimum orbit intersection distance (MOID) of 0.00111549 AU is among the smallest values for asteroids larger than 100 m (JPL/SSD).

Numerical modeling of NEA orbital evolution indicates that Ryugu originated in the inner asteroid belt. According to a model by Bottke et al. (2002), Ryugu is statistically most likely to have followed the ν_6 secular resonance pathway from the main asteroid belt (Campins et al. 2013). This pathway marks the inner boundary of the main asteroid belt at ~ 2.1 AU and is the most efficient way to deliver bodies into near-Earth space, accounting for most of those that reach very Earth-like orbits characterized by low-MOID values and low delta-V space mission trajectories. Several models suggest that Ryugu likely originated in the inner main asteroid belt, between ~ 2.1 and 2.5 AU, and reached ν_6 by inward Yarkovsky drift (Campins et al. 2013, Walsh et al. 2013, Bottke et al. 2015).

Table 1 Summary of important orbital and global properties. The orbital data are taken from JPL's SSD website (NASA's Jet Propulsion Laboratory Solar Systems Dynamics Group - HORIZONS at http://ssd.jpl.nasa.gov/). Rotation period is from numerous references (Kim et al. 2013; Müller et al. 2017; Perna et al. 2017)

Eccentricity	$0.190225753 \pm 4.1 \times 10^{-8}$
Semi-major axis (AU)	$1.1896008220 \pm 3.3 \times 10^{-9}$
Inclination (°)	$5.8837424 \pm 4.0 \times 10^{-6}$
Period (day)	$473.9148422 \pm 2.0 \times 10^{-6}$
Perihelion (AU)	$0.963308109 \pm 4.8 \times 10^{-8}$
Aphelion (AU)	$1.4158935346 \pm 4.0 \times 10^{-9}$
Rotation period (h)	7.625 ± 0.003

Working from the scenario of an inner main belt origin for Ryugu, it is possible to seek relationships with known asteroid families. Campins et al. (2013) suggested that the spectral taxonomy of Ryugu prevents a solid match with known families, while Bottke et al. (2015) studied the dynamical possibilities of delivery from either the New Polana or Eulalia families. This work found that Ryugu was more likely to have originated from the ~ 1 Gyr old New Polana family that is centered at 2.42 AU. Surveys of these two families in visible and near-infrared wavelengths find no significant differences between them, leaving Ryugu's origin uncertain (Pinilla-Alonso et al. 2016; de León et al. 2016).

Size, shape, and spin The reference Ryugu shape model was reconstructed from the inversion of optical and thermal infrared data by Müller et al. (2017). Because of the small amplitude of the obtained light curves and low quality of the optical light curves, the inversion did not lead to a unique solution for the pole, shape, and period. Formally, there were different sets of these parameters that fit the available data equally well. However, after a careful analysis, Müller et al. (2017) concluded that the most likely pole direction in ecliptic coordinates is lambda = 310–340° and beta = – 40° ± 15°. The corresponding shape model is roughly spherical with a volume-equivalent diameter of 850–880 m (see Fig. 1).

Ryugu has a rotation period of approximately 7.625 ± 0.003 (Kim et al. 2013; 7.63 h with no uncertainties in Perna et al. (2017). Müller et al. (2017) provided more possible values for the sidereal rotation period (7.6300, 7.6311, 7.6326 h), but from the analysis of the new data from Very Large Telescope (VLT) (Perna et al. 2017) and other observatories (Kim et al. 2016), the sidereal rotation period of 7.6326 h is preferred (Durech, pers. comm.). As shown in Fig. 2, the light curve has an asymmetric behavior with two different maxima and minima. The asymmetric trend of the light curve is connected to either the irregular shape or variations in albedo.

Phase function As discussed in Le Corre et al. (2018), the spherical albedo and phase integral of Ryugu are useful quantities for deriving the bolometric Bond albedo map. These quantities allow us to make temperature prediction across the surface of this asteroid during landing site selection events of the Hayabusa2 mission. These quantities can also be useful to calculate Ryugu's thermal inertia and constrain the Yarkovsky effect.

Le Corre et al. (2018) used ground-based photometric data of Ryugu, which were compiled by Ishiguro et al. (2014) and references therein, to constrain the average

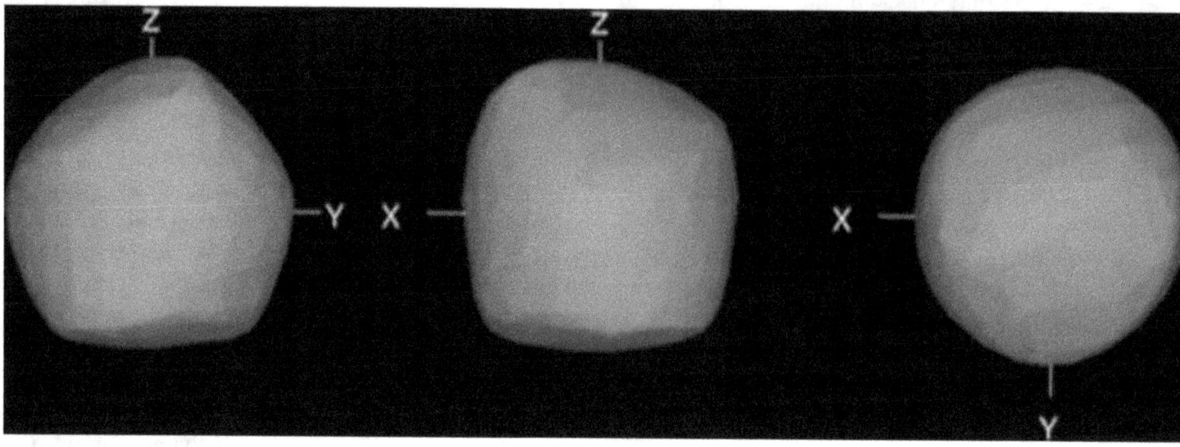

Fig. 1 Best fit shape model for Ryugu. Image taken from Müller et al. 2017

disk-resolved brightness across Ryugu's surface. The radiance factor (RADF) is the ratio of the bidirectional reflectance of a surface to that of a perfectly diffuse surface illuminated at the incident angle $i = 0$ (Hapke 2012). Reflectance, $r_{LS}(i, e, \alpha)$, is directly related to RADF(i, e, α) or $[I/\mathscr{F}](i, e, \alpha)$ as described in:

$$[I/\mathscr{F}]\,(i, e, \alpha) = \mathrm{RADF}(i, e, \alpha) = \pi r_{\mathrm{LS}}(i, e, \alpha), \qquad (1)$$

where i, e, and α being the incidence angle, the emission angle, and the phase angle, respectively. I is the radiance and has units of W m^{-2} nm^{-1} sr^{-1}. To dispel confusion, the value $J = \pi\mathscr{F}$ is the collimated light (irradiance) and has units of W m^{-2} nm^{-1}.

Under the assumption that the photometric properties of the surface are well described by a Lommel-Seeliger law, we can write:

$$[I/\mathscr{F}]\,(i, e, \alpha) = \frac{w_o}{4}\,\frac{\mu_o f(\alpha)}{\mu_o + \mu}, \qquad (2)$$

where $\mu_o = \cos(i)$, $\mu = \cos(e)$, $A_{LS} = \frac{w_o}{4\pi}$ is the Lommel-Seeliger albedo, $f(\alpha)$ is the phase function, and w_o is the average particle single scattering albedo. We approximate the phase function with an exponential-polynomial function of the form $f(\alpha) = e^{\beta\alpha+\gamma\alpha^2+\delta\alpha^3}$ (Takir et al. 2015), the coefficients of which β, γ, and δ are fitted to the data.

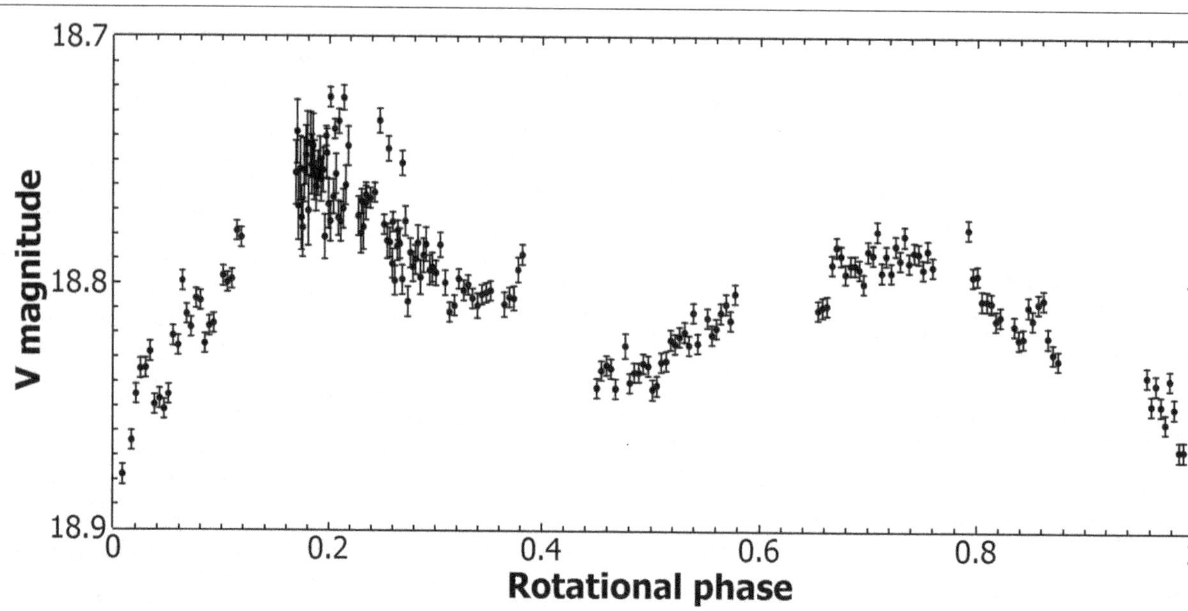

Fig. 2 Composite lightcurve of Ryugu, as observed at the Very Large Telescope of the European Southern Observatory (ESO-VLT) on 12 Jul. 2016 by Perna et al. (2017). The rotational phase corresponds to a synodic period of 7.63 h. The zero phase time corresponds to 57,581.0 MJD

Figure 3 shows Lommel-Seeliger models that capture low and high phase angle behavior and the scatter in the moderate phase angle ground-based observations of Ishiguro et al. (2014). Le Corre et al. (2018) computed models for nominal, maximum, and minimum predicted brightness of Ryugu at 550 nm (Table 2). Although this photometric model fitted the disk-integrated data of Ryugu well, it may not be an appropriate model for this asteroid's disk-resolved data especially at larger phase angles (> 100°) (e.g., Schröder et al. 2017). This Lommel-Seeliger model is a preliminary model and can be updated when the spacecraft arrives at Ryugu.

Table 3 includes the albedo quantities of Ryugu, computed with the Lommel-Seeliger model, and the quantities computed by Ishiguro et al. (2014). The two sets of quantities are consistent with each other.

Table 2 Lommel-Seeliger functions that predict $[I/\mathcal{F}]$ (i, e, α) (reflectance) of Ryugu at 550 nm. A_{LS} is Lommel-Seeliger Albedo and $f(\alpha) = e^{\beta\alpha+\gamma\alpha^2+\delta\alpha^3}$. Table from Le Corre et al. (2018)

	A_{LS}	β	γ	δ
Nominal	0.027	-4.14×10^{-2}	3.22×10^{-4}	19.16×10^{-7}
Maximum	0.030	-3.99×10^{-2}	3.27×10^{-4}	22.15×10^{-7}
Minimum	0.024	-4.46×10^{-2}	3.77×10^{-4}	21.33×10^{-7}

Spectral properties Ryugu is a primitive Apollo NEA classified as a C-type asteroid. At present, there are eight ground-based spectroscopic observations of this asteroid, three of which extend toward the near-IR (Table 4). In nearly all the observations, the spectrum is featureless, with a slight reddening in the IR and a slight UV. The best meteorite spectral counterpart is a heated CM chondrite or possibly a CI chondrite (Tonui et al. 2014); however, this comparison is complicated by the poorly known effects of space weathering on Ryugu.

The spectral analysis by Vilas (2008) at visible wavelengths showed a feature at 0.7 μm attributed to the Fe^{2+}- > Fe^{3+} charge transfer transition in oxidized phyllosilicates. This feature has not been detected in any other observations of Ryugu, and the author suggested a spatial heterogeneity in the asteroid surface composition, as already observed in other main-belt C-type asteroid spectra (see Rivkin et al. 2002).

The observations by Lazzaro et al. (2013) in the 0.4–0.85 μm spectral range covered 70% of Ryugu's surface and showed an almost uniform C-type featureless spectrum, with no direct evidence of hydration features.

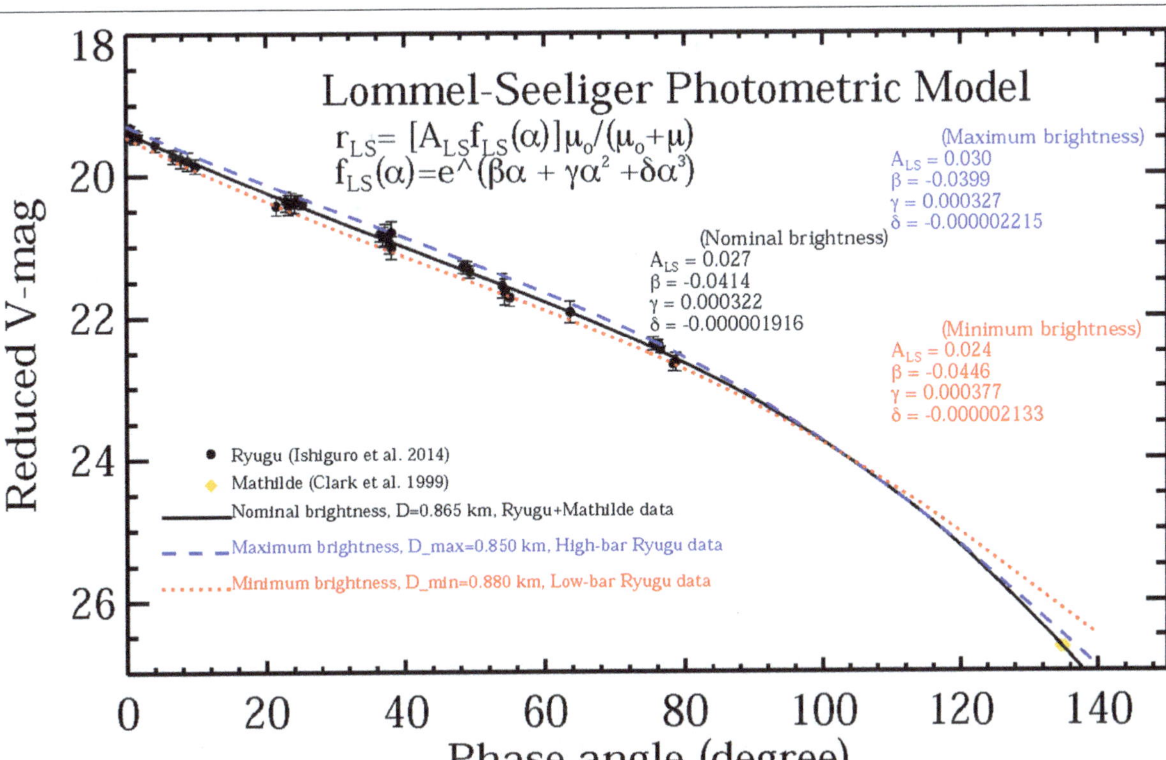

Fig. 3 The reduced V magnitude of Ryugu as a function of phase angle predicted by the Lommel-Seeliger model is shown compared with the ground-based measurements of Ishiguro et al. (2014) and references therein. Shown are the minimum (red dots), maximum (blue dashes), and nominal (black solid line) models. Our minimum and maximum models do not include the Mathilde data; however, our nominal model does. Reflectance r_{LS} is in units of sr^{-1}. Figure from Le Corre et al. (2018)

Table 3 Albedo quantities of Ryugu computed using the Lommel-Seeliger model and by Ishiguro et al. (2014). p_V is the geometric albedo, q is the phase integral, and A_B is the spherical Bond albedo. Table from Le Corre et al. (2018)

	p_V	q	A_B
Lommel-Seeliger	$0.042^{+0.005}_{-0.004}$	$0.34^{+0.01}_{-0.01}$	$0.014^{+0.001}_{-0.001}$
Ishiguro et al. (2014)	$0.047^{+0.003}_{-0.003}$	$0.32^{+0.03}_{-0.03}$	$0.014^{+0.002}_{-0.002}$

Few variations have been observed at wavelengths shorter than 0.5 μm, which lie in a UV drop-off as found in previous observations (Binzel et al. 2001). The UV drop-off is a common spectral feature of asteroids whose taxonomy is associated with carbonaceous chondrites, particularly for G and F asteroid classes (Tholen and Barucci 1989) or Cg in the Bus-DeMeo system (Bus and Binzel 2002; DeMeo et al. 2009). This spectral behavior was investigated and widely discussed by Hiroi et al. (1996, 2003) and Cloutis et al. (2012). These authors noticed decreased reflectance in the UV after heating or laser bombardment processing to simulate micrometeorite bombardment. These hypervelocity impacts cause localized melting and the production of spherical glassy droplets. The slight UV drop-off observed in different regions of Ryugu's surface could thus be explained by the effect of heterogeneous processing due to space weathering and/or episodes of substantial heating via, presumably, closer perihelion passages.

The IR observations by Abe et al. (2008) and Pinilla-Alonso et al. (2013) show a featureless/flat spectrum and confirm that Ryugu is a C-type asteroid in line with all the other observations. A very slight positive slope is observed. The two studies derive different slope values, which could be indicative of a heterogeneous surface; however, the visible observations by Moskovitz et al. (2013) reported a homogenous and flat spectrum in different observations of Ryugu, suggesting that possible heterogeneities should be constrained to 5% of the entire asteroid surface. This work proposed that Ryugu was best represented by a thermally altered sample of the Murchison (CM2) meteorite and by the thermally metamorphosed CI chondrite Yamato 86029. The most recent work by Perna et al. (2017) confirms all the previous observations with a slight UV drop-off and a featureless/flat spectrum in both the visible and near-IR (Fig. 4). Le Corre et al. (2018) observed Ryugu during its close flyby of the Earth in July 2016. Their spectrum differs from those presented in Moskovitz et al. (2013) and Perna et al. (2017), with a stronger red spectral slope shortward of 1.6 μm. However, like the previous observations, Ryugu's spectrum does not show any well-defined absorption bands. Le Corre et al. (2018) suggested the possible presence of two broad absorption bands centered around 1 and 2.2 μm in their data. They proposed that if both are present, these bands could be indicative of a CO or CV chondrite-like composition. These meteorites are typically too bright to match Ryugu's reflectance and would require darkening effects of space weathering to provide a match to the asteroid (Le Corre et al. 2018). These authors used curve fitting techniques to confirm that CM chondrites remain the best match. They also found two small asteroids with very similar spectra (showing a pronounced red slope) to Ryugu: NEA (85275) 1994 LY and Mars-crossing asteroid (316720) 1998 BE7. In addition, this work showed that the newly observed

Table 4 Spectroscopic observations of asteroid Ryugu since its discovery in 2001. Spectral and compositional characteristics are listed

Reference	Spectral interval (μm)	Taxonomy type (T = Tholen; BD = Bus-De Meo)	Meteoritical counterpart	Spectral characteristics
Binzel et al. 2001	0.4–0.9	Cg (BD)	N.A.	UV drop-off: red (0.4–0.65 μm); flat (0.65–0.9 μm)
Vilas 2008	0.42–0.93	C (T)	CM2	Band at 0.7 μm; flat with very weak band at 0.6 μm
Abe et al. 2008	0.4–2.4	C	N.A.	Flat and featureless spectral slope (0.85–2.2 μm) 0.89 ± 0.03%/1000 Å
Lazzaro et al. 2013	0.4–0.85	C	N.A.	Flat and featureless UV drop-off (< 0.45 μm)
Pinilla-Alonso et al. 2013	0.85–2.2	C	N.A.	Featureless spectral slope (0.85–2.2 μm) 0.37 ± 0.28%/1000 Å
Sugita et al. 2013	0.47–0.8	163 Erigone family	CM (heated Murchison)	Flat and featureless
Moskovitz et al. 2013	0.44–0.94	C (BD)	CM (heated Murchison) CI (Y-86029)	Spectrally flat
Perna et al. 2017	0.35–2.15	C (BD)	heated Murchison CM unusual/heated CI	Featureless UV drop-off (< 0.45 μm); spectral slope (0.5–0.8 μm)
Le Corre et al. 2018	0.8–2.4	C	CM2 carbonaceous chondrites, Mighei (grain size < 40 μm) and ALH83100 (grain size < 100 μm)	Near-perfect match with NEA (85275) 1994 LY and Mars-crossing asteroid (316720) 1998 BE7

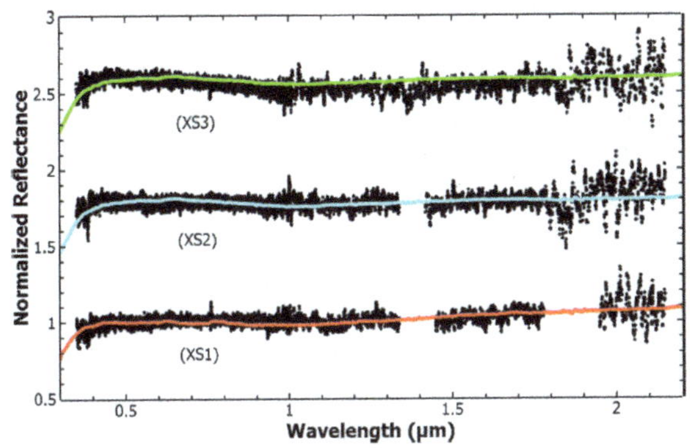

Fig. 4 The featureless/flatness nature of the Ryugu spectrum is evident in this series of three Vis-IR spectra. Superimposed are the spectra of thermally altered samples of the CM carbonaceous chondrite Murchison, in red (heated at 900 °C) and in cyan and green (heated at 1000 °C) (from Perna et al. 2017)

spectrum of main belt asteroid (302) Clarissa, suggested as a possible source family for Ryugu (Campins et al. 2013), does not match their spectrum of Ryugu. It is however a closer match with the spectrum from Moskovitz et al. (2013).

Derived quantities

Determining the composition of Ryugu by remote spectroscopic observations requires accounting for viewing geometry and modeling of the asteroid's surface properties. Reflectance properties of asteroids may differ from laboratory analogs due to grain size, phase angle, and space weathering (Binzel et al. 2015; Sanchez et al. 2012; Reddy et al. 2012). Laboratory spectral measurements of meteorites relevant to Ryugu can improve the interpretation of future hyperspectral observations obtained by Hayabusa2 NIRS3 instrument by linking absorption bands to a well-characterized mineralogical composition. Here, we summarize various authors' interpretations of their Ryugu spectra, most of which are included in Table 4 (see Le Corre et al. 2018 for a lengthy discussion of spectral interpretations of Ryugu).

Composition Generally, ground-based observations of asteroid Ryugu show a featureless spectrum, with a slight reddening in the near-IR and with a slight drop-off in the UV range. Based on laboratory data, the best available meteoritic spectral analogs are, to first order, a heated CM or possibly CI chondrites. However, a list of possible surface analogs for Ryugu should contain at least components of CM and CI meteorites (e.g., Murchison, Orgueil, Jbilet Winselwan, Y82182, Y86029, and so on), plus a suite of serpentines, saponite, phyllosilicates, magnetite, sulfates, sulfides, pyroxenes, and carbonates.

Laboratory measurements of reflectance spectra of a series of 23 carbonaceous chondrites belonging to the CM and CR groups were obtained between 0.3 and 100 μm in support of the efforts to be able to interpret and relate measurements of the surface of Ryugu (Beck et al. 2018). The major results are (i) the presence of 0.7 and 0.9 μm features correlated to the amount of phyllosilicates (confirming previous work such as Vilas and Sykes 1996); (ii) reporting for the first time of the presence of 0.7 and 0.9 μm features in spectra of a CR chondrites (Grosvenor Mountains 95577), these data suggest that Cgh-type asteroids (which make up about 1/3 of the main-belt C-type asteroids) could be the parent bodies of CM but also of some CR chondrites; (iii) the presence of a goethite-like 3 μm band for some CR chondrites (Beck et al. 2018).

Predicted quantities

Moving from derived quantities, where measured properties are used to derive new values for the properties of Ryugu, here, we use the best evidence regarding Ryugu to predict properties for which there are no measurements of any kind. This section will often rely on the knowledge gained from previous explorations and general estimates for C-type and NEAs. Some are explained in depth, notably block sizes, due to the importance they could have for mission operations planning regarding spacecraft safety.

Satellites, dust, and the local environment There are no directly observed, or inferred, satellites at Ryugu, so this feature, along with dust floating around the asteroid, has fallen into the "predicted" category of this work. What would drive a prediction of the existence of a satellite? Satellites among NEAs are common—it is expected that 15% of the population have satellites on relatively close orbits (Walsh and Jacobson 2015; Margot et al. 2015). However, nearly all NEAs with satellites are very rapidly rotating, typically near the critical rotation rate of ~ 2.2 h, and nearly all have small lightcurve amplitudes, and where available, their shapes are often found to be "top-shapes" or nearly spherical with equatorial bulges. Given that small bodies can be spun up or down by the Yarkovsky-O'Keefe-Radzievskii-Paddack effect (YORP), but this is highly uncertain, the current spin rate of Ryugu certainly does not indicate that it was ever a rapid rotator in the past, but at first glance, there is no strong indication that Ryugu will have a satellite or significant floating dust.

Boulders, craters, and surface roughness Safely delivering the spacecraft to the asteroid surface and effectively obtaining a sample are mission critical events that require careful evaluation of the asteroid surface properties. The distribution and abundance of boulders and craters pose the greatest risk to spacecraft flight that must be carried out autonomously. The surface roughness or regolith grain size distribution is irrelevant to safety but critical to the ability to successfully obtain a sample. Here, we summarize spacecraft observations of asteroids that provide useful guidance for what we can expect upon arrival to asteroid Ryugu.

Asteroid Itokawa is similar in size (535 m × 294 m × 209 m; Demura et al. 2006) to Ryugu (~ 900 m), but significant differences in geological features are expected. Returned sample analyses confirmed that S-type asteroid Itokawa is composed of an ordinary chondrite-like material, whereas C-type asteroid Ryugu is likely to be more volatile-rich and weaker. It is already known that Itokawa and Ryugu have strikingly differing shapes. Itokawa is highly elongated and appears to be a rubble pile composed of fragments from the catastrophic disruption of a larger body (Michel and Richardson 2013). In contrast, Ryugu is roughly spherical, but its evolutionary history is not yet known.

The only C-type asteroid that has been previously visited by spacecraft is the main belt asteroid Mathilde—a significantly larger asteroid at ~ 53 km mean diameter (Thomas et al. 1999). The best images obtained during the NEAR flyby of Mathilde were 160 m/pixel (Thomas et al. 1999). At this scale, Ryugu would be only a few pixels across, so these images do not offer much insight into what to expect on Ryugu's surface. Several other asteroids (S-types Eros, (243) Ida, (951) Gaspra, (2867) Steins, (4179) Toutatis, (5535) Annefrank; Q-type (9969) Braille; M-type (21) Lutetia) and other small bodies (Phobos, Deimos, several comets, and moons of Saturn) have been imaged by spacecraft and provide some constraints as to what to expect from Ryugu; however, of those bodies, only Eros, Phobos, Toutatis, and 67P/Churyumov-Gerasimenko (67P) were imaged at scales sufficient to resolve small boulders, and only Eros, Phobos (highest resolution images only available in one area), and 67P have been imaged globally. The size of Eros, the size and celestial location of Phobos, and the cometary composition/nature of 67P all pose complications for using these bodies as analogs to Ryugu.

At a larger scale, images of Ida and Gaspra look broadly similar to those of Eros, perhaps indicating that Eros is representative of an S-type asteroid of its size. Aside from Itokawa, Steins (5.3 km mean diameter) and Toutatis (3 km mean diameter) are the asteroids that have been observed by spacecraft that are closest in size (both are larger) to Ryugu but have a different taxonomic type; the best global Steins images have a ~ 160 m pixel scale, again insufficient to give much insight into Ryugu. The Toutatis data are limited but have a pixel scale of ~ 3 m. There is limited, very low-resolution data available for Annefrank and Braille. (4) Vesta and Ceres are so much larger than Ryugu that they do not provide very good points of comparison for what to expect.

Size frequency distribution of boulders The cumulative size-frequency distributions (SFD) of boulders on many small bodies (where sufficient data are available) can be fit by power laws (see Table 5). Many, though not all, of the fits to global boulder populations have power-law exponents around − 3. The exponent of the power-law fit depends on several factors, including the geological context, the strength of the material, and

Table 5 Small bodies for which boulder counts have been made from spacecraft images. The minimum boulder sizes measured are directly related to the best image resolution available for a given object

Name	Mean diameter (km)	Spectral type	Min boulder size of global count (m)	Min boulder size of regional count (m)	Power law found	Data source	References
Eros	17	S	15	0.05	− 3.2 as low as − 2.3 locally	NEAR	Rodgers et al. 2016; Thomas et al. 2001
Itokawa	0.35	S	6	0.1	− 3.1 − 3.5 as low as − 1.7 or as high as − 4.1 locally	Hayabusa	Mazrouei et al. 2014; Michikami et al. 2008; Noviello et al. 2014; Rodgers et al. 2016 Tancredi et al. 2015
Toutatis	2.9	S	n/a	10	− 4.4 locally	Chang'e 2	Jiang et al. 2015; Huang et al. 2013
Lutetia	99	M	n/a	60	− 5.0	Rosetta	Küppers et al. 2012; Sierks et al. 2011
Ida	32	S	n/a	45	n/a	Galileo	Lee et al. 1996
Phobos	22	D	n/a	5	− 3.2	Viking MGS MEX MRO	Ernst et al. 2015; Rodgers et al. 2016; Thomas et al. 2000
Deimos	12	D	n/a	~ 4	− 3.2	Viking	Lee et al. 1986; C. Ernst, personal communication
Churyumov-Gerasimenko	4	Comet	7	2	− 3.6 global; local ranges − 2.2 to − 4.0	Rosetta	Pajola et al. 2015, 2016

possibly the size range of boulders. The power-law behavior of cumulative size-frequency distributions of terrestrial fragmented objects has also been observed (Turcotte 1997; see also Table 1 in Pajola et al. 2015), with the examples listed in Pajola et al. (2015) having exponents ranging from – 1.89 to – 3.54.

In the cases of Eros, Itokawa, and Phobos, extending the SFD power-law fit from large, tens-of-meter-sized blocks down to small, tens-of-centimeter-sized blocks yields reasonable estimates of small block populations (Rodgers et al. 2016, see Fig. 5). However, the geological context of an area matters for the absolute block density—if lower-resolution counts include multiple geological settings, they will not extrapolate accurately to local areas containing only one setting (Ernst et al. 2015; Rodgers et al. 2016), and there are some indications of broken power laws on Itokawa that complicate simple extrapolations (Tancredi et al. 2015).

Radar observations of some NEAs have identified large boulders on at least 14 NEAs (Benner et al. 2015), but the size of the boulders identified is limited by the resolution of the data. Toutatis provides the only "ground truth" between radar identifications of boulders and spacecraft images of boulders—the SNR of the radar observations was too low to positively identify boulders on Eros or Itokawa (Benner et al. 2015). Radar data can be used to measure the radar scattering properties of asteroids, which gives insight into surface cobbles of the scale of the radar wavelength (0.035 m for Goldstone, 0.126 m for Arecibo).

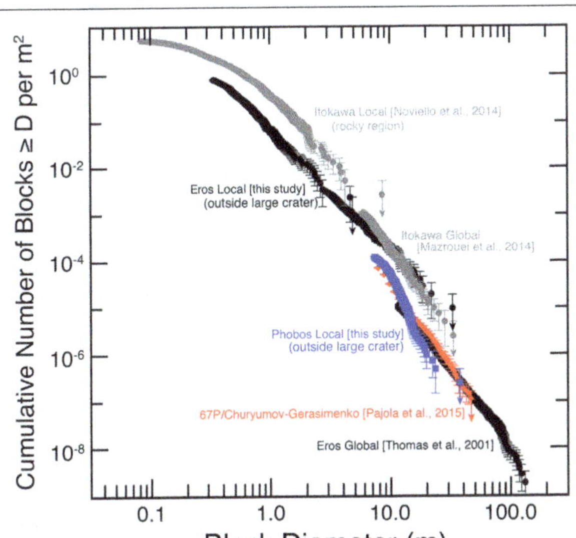

Fig. 5 Figure taken from Rodgers et al. (2016) showing the size frequency distribution of the blocks measured on Eros, Itokawa, Phobos, and comet 67P/Churyumov-Gerasimenko (see Rodgers et al. 2016, Fig. 1)

If there were radar data that suggest either a number of boulders or cobbles, an average size distribution of boulders for the asteroid could be estimated by assuming a – 3 power-law exponent. The actual distribution of boulders will be influenced by factors such as their origin and the geopotential of Ryugu. Ponds full of centimeter-sized pebbles are found on Itokawa at geopotential lows, and most boulders are located in other regions of the asteroid (Fujiwara et al. 2006; Saito et al. 2006). If this distribution is typical of a small asteroid, Ryugu should have concentrations of finer material (ponds) near the geopotential lows, and boulders will be found at higher geopotentials. Generally, the lowest potential for an asteroid of Ryugu's shape is expected to be found at or near the equator; the shape and density of the asteroid will affect the location of these geopotential lows.

If the boulders are sourced from the craters, as with Eros (Thomas et al. 2001), they are likely to be distributed around the craters as ejecta would fall on the irregularly shaped body. On the other hand, they may preferentially settle in the equatorial or near-equatorial regions.

Shapes of boulders The aspect ratios of boulders on Itokawa > 6 m in diameter were measured by Mazrouei et al. (2014), who found that most boulders in this size range were elongated, with *b/a* ratios of 0.7 (*a*, *b*, and *c* are triaxial dimensions of the boulders, defined to be $a \geq b \geq c$). It was not possible to measure the third dimension of most boulders in the Hayabusa AMICA images. Similar *b/a* ratios (0.62–0.68) were found by Michikami et al. (2010) and found to peak near 0.7 in Tancredi et al. (2015). The *c/a* ratio was measured for 21 boulders, the mean was 0.46 (Michikami et al. 2016). A compilation of fragment dimensions from several laboratory experiments is shown in Table 6. Although the measurements assume an ellipsoidal shape, the fragments are actually irregular in shape. It is not known whether the aspect ratio of rocks on a C-type asteroid would be different from those seen on Itokawa or in the strong-rock experiments of Table 6. Also unknown is the magnitude of the influence of thermal degradation on boulder fragmentation.

Craters Itokawa has a small number of identified impact craters, especially small ones (Hirata et al. 2009). Given its similar size to Itokawa, Ryugu may also not exhibit many craters. Eros, Phobos, Ida, and Gaspra all have numerous craters; on Eros, seismic shaking is suggested to have acted to erase many small (< 100 m diameter) craters (Chapman et al. 2002; Thomas and Robinson 2005); craters of 100 m size would be of significant size compared to Ryugu. These observations imply that Eros has a global loose regolith as deep as 100 m

Table 6 Compilation of fragment ratios *b/a* and *c/a* from several publications in the literature. Dimensions are defined to be $a \geq b \geq c$

Reference	Target	Projectile	Impact velocity (km/s)	*b/a*	*c/a*
Fujiwara et al. 1978	Basalt	Polycarbonate cylinders	2.6 and 3.7	0.73	0.50
Capaccioni et al. 1986	Basalt Concrete	Aluminum spheres	8.8 and 9.7	0.7	0.5
Durda et al. 2015	Basalt	Aluminum spheres	3.9–5.8	0.72 ± 0.13	0.39 ± 0.13
Michikami et al. 2016	Basalt	Nylon spheres	1.6–7.1	0.74	0.5 (catastrophic disruption) 0.2 (impact cratering)

(Robinson et al. 2002). Eros's relatively flat topography (< 55% of the surface is steeper than 30°) may also be explained by a thick global regolith (Zuber et al. 2000). Craters are more apparent on Toutatis than on Itokawa (Huang et al. 2013), with different densities of craters seen on the different lobes of the asteroid (Zou et al. 2014). As Ryugu is intermediate in size between Itokawa and Toutatis, it might be expected to have more craters per unit area than Itokawa but fewer than Toutatis. Since the size of regolith particles on Ryugu is expected to be similar to that on Itokawa (see the "Regolith thermophysical properties" section), steep cratered topography is not expected on Ryugu as well as on Itokawa. However, the different composition of Ryugu could affect the size of craters and the regolith depth and structure.

Meter-scale surface roughness Here, surface roughness at the meter-scale is described, to be distinguished from the smaller-scale roughness that is a parameter in thermophysical model fitting routines, or roughness in terms of very fine scale regolith grains—both are discussed in later sections. The surface roughness of Itokawa and Eros has been measured by rendezvous spacecraft laser altimeters. Itokawa's two major regions, the highlands and lowlands, have different surface roughnesses that are obvious even in the images. The roughest parts of the asteroid correspond to those with more boulders, which correspond to regions of higher geopotential (Barnouin-Jha et al. 2008). It is thought that finer materials may move to regions of lower geopotential and cover up larger boulders, creating a smoother (less rough) region. These areas are thought to correspond to areas with the thickest (fine) regolith. Over baselines of 8–50 m, the highlands of Itokawa are 2.3 ± 0.4 m rougher than the MUSES-C region (Barnouin-Jha et al. 2008); this difference may correspond to the expected regolith depth in this area. Alternatively, it is possible that large boulders are segregated to higher geopotential through the granular physics phenomena known as the Brazil nut effect, whereby under the action of various shaking processes, segregation occurs with larger bodies moving upward in the geopotential (Tancredi et al. 2015).

The highlands of Itokawa are at a baseline of 20 m—rougher than Eros—and the lowlands of Itokawa are significantly smoother than Eros (with the possible exception of Eros' ponds) (Barnouin-Jha et al. 2008). The surface roughness of Itokawa does not exhibit self-affine over the examined baselines (Barnouin-Jha et al. 2008).

On Eros, the surface roughness is self-affine over two orders of magnitude of baseline (~ 4–200 m), suggesting similar processes control the surface properties over these scales (Cheng et al. 2002). The corresponding Hurst exponents lie between 0.81 and 0.91 (Cheng et al. 2002). There are regional variations in surface roughness of Eros; smaller roughness values are found where more regolith is thought to have accumulated, and higher values are found along the crater walls and grooves (Susorney and Barnouin 2016).

There is a strong dependence of radar circular polarization ratio with asteroid class, suggesting a relationship between roughness at the centimeter- to decimeter-scale and composition (Benner et al. 2008). However, the mean ratios for S- and C-type NEAs are indistinguishable, perhaps indicating similar average surface properties for these populations (Benner et al. 2008).

Regolith thermophysical properties

Regolith thermophysical properties govern the exchange of radiative energy between the asteroid and its environment, and knowledge of these parameters is needed to calculate the surface and subsurface temperatures. While the surface energy balance is governed by insolation and regolith thermal inertia, heat diffusion governs temperatures in the subsurface. Heat is conducted to the subsurface according to:

$$\rho c_{\mathrm{p}} \frac{\partial T}{\partial t} = \frac{\partial}{\partial z}\left(k \frac{\partial T}{\partial z}\right), \tag{3}$$

where ρ is regolith density, c_{p} is specific heat capacity, T is temperature, t is time, z is depth, and k is thermal conductivity. Equation (3) is a second-order differential equation, which can be solved by prescribing two boundary conditions: One is usually given by zero heat flux from the interior, while the other is usually given in terms of the surface energy balance. For periodic insolation forcing, the surface energy balance takes the convenient form:

$$\sigma_{SB}\epsilon T^4 = (1\text{–}A)S + I\sqrt{\frac{\pi}{P}}\frac{\partial T}{\partial z'}\bigg|_{z'=0}, \quad (4)$$

where σ_{SB} is the Stefan-Boltzmann constant, ϵ is the surface emissivity, A is the Bond albedo, S is the total solar radiative flux including scattered radiation, P is the period of the forcing, and $z' = z/d_e$ is the depth normalized to the skin depth $d_e = \sqrt{kP/\rho c_p \pi}$. In Eq. (4), all material parameters have been absorbed in the single thermal inertia parameter I, which is defined as:

$$I = \sqrt{k\rho c_p}. \quad (5)$$

It is worth noting that Eq. (5) is only valid at the surface and provided that thermal conductivity is constant, which is not strictly true (see below). However, thermal inertia is a convenient way to describe the reaction of surface temperatures to insolation changes, and it is thus widely used. In addition, lack of data usually does not allow different parameters to be disentangled, although recent studies indicate that the radiometrically measured thermal inertia changes with heliocentric distance for an individual object (Rozitis et al. 2017; Marsset et al. 2017). Nevertheless, temperature dependence of thermal conductivity can have a significant influence when interpreting conductivity in terms of regolith grain size (e.g., Opeil et al. 2012; Gundlach and Blum 2013; Piqueux and Christensen 2011; Sakatani et al. 2017), and thermal inertia is therefore generally interpreted at a representative surface temperature (e.g., Müller et al. 2017; Gundlach and Blum 2013). In addition, care must be taken when converting thermal inertia to material parameters like thermal conductivity, since different combinations of material parameters govern the temperature at the surface (thermal inertia) and in the subsurface (thermal diffusivity).

There is evidence for regolith layering in terms of thermal conductivity from remote sensing data of main belt asteroids (MBA), and Harris and Drube (2016) argue that thermal inertia typically increases significantly as a function of thermal skin depth for these bodies. This effect may be caused by the fact that for slowly rotating asteroids surface temperature and thus surface thermal inertia is influenced by deeper layers than for fast rotators, and thus for slow rotators, they probe regolith material parameters to a greater depth. The observed increase of surface thermal inertia as a function of skin depth can thus be interpreted as thermal conductivity increasing as a function of depth, and it should therefore be kept in mind that a similar layering may be present on Ryugu.

Measured quantities

Thermal inertia and albedo Relevant thermal observations of Ryugu are summarized in Müller et al. (2017) and include the following: SUBARU-COMICS, multiple N-band (Aug 2007), AKARI IRC 15/24 μm photometry (May 2007), Herschel-PACS, 70/160 μm photometry (Apr 2013), Spitzer-IRS spectrum (5–38 μm) (May 2008), Spitzer-IRAC 3.6/4.5 μm two-epoch thermal light curves (Feb/May 2013), Spitzer-IRAC 3.6/4.5 μm multiple-epoch point-and-shoot sequence (Jan-May 2013), and new data from NEOWISE at 4.6 μm (Sep 2016; J. Masiero, priv. comm.)

A thermal inertia of 150 to 300 J m^{-2} $s^{-1/2}$ K^{-1} is required to explain the available rich data set of thermal measurements over a wide phase angle and wavelength range. The corresponding maximum surface temperatures are in the range of ~ 320 to 375 K (r_{helio} = 1.00–1.41 AU for our observational thermal data set), with T_{max} ~ 350 K at the object's semi-major axis distance (a = 1.18 AU). The corresponding shape model is very spherical with the volume-equivalent diameter 850–880 m (compare "Global properties" section). The surface has thermal inertia of 150–300 J m^{-2} $s^{-1/2}$ K^{-1}, and the roughness is small with the rms of surface slopes < 0.1, which means the surface appears very smooth in the radiometric context (connected to a very low width-to-depth ratio in the modeled spherical crater segments on Ryugu's surface). The reference model of Müller et al. (2017) has the pole direction (340°, − 40°), the volume-equivalent diameter 865 m, surface thermal inertia 200 J m^{-2} $s^{-1/2}$ K^{-1}, low surface roughness with rms of surface slopes 0.05, geometric V-band albedo 0.049, infrared emissivity 0.9, and rotation period 7.63109 h (compare "Global properties" section). These are consistent with other estimates independently conducted by Yu et al. (2014). Geometric albedo, phase integral, and the spherical bond albedo have already been discussed in the "Global properties" section, and the values presented there are compatible with the estimates from thermophyiscal models, which yield geometric V-band albedos of approximately 0.042 to 0.055. Using a phase integral of $q = 0.29 + 0.684G$, where $G = 0.13 \pm 0.02$ is the slope parameter as derived from V-band observations (Ishiguro et al. 2014), the bond albedo can be calculated using $A = qp_V$. The uncertainty in G translates into an uncertainty in the phase integral q (Bowell 1989) and combined with a 5% accuracy of the $q - G$ relation (Muinonen et al. 2010) a bond albedo of $A = 0.019 \pm 0.003$ for the full admissible thermal inertia and roughness range is obtained.

Derived quantities

Thermal conductivity The thermal conductivity of particulate media is known to be very low in a vacuum because the weak contacts between particles act as high thermal resistance points for heat conduction, and contributions of heat transfer by gas are absent (e.g., Fountain and West 1970). In fact, thermal conductivity of the lunar regolith was estimated to be less than 0.03 W m^{-1} K^{-1} even at depths below 50 cm based on the analyses of Apollo Heat Flow Experiment data (Langseth et al. 1976; Grott et al. 2010), and much lower values are expected at shallower depth. Accordingly, the thermal conductivity of Ryugu's surface is also expected to be very low if it is covered by regolith or small particles.

The thermal conductivity of the top surface layer can be constrained using Eq. (5) if the thermal inertia is assumed to be known from, for example, ground-based observations. For Ryugu, using the estimated typical value of 200 J m^{-2} K^{-1} s$^{-1/2}$ (Müller et al. 2017) and assuming a bulk density of $\rho = 1100$–1500 kg m^{-3} (typically 1270 kg m^{-3}) as well as a specific heat capacity c_p of 758 J kg^{-1} K^{-1} at 300 K (as an averaged temperature of daytime on Ryugu), thermal conductivity is constrained to be $0.020 < k < 0.108$ W m^{-1} K^{-1}, with a most likely value of $k = 0.042$ W m^{-1} K^{-1}. However, it is worth pointing out again that this estimate is only valid if density, specific heat capacity, and thermal conductivity are constant as a function of depth and time, which will in general not hold. Rather, on the actual Ryugu, these parameters may have depth-directional distributions and will vary with time due to the diurnal temperature variations.

Thermal conductivity of regolith-like powders is a complicated function of parameters, such as grain size, porosity, temperature, and particle surface energy, the latter being a measure for the attractive surface force acting between solids in close proximity (Johnson et al. 1971), and different models have been proposed for this problem (e.g., Halajian and Reichman 1969; Hütter et al. 2008; Gundlach and Blum 2012). Here, we use the integrated model by Sakatani et al. (2017) that was developed based on experimental studies. Under vacuum conditions, effective thermal conductivity is given by the sum of "solid conductivity" originating from thermal conduction through inter-particle contacts and "radiative conductivity" originating from thermal radiation through void spaces between the particles, and it is given by:

$$k = k_{\text{solid}} + k_{\text{rad}}. \tag{6}$$

The solid part of the conductivity k_{solid} can be modeled as serial and parallel connections of heat paths around the contact areas between spheres and it may be expressed as:

$$k_{\text{solid}} = \frac{4}{\pi^2} k_{\text{m}} (1-\phi) C \xi \frac{r_{\text{c}}}{R_{\text{p}}}, \tag{7}$$

where k_{m} is the thermal conductivity of the bulk material, ϕ is the porosity, C is the coordination number, r_{c} is the radius of the contacts area between two spheres, R_{p} is the radius of the grains, and ξ is a coefficient that represents a reduction of contact heat conductance due to surface roughness on the grains; $\xi = 1$ if particles are perfectly smooth spheres, and comparison with experimental data showed that $\xi \leq 1$ for natural samples including spherical glass beads. The coordination number may be expressed as a function of porosity, and according to Suzuki et al. (1980), it is given by:

$$C = \frac{2.812(1-\phi)^{-1/3}}{f^2(1+f^2)}, \tag{8}$$

where $f = 0.07318 + 2.193\phi - 3.357\phi^2 + 3.194\phi^3$. The contact radius r_{c} between two spheres depends on the force with which the particles are pressed together as well as their elastic parameters, and under the micro-gravity environment of Ryugu surface energy γ, and thus, van der Waals forces will dominate the solid conductivity contribution. Contact radius r_{c} may be expressed as:

$$r_{\text{c}} = \left[\frac{3(1-\nu^2)}{4E} \left\{ F + \frac{3}{2}\pi\gamma R_{\text{p}} + \sqrt{3\pi\gamma R_{\text{p}} F + \left(\frac{3}{2}\pi\gamma R_{\text{p}}\right)^2} \right\} \right]^{1/3}, \tag{9}$$

where ν is Poisson's ratio, E is Young's modulus, F is the external compressive force on the particle, and γ is the surface energy (Johnson et al. 1971). F is given as the compressional stress σ divided by the number of grains per unit cross-sectional area, and

$$F = \frac{2\pi R_{\text{p}}^2}{\sqrt{6}(1-\phi)} \sigma. \tag{10}$$

On planetary surfaces, the compressional stress is caused by the self-weight according to $\sigma = \rho_{\text{m}}(1-\phi)gz$ with ρ_{m} being the bulk density of the grain, g gravitational acceleration, and z depth, but on Ryugu, this contribution will be small. Overburden pressure on Ryugu is only of the order of 0.01 Pa at a depth of 10 cm, and assuming van der Waals forces to be of the order of 0.075 N/m times R_{p} (Kiuchi and Nakamura 2014); the latter will dominate F for particle sizes smaller than 1 m (compare Eq. (10)). On the other hand, the overall contribution of the solid conductivity to the total conductivity will be small for typical NEAs with thermal inertia I larger than 100 J m^{-2} K^{-1} s$^{-1/2}$ when compared to the radiative contribution, and radiative conductivity will dominate (see below).

The radiative part of the conductivity may be modeled by approximating the particle layers as multiple infinitely thin parallel slabs. Then, radiative conductivity may be written as:

$$k_{\rm rad} = \frac{4\epsilon}{2-\epsilon}\sigma_{\rm SB} L T^3, \tag{11}$$

where ϵ is the grain surface's thermal emissivity, $\sigma_{\rm SB}$ is the Stefan-Boltzmann constant, T is temperature, and L is the effective radiative distance between the slabs. Thus, L is proportional to the geometrical clearance or typical pore length between the grains, and in a homogeneous packing of equal-sized spheres, the geometrical pore length can be expressed in terms of grain size and porosity (Piqueux and Christensen 2009). Additionally, introducing a factor ζ to scale the geometrical length to the effective one, L is given by:

$$L = 2\,\zeta\left(\frac{\phi}{1-\phi}\right)^{1/3} R_{\rm p}. \tag{12}$$

Comparison with experimental data for glass beads showed that ζ increases with decreasing grain size, which may be caused by a number of effects: First, the assumption of independent scatterers may break down, or a change of the scattering characteristics may take place for particles smaller than the thermal wavelength, resulting in more forward scattering. Further, radiative transfer may be enhanced across gaps smaller than the thermal wavelength (Rousseau et al. 2009), or there may be bias in the experimentally derived $k_{\rm solid}$. The latter may be due to the fact that the temperature dependence of $k_{\rm rad}$ is not exactly T^3 due to the coupling between solid and radiative heat transfer (Singh and Kaviany 1994); thus, an extrapolation of measured data to 0 K for determining $k_{\rm solid}$ may not be perfect.

A potential shortcoming of the above theory is that natural regolith does not consist of monodispersed spherical particles. However, laboratory measurements of the thermal conductivity of lunar regolith analog material (Sakatani et al. 2018) show that measured thermal conductivity is in close agreement with theory if the median particle size (in the volume fraction sense) is used in the above equations. This is due to the fact that (a) the solid conduction only weakly depends on particle size and (b) the average pore size is close to the one calculated with the median radius using Eq. (12).

Another simplifying assumption made above is the spherical particle shape, while natural regolith is generally both non-spherical and angular. The deviation from the ideal shape can be measured by quantities called sphericity and angularity, and both shape indicators may influence the contact radii in the solid conduction contribution in complicated ways. However, their main contribution on thermal conductivity is to change regolith porosity, which is sufficiently described by the median particle radius. Thus, the current best estimate for determining ζ is an empirical relation derived from the experimental results of Sakatani et al. (2018) for a lunar regolith simulant, and it is given by:

$$\zeta = 0.68 + \frac{7.6\times10^{-5}}{R_{\rm p}}, \tag{13}$$

where $R_{\rm p}$ is in units of meter. In the above model, Thermal conductivity of the regolith depends mainly on the particle size and porosity, and ranges for these two parameters can be estimated if thermal inertia or effective thermal conductivity is known. Figure 6 shows a thermal inertia contour plot as a function of particle size and porosity. For thermal inertia values of I = 150–300 J m^{-2} K^{-1} s$^{-1/2}$ as derived from remote sensing observations (Müller et al. 2017), a typical particle size for the Ryugu regolith can be estimated to be 3–30 mm, with 6–10 mm being the most likely. For such large grain sizes, the radiative conductivity is dominant over the solid conductivity even at low temperatures. For example, assuming a particle diameter of 3 mm and a temperature of 150 K, the solid contribution to thermal conductivity is of the order of $5{\cdot}10^{-4}$ W m^{-1} K^{-1}, whereas the radiative part is larger than $3{\cdot}10^{-3}$ W m^{-1} K^{-1}. Therefore, the uncertainty of assumed parameters in the solid conductivity model, such as Young's modulus and surface energy, does not affect the grain size estimation. On the other hand, porosity cannot be effectively constrained because a major part of heat flow is made by the radiative heat transfer in this particle size range. Radiative thermal conductivity increases while the heat capacity decreases with porosity; therefore, this balance of both parameters makes thermal inertia nearly constant for various porosities.

For a verification of the determined grain size, a comparison with the thermal model developed by Gundlach and Blum (2013) can be made. This model is based on the same physical concept as the calculations described above, however deviates in some details. Basically, these differences are (i) the formulation of the network heat conductivity, (ii) the description of the radiative thermal transport process, and (iii) the irregularity parameter measured by different calibration measurements. A derivation of the regolith grain radius based on the Gundlach and Blum (2013) formulations yields radii ranging between 1.1 and 2.5 mm (for volume filling factors between 0.1 and 0.6). Thus, both models agree rather well, with the Gundlach and Blum (2013) model having slightly smaller sizes, which provides confidence in the grain size determination from thermal inertia measurements of planetary surfaces.

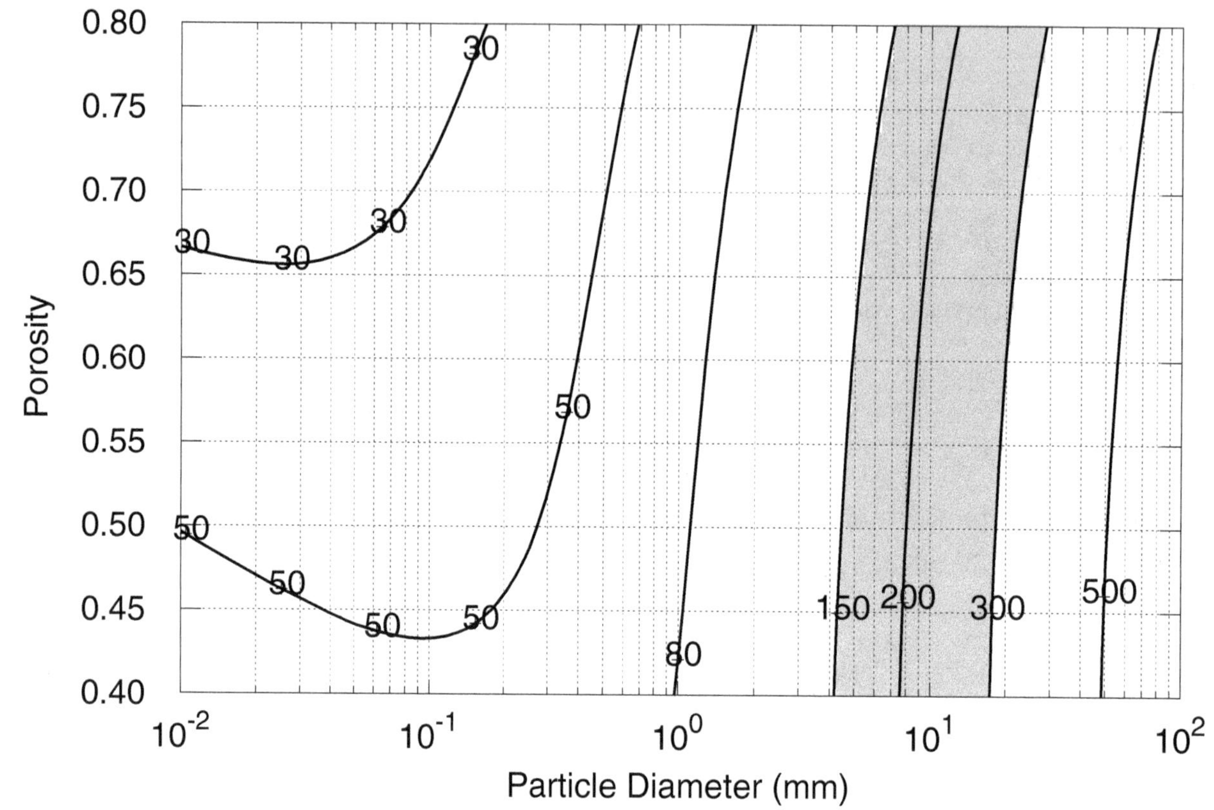

Fig. 6 Contour plot of thermal inertia (in units of J m^{-2} $s^{-1/2}$ K^{-1}) as a function of particle diameter and porosity. The gray-filled region indicates the thermal inertia range estimated for Ryugu by ground-based observations (Müller et al. 2017). The regolith parameters assumed in the calculation are $c_p = 758$ J kg^{-1} K^{-1}, $T = 300$ K, $\rho_m = 3110$ kg m^{-3}, $k_m = 0.0021\,T + 1.19$ W m^{-1} K^{-1}, $\gamma = 0.032$ J m^{-2}, $\nu = 0.25$, $E = 78$ GPa, $\epsilon = 1.0$, $\xi = 0.12$, $z = 0.01$ m, and $g = 1.8 \times 10^{-4}$ m s^{-2}, respectively. γ and ξ were optimized for experimental data of lunar regolith simulant JSC-1A (Sakatani et al. 2018)

Predicted properties

Regolith heat capacity The specific heat capacity of rocks and soils at low temperatures has been studied for lunar samples (Robie et al. 1970; Fujii and Osako 1973; Hemingway et al. 1973; Opeil et al. 2012), and a strong temperature dependence has been found. The suite of materials studied includes particulate material such as lunar fines and soils, but brecciated lunar rocks as well as basalts have also been studied. The best fit to the lunar soils data is given by Hemingway et al. (1973), and the specific heat capacity can be approximated as:

$$c_\mathrm{p} = -23.173 + 2.127T + 1.5009{\cdot}10^{-2}T^2 \\ -7.3699{\cdot}10^{-5}T^3 + 9.6552{\cdot}10^{-8}T^4, \tag{14}$$

where c_p is the specific heat capacity in units of J kg^{-1} K^{-1} and T is temperature in K. This best-fitting formula is accurate to within 2% down to temperatures of 200 K and to within 6% down to temperatures of 90 K. At higher temperatures, the expression

$$c_\mathrm{p} = 4184{\cdot}\left[0.2029 + 0.0383\left(1 - \exp\left(\frac{350-T}{100}\right)\right)\right], \tag{15}$$

by Wechsler et al. (1972) may be used (also see Ledlow et al. (1992), Stebbins et al. (1984), and Schreiner et al. (2016)). The fit to the data is shown along with the data in Fig. 7 and extends to temperatures of up to 350 K, as appropriate for Ryugu. The extrapolation applicable at higher temperatures is shown as a dashed line.

Measurements on lunar material are in good agreement with measurements performed by Winter and Saari (1969) using other geological materials as well as measurements of the specific heat capacity of meteorites performed by Yomogida and Matsui (1983) as well as Consolmagno et al. (2013). It may be worth noting that a trend exists with respect to the iron content of the considered samples, with low iron corresponding to high heat capacity (Yomogida and Matsui 1983). While heat capacity thus shows a strong temperature dependence, this is only relevant if the near surface regolith layer is considered. At depths below a few skin depths, perturbations rapidly decay such that the

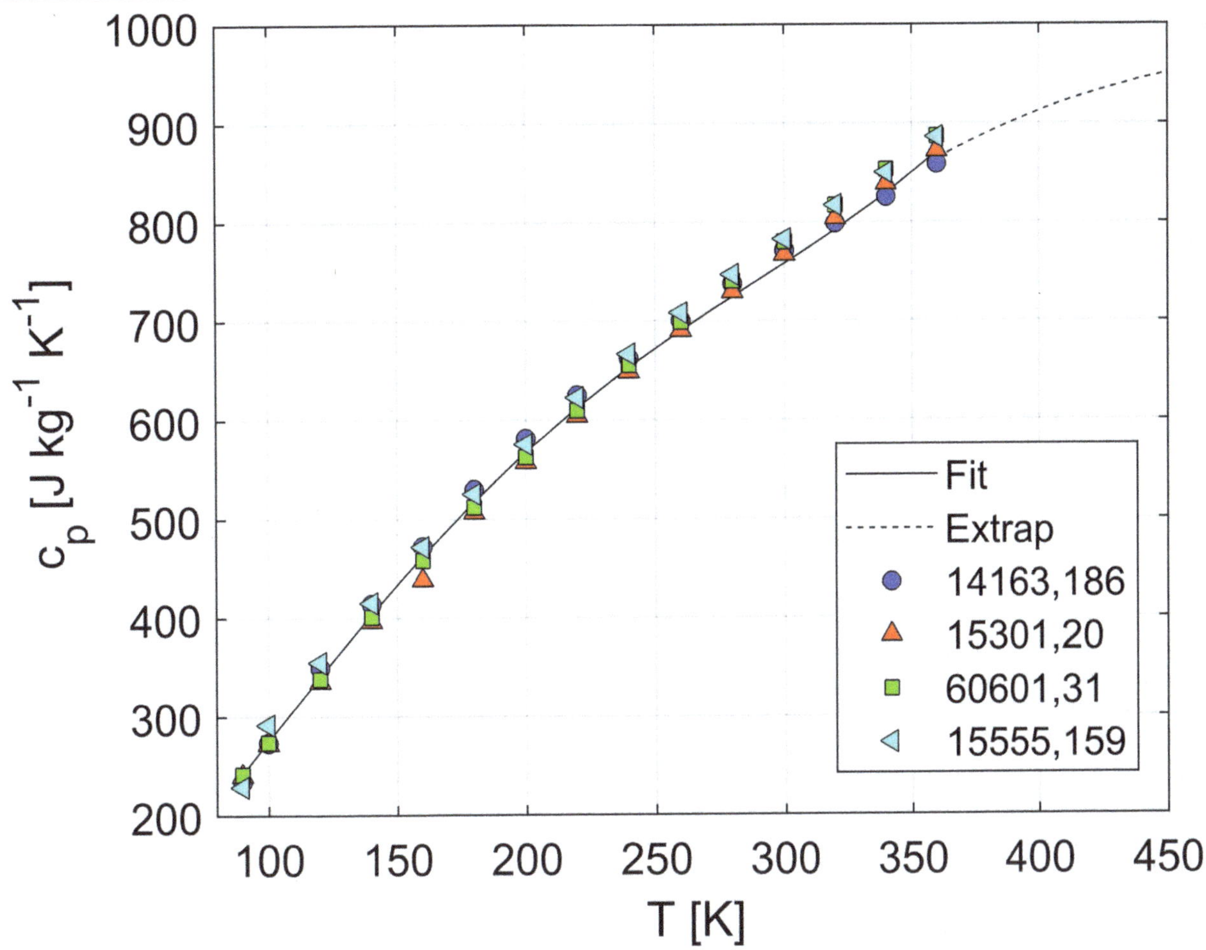

Fig. 7 Specific heat capacity of lunar samples 14,163,186 (fines > 1 mm, blue circles), 15,301,20 (soil, red triangles), 60,601,31 (soil, green squares), and 15,555,159 (basalt, cyan left triangles) as a function of temperature together with the best fitting curve (solid line) as well as an extrapolation for temperatures beyond 350 K (dashed line). Data and fit from Hemingway et al. (1973), extrapolation from Ledlow et al. (1992)

regolith can be assumed isothermal for the purpose of determining its heat capacity. Evaluated at a representative average daytime surface temperature of 300 K, a value of $c_p = 758$ J kg^{-1} K^{-1} is obtained by evaluating Eq. (14).

Surface temperatures, thermal gradients, and thermal fatigue The thermal evolution of Ryugu is simulated by solving Eq. (3) numerically. In general, the input flux at the surface is not only the heat flux from the sun but also the sum of the heat radiation from the other surface area (Rozitis and Green 2011; Davidsson and Rickman 2014; Davidsson et al. 2015). However, Ryugu is predicted to be a round-shaped body from ground-based observations. Thus, hereafter, for simplicity, it is assumed that the surface is convex, and areas to be calculated do not receive heat radiation from other areas.

It is to be noted that the thermal emissivity is not needed to be the complement of the surface albedo because the peak wavelength of the solar radiation is in the range of the visible light while that of the thermal radiation from the surface of an asteroid is around or longer than 10 μm. Thus, the thermal emissivity and the albedo are independent parameters. We adopt 0.019 for the albedo and 0.90 for the emissivity of the surface of Ryugu as nominal values. Hereafter, we will show the equilibrium temperature evolution as a function of local time, which is obtained after 100 asteroid rotations at a fixed solar distance and a sub-solar latitude. The difference between the equilibrium temperature distribution and a full numerical simulation taking into account the orbit of the body around the sun is less than a few K.

Figure 8 shows the evolution of equilibrium surface temperature as a function of local time with a constant thermal inertia (left panel, thermal inertia equal to 200 J m^{-2} s$^{-1/2}$ K^{-1}) but various local latitudes, and constant local latitude (right panel, 0°) but various thermal inertias. In these numerical simulations, we adopt a solar distance of 1.2 AU and the sub-solar

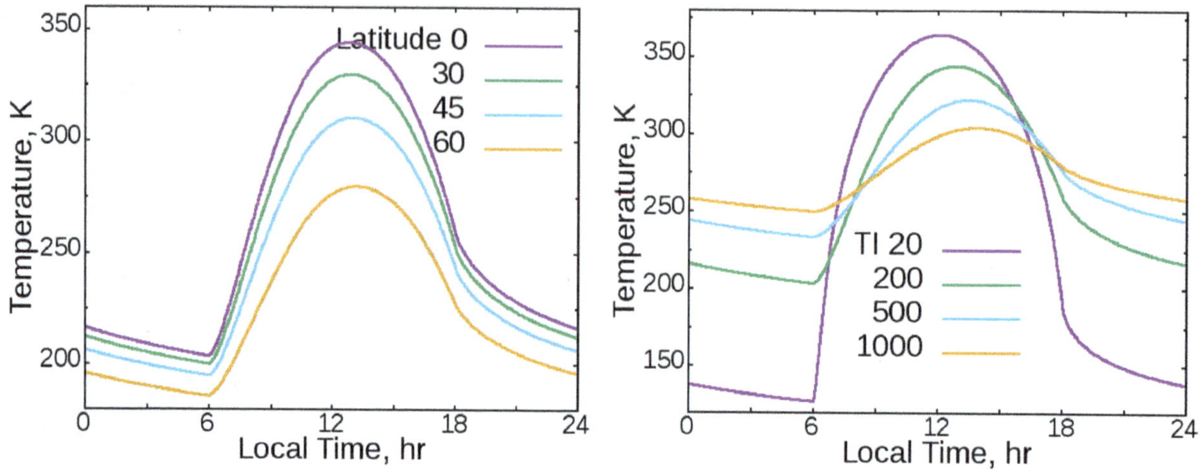

Fig. 8 Surface temperature change as a function of local time for various latitudes (left) and thermal inertia (right). The sub-solar latitude is assumed to be 0 (equator)

latitude of 0° (equator) for simplicity. As shown in Fig. 8, surface temperature changes with solar direction and decreases almost linearly during night time. Both the maximum and minimum temperatures decrease with latitude for a constant thermal inertia. The local time of the maximum temperature is achieved after noon, and the degree of phase shift relative to the noon is almost the same for each latitude. On the other hand, as shown in the right panel of Fig. 8, the local time of the maximum temperature becomes later with thermal inertia. The phase shift is a diagnostic parameter to estimate the thermal inertia from the surface temperature distribution obtained by the instrument TIR (Takita et al. 2017). The maximum temperature decreases with thermal inertia and the minimum temperature increases with thermal inertia because subsurface temperatures are kept warmer for larger thermal conductivities. It is to be noted that the total thermal radiation from the surface must balance with the solar influx in the equilibrium state. Thus, the decrease of maximum temperature is smaller than the increase of minimum temperature since the thermal radiation is proportional to the fourth power of the temperature.

Figure 9 shows results similar to Fig. 8, but for a spin axis tilted by 30°. In this case, the duration of day and night depends on the latitude. As a result, the maximum temperature is not needed to be obtained at the latitude identical to the sub-solar latitude. For the case with a thermal inertia of 1000 J m^{-2} $s^{-1/2}$ K^{-1}, the maximum temperature is obtained at the latitude of 45°. Despite the change of the length of day time, the shift of the local time of peak temperature is

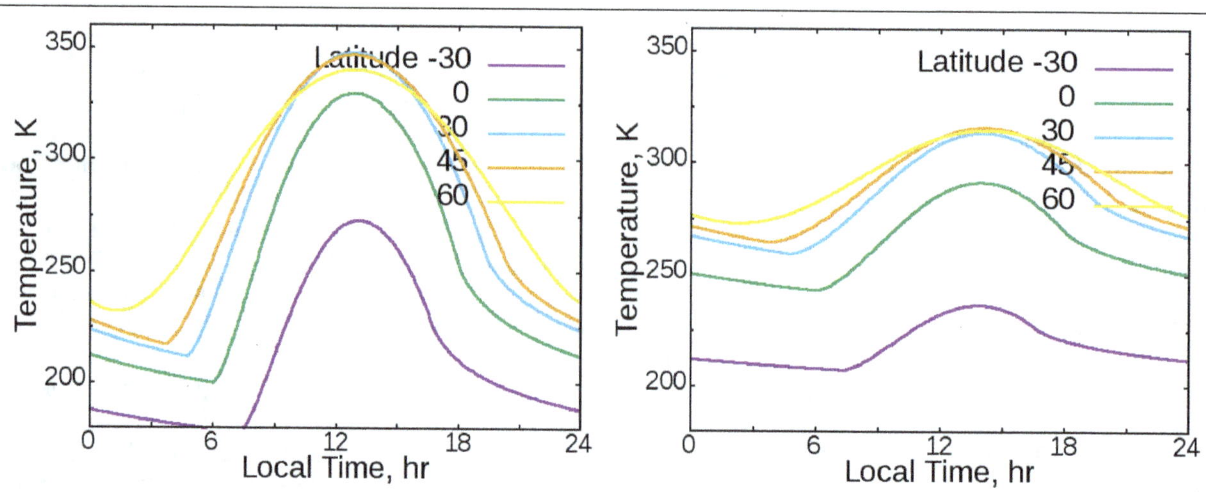

Fig. 9 Left: similar to Fig. 8, but for a case with sub-solar latitude of 30, i.e., a tilted spin axis. Left: the thermal inertia is 200 J m^{-2} $s^{-1/2}$ K^{-1}. Right: thermal inertia is 1000 J m^{-2} $s^{-1/2}$ K^{-1}

independent of the latitude. These results are similar to those obtained by Müller et al. (2017), who calculated the maximum surface temperatures in the range of 320 to 375 K by considering the derived object properties and heliocentric distances of 1.00–1.41 AU. At the object's semi-major axis distance of 1.18 AU, Müller et al. (2017) find a reference maximum temperature of 350 K.

Delbo et al. (2014) suggested that thermal fatigue could be the driver of regolith formation on asteroids with perihelion distances of less than 2.5 AU, including NEAs such as Ryugu. Thermal fatigue describes the breakdown of larger boulders due to thermal stress. The regolith is exposed to large temperature differences between night and day time temperatures causing large temperature gradients and therefore stress in the material leading to crack formation and propagation. Delbo et al. (2014) estimated that thermal fatigue erodes the surfaces of NEAs like Ryugu several orders of magnitude faster than erosion caused by micrometeoroids. Consequently, studying the temperature gradient at the surface might be essential for understanding the evolution of the surface of Ryugu. Figure 10 represents the evolution of the temperature gradient just beneath the surface as a function of the local time and latitude. The thermal inertia and the sub-solar latitude are assumed to be 1000 J m^{-2} $s^{-1/2}$ K^{-1} and 0°, respectively. We adopted a larger thermal inertia than the nominal value of Ryugu because dust creation due to the thermal fatigue takes place at the surface of consolidated boulders or monolithic rocks. As shown in Fig. 10, the maximum temperature gradient is achieved before noon. The absolute value of the thermal gradient is larger for lower latitude. This is simply because the larger thermal spatial gradient is obtained at the time when the maximum temporal gradient of the surface temperature is achieved. In fact, as shown in the right panel of Fig. 8, the maximum temporal gradient at the surface is achieved before noon. The maximum thermal spatial gradient becomes larger for smaller thermal inertia. On the other hand, the negative maximum temperature gradient is achieved at sunset. After sunset, the temperature gradient becomes smaller because the underground temperature distribution is alleviated during night time. The right-hand panel of Fig. 10 is similar to the left one but uses a spin axis tilted by 30°. In this case, the maximum thermal gradient is achieved not only at the sub-solar latitude but also at the sub-solar latitude and the equator. Thus, the determination of location at which the maximum temperature gradient is achieved is not straightforward (Hamm et al. 2019). It is to be noted that the integration of thermal gradient through a day should be zero in the equilibrium state if the thermal conductivity is independent of temperature.

We assumed above that the surface of Ryugu is smooth. If the roughness of the surface is taken into account, the apparent temperature distribution changes depending on the degree of roughness (Rozitis and Green 2011; Davidsson and Rickman 2014, Davidsson et al. 2015). The temperature of a slope is especially affected at dawn and dusk. A wall of a valley observed at a low phase angle will have a much higher temperature than the horizontal, surrounding surface. Contrarily, when observing a wall against the sun, it will be much colder than its surroundings. In addition to the case of concave areas such as craters, the absorption of radiation could raise the temperature of the

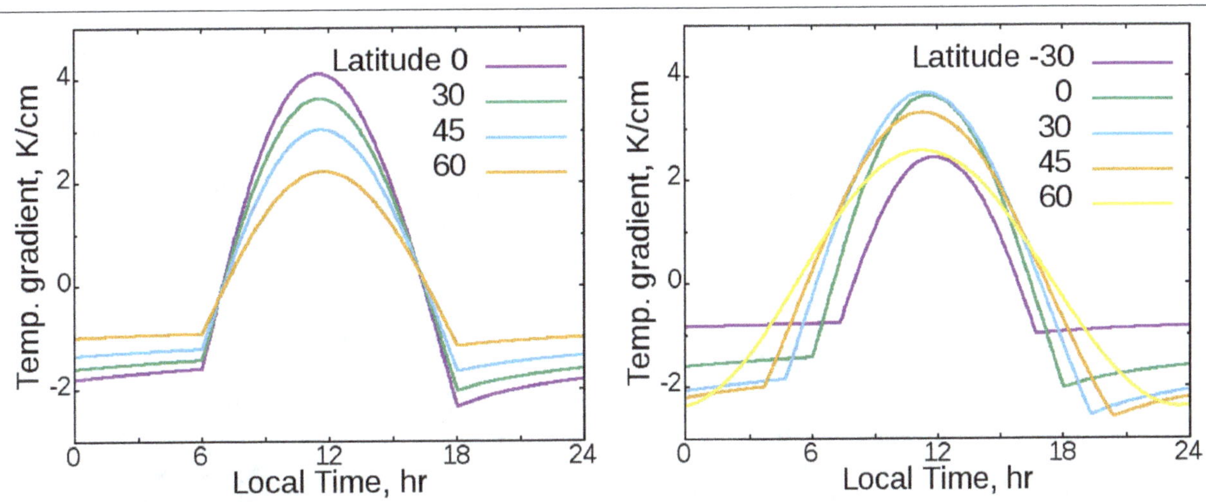

Fig. 10 Thermal gradient just beneath the surface as a function of local time. Left: sub-solar latitude of 0°. Right: sub-solar latitude of 30°. The negative maximum of the thermal gradient is achieved at sunset while the positive maximum of the thermal gradient is achieved just before noon

bottom area of the concavity depending on its size and shape. Moreover, we must point out that the bond albedo of a rough surface changes with solar incident angle (Senshu et al. in preparation). These effects need to be taken into account when analyzing the apparent temperature distribution obtained by the Hayabusa2 TIR instrument.

Regolith mechanical properties

Measured quantities

There are no directly measured quantities of the mechanical properties of the regolith on Ryugu, as for any asteroid that has not been visited by a spacecraft. Therefore, the only quantities that can be determined must be derived from current data or rely on pure assumptions.

Derived quantities

The only quantity that can be derived from observations is the dominant grain size, based on the measurement of the thermal inertia of the asteroid (see the previous section). The grain shape can also be derived from the only asteroid from which regolith particles have been returned, i.e., the asteroid Itokawa. However, we note that Itokawa belongs to the S taxonomic class, and its composition is thus very different from that expected for Ryugu. This probably also means that its mechanical properties are different, and therefore, how regolith is generated on this body, as well as its properties, are likely to be different too. But since this is the only data point that we have so far, we use it in this section.

Grain size As summarized previously in the "Regolith thermophysical properties" section, a thermal inertia of 150 to 300 J m^{-2} $s^{-0.5}$ K^{-1} is required to be compatible with remote sensing observations (Müller et al. 2017). Using the model by Sakatani et al. (2016), this translates to typical particle diameters of 3–30 mm for Ryugu, with 6–10 mm being the most likely. This is consistent with, although slightly larger than, the results by Gundlach and Blum (2013), who obtain grain diameters of 2.2 to 5 mm.

Grain shape For the grain shape as shown in Figs. 11 and 12, we note that the mean aspect ratios of particles returned from the asteroid Itokawa by the Hayabusa

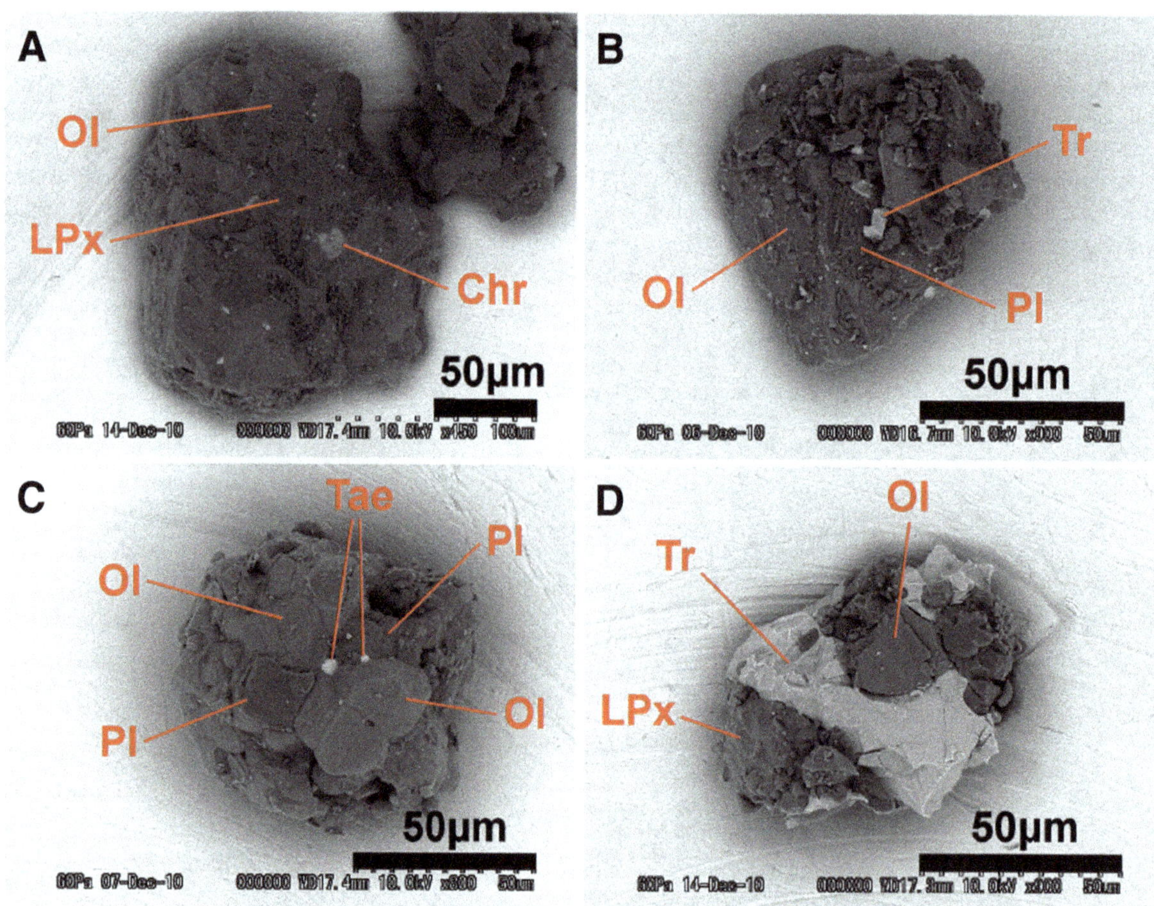

Fig. 11 Backscattered electron (BSE) images of asteroid Itokawa regolith grains RA-QD02-0030 (**a**), RA-QD02-0024 (**b**), RA-QD02-0013 (**c**), and RA-QD02-0027 (**d**) from Nakamura et al. (2011). Ol, olivine; LPx, low-Ca pyroxene; Chr, chromite; Pl, plagioclase; Tr, troilite; Tae, taenite

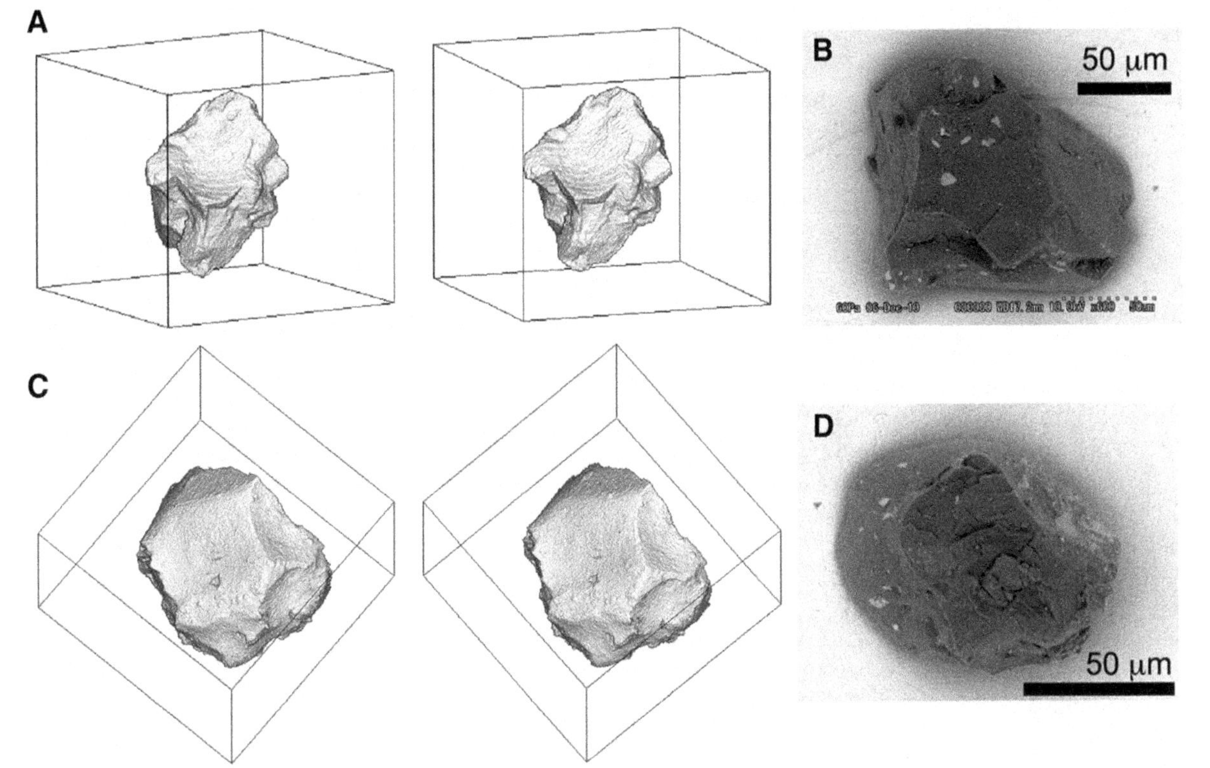

Fig. 12 The 3D external shapes of Itokawa particles. **a** Stereogram (box size 232 µm × 232 µm × 203 µm) and **b** SEM micrograph of RA-QD02-0023. **c** Stereogram (box size, 112 µm × 112 µm × 93 µm) and **d** SEM micrograph of RA-QD02-0042

mission, b/a and c/a (a is the longest axis, b is the medium axis, c is the shortest axis) (Tsuchiyama et al. 2014), are 0.72 ± 0.13 and 0.44 ± 0.15, respectively, and are similar to the mean axial ratio of fragments generated in laboratory impact experiments ($a:b:c \approx 2:\sqrt{2}:1$ or $b/a \approx 0.71$ and $c/a \approx 0.5$). Note that the bulk density of Itokawa samples is 3.4 g cm^{-3}. The average porosity of Itokawa samples is 1.5%. We do not have any data yet for a C-type asteroid like Ryugu.

Predicted quantities

Because we do not yet have any detailed image of a C-type asteroid of Ryugu's size and any experiment done on such a surface allowing us to understand how it responds to an external action, most of the information regarding regolith mechanical properties of Ryugu are predictions.

Regolith porosity Natural regolith is typically non-spherical (measured by a quantity called sphericity) and angular (measured by "angularity"). Together with the adhesion properties, the median grain size and the size distribution determine the porosity.

It is well known that random packings of monosized coarse spheres have a porosity of 0.363 (dense) to 0.42 (loose packing). Adhesion increases the porosity, to the limit of 1; the decisive parameter is the granular Bond number $B_o = F_{vdw}/F_g$, i.e., the ratio of van der Waals forces and weight acting on a grain (Kiuchi and Nakamura 2014). If the particles are polydisperse, the porosity tends to decrease with the width of the size distribution (the smaller particles filling up the voids of the larger particles). Finally, non-spherical grains tend to increase the porosity compared to packings consisting of spherical grains.

A well-known packing theory which can be used for predictions is described in Zou et al. (2011). Here, the equation expressing the "initial porosity" as a function of grain size has to be generalized to arbitrary gravity (instead of 1 G), and the assumption has to be made on the increase of initial porosity due to shape effects.

Ryugu has a very low gravity so that compaction by lithostatic pressure is not likely to play a role in the upper meter or so. This is different from the moon: the porosity of lunar regolith (as a function of depth) is relatively well known; it is 83% on the very surface (Hapke and Sato 2016) and decreases quickly to about 58% at a few centimeters and to about 38% at a few meter depth. Apart from the very upper surface, we see compaction by overburden pressure at work (Schräpler et al. 2015; Omura and Nakamura 2017).

Given these considerations, a porosity of ~ 0.4–0.5 can be assumed for the near-surface regolith, which corresponds to the random loose packing of slightly adhesive (relative to weight) centimeter-sized irregular but roundish particles with a size distribution.

Regolith subsurface properties According to Britt and Consolmagno (2001), the materials in porous asteroids may be sorted by the particle size. The large irregular pieces (and larger voids/fractures) may be located deeper inside the asteroid, and the fine particle size fractions that are observed on their surfaces are restricted to the surface regolith zone (Fig. 13). The large interior voids/fractures are preserved from infilling by the effects of friction on the smaller size fractions. Friction tends to dominate the downward pull of gravity and prevents the fine fractions from filtering into the interior of the asteroid and infilling the large fractures and voids. Friction may also play a role in allowing shattered asteroids to maintain their relief features and shape by resisting the movements of pieces within the object, in effect providing strength to non-coherent objects.

However, another mechanism, called the Brazil nut effect, may cause an opposite trend (Tancredi et al. 2012, 2015; Matsumura et al. 2014; Maurel et al. 2017) and segregate larger particles to the surface, as seen on Itokawa. Thus, coarse gravel (1 cm or greater) may be expected to exist on the surface overlaying fine-grained material with the fraction of fines increasing with depth. The combination of low surface acceleration and solar radiation pressure tends to strip off fine particles that have been generated by comminution processes and leave lags of larger, harder to move materials. On the other hand, this material may experience thermal fatigue that can turn them in smaller pieces. There is thus a competition between various processes, and we are not yet in a position to understand which one is dominant or their respective contributions to the final product. We would actually need at least one image of the surface of a C-type asteroid of Ryugu's size, which does not yet exist, and thus, we have to wait for Hayabusa2 to provide one.

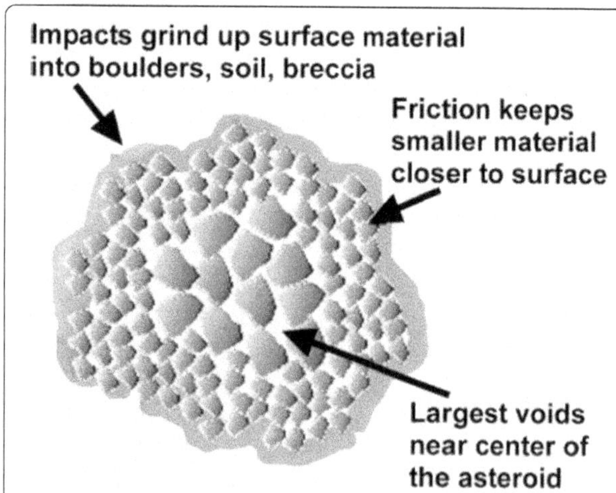

Fig. 13 Schematic of the possible structure of rubble pile asteroids. The original reaccretion of these objects is assumed to be particle size sorted with the largest pieces accreting first to form the core of the object and successively smaller sizes accreting later (Michel and Richardson 2013). The finest size fractions are generated at the surface by regolith action. These fine particles are confined to the near-surface zone by the dominance of frictional forces (from Britt and Consolmagno 2001)

Regolith cohesion, angle of friction, tensile strength, and their influence on shear strength In terrestrial environments, the shear strength of regolith is driven by inter-grain frictional forces and rotations; however, both effects are largely affected by the level of gravity. Thus, as gravitational forces decrease, other forces start to govern the behavior of granular matter. With the decrease of gravitational forces, the importance of attractive forces like van der Waals forces rise. These attractive forces may, like higher gravity, increase the frictional forces by holding the grains together and hence increasing the normal pressure. Thereby, both frictional forces and resistance torques are increased, which in turn increase shear strength. Taking these considerations into account, the behavior of granular matter gets even more complex in micro-gravity (μg). Furthermore, each particle is affected by the spin of the asteroid (particularly if it is fast) and the resulting forces, which constantly try to separate the grains and thus break the soil.

Cohesion, friction angles, and relative density have been linked for the lunar regolith as shown in Fig. 14. With an assumption of 50% porosity, the cohesion

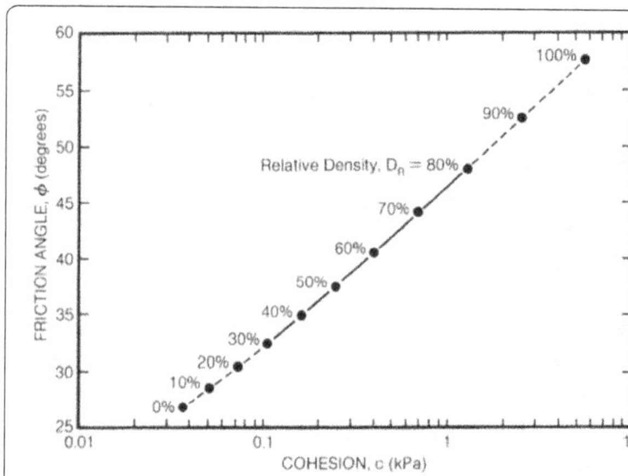

Fig. 14 Friction angle as a function of cohesion and relative density for the regolith on the Moon (same as Fig. 9.27 of Heiken et al. 1991)

should be 300 Pa, and the angle of friction about 37°. However, this is estimated for lunar regolith with fine grains. If Ryugu's regolith is coarser (1 mm to 1 cm grains), then it may have lower cohesion, say 100 Pa (30% relative density) equivalent to an angle of friction of 33°. From this, we can estimate the tensile strength from the Mohr-Coulomb curve to be 1 kPa or less (see Fig. 9.26 of Heiken et al. 1991). The effect of gravity is a complicating factor, but Kleinhans et al. (2011) have shown that if gravity decreases, the static angle of repose of regolith increases.

While cohesion in regolith is best known for the influence of water bridges in terrestrial environments, here the term cohesion covers the whole range of attractive forces. Due to the absence of water, the most important attractive forces for μg environments are van der Waals forces, electrostatic forces, magnetostatic forces, and self-gravity (Scheeres et al. 2010; Murdoch et al. 2015). In a study comparing the cohesive forces to gravitational forces in different levels of gravity, Scheeres et al. (2010) have shown that cohesion might be one of the governing effects for the shear strength of granular matter in micro-gravity. Using scaling laws, the study showed that cohesion mostly governed by van der Waals forces may be able to be the governing force even for gravel-sized particles (~centimeter size) in Ryugu's ambient gravity as shown in Fig. 15.

The non-dimensional bond number can be expressed by the relation between the gravitational weight and the cohesional forces (Scheeres et al. 2010, see Fig. 16):

$$B = \frac{F_C}{F_g}. \quad (16)$$

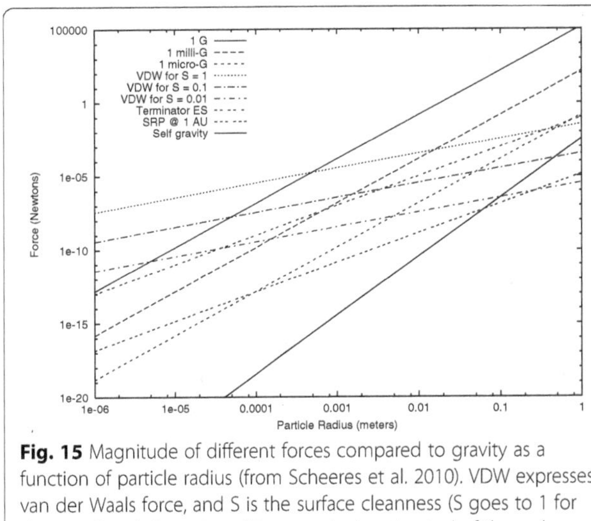

Fig. 15 Magnitude of different forces compared to gravity as a function of particle radius (from Scheeres et al. 2010). VDW expresses van der Waals force, and S is the surface cleanness (S goes to 1 for clean surfaces). Ryugu's ambient gravity is estimated of the order of 0.25 m G

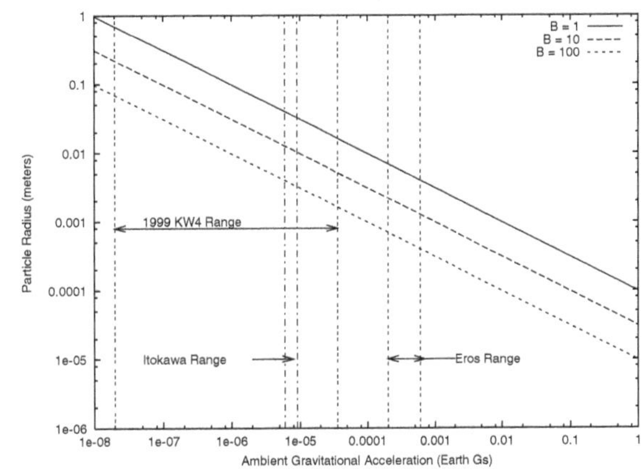

Fig. 16 Bond numbers as a function of particle radius and gravitational acceleration (from Scheeres et al. 2010). Ryugu's ambient gravity is estimated of the order of 0.25 m G

For the expected gravitational level of Ryugu, Scheeres et al. (2010) predict bond numbers of one for gain sizes up to 10th of cm. They also state that clumping and macroscopic cohesional behavior will be present even in regolith made of coarse sand portions. Other authors like Durda et al. (2012) studied the likely behavior of asteroid regolith using terrestrial laboratory simulants.

As inter-grain cohesion forces act on the grain friction level, they act similarly to overburden pressure regarding macroscopic friction and grain rolling. This may easily be shown regarding the normal F_N and tangential forces F_T on grain level:

$$F_N = F_r - F_C \quad (17)$$

$$F_T = (F_N + F_C)\mu \quad (18)$$

where F_r is the repulsive force due to contact, F_C is the total cohesion force, and μ is the micro-scale friction coefficient. The effect of cohesion lowers the repulsive effect of contact forces by holding grains together, whereas this effect does not lower the effective normal force for friction as cohesion keeps the contact in a tense condition. Regarding this issue, the magnitude of the frictional force may not only depend on the level of gravity and the bond number but also on the loading history. Higher loads that might have occurred in the past will compact the material establishing a larger number of bonds at possibly larger bond areas and thus increase the total amount of contact tension. The higher contact tension and thus micro-scale normal forces, the higher the friction forces and hence the macro-scale shear strength.

Given the expected angle of friction as stated above and the expected grain shape, inter-grain friction can also be expected to be higher than for most terrestrial sands. Thus, the inter-grain friction angle for a spherical surrogate particle is expected to be in the range of 40–50°.

Regolith bearing strength Assuming that the shape of an interacting device with the regolith is a square or a circle, the bearing capacity or strength of the regolith is defined as $\sigma = 1.3 \times c \times N_c$, where 1.3 is the shape factor (corresponding to a square or a circle), c is the regolith cohesion, and N_c is the bearing capacity factor. Assuming that cohesion is of the order of 100 Pa and that N_c is in the range 6–20 depending on the actual angle of friction (0–33°), then the bearing strength may be of the order of a few kPa or lower.

Whole rock properties We use as a reference the Master's thesis of Shigaki (2016, Kobe University) for regolith with millimeter size range particles (Table 7); for bigger rocks, strength scaling may apply, resulting in weaker strength with increasing size. Note that the tensile strength measure gives similar or slightly higher values. The compressive (unconfined) strength of sandstone is of the order of 50–100 MPa, while tensile strength is of the order of a few MPa (Ahrens 1995). However, the rock porosity of Ryugu may be higher (~ 40%) than the rock porosity of sandstone (~ 20%), and therefore, the compressive and tensile strengths may be a factor 3 lower than that of sandstone, similar to Weibern tuff (Poelchau et al. 2014). This means that we can assume that the compressive strength could be of the order of 15–30 MPa and the tensile strength about 1 MPa or less.

Presence of ponds and their properties For sample return missions, very flat areas among the wide variety of asteroid surfaces are the most sought after since they are the safest places to touch down. As revealed by asteroid surfaces imaged at the centimeter scale, ponds are good candidates (Veverka et al. 2001; Robinson et al. 2001; Cheng et al. 2002; Miyamoto et al. 2007). Since it is likely ponds are present on Ryugu's surface, it is essential to know if there is a difference between the material present in the ponds and the material in other areas. Do our chondrite samples help us solve this issue?

Table 7 Crushing strength of various materials from Shigaki (2016)

Materials	Crushing strength (MPa)
Chondrule from Allende (CV3)	7.7 ± 5.9
Chondrule from Saratov (L4)	9.4 ± 6.0
Glass bead (1 mm size)	223 ± 61
Dunite (a few mm)	13.1 ± 2.6
Basalt (a few mm)	16.9 ± 2.6
Sandstone (a few mm)*	3.4 ± 1.1

*Sandstone can be considered as an upper limit for the whole rock properties of Ryugu

Some fine-grained cognate lithologies in chondrites (Figs. 17 and 18) have been inferred to be sedimentary features indicating that they sample asteroid ponds. A cognate clast in the Vigarano CV3 chondrite, first identified by Johnson et al. (1990), appears to be a sample from such a pond of a C-type asteroid. The clast consists mainly of micron-sized grains of olivine, arranged into layers with varying composition and porosity. The densest (most Fe-rich) olivine grains are in sections of layers with the lowest bulk porosity, which appears bright in BSE images. The most remarkable aspect of the clast is crossbeds. Zolensky et al. (2002), Zolensky and Le (2017) proposed that this clast is formed by the processes of electrostatic grain levitation and subsequent seismic shaking. They then described similar clasts in another Vigarano sample, and in a section of Allende—all CV3 chondrites, and most recently in the LL3 chondrite NWA 8330 (Zolensky and Le 2017). These clasts consist mainly of olivine, whose compositional range is essentially identical to the host meteorite, although there is a striking predominance of heavy, Fe-rich olivine in the bulk clast relative to the bulk host. However, because of the compositional zoning within the pond deposit layers, the top-most surface is depleted in Fe-rich, heavier olivine. Based on these observations, samples collected from the surfaces of ponds on Ryugu will *significantly* differ from the bulk asteroid by being finer-grained and metal-poor, and silicates will be Fe-poor. Note that Itokawa samples were also depleted from typical LL chondrites by being metal depleted. Grain sizes in Ryugu regolith probably will extend down to the micron size, as was the case for Itokawa.

This interpretation is however challenged by a recent survey of Fe-rich secondary phases in CV chondrites (Ganino and Libourel 2017), which states that the phases constituting such similar networks of veinlets formed in reduced conditions near the iron-magnetite redox buffer at low aSiO2 (log(aSiO2) < – 1) and moderate temperature (210–610 °C). The various lithologies in CV3 chondrites, i.e., CV3Red, CV3OxA, and CV3OxB, including those present in Figs. 17 and 18, and their diverse porosity and permeability (MacPherson and Krot 2014) are in good agreement with CV3 lithologies being variably altered crustal pieces coming from an asteroid percolated heterogeneously via a porous flow of hydrothermal fluids. These hot, reduced, and iron-rich fluids resemble pervasive, Darcy flow type, supercritical hydrothermal fluids, and consistent with textural settings of secondary phases, as scattered patches in the CV chondrite matrices or, when

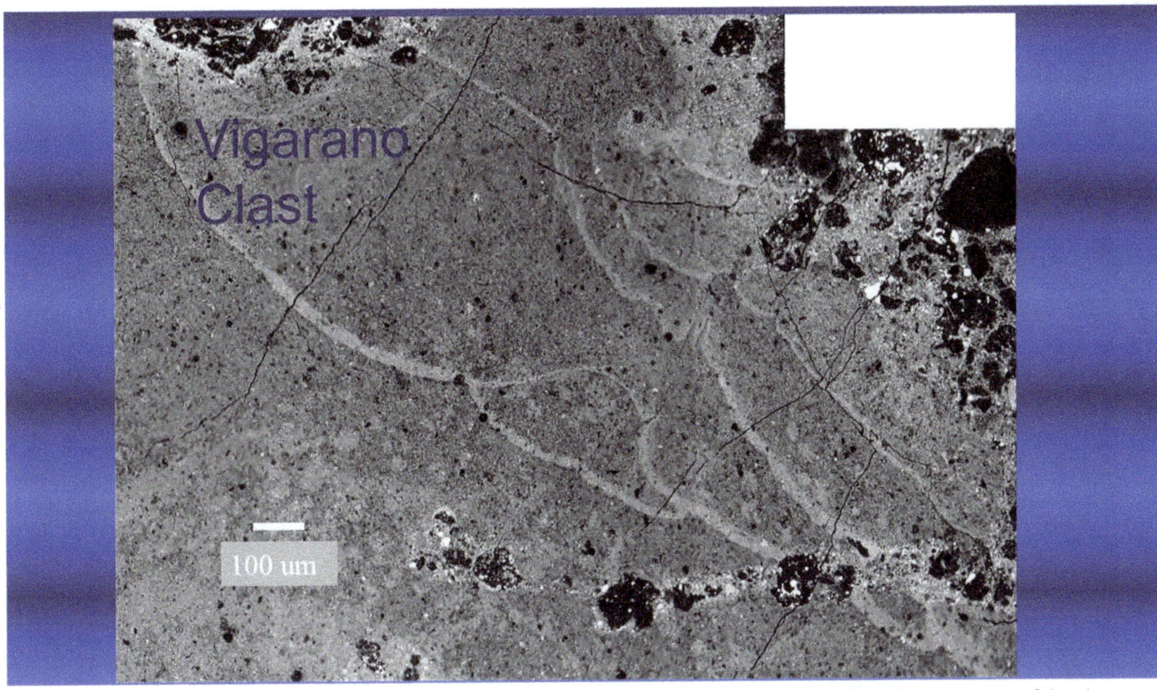

Fig. 17 Backscattered electron image of a portion of a pond deposit clast from the Vigarano CV3 chondrite. The mineralogy of the clast is mainly olivine, with Fe-rich olivine being lighter in shade. Sedimentary beds are apparent, and light gray (Fe-olivine-rich) layers are at the bottoms of the beds. Crossbeds are apparent

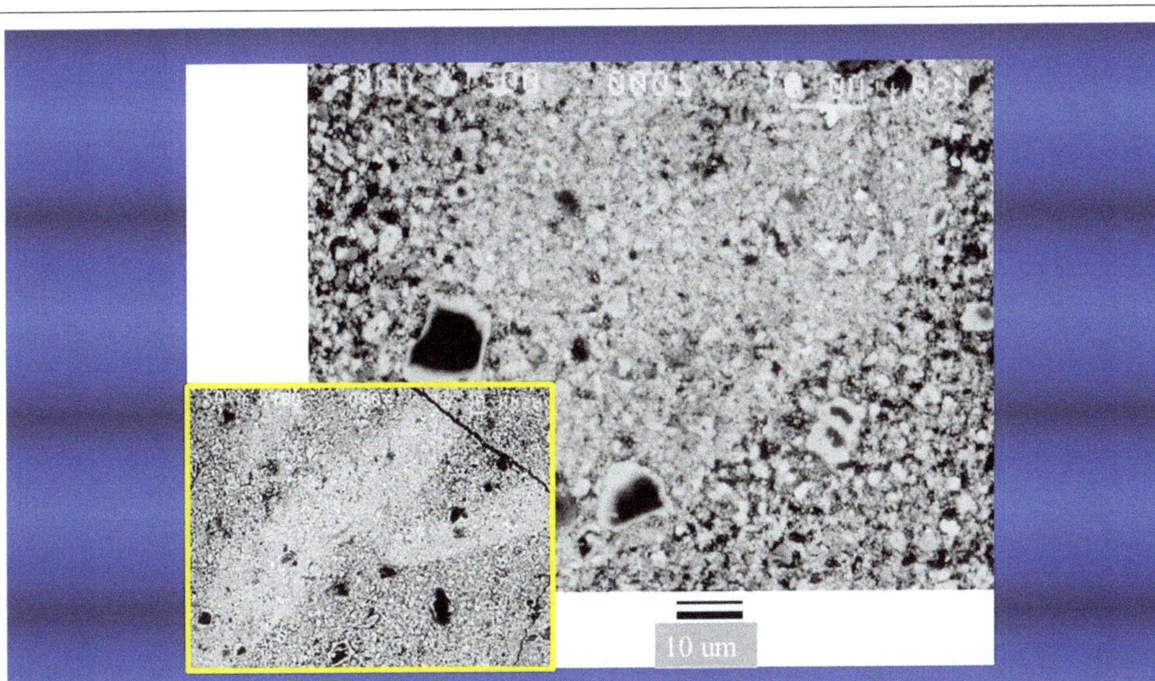

Fig. 18 Close-up on the clast in Fig. 17, showing the bottom of one bed (upper left) with a sharp contact with the top of the succeeding bed (lower right). The mineralogy of the clast is mainly olivine, with Fe-rich olivine being lighter in shade. The highest concentration of light gray, Fe-rich olivine is at the bottom of the bed. Insert image at lower left is a lower mag view of the same area, with the high mag area at the lower left of the insert. Note that the grain size extends down to the micron scale

Table 8 Properties of Ryugu, covered and discussed in the text

Global properties			
Measured	Eccentricity	0.190208	
	Semi-major axis	1.189555	AU
	Inclination	5.883556	°
	Period	473.8878	Days
	Perihelion	0.963292	AU
	Aphelion	1.415819	AU
	Rotation period	7.6326	h
	Pole direction in ecliptic coordinates (lambda, beta)	(310–340, − 40 ± 15)	°
	Shape	Almost spherical	
	Volume-equivalent diameter	850–880	m
	Phase function (Lommel -Seeliger model)	see Table 2	
	Geometric and bond albedos	see Table 3	
	Spectral type	C-type, see also Table 4	
Derived	Composition (meteorite spectral counterpart)	heated CM, CM2, or CI	
Predicted	Satellite	no	
	Dust around Ryugu	no or little	
	Boulders: power law exponent of cumulative SFD	− 3	
	Crater density	more than Itokawa, less than Toutatis	
	Surface roughness at meter scales	like Eros and Itokawa	
	Origin	Originated in the inner main asteroid belt, between ~ 2.1 and 2.5 AU, and reached the ν_6 by inward Yarkovsky drift	
Regolith thermophysical properties			
Measured	Thermal inertia	150–300, typically 200	$J\ m^{-2}\ s^{-1/2}\ K^{-1}$
	Maximum surface temperature	320–375	K
	Roughness (the rms of surface slopes)	< 0.1	
Derived	Regolith thermal conductivity	0.020–0.108, most likely 0.042	$W\ m^{-1}\ K^{-1}$
	Typical particle size in diameter	3–30 most likely 6–10 (2.2–5 by Gundlach and Blum 2013 model)	mm
Predicted	Regolith heat capacity at 300 K	758	$J\ kg^{-1}\ K^{-1}$
	Regolith emissivity	0.9	
	Regolith thermal albedo	0.019	
	Regolith bulk density	1100–1500	$kg\ m^{-3}$
Regolith mechanical properties			
Predicted	Regolith porosity	~ 0.4–0.5	
	Regolith cohesion	100	Pa
	Regolith angle of friction	33	°
	Regolith tensile strength	< 1000	Pa
	Inter-grain friction angle	40–50	°

Table 8 Properties of Ryugu, covered and discussed in the text *(Continued)*

	Regolith bearing strength	< a few 1000	Pa
	Compressive strength of rock	15–30	MPa
	Tensile strength of rock	< 1	MPa
	Presence of ponds	Possible, there are compositions being expected to be significantly different from that of the bulk asteroid	

the fluid is channelized, as subtle veinlet networks. If this interpretation is correct, the permeability/porosity/cohesiveness of the unprocessed Ryugu surface, assuming CV (or CO) chondrites are relevant proxy, will depend on the percolation of hydrothermal fluids in the Ryugu's parent body, and their efficiency to precipitate secondary phases, which will in turn control the formation and evolution of the regolith, its size distribution, and the ease of its sampling (see the "Regolith thermophysical properties" section).

Conclusions

We have reviewed a wide range of information about the C-type, NEA (162173) Ryugu from the point of view of describing the regolith layer of Ryugu: the global properties, the thermophysical properties, and the mechanical properties. Each property is categorized into three types, i.e., the measured/observed, the derived, and the predicted properties, as summarized in Table 8. This table gives a reference model of Ryugu, especially on the surface regolith.

As a whole, this paper provides a reference model of Ryugu's regolith which should be extremely useful both for science applications until Hayabusa2's arrival and for Hayabusa2 operations at Ryugu. It will be checked and refined thanks to Hayabusa2 data, and further related scientific studies will be promoted toward understanding Ryugu, other small bodies, and solar system evolution.

Abbreviations

ESO-VLT: Very Large Telescope of the European Southern Observatory; CNES: Centre National d'Etudes Spatiales; DCAM3: Deployable camera 3; DLR: German Aerospace Centre; IRSG: International Regolith Science Group; JAXA: Japan Aerospace Exploration Agency; LIDAR: Light detection and ranging; MARA: Multi-channel radiometer; MasCam: MASCOT camera with illumination unit; MasMag: Magnetometer; MBA: Main belt asteroids; MicrOmega: Near- to mid-infrared spectromicroscope; MOID: Minimum orbit intersection distance; NEA: Near-Earth asteroid; NIRS3: Near-infrared spectrometer; ONC: Optical navigation cameras; ONC-T: Optical navigation telescopic camera; ONC-W1 and W2: Optical navigation wide field of view cameras 1 and 2; PSL: Planetary Science Laboratory; RADF: Radiance factor; SCI: Small Carry-on Impactor; SFD: Cumulative size-frequency distributions; TIR: Thermal infrared imager; VLT: Very Large Telescope; VWD: van der Waals force; YORP: Yarkovsky-O'Keefe-Radzievskii-Paddack effect; μg: Micro-gravity

Acknowledgements

We thank two anonymous reviewers for the useful comments that improved this paper. We are grateful to Daniel Jay Scheeres, Paul Sanchez, and Christine Hartzell for providing figures and helpful comments. This paper is a product of IRSG in Hayabusa2 project, JAXA. The members of IRSG, except for the authors of this manuscript, are as follows: Paul Abell, Masahiko Arakawa, Young-Jun Choi, Harold C. Connolly Jr., Carsten Güttler, Naru Hirata, Tra-Mi Ho, Trevor R. Ireland, Masato Kiuchi, Goro Komatsu, Dante Lauretta, Deborah Domingue Lorin, Clara Maurel, Tomokatsu Morota, Sho Sasaki, Hirotaka Sawada, Daniel Jay Scheeres, Nicole Schmitz, Stephen R. Schwartz, Seiji Sugita, Shogo Tachibana, Satoshi Tanaka, Eri Tatsumi, Florian Thuillet, Yuichi Tsuda, Faith Vilas, Sei-ichiro Watanabe, Hikaru Yabuta, Hajime Yano, and Makoto Yoshikawa.

Funding

KW is supported by JSPS KAKENHI Grant Number 15K05273. PM and MAB acknowledge financial support from the French space agency CNES. MEZ is supported by the NASA Emerging Worlds and Hayabusa2 Participating Scientist Programs, and the SERVII Center for Lunar Science and Exploration. TM has received funding from the European Union's Horizon 2020 Research and Innovation Programme, under Grant Agreement no 687378. This material is based partly upon work supported by NASA under Contract NNM10AA11C issued through the New Frontiers Program. CME and DT are supported by the NASA Hayabusa2 Participating Scientist Program.

Authors' contributions

KW led this study. MG, PM, and KJW proposed the topic and designed each section of this study. The other authors contributed to reviewing related works, writing parts of the manuscript, and proposing some critical ideas on this study, based on their own expertise. All authors read and approved the final manuscript.

Authors' information

KW is a PI of the International Regolith Science Group (IRSG) which is a branch of the Interdisciplinary Science Team in the Hayabusa2 project. KW, MG, PM, KJW, and HM are core members of IRSG.

Competing interests

The authors declare that they have no competing interests.

Author details

[1]Planetary Exploration Research Center, Chiba Institute of Technology, Tsudanuma 2-17-1, Narashino, Chiba 275-0016, Japan. [2]DLR Institute of Planetary Research, Rutherfordstr. 2, 12489 Berlin, Germany. [3]Université Côte d'Azur, Observatoire de la Côte d'Azur, CNRS, Laboratoire Lagrange, CS 34229, 06304 Nice Cedex 4, France. [4]Southwest Research Institute, Boulder, USA. [5]Laboratoire d'Etudes Spatiales et d'Instrumentation en Astrophysique (LESIA), Observatoire de Paris, PSL Research University, CNRS, Université Paris Diderot, Sorbonne Paris Cité, UPMC Université Paris 06, Sorbonne Universités, 5 Place J. Janssen, 92195 Meudon Principal Cedex, France. [6]DLR RB-MSC German Aerospace Center, 51147 Cologne, Germany. [7]Institut für Geophysik und extraterrestrische Physik, Technische Universität Braunschweig, Mendelssohnstr. 3, 38106 Braunschweig, Germany. [8]Johns Hopkins University Applied Physics Laboratory, Laurel 20723, MD, USA. [9]DLR German Aerospace Center, Institute of Space Systems, System Engineering and Project Office, Robert-Hooke-Strasse 7, 28359 Bremen, Germany. [10]Department of Biological

and Environmental Sciences, University of Stirling, Stirling FK9 4LA, Scotland. [11]Physics Institute, University of Bern, Bern, Switzerland. [12]Center for Space Situational Awareness, Korea Astronomy and Space Science Institute, 776, Daedeokdae-ro, Yuseong-gu, Daejeon 305-348, Republic of Korea. [13]Planetary Science Institute, 1700 East Fort Lowell, Suite 106, Tucson, AZ 85719, USA. [14]DLR Institute of System Dynamics and Control, Oberpfaffenhoffen, Germany. [15]Robert M Walker for Space Sciences, Astromaterials Research and Exploration Science, NASA Johnson Space Center, Houston, TX 77058, USA. [16]Faculty of Engineering, Kindai University, Hiroshima Campus, 1 Takaya Umenobe, Higashi-Hiroshima, Hiroshima 739-2116, Japan. [17]Department of Systems Innovation, University of Tokyo, Tokyo 113-8656, Japan. [18]Max-Planck Institute for Extraterrestrial Physics, Garching, Germany. [19]Department of Planetology, Graduate School of Science, Kobe University, 1-1 Rokkodai, Nada-ku, Kobe 657-8501, Japan. [20]Department of Terrestrial Magnetism, Carnegie Institution of Washington, 5241 Broad Branch Rd NW, Washington, D.C. 20015, USA. [21]Institute of Space and Astronautical Science, Japan Aerospace Exploration Agency, Sagamihara, Japan. [22]INAF, Istituto di Astrofisica e Planetologia Spaziali, via Fosso del Cavaliere, 00133 Rome, Italy. [23]JETS/Astromaterials Research and Exploration Science, NASA Johnson Space Center, Houston 77058, TX, USA.

References

Abe M et al (2008) Ground-based Observational Campaign for Asteroid 162173 1999 JU3, 39th lunar planet. Sci, p 1594

Ahrens,TJ. (1995) Rock physics and phase relations: a handbook of physical constants

Arai T, Nakamura T, Tanaka S, Demura H, Ogawa Y, Sakatani N, Horikawa Y, Senshu H, Fukuhara T, Okada T (2017) Thermal imaging performance of TIR onboard the Hayabusa2 spacecraft. Space Sci Rev 208:239–254

Arakawa M, Wada K, Saiki T, Kadono T, Takagi Y, Shirai K, Okamoto C, Yano H, Hayakawa M, Nakazawa S, Hirata N, Kobayashi M, Michel P, Jutzi M, Imamura H, Ogawa K, Sakatani N, Iijima Y, Honda R, Ishibashi K, Hayakawa H, Sawada H (2017) Scientific objectives of Small Carry-on Impactor (SCI) and deployable camera 3 digital (DCAM3-D): observation of an ejecta curtain and a crater formed on the surface of Ryugu by an artificial high-velocity impact. Space Sci Rev 208:187_212

Barnouin-Jha OS, Cheng AF, Mukai T, Abe S, Hirata N, Nakamura R, Gaskell RW, Saito J, Clark BE (2008) Small-scale topography of 25143 Itokawa from the Hayabusa laser altimeter. Icarus 198:108–124. https://doi.org/10.1016/j.icarus.2008.05.026

Beck P, Maturilli A, Garenne A, Vernazza P, Helbert J, Quirico E, Schmitt B (2018) What is controlling the reflectance spectra (0.35-150 µm) of hydrated (and dehydrated) carbonaceous chondrites?. Icarus 313:124–138. https://doi.org/10.1016/j.icarus.2018.05.010

Benner LAM, Busch MW, Giorgini JD, Taylor PA, Margot J-L (2015) Radar observations of near-Earth and main-belt asteroids. In: Asteroids IV

Benner LAM, Ostro SJ, Magri C, Nolan MC, Howell ES, Giorgini JD, Jurgens RF, Margot J-L, Taylor PA, Busch MW, Shepard MK (2008) Near-Earth asteroid surface roughness depends on compositional class. Icarus 198:294–304. https://doi.org/10.1016/j.icarus.2008.06.010

Bibring J-P, Hamm V, Langevin Y, Pilorget C, Arondel A, Bouzit M, Chaigneau M, Crane B, Darié A, Evesque C, Hansotte J, Gardien V, Gonnod L, Leclech J-C, Meslier L, Redon T, Tamiatto C, Tosti S, Thoores N (2017) The MicrOmega investigation onboard Hayabusa2. Space Sci Rev 208:401–412

Binzel RP, DeMeo FE, Burt BJ, Cloutis EA, Rozitis B, Burbine TH, Campins H, Clark BE, Emery JP, Hergenrother CW, Howell ES, Lauretta DS, Nolan MC, Mansfield M, Pietrasz V, Polishook D, Scheeres DJ (2015) Spectral slope variations for OSIRIS-REx target asteroid (101955) Bennu: possible evidence for a fine-grained regolith equatorial ridge. Icarus 256:22–29

Binzel RP, Harris AW, Bus SJ, Burbine TH (2001) Spectral properties of near-Earth objects: Palomar and IRTF results for 48 objects including spacecraft targets (9969) Braille and (10302)1989 ML. Icarus 151:139–149

Bottke WF, Morbidelli A, Jedicke R, Petit J-M, Levison H, Michel P, Metcalfe T (2002) Debiased orbital and absolute magnitude distribution of the near-Earth objects. Icarus 156:399

Bottke WF, Vokrouhlický D, Walsh KJ, Delbo M, Michel P, Lauretta DS, Campins H, Connolly HC Jr, Scheeres DJ, Chelsey SR (2015) In search of the source of asteroid (101955) Bennu: applications of the stochastic YORP model. Icarus 247:191

Bowell E (1989) Application of photometric models to asteroids. In: Asteroids II, Proceedings of the Conference, Tucson: University of Arizona press, AZ, Mar. 8–11, 1988 (A90-27001 10-91). University of Arizona Press, pp 524–556

Britt DT, Consolmagno GJ (2001) Modeling the structure of high porosity asteroids. Icarus 152(1):134–139

Bus SJ, Binzel RP (2002) Phase II of the small main-belt asteroid spectroscopic survey: a feature-based taxonomy. Icarus 158:146–177

Campins H, de Leon J, Morbidelli A, Licandro J, Gayon-Markt J, Delbo M, Michel P (2013) The origin of asteroid 162173 (1999 JU3). AJ 146:26

Capaccioni F, Cerroni P, Coradini M, Di Martino M, Farinella P (1986) Asteroidal catastrophic collisions simulated by hypervelocity impact experiments. Icarus 66:487–514. https://doi.org/10.1016/0019-1035(86)90087-4 ISSN 0019-1035

Chapman CR, Merline WJ, Thomas PC, Joseph J, Cheng AF, Izenberg N (2002) Impact history of Eros: craters and boulders. Icarus 155:104–118. https://doi.org/10.1006/icar.2001.6744

Cheng AF (2002) Near Earth asteroid rendezvous: mission summary. In: Bottke WF, Cellino A, Paolicchi P, Binzel RP (eds) Asteroids III. Univ. Arizona Press, Tucson, pp 351–366

Cheng AF, Barnouin-Jha O, Prockter L, Zuber MT, Neumann G, Smith DE, Garvin J, Robinson M, Veverka J, Thomas PC (2002) Small-scale topography of 433 Eros from laser altimetry and imaging. Icarus 155:51–74. https://doi.org/10.1006/icar.2001.6750

Cloutis EA, Hudon P, Hiroi T, Gaffey MJ (2012) Spectral reflectance properties of carbonaceous chondrites 4: aqueously altered and thermally metamorphosed meteorites. Icarus 220:586–617

Consolmagno GJ, Schaefer MW, Schaefer BE, Britt DT, Macke RJ, Nolan MC, Howell ES (2013) The measurement of meteorite heat capacity at low temperatures using liquid nitrogen vaporization. Planet Space Sci 87:146–156

Davidsson BJR, Rickman H (2014) Surface roughness and three-dimensional heat conduction in thermophysical models. Icarus 243:58–77

Davidsson BJR, Rickman H, Bandfield JL, Groussin O, Gutiérrez PJ, Wilska M, Capria MT, Emery JP, Helbert J, Jorda L, Maturilli A, Mueller TG (2015) Interpretation of thermal emission. I. The effect of roughness for spatially resolved atmosphereless bodies. Icarus 252:1–21

de León J, Pinilla-Alonso N, Delbo M, Campins H, Cabrera-Lavers A, Tanga P, Cellino A, Bendjoya P, Gayon-Markt J, Licandro J, Lorenzi V (2016) Visible spectroscopy of the Polana-Eulalia family complex: spectral homogeneity. Icarus 266:57

Delbo M, Libourel G, Wilkerson J, Murdoch N, Michel P, Ramesh KT, Ganino C, Verati C, Marchi S (2014) Thermal fatigue as the origin of regolith on small asteroids. Nature 508:233–236

DeMeo FE, Binzel RP, Slivan SM, Bus SJ (2009) An extension of the Bus asteroid taxonomy into the near-infrared. Icarus 202:160–180

Demura H, Kobayashi S, Nemoto E, Matsumoto N, Furuya M, Yukishita A, Muranaka N, Moritz H, Shirakawa K, Maruya M, Ohyama H, Uo M, Kubota T, Hashimoto T, Kawaguchi J, Fujiwara A, Saito J, Sasaki S, Miyamoto H, Hirata N (2006) Pole and global shape of 25143 Itokawa. Science 312:1347–1349. https://doi.org/10.1126/science.1126574

Durda DD, Campo Bagatin A, Alemañ RA, Flynn GJ, STRAIT MM, Clayton AN, Patmore EB (2015) The shapes of fragments from catastrophic disruption events: effects of target shape and impact speed. Planet Space Sci 107:77–83. https://doi.org/10.1016/j.pss.2014.10.006

Durda DD, Scheeres DJ, Roark SE, Sanchez DP (2012) Asteroid regolith mechanical properties: laboratory experiments with cohesive powders. In: American Astronomical Society, DPS meeting #44, id.105.04

Ernst CM, Rodgers DJ, Barnouin OS, Murchie SL, Chabot NL (2015) Evaluating small body landing hazards due to blocks. In: 46th Lunar and Planetary Science Conference, vol 46, p 2095

Fountain JA, West EA (1970) Thermal conductivity of particulate basalt as a function of density in simulated lunar and martian environment. J Geophys Res 75:4063–4069

Fujii N, Osako M (1973) Thermal diffusivity of lunar rocks under atmospheric and vacuum conditions. Earth Planet Sci Lett 18:65–71

Fujiwara A, Kamimoto G, Tsukamoto A (1978) Expected shape distribution of asteroids obtained from laboratory impact experiments. Nature 272:602. https://doi.org/10.1038/272602a0

Fujiwara A, Kawaguchi J, Yeomans DK, Abe M, Mukai T, Okada T, Saito J, Yano H, Yoshikawa M, Scheeres DJ, Barnouin-Jha O, Cheng AF, Demura H, Gaskell RW, Hirata N, Ikeda H, Kominato T, Miyamoto H, Nakamura AM, Nakamura R,

Sasaki S, Uesugi K (2006) The rubble-pile asteroid Itokawa as observed by Hayabusa. Science 312:1330–1334. https://doi.org/10.1126/science.1125841

Ganino C, Libourel G (2017) Reduced and unstratified crust in CV chondrite parent body. Nat Commun 8:261. https://doi.org/10.1038/s41467-017-00293-1 Published online 2017 Aug 15

Grott M, Knollenberg J, Borgs B, Hänschke F, Kessler E, Helbert J, Maturilli A, Müller N (2017) The MASCOT radiometer MARA for the Hayabusa 2 mission. Space Sci Rev 208:413–431

Grott M, Knollenberg J, Krause C (2010) Apollo Lunar heat flow experiment revisited: a critical reassessment of the in situ thermal conductivity determination. J Geophys Res 115:E11005

Gundlach B, Blum J (2012) Outgassing of icy bodies in the solar system - II: heat transport in dry, porous surface dust layers. Icarus 219:618–629

Gundlach B, Blum J (2013) A new method to determine the grain size of planetary regolith. Icarus 223(1):479–492

Halajian JD, Reichman J (1969) Correlation of mechanical and thermal properties of the lunar surface. Icarus 10:179–196

Hamm M, Senshu H, Grott M (2019) Latitudinal dependence of asteroid regolith formation by thermal fatigue. Icarus 319:308-311

Hapke B (2012) Theory of reflectance and emittance spectroscopy. Cambridge: Cambridge University Press

Hapke B, Sato H (2016) The porosity of the upper lunar regolith. Icarus 273:75–83

Harris AW, Drube L (2016) Thermal tomography of asteroid surface structure. Astrophys J 832(2):127

Heiken G, Vaniman D, French B (eds) (1991) Lunar sourcebook. Cambridge Univ. Press, Cambridge

Hemingway BS, Robie RA, Wilson WH (1973) Specific heats of lunar soils, basalt, and breccias from the Apollo 14, 15, and 16 landing sites, between 90 and 350 K. In: Proceedings of the fourth lunar science conference, vol 3, pp 2481–2487

Herčík D, Auster H-U, Blum J, Fornaçon K-H, Fujimoto M, Gebauer K, Güttler C, Hillenmaier O, Hördt A, Liebert E, Matsuoka A, Nomura R, Richter I, Stoll B, Weiss BP, Glassmeier K-H (2017) The MASCOT magnetometer. Space Sci Rev 208:433–449

Hirata N, Barnouin-Jha OS, Honda C, Nakamura R, Miyamoto H, Sasaki S, Demura H, Nakamura AM, Michikami T, Gaskell RW, Saito J (2009) A survey of possible impact structures on 25143 Itokawa. Icarus 200:486–502. https://doi.org/10.1016/j.icarus.2008.10.027

Hiroi T, Moroz LV, Shingareva TV, Basilevsky AV, Pieters CM (2003) Effects of microsecond pulse laser irradiation on vis–NIR reflectance spectrum of carbonaceous chondrite stimulant: implications for Martian moons and primitive asteroids. In: 34th lunar planet. Sci, p 1324

Hiroi T, Zolensky ME, Pieters CM, Lipschutz ME (1996) Thermal metamorphism of the C, G, B, and F asteroids seen from the 0.7 μm, 3 μm, and UV absorption strengths in comparison with carbonaceous chondrites. Meteorit Planet Sci 31(3):321–327

Ho T-M, Baturkin V, Grimm C, Grundmann JT, Hobbie C, Ksenik E, Lange C, Sasaki K, Schlotterer M, Talapina M, Temtanasombat N, Wejmo E, Witte L, Wrasmann M, Wübbles G, Rößler J, Ziach C, Findlay R, Biele J, Krause C, Ulamec S, Lange M, Mierheim O, Lichtenheldt R, Maier M, Reill J, Sedlmayr H-J, Bousquet P, Bellion A, Bompis O, Cenac-Morthe C, Deleuze M, Fredon S, Jurado E, Canalias E, Jaumann R, Bibring J-P, Glassmeier KH, Herčík D, Grott M, Celotti L, Cordero F, Hendrikse J, Okada T (2017) MASCOT—the Mobile Asteroid Surface Scout onboard the Hayabusa2 mission. Space Sci Rev 208:339–374

Huang J, Ji J, Ye P, Wang X, Yan J, Meng L, Wang S, Li C, Li Y, Qiao D, Zhao W, Zhao Y, Zhang T, Liu P, Jiang Y, Rao W, Li S, Huang C, Ip W-H, Hu S, Zhu M, Yu L, Zou Y, Tang X, Li J, Zhao H, Huang H, Jiang X, Bai J (2013) The ginger-shaped asteroid 4179 Toutatis: new observations from a successful flyby of Chang'e-2, Nature scientific reports 3, p 3411. https://doi.org/10.1038/srep03411

Hütter ES, Koemle NI, Kargl G, Kaufmann E (2008) Determination of the effective thermal conductivity of granular materials under varying pressure conditions. J Geophys Res 113:E12004. https://doi.org/10.1029/2008JE003085

Ishibashi K, Shirai K, Ogawa K, Wada K, Honda R, Arakawa M, Sakatani N, Ikeda Y (2017) Performance of Hayabusa2 DCAM3-D camera for short-range imaging of SCI and ejecta curtain generated from the artificial impact crater formed on asteroid 162137 Ryugu (1999 JU_3). Space Sci Rev 208:213–238

Ishiguro M, Kuroda D, Hasegawa S et al (2014) Optical properties of (162173) 1999 JU3: in preparation for the JAXA Hayabusa 2 sample return mission. Astrophys J 792:74

Iwata T, Kitazato K, Abe M, Ohtake M, Arai T, Arai T, Hirata N, Hiroi T, Honda C, Imae N, Komatsu M, Matsunaga T, Matsuoka M, Matsuura S, Nakamura T, Nakato A, Nakauchi Y, Osawa T, Senshu H, Takagi Y, Tsumura K, Takato N, Watanabe S, Barucci M-A, Palomba E, Ozaki M (2017) NIRS3: the near infrared spectrometer on Hayabusa2. Space Sci Rev 208:317–337

Jaumann R, Schmitz N, Koncz A, Michaelis H, Schroeder SE, Mottola S, Trauthan F, Hoffmann H, Roatsch T, Jobs D, Kachlicki J, Pforte B, Terzer R, Tschentscher M, Weisse S, Müller U, Perez-Prieto L, Broll B, Kruselburger A, Ho T-M, Biele J, Ulamec S, Krause C, Grott M, Bibring J-P, Watanabe S, Sugita S, Okada T, Yoshikawa M, Yabuta H (2017) The camera of the MASCOT asteroid lander onboard Hayabusa2. Space Sci Rev 208:375–400

Jiang Y, Ji J, Huang J, Marchi S, Li Y, Ip W-H (2015) Asteroid 4179 Toutatis: boulders distribution as closely flew by Chang'e-2, IAU General Assembly, p 22

Johnson CA, Prinz M, Weisberg MK, Clayton RN, Mayeda TK (1990) Dark inclusions in Allende, Leoville, and Vigarano - evidence for nebular oxidation of CV3 constituents. Geochim Cosmochim Acta 54:819–830

Johnson KL, Kendall K, Roberts AD (1971) Surface energy and the contact of elastic solids. Proc R Soc Lond A Meth Phys Sci 324:301–313

Kameda S, Suzuki H, Cho Y, Koga S, Yamada M, Nakamura T, Hiroi T, Sawada H, Honda R, Morota T, Honda C, Takei A, Takamatsu T, Okumura Y, Sato M, Yasuda T, Shibasaki K, Ikezawa S, Sugita S (2015) Detectability of hydrous minerals using ONC-T camera onboard the Hayabusa2 spacecraft. Adv Space Res 56:1519–1524

Kameda S, Suzuki H, Takamatsu T, Cho Y, Yasuda T, Yamada M, Sawada H, Honda R, Morota T, Honda C, Sato M, Okumura Y, Shibasaki K, Ikezawa S, Sugita S (2017) Preflight calibration test results for optical navigation camera telescope (ONC-T) onboard the Hayabusa2 spacecraft. Space Sci Rev 208:17–31

Kim M-J, Choi Y-J, Moon H-K, Ishiguro M, Mottola S, Kaplan M, Kuroda D, Warjurkar DS, Takahashi J, Byun Y-I (2013) Optical observations of NEA 162173 (1999 JU3) during the 2011-2012 apparition. Astron Astrophys 550:L11

Kim M-J, Choi Y-J, Moon H-K, Krugly YN, Greenstreet S, Lister T, Kaynar S, Rumyantsev W, Molotov IE, Donchev Z (2016) Optical observations of NEA 162173 Ryugu (1999 JU3) during the 2016 apparition. In: American Astronomical Society, DPS meeting #48, 326.12

Kiuchi M, Nakamura AM (2014) Relationship between regolith particle size and porosity on small bodies. Icarus 239:291–293

Kleinhans MG, Markies H, de Vet SJ et al (2011) Static and dynamic angles of repose in loose granular materials under reduced gravity. J Geophys Res Planets 116: E11004

Küppers M, Moissl R, Vincent J-B, Besse S, Hviid SF, Carry B, Grieger B, Sierks H, Keller HU, Marchi S, Team O (2012) Boulders on Lutetia. Planet Space Sci 66: 71–78. https://doi.org/10.1016/j.pss.2011.11.004

Langseth MG, Keihm SJ, Peters K (1976) Revised lunar heat-flow values. Proc Lunar Sci Conf 7:3143–3171

Lauretta DS, Balram-Knutson SS, Beshore E, Boynton WV, Drouet d'Aubigny C, DellaGiustina DN, Enos HL, Gholish DR, Hergenrother CW, Howell ES, Johnson CA, Morton ET, Nolan MC, Rizk B, Roper HL, Bartels AE, Bos BJ, Dworkin JP, Highsmith DE, Moreau MC, Lorenz DA, Lim LF, Mink R, Nuth JA, Reuter DC, Simon AA, Bierhaus EB, Bryan BH, Ballouz R, Barnouin OS, Binzel RP, Bottke WF, Hamilton VE, Walsh KJ, Chesley SR, Christensen PR, Clark BE, Connolly HC, Crombie MK, Daly MG, Emery JP, McCoy TJ, McMahon JW, Scheeres DJ, Messenger S, Nakamura-Messenger K, Righter K, Sandford SA (2017) OSIRIS-REx: sample return from asteroid (101955) Bennu. Space Sci Rev 212(1–2):925–984

Lazzaro D, Barucci MA, Perna D, Jasmim FL, Yoshikawa M, Carvano JMF (2013) Rotational spectra of (162173) 1999 JU3, the target of the Hayabusa2 mission, Astron. Astrophys 549:L2

Le Corre L, Sanchez JA, Reddy V, Takir D, Cloutis EA, Thirouin A, Becker KJ, Li J-Y, Sugita S, Tatsumi E (2018) Ground-based characterization of Hayabusa2 mission target asteroid 162173 Ryugu: constraining mineralogical composition in preparation for spacecraft operations. Mon Not R Astron Soc 475:614–623

Ledlow MJ, Burns JO, Gisler GR, Zhao J-H, Zeilik M, Baker DN (1992) Subsurface emission from mercury: VLA radio observations at 2 and 6 centimeters. ApJ 384:640–655

Lee P, Veverka J, Thomas PC, Helfenstein P, Belton MJS, Chapman CR, Greeley R, Pappalardo RT, Sullivan R, Head JWI (1996) Ejecta blocks on 243 Ida and on other asteroids. Icarus 120:87–105. https://doi.org/10.1006/icar.1996.0039

Lee SW, Thomas P, Veverka J (1986) Phobos, Deimos, and the Moon: size and distribution of crater ejecta blocks. Icarus 68:77–86. https://doi.org/10.1016/0019-1035(86)90075-8

MacPherson GJ, Krot AN (2014) The formation of Ca-, Fe-rich silicates in reduced and oxidized CV chondrites: the roles of impact-modified porosity and permeability, and heterogeneous distribution of water ices. Meteorit Planet Sci 49:1250–1270

Margot J-L, Pravec P, Taylor P, Carry B, Jacobson S (2015) Asteroid systems: binaries, triples, and pairs. In: Bottke WF, Cellino A, Paolicchi P, Binzel RP (eds) Asteroids III. Univ. Arizona Press, Tucson, pp 355–374

Marsset M, Carry B, Dumas C, Hanuš J, Viikinkoski M, Vernazza P, Müller TG, Delbo M, Jehin E, Gillon M, Grice J, Yang B, Fusco T, Berthier J, Sonnett S, Kugel F, Caron J, Behrend R (2017) 3D shape of asteroid (6) Hebe from VLT/SPHERE imaging: implications for the origin of ordinary H chondrites. Astron Astrophys 604:A64. https://doi.org/10.1051/0004-6361/201731021

Matsumura S, Richardson DC, Michel P, Schwartz SR, Ballouz R-L (2014) The Brazil nut effect and its application to asteroids. Mon Not R Astron Soc 443:3368–3380

Maurel C, Ballouz RL, Richardson DC, Michel P, Schwartz SR (2017) Numerical simulations of oscillation-driven regolith motion: Brazil-nut effect. Mon Not R Astron Soc 464:2866–2881

Mazrouei S, Daly MG, Barnouin OS, Ernst CM, DeSouza I (2014) Block distributions on Itokawa. Icarus 229:181–189. https://doi.org/10.1016/j.icarus.2013.11.010

Michel P, Richardson DC (2013) Collision and gravitational reaccumulation: possible formation mechanism of the asteroid Itokawa. Astron Astrophys 554:L1–L4

Michikami T, Hagermann A, Kadokawa T, Yoshida A, Shimada A, Hasegawa S, Tsuchiyama A (2016) Fragment shapes in impact experiments ranging from cratering to catastrophic disruption. Icarus 264:316–330. https://doi.org/10.1016/j.icarus.2015.09.038

Michikami T, Nakamura AM, Hirata N (2010) The shape distribution of boulders on asteroid 25143 Itokawa: comparison with fragments from impact experiments. Icarus 207:277–284. https://doi.org/10.1016/j.icarus.2009.10.008

Michikami T, Nakamura AM, Hirata N, Gaskell RW, Nakamura R, Honda T, Honda C, Hiraoka K, Saito J, Demura H, Ishiguro M, Miyamoto H (2008) Size-frequency statistics of boulders on global surface of asteroid 25143 Itokawa. Earth Planets Space 60:13–20

Miyamoto H, Yano H, Scheeres DJ, Abe S, Barnouin-Jha O, Cheng AF, Demura H, Gaskell RW, Hirata N, Ishiguro M, Michikami T, Nakamura AM, Nakamura R, Saito J, Sasaki S (2007) Regolith migration and sorting on asteroid Itokawa. Science 316:1011–1014

Mizuno T, Kase T, Shiina T, Mita M, Namiki N, Senshu H, Yamada R, Noda H, Kunimori H, Hirata N, Terui F, Mimasu Y (2017) Development of the laser altimeter (LIDAR) for Hayabusa2. Space Sci Rev 208:33–47

Moskovitz NA, Ane S, Pan K, Osip DJ, Pefkou D, Melita MD, Elias M, Kitazato K, Bus SJ, DeMeo FE, Binzel RP, Abell PA (2013) Rotational characterization of Hayabusa II target asteroid (162173) 1999 JU3. Icarus 224:24–31

Muinonen K, Belskaya IN, Cellino A, Delbò M, Levasseur-Regourd AC, Penttilä A, Tedesco EF (2010) A three-parameter magnitude phase function for asteroids. Icarus, 209, 542–555

Müller TG, Ďurech J, Ishiguro M, Müller M, Krühler T, Yang H, Kim M-J, O'Rourke L, Usui F, Kiss C, Altieri B, Carry B, Choi Y-J, Delbo M, Emery JP, Greiner J, Hasegawa S, Hora JL, Knust F, Kuroda D, Osip D, Rau A, Rivkin A, Schady P, Thomas-Osip J, Trilling D, Urakawa S, Vilenius E, Weissman P, Zeidler P (2017) Hayabusa-2 mission target asteroid 162173 Ryugu (1999 JU3): searching for the object's spin-axis orientation★. Astron Astrophys 599:A103

Murdoch N, Sanchez P, Schwartz SR, Miyamoto H (2015) Asteroid surface geophysics. In: Asteroids IV, pp 767–792

Nagao K, Okazaki R, Nakamura T, Miura YN, Osawa T, Bajo K, Matsuda S, Ebihara M, Ireland TR, Kitajima F, Naraoka H, Noguchi T, Tsuchiyama A, Yurimoto H, Zolensky ME, Uesugi M, Shirai K, Abe M, Yada T, Ishibashi Y, Fujimura A, Mukai T, Ueno M, Okada T, Yoshikawa M, Kawaguchi J (2011) Irradiation history of Itokawa regolith material deduced from noble gases in the Hayabusa samples. Science 333:1128–1131

Nakamura T, Noguchi T, Tanaka M, Zolensky ME, Kimura M, Tsuchiyama A, Nakato A, Ogami T, Ishida H, Uesugi M, Yada T, Shirai K, Fujimura A, Okazaki R, Sandford SA, Ishibashi Y, Abe M, Okada T, Ueno M, Mukai T, Yoshikawa M, Kawaguchi J (2011) Itokawa dust particles: a direct link between S-type asteroids and ordinary chondrites. Science 333(6046): 1113–1116

Namiki N, Mizuno T, Hirata N, Noda H, Senshu H, Yamada R, Ikeda H, Abe S, Matsumoto K, Oshigami S, Shizugami M, Yoshida F, Hirata N, Miyamoto H, Sasaki S, Araki H, Tazawa S, Ishihara Y, Kobayashi M, Wada K, Demura H, Kimura J, Hayakawa M, Kobayashi N (2014) Scientific use of LIDAR data of Hayabusa-2 mission, Proceeding of an International CJMT-1, Workshop on Asteroidal Science, pp 74–96

Noviello JL, Ernst CM, Barnouin OS, Daly M (2014) Block distribution on Itokawa: implications for asteroid surface evolution. In: 44th lunar and planetary science conference, vol 45, p 1587

Ogawa K, Shirai K, Sawada H, Arakawa M, Honda R, Wada K, Ishibashi K, Iijima Y, Sakatani N, Nakazawa S, Hayakawa H (2017) System configuration and operation plan of Hayabusa2 DCAM3-D camera system for scientific observation during SCI impact experiment. Space Sci Rev 208:125–142

Okada T, Fukuhara T, Tanaka S, Taguchi M, Imamura T, Arai T, Senshu H, Ogawa Y, Demura H, Kitazato K, Nakamura R, Kouyama T, Sekiguchi T, Hasegawa S, Matsunaga T, Wada T, Takita J, Sakatani N, Horikawa Y, Endo K, Helbert J, Müller TG, Hagermann A (2017) Thermal infrared imaging experiments of C-type asteroid 162173 Ryugu on Hayabusa2. Space Sci Rev 208:255–286

Omura T, Nakamura AM (2017) Experimental study on compression property of regolith analogues. Planet Space Sci 149:14–22

Opeil CP, Consolmagno GJ, Safarik DJ, Britt DT (2012) Stony meteorite thermal properties and their relationship to meteorite chemical and physical states. Meteorit Planet Sci 47:319–329

Pajola M, Oklay N, La Forgia F, Giacomini L, Massironi M, Bertini I, El-Maarry MR, Marzari F, Preusker F, Scholten F, Höfner S, Lee J-C, Vincent J-B, Groussin O, Naletto G, Lazzarin M, Barbieri C, Sierks H, Lamy P, Rodrigo R, Koschny D, Rickman H, Keller HU, Agarwal J, A'Hearn MF, Barucci MA, Bertaux J-L, Cremonese G, Da Deppo V, Davidsson B, De Cecco M, Debei S, Ferri F, Fornasier S, Fulle M, Güttler C, Gutiérrez PJ, Hviid SF, Ip W-H, Jorda L, Knollenberg J, Kramm JR, Küppers M, Kürt E, Lara LM, Lin Z-Y, Lopez Moreno JJ, Magrin S, Michalik H, Mottola S, Thomas N, Tubiana C (2016) Aswan site on comet 67P/Churyumov-Gerasimenko: morphology, boulder evolution, and spectrophotometry. Astron Astrophys 592:A69. https://doi.org/10.1051/0004-6361/201527865

Pajola M, Vincent J-B, Güttler C, Lee J-C, Bertini I, Massironi M, Simioni E, Marzari F, Giacomini L, Lucchetti A, Barbieri C, Cremonese G, Naletto G, Pommerol A, El-Maarry MR, Besse S, Küppers M, La Forgia F, Lazzarin M, Thomas N, Auger A-T, Sierks H, Lamy P, Rodrigo R, Koschny D, Rickman H, Keller HU, Agarwal J, A'Hearn MF, Barucci MA, Bertaux J-L, Da Deppo V, Davidsson B, De Cecco M, Debei S, Ferri F, Fornasier S, Fulle M, Groussin O, Gutiérrez PJ, Hviid SF, Ip W-H, Jorda L, Knollenberg J, Kramm JR, Kürt E, Lara LM, Lin Z-Y, Lopez Moreno JJ, Magrin S, Marchi S, Michalik H, Moissl R, Mottola S, Oklay N, Preusker F, Scholten F, Tubiana C (2015) Size-frequency distribution of boulders ≥7 m on comet 67P/Churyumov-Gerasimenko. Astron Astrophys 583:A37. https://doi.org/10.1051/0004-6361/201525975

Perna D, Barucci MA, Ishiguro M, Alvarex-Candal A, Kuroda D, Yoshiwaka M, Kim M, Fornasier S, Hasegawa S, Roh D, Müller TG, Kim Y (2017) Spectral and rotational properties of near-Earth asteroid (162173) Ryugu, target of the Hayabusa2 sample return mission. Astron Astrophys 599:L1

Pinilla-Alonso N, de León J, Walsh KJ, Campins H, Lorenzi V, Delbo M, DeMeo F, Licandro J, Landsman Z, Lucas MP, Alí-Lagoa V (2016) Portrait of the Polana-Eulalia family complex: surface homogeneity revealed from near-infrared spectroscopy. Icarus 274:231

Pinilla-Alonso N, Lorenzi V, Campins H, de Leon J, Licandro J (2013) Near-infrared spectroscopy of 1999 JU3, the target of the Hayabusa 2 mission. Astron Astrophys 552:A79

Piqueux S, Christensen PR (2009) A model of thermal conductivity for planetary soils: 1. Theory for unconsolidated soils. J Geophys Res 114:E09005

Piqueux S, Christensen PR (2011) Temperature-dependent thermal inertia of homogeneous Martian regolith. J Geophys Res 116:E07004. https://doi.org/10.1029/2011JE003805

Poelchau MH, Kenkmann T, Hoerth T, Schafer F, Rudolf M, Thoma K (2014) Impact cratering experiments into quartzite, sandstone and tuff: the effects of projectile size and target properties on spallation. Icarus 242:211–224

Reddy V, Sanchez JA, Nathues A, Moskovitz NA, Li JY, Cloutis EA, Archer K, Tucker RA, Gaffey MJ, Paul Mann J, Sierks H, Schade U (2012) Photometric, spectral phase and temperature effects on 4 Vesta and HED meteorites: Implications for the Dawn mission, Icarus 217, 153–168

Rivkin AS, Howell ES, Vilas F, Lebofksy LA (2002) Hydrated minerals on asteroids: the astronomical record. In: Bottke WF, Cellino A, Paolicchi P, Binzel RP (eds) Asteroids III. Univ. Arizona Press, Tucson, pp 235–253

Robie RA, Hemingway BS, Wilson WH (1970) Specific heats of lunar surface materials from 90 to 350K, Proceedings of the Apollo 11 lunar science conference, vol 3, pp 2361–2367

Robinson MS, Thomas PC, Veverka J, Murchie S, Carcich B (2001) The nature of ponded deposits on Eros. Nature 413:396–400

Robinson MS, Thomas PC, Veverka J, Murchie SL, Wilcox BB (2002) The geology of 433 Eros. Meteorit Planet Sci 37:1651–1684. https://doi.org/10.1111/j.1945-5100.2002.tb01157.x

Rodgers DJ, Ernst CM, Barnouin OS, Murchie SL, Chabot NL (2016) Methodology for finding and evaluating safe landing sites on small bodies. Planet Space Sci:1–37. https://doi.org/10.1016/j.pss.2016.10.010

Rousseau E, Siria A, Jourdan G, Volz S, Comin F, Chevrier J, Greffet J-J (2009) Radiative heat transfer at the nanoscale. Nat Photonics 3(9):514–517

Rozitis B, Green SF (2011) Directional characteristics of thermal-infrared beaming from atmosphereless planetary surfaces - a new thermophysical model. Mon Not R Astron Soc 415:2042–2062

Rozitis B, Green SF, MacLennan E, Emery JP (2017) The variations of asteroid thermal inertia with heliocentric distance, asteroids, comets, meteors, Montevideo

Saiki T, Imamura H, Arakawa M, Wada K, Takagi Y, Hayakawa M, Shirai K, Yano H, Okamoto C (2017) The small carry-on impactor (SCI) and the Hayabusa2 impact experiment. Space Sci Rev 208:165–186

Saito J, Miyamoto H, Nakamura R, Ishiguro M, Michikami T, Nakamura AM, Demura H, Sasaki S, Hirata N, Honda C, Yamamoto A, Yokota Y, Fuse T, Yoshida F, Tholen DJ, Gaskell RW, Hashimoto T, Kubota T, Higuchi Y, Nakamura T, Smith P, Hiraoka K, Honda T, Kobayashi S, Furuya M, Matsumoto N, Nemoto E, Yukishita A, Kitazato K, Dermawan B, Sogame A, Terazono J, Shinohara C, Akiyama H (2006) Detailed images of asteroid 25143 Itokawa from Hayabusa. Science 312:1341–1344. https://doi.org/10.1126/science.1125722

Sakatani N, Ogawa K, Arakawa M, Tanaka S (2018) Thermal conductivity of lunar regolith simulant JSC-1A under vacuum. Icarus 309:13–24. https://doi.org/10.1016/j.icarus.2018.02.027

Sakatani N, Ogawa K, Iijima Y, Arakawa M, Honda R, Tanaka S (2017) Thermal conductivity model for powdered materials under vacuum based on experimental studies. AIP Adv 7:015310

Sakatani N, Ogawa K, Iijima Y, Arakawa M, Tanaka S (2016) Compressional stress effect on thermal conductivity of powdered materials: measurements and their implication to lunar regolith. Icarus 267:1–11. https://doi.org/10.1016/j.icarus.2015.12.012

Sanchez JA, Reddy V, Nathues A, Cloutis EA, Mann P, Hiesinger H (2012) Phase reddening on near-Earth asteroids: Implications for mineralogical analysis, space weathering and taxonomic classification, Icarus 220, 36–50

Sawada H, Ogawa K, Shirai K, Kimura S, Hiromori Y, Mimasu Y (2017) (DCAM3 development team), deployable camera (DCAM3) system for observation of Hayabusa2 impact experiment. Space Sci Rev 208:143–164

Scheeres DJ, Hartzell CM, Sanchez P (2010) Scaling forces to asteroid surfaces: the role of cohesion. Icarus 210(2):968–984

Schräpler R, Blum J, von Borstel I, Güttler C (2015) The stratification of regolith on celestial objects. Icarus 257:33–46

Schreiner SS, Dominguez JA, Sibille L, Hoffman JA (2016) Thermophysical property models for lunar regolith. Adv Space Res 57(5):1209–1222

Schröder SE, Mottola S, Carsenty U, Ciarniello M, Juamann R, Li J-Y, Longobardo A, Plamer E, Pierters C, Preusker F, Raymond CA, Russel CT (2017) Resolved spectrophotometric properties of the Ceres surface from dawn framing camera images. Icarus 288:201–225

Senshu H, Oshigami S, Kobayashi M, Yamada R, Namiki N, Noda H, Ishihara Y, Mizuno T (2017) Dust detection mode of the Hayabusa2 LIDAR. Space Sci Rev 208:65–79

Shigaki S (2016) Compression process of chondrite parent bodies inferred from the strength of chondrules (in Japanese), Master's thesis. Kobe University

Sierks H, Barbieri C, Lamy PL, Rodrigo R, Koschny D, Rickman H, Keller HU, Agarwal J, A'Hearn MF, Angrilli F, Auger A-T, Barucci MA, Bertaux J-L, Bertini I, Besse S, Bodewits D, Capanna C, Cremonese G, Da Deppo V, Davidsson B, Debei S, De Cecco M, Ferri F, Fornasier S, Fulle M, Gaskell R, Giacomini L, Groussin O, Gutierrez-Marques P, Gutiérrez PJ, Güttler C, Hoekzema N, Hviid SF, Ip W-H, Jorda L, Knollenberg J, Kovacs G, Kramm JR, Kührt E, Küppers M, La Forgia F, Lara LM, Lazzarin M, Leyrat C, Lopez Moreno JJ, Magrin S, Marchi S, Marzari F, Massironi M, Michalik H, Moissl R, Mottola S, Naletto G, Oklay N, Pajola M, Pertile M, Preusker F, Sabau L, Scholten F, Snodgrass C, Thomas N, Tubiana C, Vincent J-B, Wenzel K-P, Zaccariotto M, Pätzold M (2015) On the nucleus structure and activity of comet 67P/Churyumov-Gerasimenko. Science 347:1044

Sierks H, Lamy P, Barbieri C, Koschny D, Rickman H, Rodrigo R, A'Hearn MF, Angrilli F, Barucci MA, Bertaux J-L, Bertini I, Besse S, Carry B, Cremonese G, Da Deppo V, Davidsson B, Debei S, De Cecco M, De Leon J, Ferri F, Fornasier S, Fulle M, Hviid SF, Gaskell RW, Groussin O, Gutierrez P, Ip W, Jorda L, Kaasalainen M, Keller HU, Knollenberg J, Kramm R, Kührt E, Küppers M, Lara L, Lazzarin M, Leyrat C, Lopez Moreno JJ, Magrin S, Marchi S, Marzari F, Massironi M, Michalik H, Moissl R, Naletto G, Preusker F, Sabau L, Sabolo W, Scholten F, Snodgrass C, Thomas N, Tubiana C, Vernazza P, Vincent J-B, Wenzel K-P, Andert T, Pätzold M, Weiss BP (2011) Images of asteroid 21 Lutetia: a remnant planetesimal from the early solar system. Science 334:487–490

Singh BP, Kaviany M (1994) Effect of solid conductivity on radiative heat transfer in packed beds. Int J Heat Mass Transf 37(16):2579–2583

Stebbins J, Carmichael I, Moret L (1984) Heat capacities and entropies of silicate liquids and glasses. Contrib Mineral Petrol 86(2):131–148

Sugita S et al (2013) Visible spectroscopic observations of asteroid 162173 (1999 JU3) with the Gemini-S telescope. In: 44th lunar planet. Sci, p 2591

Susorney HC, Barnouin OS (2016) The global surface roughness of 433 Eros from the NEAR-shoemaker laser altimeter (NLR). In: AAS/Division for Planetary Sciences Meeting Abstracts 48

Suzuki M, Makino K, Yamada M, Iinoya K (1980) A study of coordination number in a random packed system of monosized sphere particles (in Japanese). Kagaku Kogaku Ronbunshu 6:59–64

Takir D, Clark BE, d'Aubigny CD, Hergenrother CW, Li JY, Lauretta DS, Binzel RP (2015) Photometric models of disk-integrated observations of the OSIRIS-REx target asteroid (101955) Bennu. Icarus 252:393–399

Takita J, Senshu H, Tanaka S (2017) Feasibility and accuracy of thermophysical estimation of asteroid 162173 Ryugu (1999 JU3) from the Hayabusa2 thermal infrared imager. Space Sci Rev 208:287–315

Tancredi G, Maciel A, Heredia L, Richeri P, Nesmachnow S (2012) Mon Not R Astron Soc 420:3368–3380. https://doi.org/10.1111/j.1365-2966.2011.20259.x

Tancredi G, Roland S, Bruzzone S (2015) Distribution of boulders and the gravity potential on asteroid Itokawa. Icarus 247:279–290. https://doi.org/10.1016/j.icarus.2014.10.011

Tholen D, Barucci MA (1989) In: Binzel RP, Gehrels T, Matthews MS (eds) Asteroids II. Tucson: Univ. of Arizona Press, p 298

Thomas PC, Robinson MS (2005) Seismic resurfacing by a single impact on the asteroid 433 Eros. Nature 436:366–369. https://doi.org/10.1038/nature03855

Thomas PC, Veverka J, Bell JF, Clark BE, Carcich B, Joseph J, Robinson M, McFadden LA, Malin MC, Chapman CR, Merline W, Murchie S (1999) Mathilde: size, shape, and geology. Icarus 140:17–27. https://doi.org/10.1006/icar.1999.6121

Thomas PC, Veverka J, Robinson MS, Murchie S (2001) Shoemaker crater as the source of most ejecta blocks on the asteroid 433 Eros. Nature 413:394–396

Thomas PC, Veverka J, Sullivan R, Simonelli DP, Malin MC, Caplinger M, Hartmann WK, James PB (2000) Phobos: regolith and ejecta blocks investigated with Mars orbiter camera images. J Geophys Res 105:15091–15106. https://doi.org/10.1029/1999JE001204

Thuillet F, Michel P, Maurel C, Ballouz R-L, Zhang Y, Richardson DC, Biele J, Tastumi E, Sugita S (2018) Numerical modeling of lander interaction with a low-gravity asteroid regolith surface: application to MASCOT onboard Hayabusa2. Astron Astrophys 615, A41. https://doi.org/10.1051/0004-6361/201832779

Tonui E, Zolensky M, Hiroi T, Nakamura T, Lipschutz M, Wang M-S, Okudaira K (2014) Petrographic, chemical and spectroscopic evidence for thermal metamorphism in carbonaceous chondrites I: CI and CM chondrites. Geochim Cosmochim Acta 126:284–306

Tsuchiyama A, Uesugi M, Uesugi K, Nakano T, Noguchi R, Matsumoto T, Matsuno J, Nagano T, Imai Y, Shimada A, Takeuchi A, Suzuki Y, Nakamura T, Noguchi T, Abe M, Yada T, Fujimura A (2014) Three-dimensional microstructure of samples recovered from asteroid 25143 Itokawa: comparison with LL5 and LL6 chondrite particles. Meteorit Planet Sci 49(2):172–187

Tsuda Y, Yoshikawa M, Abe M, Minamino H, Nakazawa S (2013) System design of the Hayabusa2—asteroid sample return mission to Ryugu. Acta Astronaut 91:356

Turcotte DL (1997) Fractals and chaos in geology and geophysics. Cambridge University Press, Cambridge, p 398

Veverka J, Thomas PC, Robinson M, Murchie S, Chapman C, Bell M, Harch A, Merline WJ, Bell JF, Bussey B, Carcich B, Cheng A, Clark B, Domingue D, Dunham D, Farquhar R, Gaffey MJ, Hawkins E, Izenberg N, Joseph J, Kirk R, Li H, Lucey P, Malin M, McFadden L, Miller JK, Owen WM, Peterson C, Prockter L, Warren J, Wellnitz D, Williams BG, Yeomans DK (2001) Imaging of small-scale features on 433 Eros from NEAR: evidence for a complex regolith. Science 292:484–488

Vilas F (2008) Spectral characteristics of Hayabusa 2 near-Earth asteroid targets 162173 1999 JU3 and 2001 QC34. Astron J 135:1101–1105
Vilas F, Sykes MV (1996) Are Low-Albedo Asteroids Thermally Metamorphosed? Icarus 124, 483–489
Walsh KJ, Jacobson SA (2015) Formation and evolution of binary asteroids. In: Michel P, DeMeo FE, Bottke WF (eds) Asteroids IV. Univ. Arizona Press, Tucson, pp 375–393
Walsh KJ, Delbo M, Bottke WF, Vokrouhlicky D, & Lauretta D (2013) Introducing the Eulalia and new Polana asteroid families: Re-assessing primitive asteroid families in the inner Main Belt, Icarus 225, 283–297
Watanabe S, Tsuda Y, Yoshikawa M, Tanaka S, Saiki T, Nakazawa S (2017) Hayabusa2 mission overview. Space Sci Rev 208:3–16. https://doi.org/10.1007/s11214-017-0377-1
Wechsler AE, Glaser PE, Fountain JA (1972) Thermal properties of granulated materials. Prog Aero Astron 28:215
Winter DF, Saari JM (1969) A particulate thermophysical model of the lunar soil. Astrophys J 156:1135–1151
Yamada R, Senshu H, Namiki N, Mizuno T, Abe S, Yoshida F, Noda H, Hirata N, Oshigami S, Araki H, Ishihara Y, Matsumoto K (2017) Albedo observation by Hayabusa2 LIDAR: instrument performance and error evaluation. Space Sci Rev 208:49–64
Yomogida K, Matsui T (1983) Physical properties of ordinary chondrites. J Geophys Res 88:9513–9533. https://doi.org/10.1029/JB088iB11p09513
Yoshikawa M, Kawaguchi J, Fujiwara A, Tsuchiyama A (2015) Hayabusa sample return mission. In: Michel P, DeMeo FE, Bottke WF (eds). Asteroids IV, Univ. Arizona Press, Tucson, pp 397–418
Yu L, Ji J, Wang S (2014) Investigation of thermal inertia and surface properties for near-Earth asteroid (162173) 1999 JU3. Chin Astron Astrophys 42:317–331
Zolensky M, Le L (2017) Asteroid pond mineralogy: view from a cognate clast in LL3 northwest Africa 8330. In: Abstracts of the 2017 Meteoritical Society Meeting, #6164
Zolensky ME, Nakamura K, Cheng AF, Cintala MJ, Hörz F, Morris RV, Criswell D (2002) Meteoritic evidence for the mechanism of pond formation on asteroid Eros, Lunar and Planetary Science Conference, vol 33
Zou R, Gan M, Yu A (2011) Prediction of the porosity of multi-component mixtures of cohesive and non-cohesive particles. Chem Eng Sci 66(20):4711–4721
Zou X, Li C, Liu J, Wang W, Li H, Ping J (2014) The preliminary analysis of the 4179 Toutatis snapshots of the Chang'E-2 flyby. Icarus 229:348–354. https://doi.org/10.1016/j.icarus.2013.11.002
Zuber MT, Smith DE, Cheng AF, Garvin JB, Aharonson O, Cole TD, Dunn PJ, Guo Y, Lemoine FG, Neumann GA, Rowlands DD, Torrence MH (2000) The shape of 433 Eros from the NEAR-shoemaker laser rangefinder. Science 289:2097–2101. https://doi.org/10.1126/science.289.5487.2097

13

Regolith behavior under asteroid-level gravity conditions: low-velocity impact experiments

Julie Brisset*, Joshua Colwell, Adrienne Dove, Sumayya Abukhalil, Christopher Cox and Nadia Mohammed

Abstract

The dusty regolith covering the surfaces of asteroids and planetary satellites differs in size, shape, and composition from terrestrial soil particles and is subject to environmental conditions very different from those found on Earth. This regolith evolves in a low ambient pressure and low-gravity environment. Its response to low-velocity impacts, such as those that may accompany human and robotic exploration activities, may be completely different than what is encountered on Earth. Experimental studies of the response of planetary regolith in the relevant environmental conditions are thus necessary to facilitate future Solar System exploration activities.

We combined the results and provided new data analysis elements for a series of impact experiments into simulated planetary regolith in low-gravity conditions using two experimental setups and a range of microgravity platforms. The Physics of Regolith Impacts in Microgravity Experiment (PRIME) flew on several parabolic aircraft flights, enabling the recording of impacts into granular materials at speeds of ~ 4–230 cm/s. The COLLisions Into Dust Experiment (COLLIDE) is conceptually close to the PRIME setup. It flew on the Space Shuttle in 1998 and 2001 and more recently on the Blue Origin New Shepard rocket, recording impacts into simulated regolith at speeds between 1 and 120 cm/s. Results of these experimental campaigns found that there is a significant change in the regolith behavior with the gravity environment. In a $10^{-2}g$ environment (with g being the gravity acceleration at the surface of the Earth), only embedding of the impactor was observed and ejecta production was produced for most impacts at > 20 cm/s. Once at microgravity levels ($< 10^{-4}g$), the lowest impact energies also produced impactor rebound. In these microgravity conditions, ejecta started to be produced for impacts at > 10 cm/s. The measured ejecta speeds were somewhat lower than the ones measured at reduced-gravity levels, but the ejected masses were higher. In general, the mean ejecta velocity shows a power-law dependence on the impact energy with an index of ~ 0.5. When projectile rebound occurred, we observed that its coefficients of restitution on the bed of regolith simulant decrease by a factor of 10 with increasing impact speeds from ~ 5 up to 100 cm/s. We could also observe an increased cohesion between the JSC-1 grains compared to the quartz sand targets.

Keywords: Regolith, Low-velocity impacts, Microgravity experiments

Introduction

Small airless bodies of the Solar System are known to be covered in a layer of regolith, composed of grains of varying sizes. Created by the bombardment of micrometeoroids or thermal disintegration, micron-sized and larger dust particles collect on the surfaces of asteroids, comets, and moons (Housen et al. 1979; Housen and Wilkening 1982). Any space mission probing or landing on the surfaces of these small bodies has to interact with this regolith, at the very low surface gravity levels induced by the body's small mass. The first soft landings on such small bodies were performed by the NEAR-Shoemaker spacecraft on the asteroid Eros (Veverka et al. 2001), Hayabusa on Itokawa (Kawaguchi et al. 2008), and Philae on the comet 67P/Churyumov-Gerasimenko (Biele et al. 2015). In the latter instances, the landing resulted in an unexpected response of the surface to the low-velocity impact of the lander, in particular the

*Correspondence: julie.brisset@ucf.edu
Florida Space Institute, University of Central Florida, 12354 Research Parkway, Partnership 1 Building, Suite 211, Orlando 32826, FL, USA

remarkable recurrent bouncing of Philae on the comet's surface after a first touchdown at around 1 m/s.

Currently, two missions are en route to asteroids: Hayabusa-2 carrying the MASCOT lander (Tsuda et al. 2013; Jaumann et al. 2016), due to arrive at Ryugu in 2018, and the OSIRIS-REx mission arriving at Bennu in 2018, planning to sample its surface using a Touch-And-Go mechanism (Lauretta et al. 2017). In addition, the increased interest in In-Situ Resource Utilization (ISRU) and mining prospects will soon lead to more reconnaissance and pathfinder missions to small bodies of the Solar System. Understanding interactions with surface regolith at very low gravity levels has therefore become an imperative. Impacts at a few meters per second down to a few centimeters per second into regolith layers are of particular interest to these current and future missions.

Motivated by these upcoming missions, theoretical and numerical work has recently been performed in order to investigate the behavior of granular material at the surface of small asteroids, such as Ryugu and Bennu (Scheeres et al. 2010; Sanchez and Scheeres 2014; Hirabayashi et al. 2015; Thuillet et al. 2017). In particular, the response of a layer of regolith to a low-velocity impact is investigated in, e.g., Maurel et al. (2017) for MASCOT and Sánchez et al. (2013). Experimental works supporting these efforts are scarce and often performed under conditions very different from what can be expected during a landing/contact at the surface of small bodies (1*g*, hypervelocity impacts). For example, the experiment campaign performed by Housen and Holsapple (2003) to investigate crater formation in porous materials studied impacts at >2 km/s (see Holsapple et al. (2002) and Housen and Holsapple (2011), for a review of other high-speed impact experiments in porous materials). In order to investigate the relative influence of gravity on the impact cratering and ejecta production processes in very large impact events, Housen et al. (1999) performed high-speed impacts into porous materials in a centrifuge at gravity levels up to 500*g*. While these experiments at high impact velocities are relevant for meteorite and inter-asteroid impacts, the velocity ranges are not applicable to the landing or sample collection on a surface in low gravity.

Theoretical and experimental studies that are more relevant to an asteroid landing situation are found in the study of impactor deceleration in a granular medium: the Katsuragi and Durian (2007, 2013) and Katsuragi and Blum (2017) series studies the drag force exerted by a granular material on a penetrating impactor; Clark et al. (2012) and Clark and Behringer (2013) study the impactor stopping time and the energy transfer from the impactor to the granular material at the grain level in quasi-2D setups; Seguin et al. (2008) studies the influence of container walls on the impactor penetration depth; and Machii et al. (2013) measures the influence of the target strength on the impact outcome. All these experiments were performed at impact speeds ranging from a few 10 cm/s to a few meters per second and provide therefore relevant data for the exploration of asteroid surfaces. However, their experiments were performed in 1*g*, and the application of their findings to small body surfaces, which can have gravity levels $< 10^{-5}g$, remains unexplored.

The study of low-velocity impacts into granular material in reduced gravity was started in 1998 with the Shuttle experiment COLLisions Into Dust Experiment (COLLIDE) (Colwell and Taylor 1999; Colwell 2003). Following experiments were performed during parabolic flight campaigns in 2002 and 2003, using the Physics of Regolith Impacts in Microgravity Experiment (PRIME) hardware setup (Colwell et al. 2008). More recent data collection was obtained with these two experiments during the PRIME-3 (Colwell et al. 2016) and COLLIDE-3 campaigns, on the NASA C-9 aircraft in 2014 and Blue Origin's New Shepard suborbital rocket in 2016, respectively. Both these experiment setups generate impacts of approximately centimeter-sized spherical projectiles onto beds of granular material at speeds of 1–230 cm/s and gravity levels ranging from reduced gravity ($\sim 10^{-2}g$) to microgravity ($< 10^{-4}g$). The target materials used were quartz sand and JSC-1 lunar and JSC Mars-1 simulants, sieved at particle size $< 250\mu$m. While the results of each of these experiment campaigns have been published separately (except for the most recent COLLIDE-3 campaign), the collected data set has now reached a size of ~ 130 data points and presents the opportunity for a new and combined analysis, with the perspective of applying these experimental results to interactions with the regolith-covered surfaces of small Solar System bodies.

This is the undertaking of the present paper. We combine the data sets of all the COLLIDE and PRIME campaigns and show the collision outcomes, coefficients of restitution, ejecta speeds, and estimated mass in dependance of the impact velocity and environmental *g*-level. "Methods/Experimental" section describes the PRIME and COLLIDE hardware setups and data collected. The "Results" section describes the data analysis and results, which we then discuss in the "Discussion" section.

Methods/Experimental

The data results presented here were collected over almost two decades during microgravity experimental campaigns on a variety of flight platforms. The hardware used to collect data can be divided into two families: (1) the COLLIDE experiment series, which flew on the Space Shuttle and a suborbital rocket, and (2) the PRIME experiment series, which flew on parabolic aircrafts.

COLLIDE

Flight campaigns The first two COLLIDE flight campaigns were performed in 1998 and 2001, onboard the Space Transportation System (Space Shuttle) STS-90 and STS-108 missions, respectively (COLLIDE-1 and COLLIDE-2). The experiment hardware and performance were described in Colwell and Taylor (1999) and Colwell (2003). During these flights, a total of nine impacts into regolith simulant were recorded.

More recently, in April 2016, a modified setup was flown on Blue Origin's New Shepard suborbital rocket (COLLIDE-3). Parts of the Space Shuttle COLLIDE hardware were repurposed for this new campaign, and a total of four impactor boxes were accommodated into the crew capsule of New Shepard. Three of these boxes were similar to the Space Shuttle hardware, launching one projectile into a bed of regolith simulant. The fourth box was outfitted with a multi-launcher instead of the single-launcher configuration, now allowing for three impactors to be sequentially launched onto the bed of regolith simulant during the same experiment run.

Hardware For COLLIDE-3, four individual Impactor Box Systems (IBSs) of COLLIDE-2 were modified for flight on suborbital launch vehicles. We manufactured a vacuum chamber for each IBS in order to perform the impacts under vacuum conditions, because the New Shepard crew capsule stays at atmospheric pressure during the flight. Each vacuum chamber was outfitted with a viewport allowing for a camera outside of the chamber to record the impact. The IBS electronics were re-wired and connected to custom electronics outside the chamber via an electrical feedthrough. Both the chamber and IBS can be seen in Fig. 1.

The camera system was entirely re-designed for the COLLIDE-3 campaign. While COLLIDE-1 and COLLIDE-2 used camcorders recording on digital video tapes, the new design used a GoPro Hero 3+ black edition recording data files on a microSD card. Both configurations required the retrieval of the tapes/memory card after the experiment run.

Three of the IBSs were used as single-launchers (one projectile per experiment run, like in the COLLIDE-1 and COLLIDE-2 campaigns), while the fourth one was modified to allow for the launching of three projectiles during the same experiment run. This modified setup was named multi-launcher. In a single-launcher IBS, the projectile is kept on a compressed spring by an aluminum bar. This bar is held in its closed position by a pin attached to two wires of memory shape alloy. When activated, these wires retract the pin and the released aluminum bar swings in its open position via a rotary spring. More details on the single-launcher setup can be found in Colwell and Taylor (1999) and Colwell (2003).

The multi-launcher is composed of three individual launchers (slightly smaller than a single-launcher), a central stepper motor, and a circular bottom plate. The stepper motor allows for the rotation of the bottom plate, which is outfitted with a drop hole. The projectiles inside individual launchers are kept on a compressed linear spring. They are held in place by an aluminum bar forced in its closed position by the presence of the bottom plate. This bar is held under tension by a rotary spring pressing it towards an open position. When the drop hole aligns

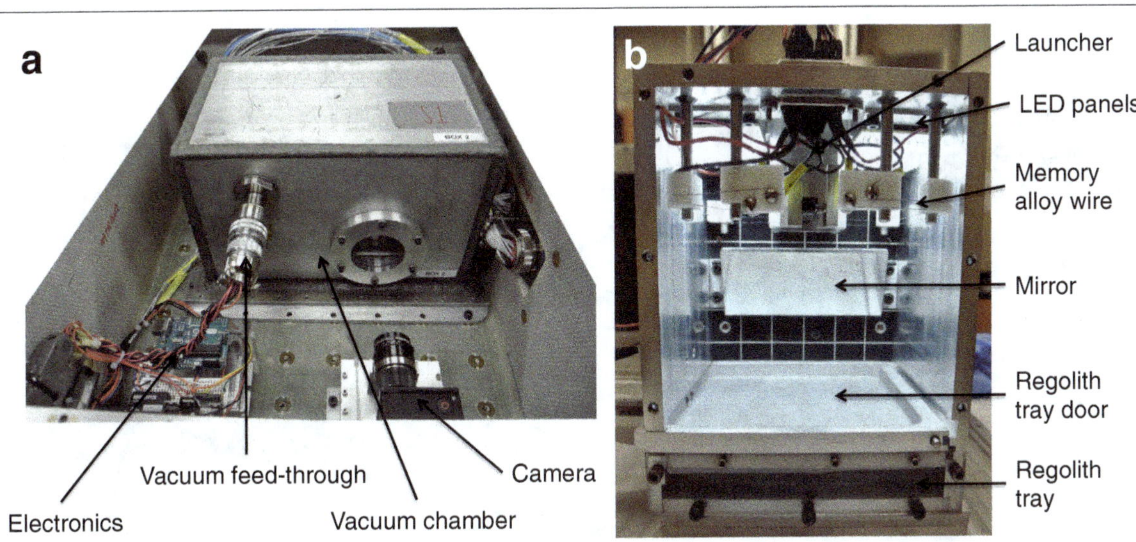

Fig. 1 COLLIDE-3 hardware. **a** The general setup showing the vacuum chamber with its viewport, the camera for data collection, and the experiment electronics for autonomous experiment run. **b** The Impactor Box System (IBS) located inside the vacuum chamber with the launcher and regolith tray

with the launcher, the pin is freed thus allowing for the linear spring of the launcher to push the projectile outwards. Figure 2 shows the multi-launcher setup.

Projectiles and simulants Table 1 details which projectiles and simulants were used for all three experiment campaigns. The projectiles used in the COLLIDE campaigns were Teflon and quartz spheres of diameters between 1 and 2 cm. These projectiles were chosen in order to vary the impact energies over the largest range of values possible while using the same hardware setup. For a similar projectile size, 1 cm projectiles had masses about one order of magnitude smaller than 2 cm projectiles.

The target materials used were quartz sand, the lunar mare basalt simulant JSC-1, and the Mars soil simulant JSC Mars-1. The choice of these materials was not motivated by their mineralogic similarity to asteroid regolith, but rather by the shape of their grains. Indeed, from the relatively long history of inter-grain collision experiments (see Blum and Wurm (2008), for a review), it appears that the actual mineral composition of the grains plays a comparatively negligible role in granular material cohesion compared to the size and shape of the grains. A notable exception to this observation takes place if the regolith is composed of ices (in particular water ice) or covered with organics, which are much stickier than rocky regolith and therefore change the behavior of grain layers. Ices and organic materials are outside the scope of the present work, and we therefore only considered grain shape and sizes. The main difference between the chosen simulants is the grain shapes and surface texture, which are increasingly angular and irregular from quartz sand to JSC Mars-1 and to JSC-1 lunar simulants (Fig. 3). Note that quartz sand and JSC Mars-1 are more similar in shape compared with JSC-1.

The target materials were sieved to sizes between 75 and 250 μm, which provides for a normal distribution between these two sizes, and filled into the target tray of dimensions of $13 \times 13 \times 2$ cm^3. Grain sizes on the surface of asteroids are known to have a size distribution ranging from micrometers to 10 m. The population of grains smaller than 1 cm is difficult to infer from available imaging of surfaces, such as for Vesta (Hiroi et al. 1994) and Eros (Dombard et al. 2010), as the image resolutions are usually >1 cm. However, from thermal inertia measurements, Gundlach and Blum (2013) has inferred average regolith grain sizes covering the surfaces of 26 asteroids. For 15 of these 26 bodies, they determined an average grain size of < 1 mm, which corresponds to the grain size distributions we used in the COLLIDE and PRIME experiments.

Seguin et al. (2008) studied the influence of the grain container size on the penetration depth of projectiles dropped into the container at speeds of < 3 m/s. They showed that for a granular material target composed of 300–400 μm-sized grains and projectile sizes ranging from 5 to 40 mm, the container bottom had an influence on the projectile penetration depth only if located at a few grain diameters from the surface. They also showed that the container side wall location had a vanishing influence on the impactor penetration for a ratio between the tray and projectile of > 5. Given their impact speeds and the size of their projectiles and target grains, their results are applicable to the COLLIDE and PRIME experiments. As the target trays were 13×13 cm^2 ($13 > 5 \times 2$ cm for a projectile of 2 cm in diameter) in lateral size and 2 cm

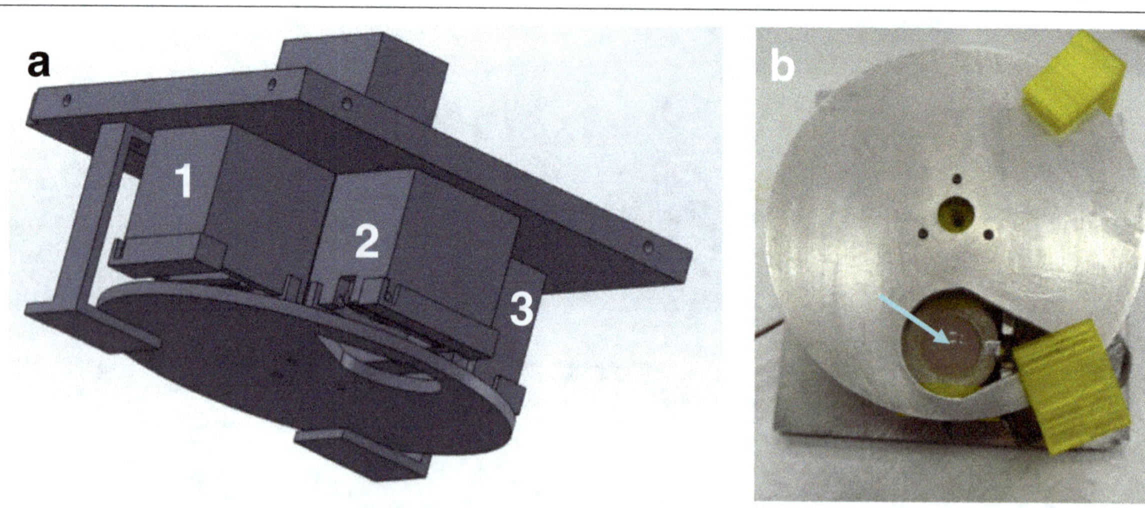

Fig. 2 COLLIDE-3 multi-launcher system. **a** CAD side view showing the individual launchers (three, numbered) and the rotating plate at the bottom. **b** The multi-launcher from the bottom showing the rotating plate and a projectile (marked by an arrow) in its individual launcher through the plate's drop hole

Table 1 Projectiles and target materials of the three COLLIDE and the PRIME-3 campaigns

COLLIDE campaign	IBS	Projectile				Target material	Outcome	Coefficient of restitution
		Diameter [cm]	Mass [g]	Material	Speed [cm/s]			
1	1	0.96	0.99	Teflon	14.80	JSC-1	R	0.028
1	4	0.96	0.99	Teflon	17.10	JSC-1	R	0.022
1	6	0.96	0.99	Teflon	90.00	JSC-1	R	0.03
2	1	2.00	10.71	Quartz	110.00	Quartz sand	E	0.01
2	2	2.00	9.64	Quartz	3.60	Quartz sand	E	N/A
2	3	2.00	10.29	Quartz	1.29	Quartz sand	E	N/A
2	4	2.00	8.66	Quartz	81.00	Quartz sand	E	0.015
2	5	2.00	8.98	Quartz	12.20	JSC-1	E	N/A
2	6	2.00	10.62	Quartz	28.00	Quartz sand	E	0.02
3	2	2.05	10.55	Quartz	26.5	Quartz sand	R	9.3×10^{-3}
3	3	2.05	10.55	Quartz	29.3	JSC Mars-1	E	N/A
PRIME-3 flight	Box							
1	1	2.05	9.82	Quartz	48.45	Quartz sand	E	N/A
1	3	2.05	11.76	Quartz	49.72	Quartz sand	E	N/A
1	4	2.05	30.75	Brass	43.19	Quartz and	A	N/A
1	5	2.05	9.9	Quartz	43.75	Quartz and	E	N/A
1	6	2.05	28.2	Brass	19.87	Quartz sand	X	N/A
1	7	2.05	9.82	Quartz	33.32	JSC-1	R	0.078
1	8	2.05	28.2	Brass	19.56	JSC-1	X	N/A
2	2	2.05	28.2	Brass	16.21	Quartz sand	R	0.046
2	3	2.05	11.76	Quartz	31.52	Quartz sand	A	N/A
2	4	2.05	30.75	Brass	38.4	Quartz sand	A	N/A
2	5	2.05	9.9	Quartz	38.73	Quartz sand	E	0.025
2	6	2.05	28.2	Brass	13.6	Quartz sand	X	N/A
2	7	2.05	9.82	Quartz	5.55	JSC-1	R	0.276
2	8	2.05	28.2	Brass	14.45	JSC-1	R	0.091
3	1	2.05	9.82	Quartz	24.75	Quartz sand	E	N/A
3	3	2.05	11.76	Quartz	26.38	Quartz sand	E	N/A
3	4	2.05	30.75	Brass	31.84	Quartz sand	A	N/A
3	5	2.05	9.9	Quartz	48.18	Quartz sand	A	N/A
3	6	2.05	28.2	Brass	15.2	Quartz sand	X	N/A
3	8	2.05	28.2	Brass	4.02	JSC-1	R	0.229
4	1	2.05	9.82	Quartz	25.85	JSC-1	E	0.075
4	3	2.05	11.76	Quartz	20.94	Quartz sand	E	0.020
4	4	2.05	30.75	Brass	27.77	JSC-1	A	N/A
4	5	2.05	9.9	Quartz	52.93	Quartz sand	A	N/A
4	6	2.05	28.2	Brass	10.02	Quartz sand	X	N/A
4	8	2.05	28.2	Brass	13.38	JSC-1	R	0.055

The table of the 108 PRIME-1 campaign impacts can be found in Colwell et al. (2008). No coefficients of restitution could be measured for the PRIME-1 impacts. *R* rebound with no ejecta, *E* ejecta production, *A* asteroid-level parabola (0.05*g*), *X* no data available (regolith pillar), *N/A* no rebound from the projectile on the target could be observed

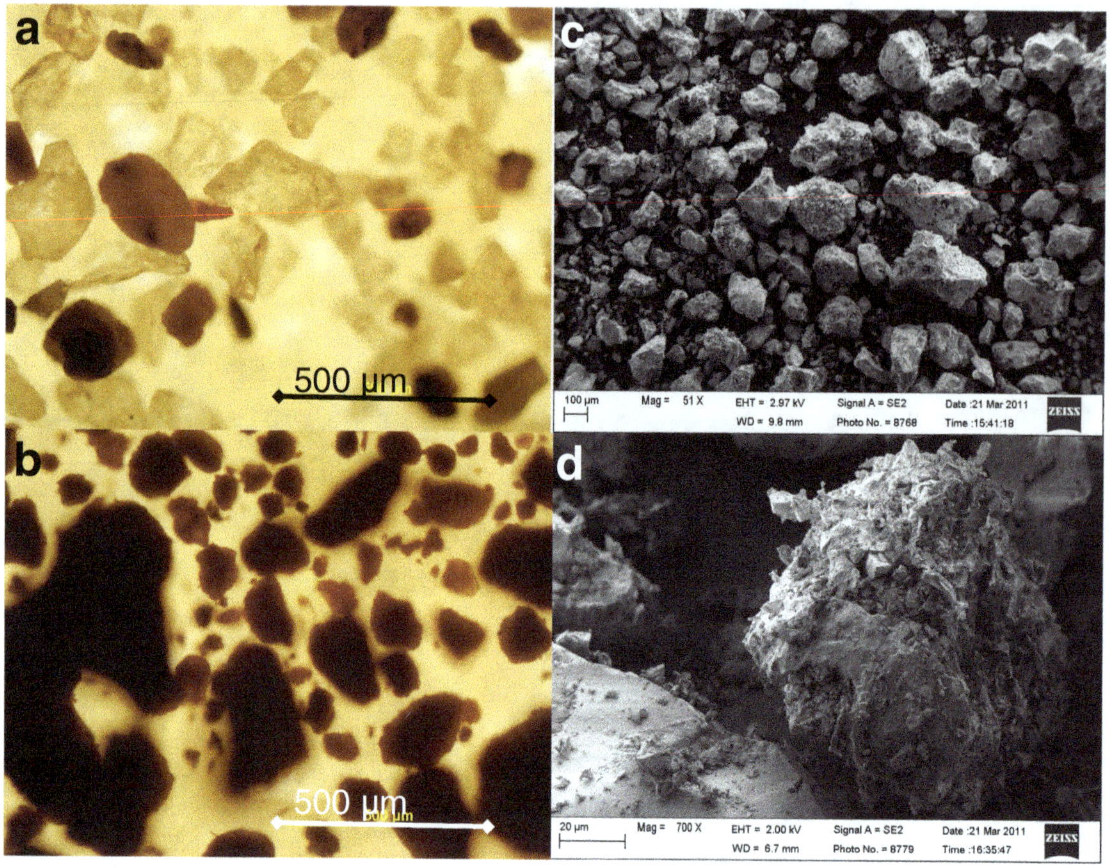

Fig. 3 Optical and scanning electron microscope (SEM) pictures of the simulants used in COLLIDE and PRIME experiments. **a** Quartz sand and **b** JSC Mars-1 and **c** JSC-1 lunar simulants. The samples were prepared in the same manner as for the experiments, by sieving the different grains to below 250 µm. **d** shows the details of the surface texture of JSC-1 grains

deep (>> 250 µm grain diameter), we expect that the tray had no influence on the target response to the impacts performed in COLLIDE and PRIME.

The trays were consistently filled using the same procedure, in order to reproduce the same target density. This filling procedure consisted in the pouring of simulant material into the tray using a funnel, followed by the leveling of the surface using a flat metal ruler. We measured the porosities of targets produced in this manner in the laboratory (30 times) by filling the tray using this procedure, then measuring the mass of the target material using a scale. Knowing the container volume and the bulk density of the grains, we determined target porosities consistently ranging from 0.4 to 0.5. While the compaction level of fine grains at the surface of asteroids is not well determined, the bulk porosity of "rubble-pile" asteroids was estimated to be of this order by Britt et al. (2002), indicating that such porosity levels are relevant for the study of low-velocity impacts on asteroid surfaces.

In the COLLIDE experiment setup, the target was contained by a sliding door, with no particular measures to prevent further compaction through post-preparation handling and launch vibrations. However, the COLLIDE-3 videos, which have a high enough resolution to show details of the target tray filling after the door opened during the rocket flight, show no volume reduction of the target sample. The target compaction during COLLIDE-1 and COLLIDE-2 could not be quantified due to the low resolution of the video data, but the projectile impact location indicates that the target surface was not significantly shifted from its original position, so that we do not estimate porosities much below 0.4 during the experiment runs.

PRIME

Flight campaigns PRIME flew on three experiment campaigns on the NASA KC-135 and C-9 parabolic airplanes. The first flight campaign induced five flight weeks between July 2002 and May 2003 and is described in

Colwell et al. (2008). The PRIME-1 campaign produced 108 successful impact recordings.

From the results of this first campaign, it was clear that the regolith was very sensitive to the residual accelerations of the aircraft. Therefore, the PRIME-2 hardware was modified to be partially free-floating. In this setup, the individual impact boxes could be pulled out of the overall payload structure, which was bolted to the airplane floor. The boxes were still attached to the structure by an electrical umbilical. Unexpectedly, the residual accelerations of the plane during parabolas were still high enough to introduce torques onto the impact boxes via this umbilical, generating a worse acceleration environment than during PRIME-1, and no scientific data could be collected.

The PRIME-3 hardware was designed to be entirely free-floating. Each impact box had its own electronics attached to it, including a battery pack for power. The following paragraph describes the hardware flown during the PRIME-3 1-week flight campaign in August 2014. The PRIME-3 campaign produced 25 impact recordings. During the flight campaign, the pilots flew a number of parabolas at 0.05*g* (named "asteroid-level" parabolas by the NASA flight support team) in addition to the usual microgravity ones. Out of the 25 recorded impacts, 7 were at asteroid level. Details on the PRIME-3 campaign can be found in Colwell et al. (2016).

Hardware Figure 4 shows the PRIME-3 hardware. It is composed of eight individual impact boxes that are kept in a structure bolted to the airplane floor. Each box can be slid out of the structure and used to perform an experiment run while completely free-floating during a parabola. The impact boxes were inherited from PRIME-1 and PRIME-2 and modified to allow for an autonomous experiment performance: while the regolith tray and launch mechanisms are similar, the electronics were re-designed to actuate them externally using a wireless signal. One camera was used to record the impacts in the individual chambers as it was attached to a bracket that could be switched from box to box. The camera used was also a GoPro Hero 3+ black edition. More details on the inside of the impact boxes can be found in Colwell et al. (2008).

During asteroid-level parabolas, the impactor box was placed on the floor of the airplane and the projectile launch initiated at the beginning of the parabola.

Projectiles and simulants The list of projectiles and simulants for the 108 impacts recorded during the PRIME-1 campaign can be found in Colwell et al. (2008). Table 1 summarizes the projectiles and simulants used during the PRIME-3 campaign.

The target materials and preparation methods were the same as for COLLIDE (see "Projectiles and simulants" section for details). The only difference with the COLLIDE targets is an additional contraption on the tray door that slightly compresses the target after preparation and through the pre-flight handling process. As the PRIME tray doors open by rotation rather than sliding, a plate

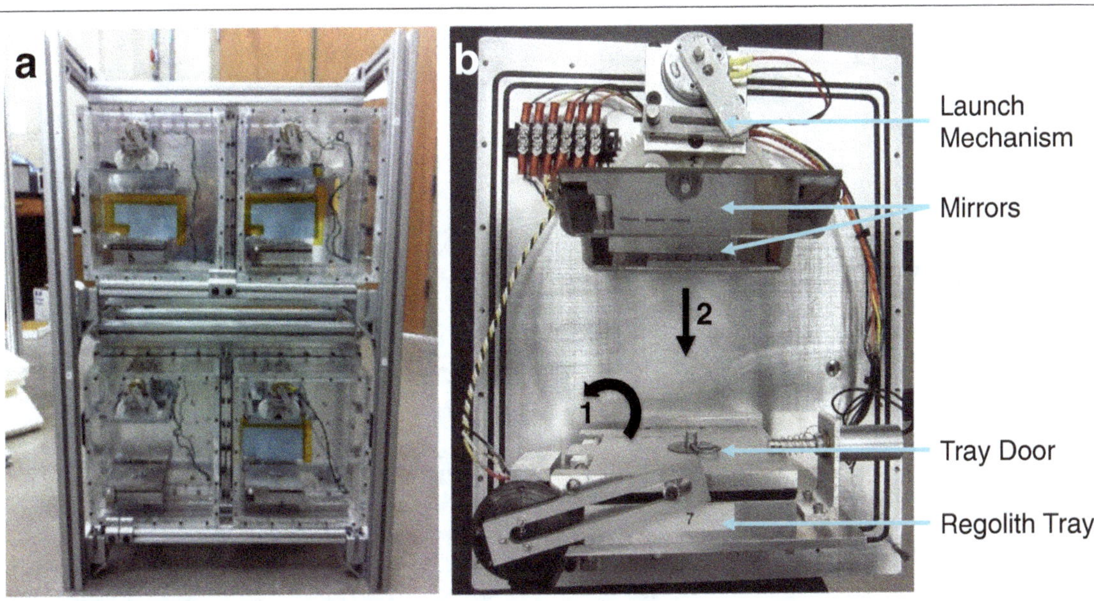

Fig. 4 PRIME-3 hardware. **a** The payload structure that is bolted to the aircraft floor and holds the eight impact boxes (only four are visible in this front view). **b** Interior of an impact box showing the regolith tray and the launch mechanism. During an experiment run, the tray door rotates to open (1) before the projectile gets launched towards the regolith bed (2)

was attached on the inner side facing the sample. Using four weak springs, this plate applies pressure on the target sample when the tray door is closed. In this way, the sample is jammed and its porosity remains unchanged (0.4 to 0.5 as described in "Projectiles and simulants" section) during loading of the experiment boxes onto the aircraft.

Results

As the target size distribution and porosity were the same for all the COLLIDE and PRIME campaigns, we combined the collected data sets. This section presents how we analyzed the impacts and the results we obtained, integrating all the data collected.

The platforms used to collect data in microgravity induced residual accelerations on the experiment hardware attached to them. These accelerations were around $10^{-3}g$ in the Space Shuttle (Colwell and Taylor 1999) and $10^{-2}g$ on the parabolic aircraft (Colwell 2003) and on Blue Origin's New Shepard rocket (Wagner and DeForest 2016). The free-floating experiment boxes of PRIME-3 offered the best microgravity quality: with only air drag acting on the box moving at very low speeds (< 1 mm/s), no residual acceleration could be detected from our video data. This implies that these experiment runs were performed at residual accelerations under our detection threshold, $10^{-4}g$. In addition to these microgravity experiment runs, the PRIME-3 flights also allowed us to perform seven experiment runs at 0.05g.

Across all the data collected, four types of collision outcome were observed:

- Collisions producing ejecta with an embedded projectile (squares in Fig. 5): in these collisions, no motion of the projectile after the impact is detected. Either the ejecta blanket or individual particles could be tracked and their speed determined.

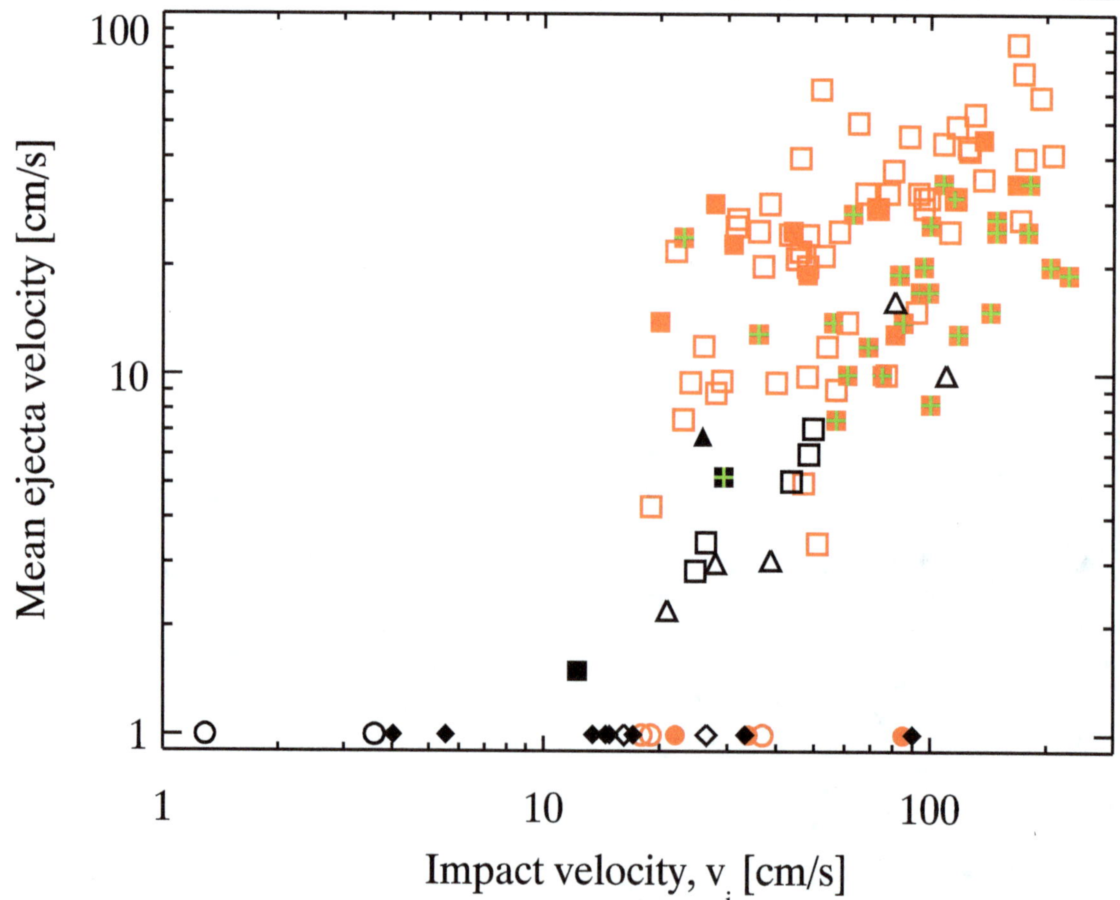

Fig. 5 Outcomes of all the low-velocity impacts observed in PRIME-1, PRIME-3, and the COLLIDE campaigns as a function of the impact velocity. Circles: projectile embedded, no ejecta; diamonds: projectile rebound, no ejecta; squares: projectile embedded, ejecta production; triangles: projectile rebound, ejecta production; open symbols: quartz sand target; filled symbols: JSC-1 target; green plus sign: JSC Mars-1 target; red: reduced gravity (~ $10^{-2}g$); black: microgravity (< $10^{-4}g$). All impacts producing no ejecta are shown at a mean ejecta velocity of 1 cm/s for the clarity of the graph, even though no speed was measured as no ejecta was observed

- Collisions producing ejecta with projectile rebound (triangles in Fig. 5): the projectile can be tracked after the collision. In addition to the ejecta speeds, a coefficient of restitution could be measured for the projectile.
- Collisions with no ejecta but projectile rebound (diamonds in Fig. 5). The coefficient of restitution for the projectile could be measured.
- Collisions with no ejecta and embedded projectile (circles in Fig. 5).

Figure 5 shows these outcomes as a function of the impact velocity. In this graph, all impacts not producing any ejecta were placed at a mean ejecta velocity of 1 cm/s. This is for the purpose of representing all the data on the same graph only, as no speed could be measured on absent ejecta (and a representation at 0 is not possible on a logarithmic scale). The following paragraphs give some additional details on these results for quartz sand and JSC-1 target materials.

JSC Mars-1 behaved similarly to quartz sand in reduced gravity (the JSC Mars-1 data points seen in Fig. 5 are mostly from the PRIME-1 campaign; see Table 1 in Colwell et al. (2008)). This can be explained by the similar grain shape between these two simulants (Fig. 3a, b). In microgravity, only one impact was available for data analysis, which was not enough for analyzing this set of data.

Ejecta production

Impacts in COLLIDE-2, COLLIDE-3, PRIME-1, and PRIME-3 resulted in the production of an ejecta blanket. In the PRIME-1 and most of the COLLIDE-2 data (Colwell 2003), individual ejecta particles could not be distinguished due to the low camera resolution. However, features of the ejecta blanket, in particular the fastest particles forming the upper edge of the blanket (named "corner" in Colwell et al. (2008)), were identified and allowed for the measurement of the upper end of the ejecta velocity distribution.

In PRIME-3, the camera resolution allowed for the individual tracking of ejecta particles. For these experiments, we were able to measure a mean ejecta velocity by directly tracking 30 particles for each impact, using the Spotlight software developed at NASA Glenn Research Center (Klimek and Wright 2006). However, due to the nature of the video data collected, the tracking of individual particles is limited to optically thin portions of the ejecta blanket. This includes but is not restricted to the ejecta "corner" mentioned above. Particle tracking was performed by three people independently in order to reduce the measurement errors. The ejecta grain velocity distribution obtained for each impact was normal and could be fit by a Gaussian curve. Details on the tracking methods can be found in Colwell et al. (2016) (see their Figs. 5 and 6). For PRIME-3 and COLLIDE-3, the characteristic ejecta velocity was chosen to be the Gaussian mean of velocity distribution of the tracked particles.

Figure 6 shows the results of the ejecta speed measurements for PRIME-1 and PRIME-3. The 1-σ error bars for the Gaussian mean determination are shown for the PRIME-3 data points. COLLIDE-3 produced ejecta for only one impact, and the target material was JSC Mars-1. As stated above, JSC Mars-1 behaved similarly to quartz sand, and we are not including JSC Mars-1 results here. The plateau formed by the PRIME-1 microgravity data points (black asterisks representing quartz sand impacts) around 10 cm/s for impact energies below 10^4 ergs is due to the residual accelerations induced by the parabolic aircraft on the experiment boxes (on the order of $10^{-2}g$). Only particles ejected with energies able to overcome the effect of this ambient gravity field were lifted from the target. For this reason, the slowest moving ejecta has a minimum speed, set by this g-level, and the measured velocities follow a plateau. This plateau also formed in PRIME-3 microgravity data, but at ejecta velocities around a few centimeters per second, indicating that residual accelerations were successfully reduced by the free-floating hardware configuration, compared to PRIME-1, which was attached to the aircraft frame.

The scatter of the PRIME-1 data points towards higher ejecta velocities was induced by the observational bias produced by the tracking method used: tracking the ejecta "corner" described above limits the recorded speeds to the fastest moving particles and is therefore only representative of the upper end of the ejecta velocity distribution: there is therefore a tendency to measure higher ejecta velocities at the same impact energies. However, the data points show a trend followed by the lower limit of the measured velocities: to guide the eye, we fit a power law to the impacts into quartz sand in microgravity. The index obtained is 1.35. For all impacts into JSC-1, a fitted power law index is at 0.50. We collected only one data point for JSC-1 ejecta at microgravity levels, which lies below this power law fit. Further data collection on JSC-1 ejecta-producing impacts in microgravity will be required in order to evaluate if this point is indicative of a different ejecta behavior between the two gravity levels.

The ejecta speeds measured for impacts at $0.05g$ during the PRIME-3 campaign are of the same range as the ones measured during the PRIME-1 campaign, where microgravity parabolas were flown, but the experiment hardware was fixed to the aircraft frame. This demonstrates that the residual accelerations produced on the experiment when attached to the aircraft are comparable to $10^{-2}g$ and that our method for determining the average ejecta speeds provides similar values than the method used in Colwell et al. (2008).

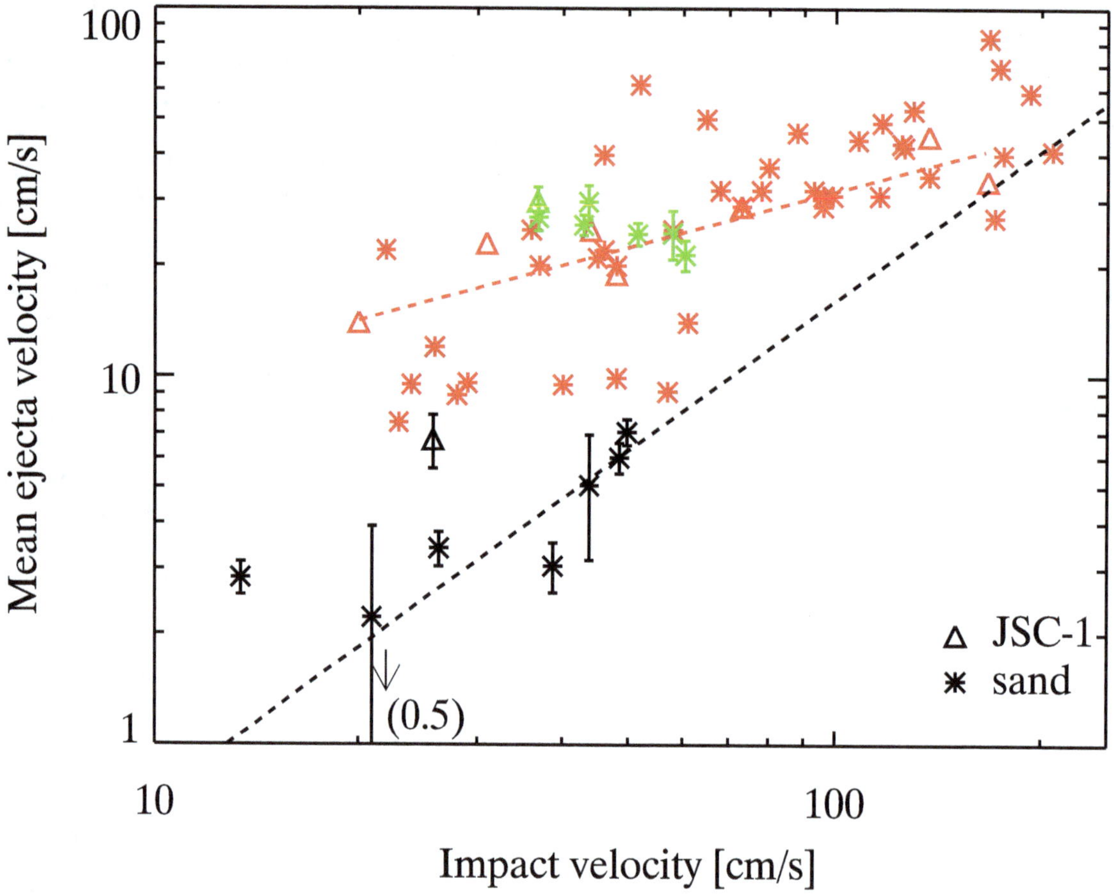

Fig. 6 Combined PRIME-1 and PRIME-3 ejecta velocity measurements as a function of the impact velocity. PRIME-1 data is shown by red symbols. For PRIME-3, black symbols show impacts in the free-floating box ($< 10^{-4}g$), while green symbols show impacts at $0.05g$. Impacts into quartz sand are marked by asterisks, and impacts into JSC-1 by triangles. Error bars are not shown for PRIME-1: they were smaller than the symbol size (Colwell 2003). PRIME-3 error bars represent the 1-σ error of the Gaussian fit to the velocity distribution of tracked particles. The dashed black line is a power law fit to the microgravity impacts in sand (index 0.50), and the red dashed line is a power law fit to all the impacts into JSC-1 (index 1.35)

We defined the effective coefficient of restitution of the impacts as the ratio between the average ejecta velocity and the projectile impact velocity and investigated its relationship to the impact velocity. These effective coefficients of restitution showed no correlation and a uniform distribution over the available range of impact velocities (Fig. 7). We calculated the overall average effective coefficient of restitution to be 0.39 ± 0.15. This value is about twice the one derived by Deboeuf et al. (2009) for impacts of centimeter-sized spheres into 100 μm glass bead beds. Deviations to our mean value were much higher than for the measurements performed in Deboeuf et al. (2009). When separating the impacts into quartz sand from the ones into JSC-1, we obtained average effective coefficients of restitution of 0.38 ± 0.15 and 0.45 ± 0.16, respectively. When separating between impacts in low gravity and microgravity, we obtained values of 0.43 ± 0.14 and 0.15 ± 0.04, respectively.

Projectile rebound

A number of impacts during the COLLIDE-1, COLLIDE-3, and PRIME-3 campaigns showed a rebound of the projectile after impact on the target and allowed for the measurement of a coefficient of restitution $\epsilon = \frac{v_f}{v_i}$, v_i and v_f being the projectile velocities before (initial) and after (final) the impact, respectively (see Table 1). Figure 8 shows the measured coefficients of restitution as a function of the impact energy. For speeds above 30 cm/s, impacts systematically produced an ejecta blanket and coefficients of restitution of the projectile could either not be measured or were of the order of 10^{-2}.

At impact speeds between about 20 and 30 cm/s, both ejecta production and projectile rebound without ejecta were observed. Below 20 cm/s, only rebounds without ejecta were observed. In order to compare our results to the experimental work of Katsuragi and Blum (2017), we fit a power law to this set of data. In Katsuragi and

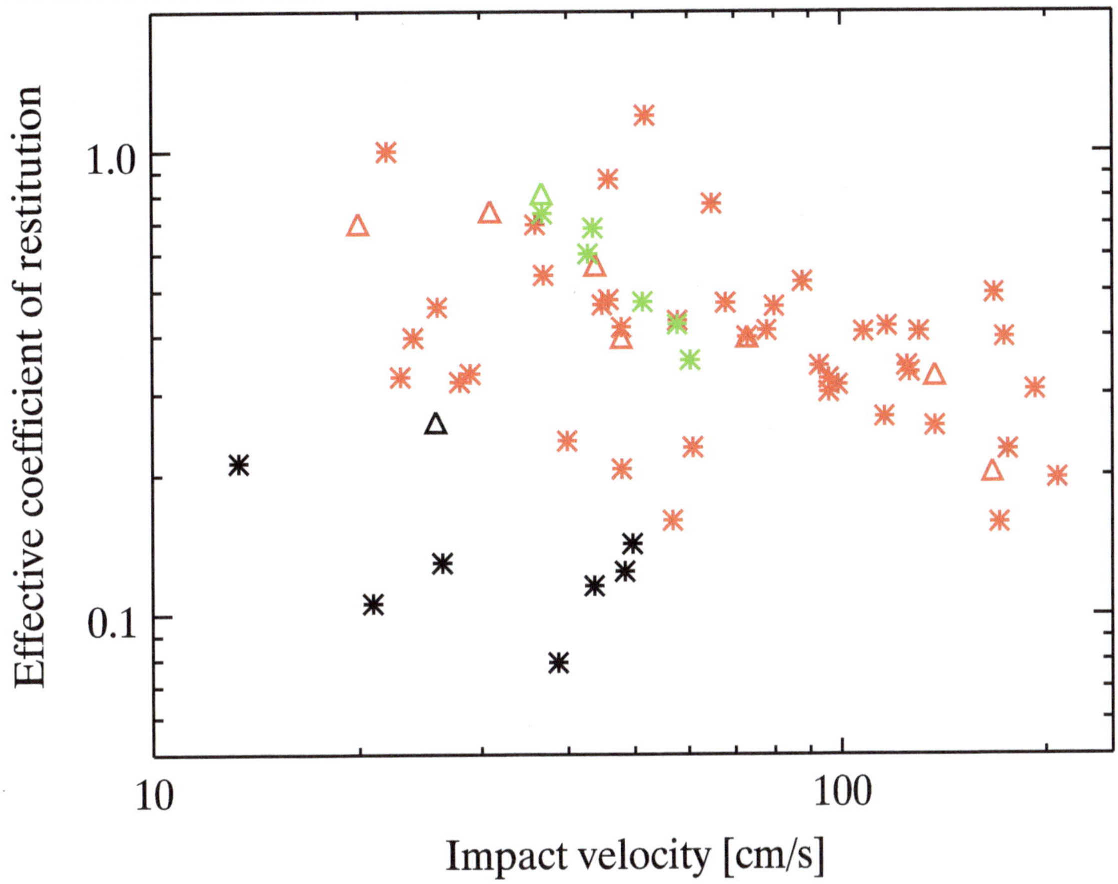

Fig. 7 Effective coefficient of restitution of the impacts depicted in Fig. 6 (the same colors and symbols were used). This coefficient of restitution is defined as the ratio between the average ejecta velocity and the projectile impact velocity

Blum (2017), spherical projectiles of about 1 cm in diameter are dropped into aggregates composed of 750 μm SiO_2 monomer grains with a packing density of $\Phi = 0.35$. The index we obtain with our data is -0.27, similar to the value of $-1/4$ found by Katsuragi and Blum (2017). We note, however, that if we separate the data sets from COLLIDE and PRIME, we obtain indexes of -0.10 and -0.50, respectively, a difference mostly due to the two very low-energy impacts into JSC-1 of the COLLIDE-1 flight. As mentioned in Katsuragi and Blum (2017), this overall index value is twice as high as expected from the theory of an elastic sphere impacting a plane surface (Johnson 1985; Thornton and Ning 1998). This indicates that energy absorption in a bed of granular material is not entirely captured by the mechanics of elastic surfaces.

Projectile penetration depth

In the PRIME-3 video data, the resolution is high enough to determine the maximum penetration depth of the projectile into the target for 21 impacts. This maximum penetration depth is defined as the distance between the bottom of the projectile and the target surface. Figure 9 shows these results as a function of the equivalent total drop distance H. This distance was derived to be able to compare our results with the ones obtained by Katsuragi and Blum (2017): $H = h + z_{max}$, where h is the drop height (experiments in Katsuragi and Blum (2017) are performed in $1g$, and the drop height is used to control the impact velocity), and z_{max} is the maximum penetration depth of the projectile into the target. As the impact velocity was not determined by the drop height in PRIME experiments, but by the stored potential energy in the spring of the projectile launcher, we calculate an equivalent h from the impact velocity v_i by equating the $1g$ potential energy with our kinetic impact energy as follows:

$$\begin{aligned} E_{1g} &= E_{0g} \\ mgh &= \frac{1}{2}mv_i^2 \\ h &= \frac{v_i^2}{2g} \end{aligned} \tag{1}$$

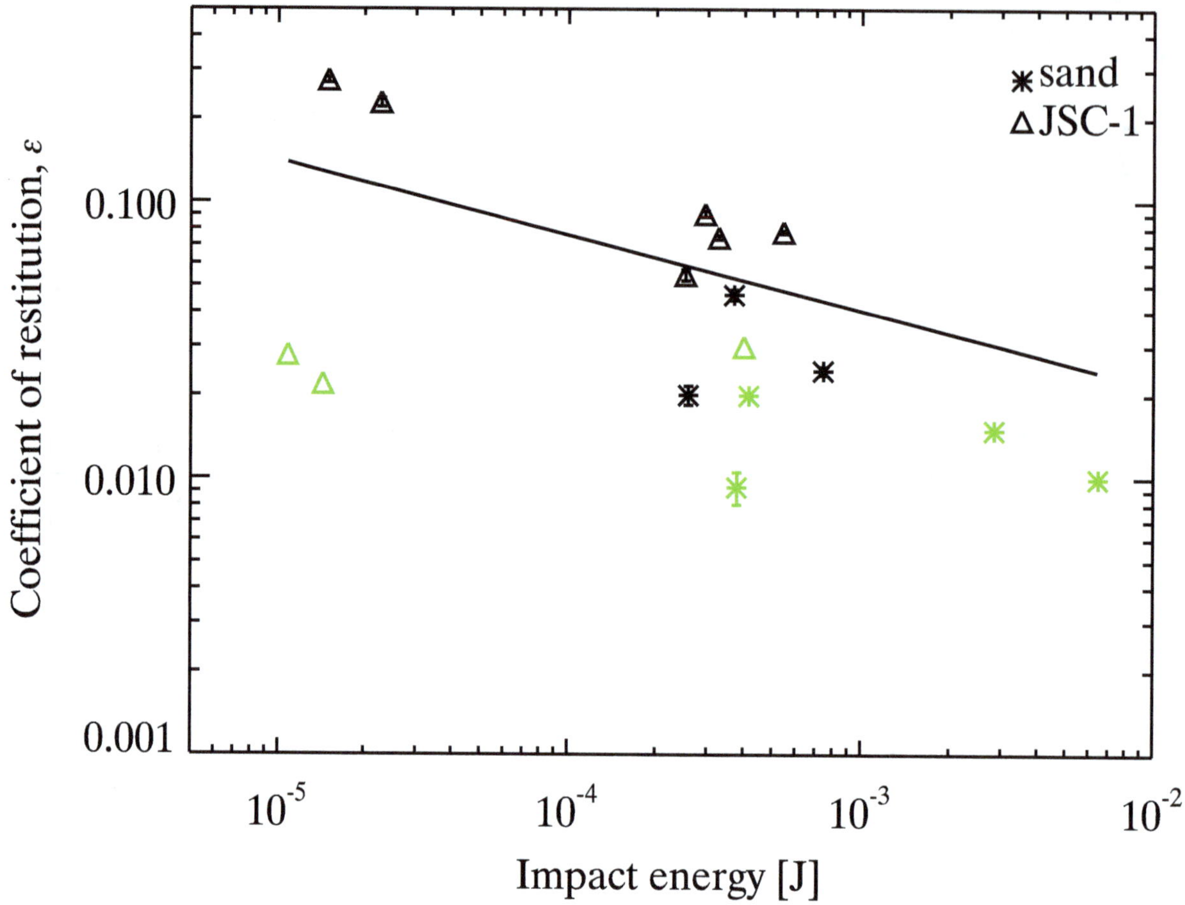

Fig. 8 Combined PRIME-3 (black) and COLLIDE (green) coefficient of restitution measurements. This coefficient of restitution is defined as the ratio of the projectile speed after and before the impact with the regolith bed. Impacts into quartz sand are marked by asterisks, and impacts into JSC-1 by triangles. Error bars are shown for the PRIME-3 and COLLIDE-3 measurements. For COLLIDE-1 and COLLIDE-2, they can be found in Colwell (2003). The straight line represents the power law fit to all data points shown, which has an index of − 0.27

with $g = 9.81$ m/s. As shown by the index 1 power law in Fig. 9, H is dominated by z_{max} in half of the impacts. This is due to the very low impact velocities, inducing equivalent drop heights $h \ll z_{max}$, so that $H \sim z_{max}$. We do not observe any correlation between the maximum penetration depth and the computed total equivalent drop distance. This is in contrast with the relation $z_{max} \propto H^{1/3}$ observed in Katsuragi and Blum (2017), representative of a scaling by impact energy. A scaling by momentum of the form $z_{max} = Am_p^{\alpha} v_i^{\beta}$, with m_p and v_i being the projectile mass and impact velocity, respectively, and A, α, and β fit parameters, as performed in Güttler et al. (2009), was not successful either.

Grain behavior with decreasing gravity level

From the experimental results presented here, we observed differences in the response of the target to low-velocity impacts in reduced gravity ($\sim 10^{-2}g$) and micro-gravity ($< 10^{-4}g$). Figure 5 shows that compared to the total number of impacts in reduced gravity ($\sim 10^{-2}g$), only a few of them did not produce ejecta: the vast majority of impacts generated an ejecta curtain. In addition, none of the reduced-gravity impacts lead to a rebound of the projectile (no red triangles or diamonds). In particular, when no ejecta was produced around the lowest impact speeds observed ($\sim$20 cm/s), only projectile embedding in the target could be seen. For the same impact speeds, a microgravity environment ($< 10^{-4}g$) lead to the rebound of the projectile. In fact, embedding into the target was only seen in the two impacts under 4 cm/s. All other impacts than those without ejecta showed a rebound of the projectile.

Projectile rebound was also observed in combination with an ejecta curtain in half of the impacts observed in microgravity (black triangles). Even though Fig. 8 shows that the coefficient of restitution of these rebounds is much smaller compared to the ones with no ejecta production (diamonds), this behavior demonstrates the

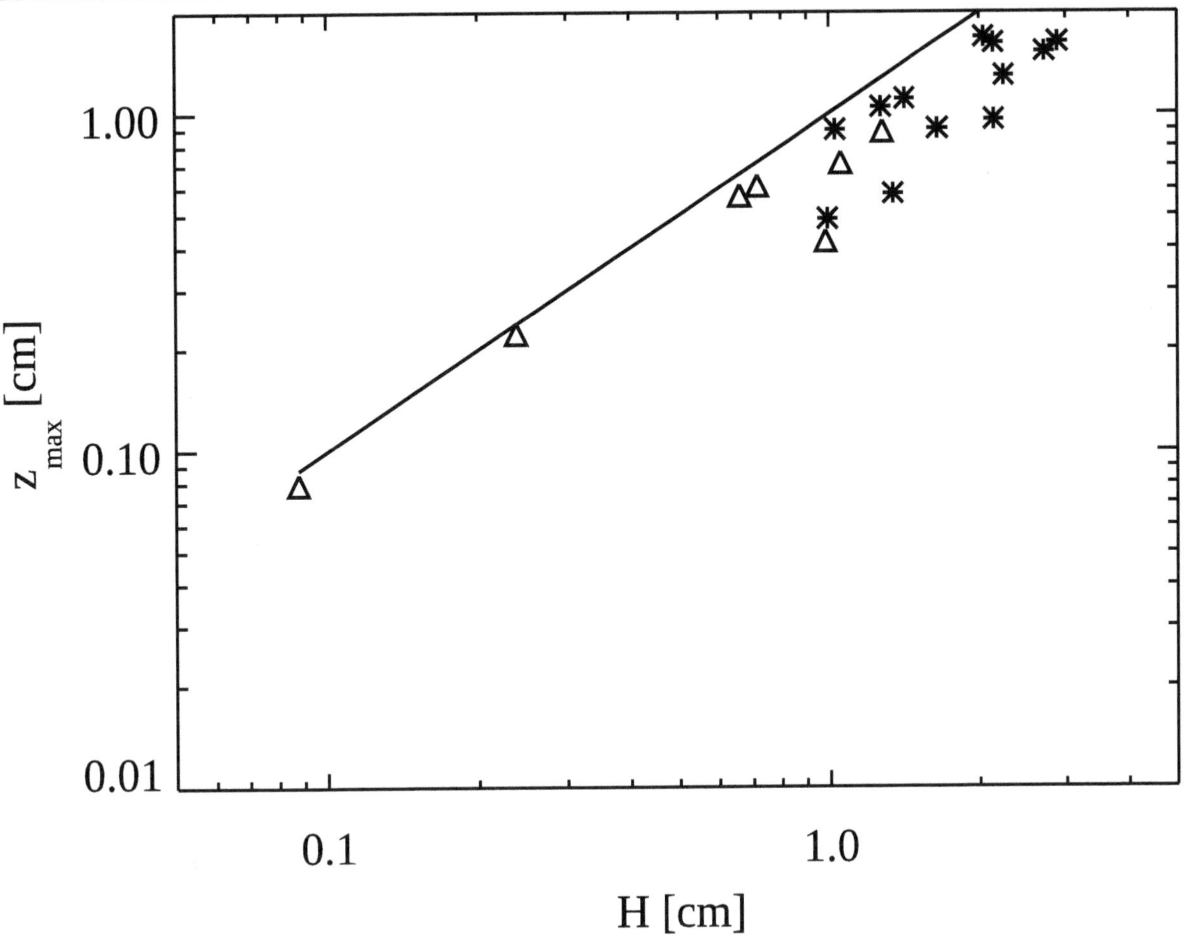

Fig. 9 Maximum penetration depth of the projectile into the target for PRIME-3 impacts. *H* is the equivalent total drop distance as described in the text. Impacts into sand are marked by asterisks, and impacts into JSC-1 by triangles. The line represents an index 1 power law

differences in target response in reduced- and microgravity environments.

It can also be noted that for the same impact speed, ejecta is faster on average in reduced gravity compared to microgravity (red symbols are above black symbols in Fig. 5). In addition, less mass is ejected than in microgravity. Figure 10 shows the average ejecta speeds for four levels of ejected mass: as the ejecta mass cannot be measured after each experiment run (the return to 1*g* mixes the target material from the tray and the ejected particles), an estimation was performed from the video images. Four levels of ejecta mass were identified: no ejecta (0), individual particles detaching from the target (1), ejected mass of the order of the projectile mass (2), and ejected mass much higher than the projectile mass (3) (we followed the same numbering as in Colwell (2003)). In Fig. 10, the red symbols show reduced-gravity impacts ($\sim 10^{-2}g$), and black symbols microgravity impacts ($< 10^{-4}g$). The impact velocity given for each target type and ejecta mass level is the average of all impacts that created the same amount of ejecta. From the sand targets, we can see that the same average impact speed results in higher ejecta masses in microgravity compared to reduced gravity. Together with Fig. 5, this shows that fewer target particles are ejected in reduced gravity, but their ejection speed is higher.

Figure 10 also shows that no ejecta blanket was observed in microgravity with JSC-1 targets. While this is also due to the fact that only very few impacts were performed in JSC-1 in microgravity at speeds > 50 cm/s, the nature of the target plays a role in the ejecta mass produced: compared to quartz sand particles, which are rounded and considered cohesionless in vacuum (once the air humidity is removed), JSC-1 particles are more angular and behave like a cohesive powder. Figure 5 shows that only two out of eight impacts into JSC-1 at > 10 cm/s resulted in ejecta production, and these were only individual particles detaching instead of an ejecta blanket. As we can see in Fig. 10, the impact speed does not seem to influence the production if these individual particles in JSC-1 targets,

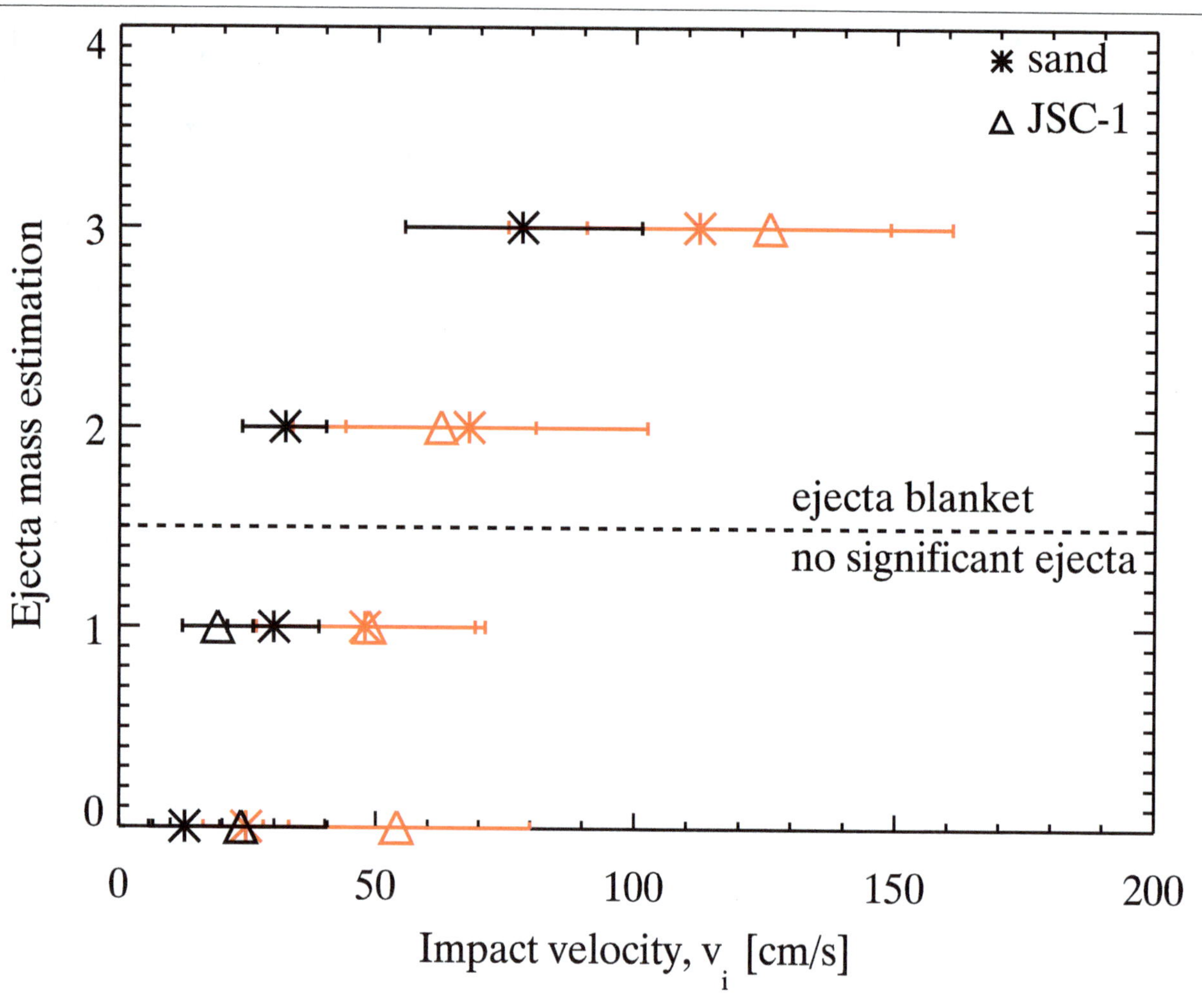

Fig. 10 Arbitrary quantified amount of ejecta for all collected data on sand and JSC-1 targets. The amount of ejecta mass was quantified as follows: 0 = no ejecta; 1 = ejection of single particles; 2 = ejecta blanket, mass of the order of the projectile mass; and 3 = ejecta blanket, mass much higher than the projectile mass. The impact velocity given for each target type is the average of all impacts that created the same amount of ejecta. Asterisk = quartz sand target, triangle = JSC-1 lunar simulant target, black = microgravity impacts, red = reduced-gravity impacts. The dotted line marks the limit between no significant ejecta produced (either none at all or only individual particles detaching) and an ejecta blanket

compared to quartz sand targets, which display a more consistent trend of increasing ejecta mass with increasing impact velocity. When an ejecta blanket is produced though, we can see that higher impact velocities result in higher ejected masses for both quartz sand and JSC-1 targets.

Discussion

Observing ejecta

Due to the hyper-*g* levels experienced after the low-gravity phase, the state of the target and projectile after each impact is destroyed and the PRIME and COLLIDE data collection relies solely on video recordings. In particular, it is not possible to measure the ejected mass, nor the trajectories of all ejected particles in the case of an ejecta blanket forming. As described in "Ejecta production" section, ejected particles could only be tracked in thin parts of the ejecta blanket or in the case of single particles being lifted rather than a blanket. Figure 11a shows an example of particle tracks superposed with the ejecta blanket: even though we can get particle speeds from these partial tracks, there is a high uncertainty in placing the origin of each track with respect to the surface of the target material and embedded projectile, which location can only be estimated as it is obscured by the ejecta blanket. For this reason, it is not possible to generate mass vs. speed or speed vs. launch position diagrams for ejecta, in order to compare them to 1-*g* data and the scaling laws derived in Housen and Holsapple (2011).

Observing projectile rebound

As described in the previous section, a rebound of the projectile from the target was never observed in reduced-gravity conditions, compared to microgravity. This could

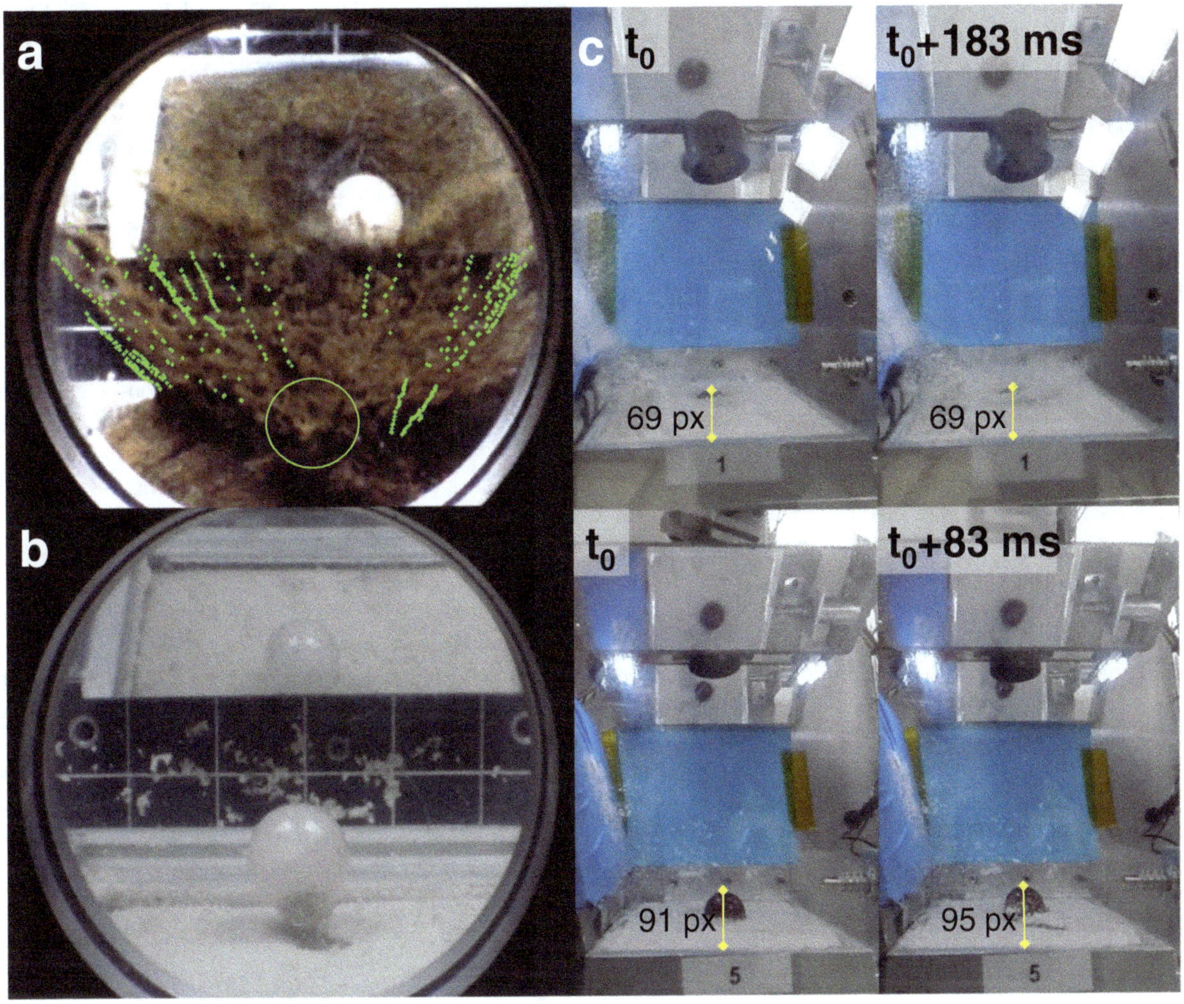

Fig. 11 Data analysis examples. **a** COLLIDE-3 IBS 3 impact into JSC Mars-1 at 29.3 cm/s. Particle tracks are superposed in green, and the approximate projectile position is shown with a green circle, as the ejecta curtain hides it entirely. The mirror above the target tray shows a top view of the impact, in which the projectile is visible. **b** COLLIDE-3 IBS 2 impact into quartz sand at 26.3 cm/s. Note the mass transfer onto the projectile after rebound. **c** PRIME-3 rebound observation: t_0 indicates the moment of deepest penetration of the projectile into the target. (Top) Case of ejecta production without projectile rebound (flight 1, box 1): no marble motion can be detected for 183 ms. In the following frames, the ejecta curtain hides the projectile, which cannot be tracked anymore. (Bottom) Case of ejecta production with projectile rebound (flight 2, box 5): we can track the motion of the projectile over 10 frames and 4 pixels

be due to two factors: either no rebound took place or the rebound was too small to be observed at the space and time resolution of the video recordings of the impacts. In the case that the absence of observed rebound in reduced gravity is only due to resolution limitations of the experiment cameras, we can assume that the rebound takes place with the same coefficient of restitution as in microgravity and calculate the distance and time travelled by the projectile after the impact before it gets stopped by gravity. As the only force on the projectile after it leaves the target again is gravity, simple Newtonian mechanics give us the time t_0 at the apogee of its trajectory and the height h above the target it reaches at t_0:

$$t_0 = \frac{v_r}{g_l} \quad ; \quad h = \frac{1}{2}\frac{v_r^2}{g_l} \tag{2}$$

where v_r is the rebound speed and g_l the value of the local gravity acceleration.

For impact speeds around 40 cm/s, we measure coefficients of restitution around 0.02 (see Fig. 8). This would set v_r to 8 mm/s. Reduced-gravity environments had gravity levels around $10^{-2}g$; thus, $g_l \sim 0.1$ m/s^2. This leads to a travel time of $t_0 = 80$ ms and a rebound travel height of $h = 320$ μm. During the PRIME-3 campaign, the camera resolution was around 80 μm/pixel and the recording was performed at 120 fps (about 8 ms between frames). In addition, reduced-gravity environments during this campaign lead to short parabolic trajectories of the ejected particles consistent with a $\sim 0.05g$ acceleration environment, leaving a clear view of the projectile during the entire recording. This means that a rebound would have been observed and a rebound motion could have been measured for 10 frames and over 4 pixels. As no rebound was observed, it appears that they were due to a different behavior of the target and no rebound actually took place in reduced gravity. To illustrate this, we show two PRIME-3 impacts with different outcomes in Fig. 11c and their measured projectile positions, at deepest penetration (t_0) and after 22 (no rebound) and 4 (rebound) frames. Some impacts generating few ejecta allow for a long-term tracking of the projectile after the impact, allowing for the detection of rebound speeds as low as a few millimeters per second. However, large ejecta blankets obstruct the view of the projectile (see Fig. 11a), thus limiting the observation of very low rebound speeds in these cases.

Scaling the penetration depth

As shown in Fig. 9, scaling the maximum penetration depth of the projectile into the target (z_{max}) with either the impact energy or momentum was not possible. Here, the difference in gravity levels (reduced or microgravity) did not appear in the distribution of the data points. Following Katsuragi and Durian (2013), Fig. 12a shows z_{max} as a function of the quantity $\mu^{-1}(\rho_p/\rho_g)^{1/2}D_p^{2/3}H^{1/3}$. In this quantity, μ is the coefficient of friction of the target material. From angle of repose measurements, we can estimate that $\mu = 0.84$ for JSC-1 and 0.67 for quartz sand (Brisset et al. 2016). ρ_p and ρ_g are the densities of the projectile and target material, respectively; D_p is the projectile diameter; and H is the equivalent total drop distance as defined in the paragraph on penetration depth results. Figure 12a allowed us to recognize two groups of data points, which we distinguished by color: impacts with projectiles of masses around 10 g (see the PRIME-3 data in Table 1: 9.82 to 11.76 g) in black and impacts with ~ 30 g projectiles (28.2 to 30.75 g). Power law fits to each population yielded similar indexes of about 2.5 and a factor of about 3 between both. We concluded that our data set scales with $\mu^{-1}(\rho_p/\rho_g)^{0.1}D_p^{2/3}H^{1/3}$, the exponent for ρ_p/ρ_g being 0.1 rather than 0.5 (all PRIME-3 projectiles have the same diameter of 2 cm). The result of this scaling is shown in Fig. 12b.

This scaling and the associated power law index are in contrast with the results by Katsuragi and Durian (2013). In particular, the contribution of the projectile to target densities ratio is reduced from an index 0.5 to 0.25, while the projectile diameter and impact energy have increased contributions from indexes of 0.67 to 1.67 and 0.33 to 0.83, respectively. The coefficient of friction has an increased contribution from an index -1 to -2.5 ($\mu < 1$). This means that in low gravity (no distinction between reduced gravity and microgravity), the diameter and kinetic energy of the projectiles play a more significant role in the penetration depth than the density ratio with the target material than in 1g. This reflects the fact that the gravity pull

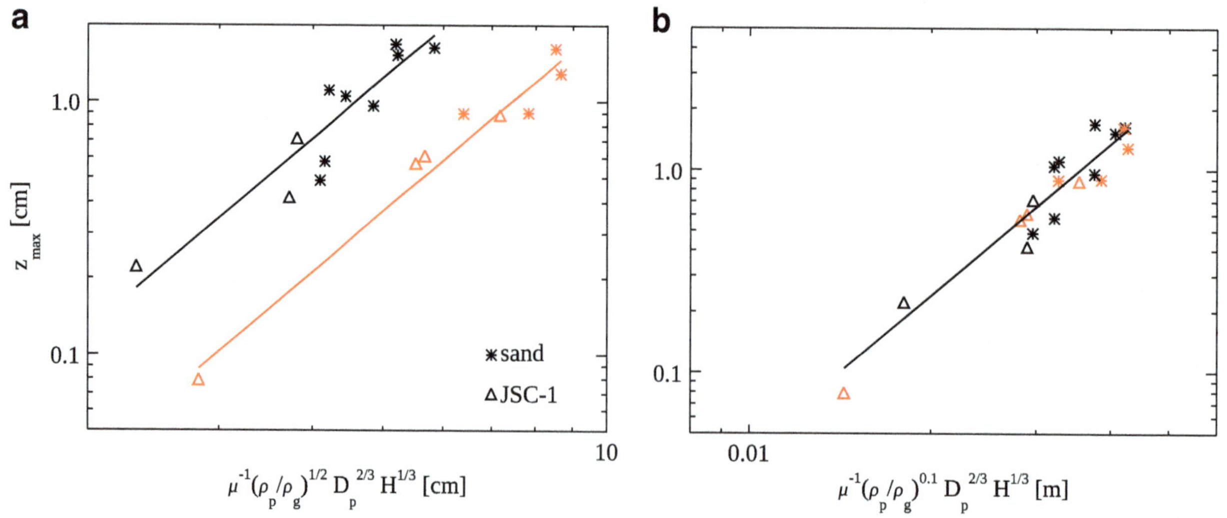

Fig. 12 Scaling of the maximum penetration depth during PRIME-3 impacts. **a** With the quantity $\mu^{-1}(\rho_p/\rho_g)^{1/2}D_p^{2/3}H^{1/3}$ (see text for details) and **b** with $\mu^{-1}(\rho_p/\rho_g)^{0.1}D_p^{2/3}H^{1/3}$. Asterisks and triangles represent quartz sand and JSC-1 targets, respectively. Impacts performed with projectiles of masses ~ 10 g are marked by black symbols, and ~ 30 g by red symbols. The black and red lines in **a** are power law fits to these two types of impacts (10 g and 30 g projectiles), both with indexes ~ 2.5. The line in **b** shows the power law fit to the entire data set, with an index of 2.5

on the projectile during penetration is much reduced, and the density ratio between projectile and target material is a less significant factor during penetration. Large and high-velocity projectiles will penetrate the target easier in reduced gravity. The role of the cohesion between the target grains (which is implicit in the coefficient for friction μ) is increased by the reduced-gravity environment: cohesive targets are more efficiently stopping a projectile than in 1*g*. This supports the notion that cohesive forces can become significantly more important in low gravity compared to the weight of the target particles (see following paragraph).

Gravity-cohesion competition

In Fig. 13, we compare the ambient gravity forces to inter-particle cohesion forces (van der Waals), following the example of Scheeres et al. (2010), using measured inter-particle force values rather than model estimations. The straight black lines in this figure represent the cohesion force between two grains, which is linearly proportional to the particle radius, as measured for two different types of particles. We are using the Johnson-Kendall-Roberts theory (Johnson et al. 1971), estimating the force between two grains to be the pull-off force, $F_{po} = 3\pi\gamma r$, γ being the surface energy of a grain and r its radius. The surface energies for quartz sand was measured in Kendall et al. (1987) and in Brisset et al. (2016) for polydisperse SiO_2 particles in similar experiment conditions (vacuum levels of around 10 mTorr at room temperature). These polydisperse SiO_2 grains were similar to JSC-1 lunar simulant in shape, while quartz sand has more rounded, less cohesive grains. The red dashed lines mark the force of gravity on individual grains for three *g*-levels, and the green vertical line marks the largest particles present in the target materials used in the present work.

In this figure, we can see that, for quartz sand, the forces induced on the target grains due to gravity are several orders of magnitude larger than inter-particle adhesive

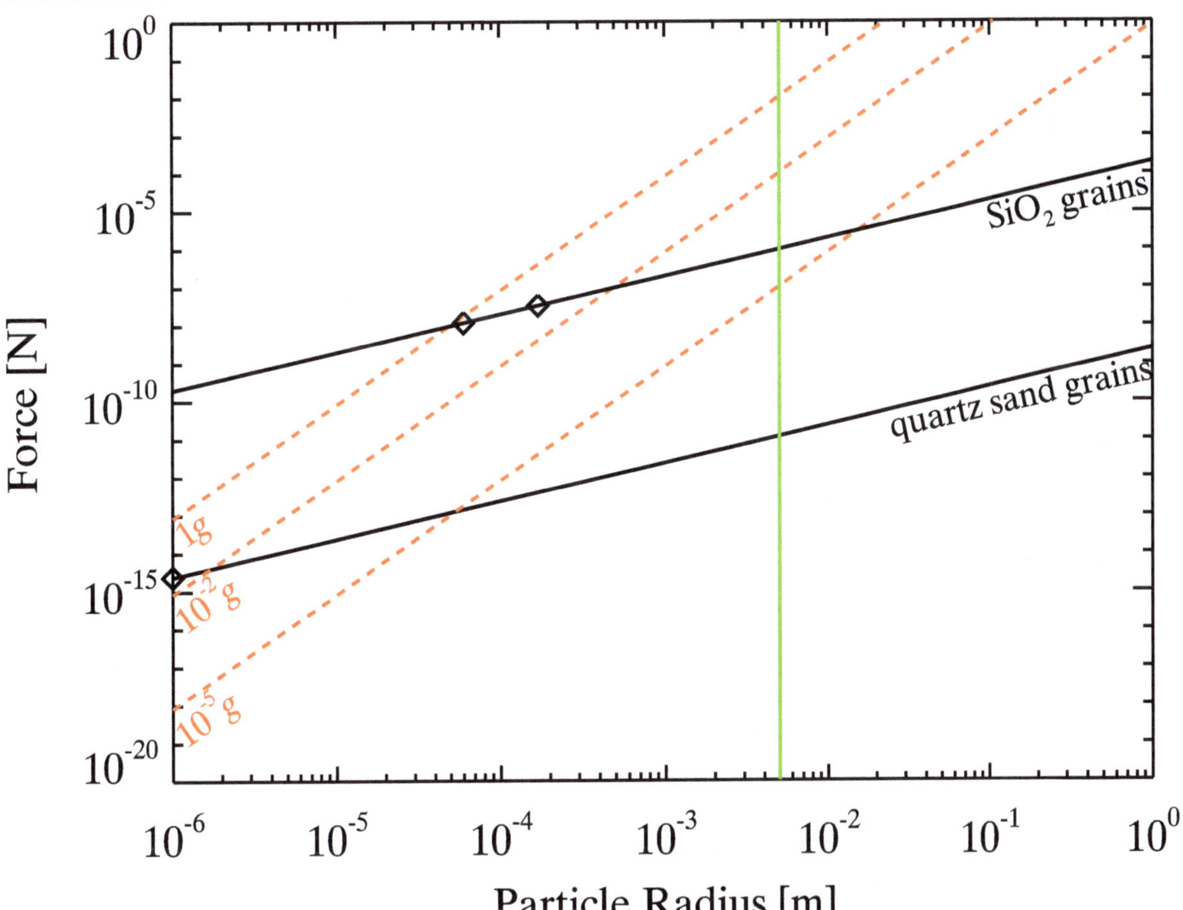

Fig. 13 Comparison between gravitational and cohesive forces depending on the grain radius. The black lines correspond to the measured pull-off force between SiO_2 polydisperse (Brisset et al. 2016) and quartz sand grains (Kendall et al. 1987). The grain sizes at which the measurements were performed are marked as diamond symbols. The dashed red lines show the gravitational forces on the regolith grains for different *g*-level environments. The green line marks the largest particles of our target size distributions

forces at $1g$ and $10^{-2}g$. Only at $10^{-5}g$, these two forces are of the same order for the smaller target particles. This can explain the reason for lower ejected masses and larger average ejecta speeds (at the same impact speed) observed in reduced gravity compared to microgravity: in reduced gravity, only the fastest, lighter grains can overcome the gravity field and be ejected from the target. In microgravity, the gravity force is low enough to allow for most grains to get lifted, even the slower ones, and the average ejecta speed is accordingly lower. In addition, as the gravity and cohesion forces are of the same order in microgravity, the target material displays a more elastic behavior: cohesion forces between the particles seem to allow for a partially elastic response of the surface to the impact, thus leading to a rebound of the projectile (see Fig. 5, open triangles and diamonds), while in reduced-gravity environments, the presence of a constant downward force locks the particles in place (e.g., jamming; (Cates et al. 1998)) and increases the inter-particle forces due to friction inside the material (Murdoch et al. 2013), thus preventing projectile rebound.

In Fig. 13, we can also see that JSC-1 targets (which approximately behave like SiO_2 grains) have stronger cohesion forces between grains than quartz sand: the cohesion and gravity forces are of the same order already in reduced gravity, while cohesion dominates by several orders of magnitude in microgravity. The strong dominance of the cohesive forces can explain why none of the impacts in JSC-1 in microgravity created an ejecta curtain. While most impacts in JSC-1 were performed at speeds below 40 cm/s and higher impact speeds may have possibly triggered larger ejecta masses, impacts at similar speeds and in the same microgravity environment generated an ejecta curtain with sand target (Fig. 10). The predominance of cohesive forces between JSC-1 grains kept the target material from escaping.

We can also see a difference in behavior of the JSC-1 targets in reduced- and microgravity environments: for the same impact velocities, rebound is observed in microgravity, while the projectile embeds into the target in reduced gravity (filled black diamonds and red circles in Fig. 5). Even though JSC-1 displays a more cohesive behavior than sand in reduced gravity, the data indicates that these cohesion forces do not dominate gravity enough at $10^{-2}g$ to lead to the rebound of the projectile.

Interacting with small body surfaces

The experiments presented here are limited by the size of the target tray and projectile. However, it is possible to learn from their results about possible outcomes of surface interactions by spacecraft landers or sampling mechanisms. In particular, impact speeds under 1 m/s are typical for past and planned missions interacting with the surface of a small body; Philae touched down on the surface of 67P/Churyumov-Gerasimenko for the first time at ~1 m/s, OSIRIS-REx's Touch-And-Go (TAG) mechanism is planned to impact the surface of Bennu at around 20 cm/s, and Hayabusa-2's MASCOT will land on Ryugu also at several 10 cm/s. The present experiments are therefore in a relevant impact speed regime. While the regolith size distribution on asteroid surfaces seems to heterogeneously include large and fine particles, images of the surfaces of asteroids visited by spacecraft show the presence of regions of very smooth terrain covered in grains of sizes below a millimeter (e.g., the smooth ponds on Eros; Cheng et al. (2002)). These smooth terrains also represent ideal surface interaction sites as they reduce the risks related to landing on uneven surfaces. Therefore, the target size distributions of our experiments are relevant to understanding spacecraft interactions with asteroid surfaces.

At these relevant impact speeds and grain sizes, we observed the new phenomenon of projectile bouncing off of granular material surfaces in microgravity. While numerical simulations are currently the only way of studying low-velocity impacts of spacecraft landers at the correct size scales, those simulations are mostly using hyper-velocity or $1g$ experiment impact data for calibration. The only simulations predicting and reproducing projectile bouncing from the surface (Maurel et al. 2017), for example, are using Earth gravity impact experiments at > 10 m/s into granular material (Yamamoto et al. 2006) or impacts on hard surfaces as input parameters (Biele et al. 2017). Maurel et al. (2017) specifically indicate the need for parameter refinement based on microgravity data from experiments.

It can be noted that the experimental conditions, in particular the level of residual accelerations, have a significant impact on the target behavior. Even for irregular grains that behave like cohesive powders, the granular material displayed a different behavior in reduced gravity and microgravity. This means that the behavior of granular material at the g-levels present on the surface of small asteroids ($10^{-4}g$ and below) cannot be extrapolated from its behavior at lunar g-level ($\sim 10^{-2}g$). Therefore, experiments relevant to the interaction with surfaces of bodies like Ryugu and Bennu will require the high-quality microgravity environment of free-fall drop towers or free-floating experiment hardware.

Conclusions

We have combined and analyzed low-velocity impacts into regolith from five experimental campaigns. The PRIME and COLLIDE experiment setups allowed for recording impacts of centimeter-sized spherical projectiles into a bed of regolith simulant (quartz sand and JSC-1 lunar and JSC Mars-1 simulants), sieved to sizes under 250 μm. Due to their flight platform, these experiments were run either at reduced-gravity levels of $\sim 10^{-2}g$ when fixed to the parabolic aircraft or during $0.05g$ parabolas, or at microgravity levels of $< 10^{-4}g$, when free-floating in the airplane or during Shuttle and suborbital rocket campaigns. The impacts into regolith simulant resulted into four types of outcomes, some producing ejecta and some displaying projectile rebound. These results revealed major differences in the target behavior depending on if the experiment was run in reduced- or microgravity conditions. The main results of the data analysis can be summarized as follows:

- We observed projectile rebound off the target in microgravity. None of the impacts in reduced gravity displayed a similar behavior. The coefficient of restitution of the impacts varied with the impact energy following a power law of index $-1/4$ (Fig. 8);
- The maximum penetration depth was observed to depend more on the projectile size and energy and less on its density difference with the target than in $1g$;
- For the same impact speeds, more mass was ejected in microgravity compared to reduced gravity, but the average ejecta speed was lower (Figs. 6 and 10);
- The difference in cohesive forces in quartz sand and JSC-1 could be observed from the experiment results, with quartz sand behaving like a cohesionless material and JSC-1 displaying cohesive behavior. None of the impacts into JSC-1 generated an ejecta curtain in microgravity, and most of them displayed projectile rebound.

The results of these flight campaigns show that the acceleration environment at which experiments are performed has a significant influence on their outcome. The study of regolith behavior at the surface of small bodies (<1 km-sized asteroids) will require acceleration environments below $10^{-4}g$, as data cannot be extrapolated from behavior in reduced-gravity environments ($> 10^{-3}g$).

From our current data set, it clearly appears that more low-velocity impact data is required in microgravity conditions. This is due to the fact that clean microgravity environments are difficult to obtain, with most platforms retaining residual acceleration levels around $10^{-2}g$. In addition, more data is currently available for sand targets than for JSC-1 lunar simulant. As JSC-1 is more representative of the granular material expected on asteroids (in particular due to the irregular particle shape), data on JSC-1 targets is more relevant for application to small body surfaces. We can also note that all experiments presented here were run using similar target size distribution: particles sieved to sizes $< 250\mu m$. From images returned of asteroid Itokawa (Fujiwara et al. 2006) and spectral analysis of Ryugu, we can infer that some asteroid surface regoliths have larger average grain sizes. For a future contribution of our experimental results into small body surface exploration missions, the investigation of grain behavior at sizes between 1 mm and 1 cm will be necessary. Such future experiments will be even more sensitive to the quality of the microgravity environment they will be run in due to the larger grain sizes (see Fig. 13).

Abbreviations

CMR: Center for Microgravity Research; COLLIDE: COLLisions Into Dust Experiment; IBS: Impactor Box System; ISRU: In situ resource utilization; JSC: Johnson Space Center; MASCOT: Mobile Asteroid Surface Scout; NASA: National Aeronautics and Space Administration; OSIRIS-REx: Origins, Spectral Interpretation, Resource Identification, Security, Regolith Explorer; PRIME: Physics of Regolith Impacts in Microgravity Experiment; TAG: Touch-And-Go

Acknowledgements

We would like to thank our two anonymous reviewers for their valuable input and comments.

Funding

This research is based in part upon work supported by the National Aeronautics and Space Administration under Grant No. NNX11AQ87G issued by the Planetary Geology and Geophysics Program and under grant NNX12AK43G issued by the Outer Planets Research Program, by the NASA Flight Opportunities Program, and by funding from Space Florida for the Center for Microgravity Research.

Authors' contributions

JC secured funding for all performed experiments and provided the COLLIDE-1 and COLLIDE-2, and PRIME-1 data. JC, AD, and JB performed the experimental data collection during the PRIME-3 and COLLIDE-3 campaigns. JB, SA, CC, and NM worked on the data analysis. All authors read and approved the final manuscript.

Competing interests

The authors declare that they have no competing interests.

References

Biele J, Ulamec S, Maibaum M, Roll R, Witte L, Jurado E, Muñoz P, Arnold W, Auster H-U, Casas C, Faber C, Fantinati C, Finke F, Fischer H-H, Geurts K, Güttler C, Heinisch P, Herique A, Hviid S, Kargl G, Knapmeyer M, Knollenberg J, Kofman W, Kömle N, Kührt E, Lommatsch V, Mottola S, Pardo de Santayana R, Remetean E, Scholten F, Seidensticker KJ, Sierks H, Spohn T (2015) The landing(s) of Philae and inferences about comet surface mechanical properties. Science 349(1). https://doi.org/10.1126/science.aaa9816

Biele J, Kesseler L, Grimm CD, Schröder S, Mierheim O, Lange M, Ho T-M (2017) Experimental determination of the structural coefficient of restitution of a bouncing asteroid lander. ArXiv preprint arXiv:1705.00701

Blum J, Wurm G (2008) The growth mechanisms of macroscopic bodies in protoplanetary disks. Annu Rev Astron Astrophys 46:21–56. https://doi.org/10.1146/annurev.astro.46.060407.145152

Brisset J, Colwell JE, Dove A, Rascon A, Mohammed N, Cox C (2016) Low-velocity impacts into cryogenic icy regolith. In: AAS/Division for Planetary Sciences Meeting Abstracts, vol. 48. American Astronomical Society. http://adsabs.harvard.edu/abs/2016DPS....4851805B

Brisset J, Heißelmann D, Kothe S, Weidling R, Blum J (2016) Submillimetre-sized dust aggregate collision and growth properties. Experimental study of a multi-particle system on a suborbital rocket. Astron Astrophys 593:A3. https://doi.org/10.1051/0004-6361/201527288

Britt DT, Yeomans D, Housen K, Consolmagno G (2002) Asteroid density, porosity, and structure p. 485–500

Cates ME, Wittmer JP, Bouchaud J-P, Claudin PH (1998) Jamming, force chains, and fragile matter. Phys Rev Lett 81(9):1841

Cheng AF, Izenberg N, Chapman CR, Zuber MT (2002) Ponded deposits on asteroid 433 Eros. Meteorit Planet Sci 37:1095–1105. https://doi.org/10.1111/j.1945-5100.2002.tb00880.x

Clark AH, Kondic L, Behringer RP (2012) Particle scale dynamics in granular impact. Phys Rev Lett 109(23):238302. https://doi.org/10.1103/PhysRevLett.109.238302

Clark AHM, Behringer RP (2013) Granular impact model as an energy-depth relation. EPL (Europhys Lett) 101:64001. https://doi.org/10.1209/0295-5075/101/64001

Colwell J, Brisset J, Dove A, Whizin A, Nagler H, Brown N, Rascon A, Brightwell K, Seward L (2016) Low-velocity impacts into regolith under microgravity conditions. In: 15th Biennial ASCE Conference on Engineering, Science, Construction, and Operations in Challenging Environments, volume 82 of *Earth and Space Conference*. https://ascelibrary.org/doi/10.1061/9780784479971.010

Colwell JE (2003) Low velocity impacts into dust: results from the COLLIDE-2 microgravity experiment. Icarus 164:188–196. https://doi.org/10.1016/S0019-1035(03)00083-6

Colwell JE, Sture S, Cintala M, Durda D, Hendrix A, Goudie T, Curtis D, Ashcom DJ, Kanter M, Keohane T, Lemos A, Lupton M, Route M (2008) Ejecta from impacts at 0.2 2.3 m/s in low gravity. Icarus 195:908–917. https://doi.org/10.1016/j.icarus.2007.12.019

Colwell JE, Taylor M (1999) Low-velocity microgravity impact experiments into simulated regolith. Icarus 138:241–248. https://doi.org/10.1006/icar.1998.6073

Deboeuf S, Gondret P, Rabaud M (2009) Dynamics of grain ejection by sphere impact on a granular bed. Phys Rev E 79(4):041306. https://doi.org/10.1103/PhysRevE.79.041306

Dombard AJ, Barnouin OS, Prockter LM, Thomas PC (2010) Boulders and ponds on the Asteroid 433 Eros. Icarus 210:713–721. https://doi.org/10.1016/j.icarus.2010.07.006

Fujiwara A, Kawaguchi J, Yeomans DK, Abe M, Mukai T, Okada T, Saito J, Yano H, Yoshikawa M, Scheeres DJ, Barnouin-Jha O, Cheng AF, Demura H, Gaskell RW, Hirata N, Ikeda H, Kominato T, Miyamoto H, Nakamura AM, Nakamura R, Sasaki S, Uesugi K (2006) The rubble-pile asteroid Itokawa as observed by Hayabusa. Science 312:1330–1334. https://doi.org/10.1126/science.1125841

Gundlach B, Blum J (2013) A new method to determine the grain size of planetary regolith. Icarus 223:479–492. https://doi.org/10.1016/j.icarus.2012.11.039

Güttler C, Krause M, Geretshauser RJ, Speith R, Blum J (2009) The physics of protoplanetesimal dust agglomerates. IV. Toward a dynamical collision model. Astrophys J 701:130–141. https://doi.org/10.1088/0004-637X/701/1/130

Hirabayashi M, Sánchez DP, Scheeres DJ (2015) Internal structure of asteroids having surface shedding due to rotational instability. Astrophys J 808:63. https://doi.org/10.1088/0004-637X/808/1/63

Hiroi T, Pieters CM, Takeda H (1994) Grain size of the surface regolith of asteroid 4 Vesta estimated from its reflectance spectrum in comparison with HED meteorites. Meteoritics 29:394–396

Holsapple K, Giblin I, Housen K, Nakamura A, Ryan E (2002) Asteroid impacts: laboratory experiments and scaling laws

Housen KR, Holsapple KA (2003) Impact cratering on porous asteroids. Icarus 163:102–119. https://doi.org/10.1016/S0019-1035(03)00024-1

Housen, KR, Holsapple KA (2011) Ejecta from impact craters. Icarus 211:856–875. https://doi.org/10.1016/j.icarus.2010.09.017

Housen, KR, Wilkening LL (1982) Regoliths on small bodies in the solar system. Ann Rev Earth Planet Sci 10:355–376. https://doi.org/10.1146/annurev.ea.10.050182.002035

Housen KR, Wilkening LL, Chapman CR, Greenberg R (1979) Asteroidal regoliths. Icarus 39:317–351. https://doi.org/10.1016/0019-1035(79)90145-3

Housen KR, Holsapple KA, Voss ME (1999) Compaction as the origin of the unusual craters on the asteroid Mathilde. Nature 402:155–157. https://doi.org/10.1038/45985

Jaumann R, Bibring JP, Glassmeier KH, Grott M, Ho TM, Ulamec S, Schmitz N, Auster HU, Biele J, Kuninaka H, Okada T (2016) A Mobile Asteroid Surface Scout (MASCOT) on board the Hayabusa 2 Mission to the near Earth asteroid (162173) Ryugu. in AGU Fall Meeting Abstracts

Johnson KL, Kendall K, Roberts AD (1971) Surface energy and the contact of elastic solids. Proc R Soc Lond 324:301–313

Johnson KL (1985) Contact mechanics. Cambridge university press, Cambridge

Katsuragi H, Blum J (2017) The physics of protoplanetesimal dust agglomerates. IX. Mechanical properties of dust aggregates probed by a solid-projectile impact. Astrophys J 851:23. https://doi.org/10.3847/1538-4357/aa970d

Katsuragi H, Durian DJ (2007) Unified force law for granular impact cratering. Nat Phys 3:420–423. https://doi.org/10.1038/nphys583

Katsuragi, H, Durian DJ (2013) Drag force scaling for penetration into granular media. Phys Rev E 87(5):052208. https://doi.org/10.1103/PhysRevE.87.052208

Kawaguchi J, Fujiwara A, Uesugi T (2008) Hayabusa - its technology and science accomplishment summary and Hayabusa-2. Acta Astronaut 62:639–647. https://doi.org/10.1016/j.actaastro.2008.01.028

Kendall K, McN. Alford N, Birchall JD (1987) A new method for measuring the surface energy of solids. Nature 6107:794–796. https://doi.org/10.1038/325794a0

Klimek R, Wright R (2006) Spotlight–8 Image Analysis Software. NASA TM 214084 2005. http://microgravity.grc.nasa.gov/spotlight/

Lauretta DS, Balram-Knutson SS, Beshore E, Boynton WV, Drouet d'Aubigny C, DellaGiustina DN, Enos HL, Golish DR, Hergenrother CW, Howell ES, Bennett CA, Morton ET, Nolan MC, Rizk B, Roper HL, Bartels AE, Bos BJ, Dworkin JP, Highsmith DE, Lorenz DA, Lim LF, Mink R, Moreau MC, Nuth JA, Reuter DC, Simon AA, Bierhaus EB, Bryan BH, Ballouz R, Barnouin OS, Binzel RP, Bottke WF, Hamilton VE, Walsh KJ, Chesley SR, Christensen PR, Clark BE, Connolly HC, Crombie MK, Daly MG, Emery JP, McCoy TJ, McMahon JW, Scheeres DJ, Messenger S, Nakamura-Messenger K, Righter K, Sandford SA (2017) OSIRIS-REx: sample return from Asteroid (101955) Bennu. Space Sci Rev 212(1-2):925–984. https://doi.org/10.1007/s11214-017-0405-1

Machii N, Nakamura AM, Güttler C, Beger D, Blum J (2013) Collision of a chondrule with matrix: relation between static strength of matrix and impact pressure. Icarus 226:111–118. https://doi.org/10.1016/j.icarus.2013.05.006

Maurel C, Michel P, Biele J, Ballouz R-L, Thuillet F (2017) Numerical simulations of the contact between the lander mascot and a regolith-covered surface. Adv Space Res 62(8):2099–2124

Murdoch N, Rozitis B, Nordstrom K, Green SF, Michel P, de Lophem T-L, Losert W (2013) Granular convection in microgravity. Phys Rev Lett 110(1):018307. https://doi.org/10.1103/PhysRevLett.110.018307

Sanchez P, Scheeres DJ (2014) The strength of regolith and rubble pile asteroids. Meteorit Planet Sci 49:788–811. https://doi.org/10.1111/maps.12293

Sánchez P, Scheeres DJ, Bierhaus EB, Clark B (2013) Regolith penetrometry in microgravity. Eur Planet Sci Congr 8:EPSC2013-882

Scheeres DJ, Hartzell CM, Sánchez P, Swift M (2010) Scaling forces to asteroid surfaces: the role of cohesion. Icarus 210:968–984. https://doi.org/10.1016/j.icarus.2010.07.009

Seguin A, Bertho Y, Gondret P (2008) Influence of confinement on granular penetration by impact. Phys Rev E 78(1):010301. https://doi.org/10.1103/PhysRevE.78.010301

Thornton C, Ning Z (1998) A theoretical model for the stick/bounce behaviour of adhesive, elastic-plastic spheres. Powder Technol 99(2):154–162

Thuillet F, Maurel C, Michel P, Biele J, Ballouz R-L, Richardson DC (2017) Numerical simulations of surface package landing on a low-gravity granular surface: application to the landing of MASCOT onboard Hayabusa2. In: Lunar and Planetary Science Conference, Vol. 48. 48th Lunar and Planetary Science Conference. https://www.hou.usra.edu/meetings/lpsc2017/pdf/1810.pdf

Tsuda Y, Yoshikawa M, Abe M, Minamino H, Nakazawa S (2013) System design of the Hayabusa 2 - asteroid sample return mission to 1999 JU3. Acta Astronaut 91:356–362. https://doi.org/10.1016/j.actaastro.2013.06.028

Veverka J, Farquhar B, Robinson M, Thomas P, Murchie S, Harch A, Antreasian PG, Chesley SR, Miller JK, Owen WM, Williams BG, Yeomans D, Dunham D, Heyler G, Holdridge M, Nelson RL, Whittenburg KE, Ray JC, Carcich B, Cheng A, Chapman C, Bell JF, Bell M, Bussey B, Clark B, Domingue D, Gaffey MJ, Hawkins E, Izenberg N, Joseph J, Kirk R, Lucey P, Malin M, McFadden L, Merline WJ, Peterson C, Prockter L, Warren J, Wellnitz D (2001) The landing of the NEAR-Shoemaker spacecraft on asteroid 433 Eros. Nature 413:390–393. https://doi.org/10.1038/35096507

Wagner E, DeForest CE (2016) Opportunities for suborbital space and atmospheric research facilities on Blue Origin's New Shepard crew capsule. AGU Fall Meeting Abstracts P53C-2216. http://adsabs.harvard.edu/abs/2016AGUFM.P53C2216W

Permissions

All chapters in this book were first published in PEPS, by Springer; hereby published with permission under the Creative Commons Attribution License or equivalent. Every chapter published in this book has been scrutinized by our experts. Their significance has been extensively debated. The topics covered herein carry significant findings which will fuel the growth of the discipline. They may even be implemented as practical applications or may be referred to as a beginning point for another development.

The contributors of this book come from diverse backgrounds, making this book a truly international effort. This book will bring forth new frontiers with its revolutionizing research information and detailed analysis of the nascent developments around the world.

We would like to thank all the contributing authors for lending their expertise to make the book truly unique. They have played a crucial role in the development of this book. Without their invaluable contributions this book wouldn't have been possible. They have made vital efforts to compile up to date information on the varied aspects of this subject to make this book a valuable addition to the collection of many professionals and students.

This book was conceptualized with the vision of imparting up-to-date information and advanced data in this field. To ensure the same, a matchless editorial board was set up. Every individual on the board went through rigorous rounds of assessment to prove their worth. After which they invested a large part of their time researching and compiling the most relevant data for our readers.

The editorial board has been involved in producing this book since its inception. They have spent rigorous hours researching and exploring the diverse topics which have resulted in the successful publishing of this book. They have passed on their knowledge of decades through this book. To expedite this challenging task, the publisher supported the team at every step. A small team of assistant editors was also appointed to further simplify the editing procedure and attain best results for the readers.

Apart from the editorial board, the designing team has also invested a significant amount of their time in understanding the subject and creating the most relevant covers. They scrutinized every image to scout for the most suitable representation of the subject and create an appropriate cover for the book.

The publishing team has been an ardent support to the editorial, designing and production team. Their endless efforts to recruit the best for this project, has resulted in the accomplishment of this book. They are a veteran in the field of academics and their pool of knowledge is as vast as their experience in printing. Their expertise and guidance has proved useful at every step. Their uncompromising quality standards have made this book an exceptional effort. Their encouragement from time to time has been an inspiration for everyone.

The publisher and the editorial board hope that this book will prove to be a valuable piece of knowledge for researchers, students, practitioners and scholars across the globe.

List of Contributors

Teresa Wong and Viatcheslav S Solomatov
Department of Earth and Planetary Sciences, Washington University, 1 Brookings Drive, 63130-4899 St. Louis, MO, USA

George Helffrich
Earth-Life Science Institute, Tokyo Institute of Technology, 2-12-1-IE-1 Ookayama, Meguro-ku, Tokyo 152-8550, Japan

Wieland Dietrich
Department of Applied Mathematics, University of Leeds, Woodhouse Lane, LS9 2JT Leeds, West Yorkshire, UK

Kumiko Hori
Department of Applied Mathematics, University of Leeds, Woodhouse Lane, LS9 2JT Leeds, West Yorkshire, UK
Solar-Terrestrial Environment Laboratory, Nagoya University, Furo-cho, Chikusa-ku, 464-8601 Nagoya, Japan

Johannes Wicht
Max Planck Institute for Solar System Research, Justus-von-Liebig-Weg 3, 37077 Göttingen, Germany

Elisavet Proedrou
Institute of Applied Physics, University of Bern, Sidlerstrasse 5, 3012 Bern, Switzerland
Center for Space and Habitability, University of Bern, Sidlerstrasse 5, 3012 Bern, Switzerland

Klemens Hocke
Institute of Applied Physics, University of Bern, Sidlerstrasse 5, 3012 Bern, Switzerland
Center for Space and Habitability, University of Bern, Sidlerstrasse 5, 3012 Bern, Switzerland
Oeschger Centre for Climate Change Research, University of Bern, Falkenplatz 16, 3012 Bern, Switzerland

Peter Wurz
Institute of Applied Physics, University of Bern, Sidlerstrasse 5, 3012 Bern, Switzerland
Physics Institute, University of Bern, Sidlerstrasse 5, 3012 Bern, Switzerland

Hagay Amit, Gaël Choblet, Benoit Langlais and Gabriel Tobie
CNRS UMR 6112, Université de Nantes, Laboratoire de Planétologie et de Géodynamique, 2 rue de la Houssinière, Nantes, F-44000, France

Peter Olson
Department of Earth and Planetary Sciences, Johns Hopkins University, 21218 Baltimore, MD, USA

Julien Monteux
Laboratoire Magmas et Volcans, Université Blaise Pascal - CNRS -IRD, OPGC, 5 rue Kessler, 63038 Clermont Ferrand, France

Frédéric Deschamps
Institute of Earth Sciences, Academia Sinica, 128 Academia Road Sec. 2, Nangang, Taipei 11529, Taiwan

Ryodo Hemmi and Hideaki Miyamoto
Department of Systems Innovation, School of Engineering, The University of Tokyo, 7-3-1 Hongo, Bunkyo-ku, Tokyo 113-8656, Japan

Naohiro Manago and Hiroaki Kuze
Center for Environmental Remote Sensing, Chiba University, 1-33 Yayoi-cho, Inage-ku, Chiba-shi, Chiba 263-8522, Japan

Katsuyuki Noguchi
Department of Chemistry, Biology, and Environmental Science, Faculty of Science, Nara Women's University, Kita-uoya Nishi-machi, Nara-shi, Nara 630-8506, Japan

George L. Hashimoto
Department of Earth Sciences, Okayama University, 3-1-1 Tsushimanaka, Kita-ku, Okayama-shi, Okayama 700-8530, Japan

Hiroki Senshu
Planetary Exploration Research Center, Chiba Institute of Technology, 2-17-1 Tsudanuma, Narashino-shi, Chiba 275-0016, Japan

Naohito Otobe
Department of Earth System Science, Faculty of Science, Fukuoka University, 8-19-1 Nanakuma, Jonan-ku, Fukuoka-shi, Fukuoka 814-0180, Japan

Makoto Suzuki
Institute of Space and Astronautical Science, Japan Aerospace Exploration Agency, 3-1-1 Yoshinodai, Chuo-ku, Sagamihara-shi, Kanagawa 252-5210, Japan

Svetlana Kostic
Computational Science Research Center, San Diego State University, San Diego, CA, USA

Isaac B. Smith
Planetary Science Institute, Lakewood, CO, USA

Doris Breuer, Tina Rueckriemen and Tilman Spohn
DLR, Institute of Planetary Research, Rutherfordstrasse 2, 12489 Berlin, Germany

Krista M. Soderlund
Institute for Geophysics, John A. & Katherine G. Jackson School of Geosciences, The University of Texas at Austin, J. J. Pickle Research Campus, Building 196 (ROC), 10100 Burnet Road (R2200), Austin, Texas, 78758-4445, USA

Andrey Sheyko
Institut für Geophysik, ETH Zürich, NO H 11.2, Sonneggstrasse 5 8092 Zürich, LA59 Switzerland

Eric M. King
U.S. Global Development Lab, U.S. Agency for International Development, Washington, DC, 20004, USA

Jonathan M. Aurnou
Department of Earth, Planetary, and Space Sciences, University of California, Los Angeles, 595 Charles Young Drive East, Los Angeles, California, 90095-1567, USA

Koji Wada and Hiroki Senshu
Planetary Exploration Research Center, Chiba Institute of Technology, Tsudanuma 2-17-1, Narashino, Chiba 275-0016, Japan

Matthias Grott, Maximilian Hamm, Ekkehard Kührt, Alessandro Maturilli, Stefano Mottola and Stefan E. Schröder
DLR Institute of Planetary Research, Rutherfordstr. 2, 12489 Berlin, Germany

Patrick Michel and Guy Libourel
Université Côte d'Azur, Observatoire de la Côte d'Azur, CNRS, Laboratoire Lagrange, CS 34229, 06304 Nice Cedex 4, France

Kevin J. Walsh
Southwest Research Institute, Boulder, USA

Antonella M. Barucci
Laboratoire d'Etudes Spatiales et d'Instrumentation en Astrophysique (LESIA), Observatoire de Paris, PSL Research University, CNRS, Université Paris Diderot, Sorbonne Paris Cité, UPMC Université Paris 06, Sorbonne Universités, 5 Place J. Janssen, 92195 Meudon Principal Cedex, France

Jens Biele
DLR RB-MSC German Aerospace Center, 51147 Cologne, Germany

Jürgen Blum and Bastian Gundlach
Institut für Geophysik und extraterrestrische Physik, Technische Universität Braunschweig, Mendelssohnstr. 3, 38106 Braunschweig, Germany

Carolyn M. Ernst
Johns Hopkins University Applied Physics Laboratory, Laurel 20723, MD, USA

Jan Thimo Grundmann
DLR German Aerospace Center, Institute of Space Systems, System Engineering and Project Office, Robert-Hooke-Strasse 7, 28359 Bremen, Germany

Axel Hagermann
Department of Biological and Environmental Sciences, University of Stirling, Stirling FK9 4LA, Scotland

Martin Jutzi
Physics Institute, University of Bern, Bern, Switzerland

Myung-Jin Kim
Center for Space Situational Awareness, Korea Astronomy and Space Science Institute, 776, Daedeokdae-ro, Yuseong-gu, Daejeon 305-348, Republic of Korea

Lucille Le Corre
Planetary Science Institute, 1700 East Fort Lowell, Suite 106, Tucson, AZ 85719, USA

Roy Lichtenheldt
DLR Institute of System Dynamics and Control, Oberpfaffenhoffen, Germany

Scott R. Messenger and Michael E. Zolensky
Robert M Walker for Space Sciences, Astromaterials Research and Exploration Science, NASA Johnson Space Center, Houston, TX 77058, USA

Tatsuhiro Michikami
Faculty of Engineering, Kindai University, Hiroshima Campus, 1 Takaya Umenobe, Higashi-Hiroshima, Hiroshima 739-2116, Japan

Hideaki Miyamoto
Department of Systems Innovation, University of Tokyo, Tokyo 113-8656, Japan

Thomas Müller
Max-Planck Institute for Extraterrestrial Physics, Garching, Germany

Akiko M. Nakamura and Kazunori Ogawa
Department of Planetology, Graduate School of Science, Kobe University, 1-1 Rokkodai, Nada-ku, Kobe 657-8501, Japan

Larry R. Nittler
Department of Terrestrial Magnetism, Carnegie Institution of Washington, 5241 Broad Branch Rd NW, Washington, D.C. 20015, USA

Tatsuaki Okada and Naoya Sakatani
Institute of Space and Astronautical Science, Japan Aerospace Exploration Agency, Sagamihara, Japan

Ernesto Palomba
INAF, Istituto di Astrofisica e Planetologia Spaziali, via Fosso del Cavaliere, 00133 Rome, Italy

Driss Takir
JETS/Astromaterials Research and Exploration Science, NASA Johnson Space Center, Houston 77058, TX, USA

Julie Brisset
Florida Space Institute, University of Central Florida, 12354 Research Parkway, Partnership 1 Building, Suite 211, Orlando 32826, FL, USA

Index

www.ingramcontent.com/pod-product-compliance
Lightning Source LLC
LaVergne TN
LVHW080922210926
848485LV00004B/709